Numerical Algorithms for Number Theory

Using Pari/GP

Mathematical
Surveys
and
Monographs

Volume 254

Numerical Algorithms
for Number Theory

Using Pari/GP

Karim Belabas
Henri Cohen

AMERICAN
MATHEMATICAL
SOCIETY
Providence, Rhode Island

2020 *Mathematics Subject Classification.* Primary 11-04, 11Y60, 65Y20, 11M06, 11M41, 30B70, 65B10, 65D30.

For additional information and updates on this book, visit
www.ams.org/bookpages/surv-254

Library of Congress Cataloging-in-Publication Data

Names: Belabas, Karim, author. | Cohen, Henri, author.

Title: Numerical algorithms for number theory : using Pari/GP | Karim Belabas, Henri Cohen.

Description: Providence, Rhode Island : American Mathematical Society, [2021] | Series: Mathematical surveys and monographs, 0076-5376 ; volume 254 | Includes bibliographical references and index.

Identifiers: LCCN 2021007368 | ISBN 9781470463519 (paperback) | 9781470465568 (ebook)

Subjects: LCSH: Numerical analysis–Computer programs. | Number theory. | Computer algorithms. | AMS: Number theory – Software, source code, etc. for problems pertaining to number theory. | Number theory – Computational number theory – Evaluation of number-theoretic constants. | Numerical analysis – Computer aspects of numerical algorithms – Complexity and performance of numerical algorithms. | Number theory – Zeta and L-functions: analytic theory; $\zeta(s)$ and $L(s,\chi)$. | Number theory – Zeta and L-functions: analytic theory – Other Dirichlet series and zeta functions. | Functions of a complex variable – Series expansions of functions of one complex variable – Continued fractions; complex-analytic aspects. | Numerical analysis – Acceleration of convergence in numerical analysis – Numerical summation of series. | Numerical analysis – Numerical approximation and computational geometry (primarily algorithms) – Numerical integration.

Classification: LCC QA297 .B375 2021 | DDC 518/.47–dc23

LC record available at https://lccn.loc.gov/2021007368

Contents

Preface

This book is a largely expanded version of a course that the second author gave at ICTP in Trieste in the summer 2012, preceding a conference on "hypergeometric motives", in Rennes in April 2014 at the Journées Louis Antoine, and at the NTCS workshop in Taiwan in August 2014.

The goal of this book is to present a number of analytic and arithmetic numerical methods used in number theory, with a particular emphasis on the ones which are less known than they should be, although very classical tools are also mentioned. Note that, as is very often the case in number theory, we want numerical methods giving sometimes *hundreds* if not thousands of decimal places of accuracy. The typical timing tables that we will give are in fact for 500 decimal digits.

The style of presentation is the following: we first give proofs of some of the tools, the prerequisites being classical undergraduate analysis. Note that since the emphasis is on *practicality*, the proofs are sometimes only heuristic, but valid in actual practice. We then give the corresponding Pari/GP programs, usually followed by a number of examples. These programs are also available as a unique separate archive on the authors' website at

http://www.math.u-bordeaux.fr/~kbelabas/Numerical_Algorithms/

Feel free to experiment and modify them to your heart's content. They can also serve as an introduction to the syntax and semantics of GP, since in general they are easy to understand and do not use much sophistication. Note we use rather recent features of the GP language, so we strongly advise to download the latest release (version 2.13 or more recent) from the Pari/GP website:

http://pari.math.u-bordeaux.fr/

The reader is advised to refer to the numerous books dealing with parts of the subject, such as (but of course not limited to!) the second author's four books [**Coh93**], [**Coh00**], [**Coh07a**], and [**Coh07b**].

Caveat: Neither the authors nor the AMS are liable for any damage caused by the use of the programs given in this book. Apart from this legalese, we would be happy to hear of any corrections and/or improvements. Even though we have tried to give simple and efficient programs, we do not claim that we give "the best" methods, and we would also be very glad to hear of new methods for solving the problems considered in this book.

Karim Belabas and Henri Cohen

CHAPTER 1

Introduction

1.1. Subject matter

In this book we are going to study a number of computational methods useful in number theory, which are rarely covered in usual texts, including in the second author's books. Some of these methods are purely number-theoretical in nature and do not involve floating point arithmetic: in particular methods involving Gauss and Jacobi sums, and applications of those such as point counting on varieties over finite fields.

But most of the methods that we will study are basically *numerical* in nature: numerical extrapolation, integration, summation, computation of continued fractions, Euler products, inverse Mellin transforms, and L-functions. It is clear that these methods are strongly interrelated: extrapolation can be used essentially in all of them, numerical integration can be done via numerical summation and vice versa, etc. Thus, the order in which the material is presented is a little arbitrary.

Since extrapolation can be used essentially everywhere, we present it first. Even though the methods that we suggest are quite simple and even simplistic, they work rather well in many cases. We then study in detail first numerical integration, then numerical summation. Although these subjects are extremely classical, the goal of the corresponding chapters is to show that in both cases, two methods stand out as being the "best" in a suitable sense. In the case of numerical integration, the first is the very classical *Gauss–Legendre* method, and the second is the more recent but now well-known method of *doubly-exponential integration* (DE). For numerical summation, however, the two best methods seem to be new, or at least overlooked: the first is *discrete Euler–Maclaurin* summation, and the second is a recent method due to H. Monien which is analogous to Gauss–Legendre integration, and which we naturally call *Monien summation*.

We then explain how Euler products and sums (which are over the irregular sets of prime numbers) can be computed quite simply using values of the Riemann zeta function, or more generally of L-functions. In particular, we give a detailed explanation of the computation of *Hardy–Littlewood constants* associated to polynomials.

The next chapter is a little special in that it does not involve much numerical analysis: we explore different methods for computing Gauss and Jacobi sums, which are essential in point counting of algebraic varieties and hypergeometric motives, and explain the use of the remarkable Gross–Koblitz formula from p-adic analysis.

As for extrapolation, the computation of continued fractions involves quite naïve methods. It can be done using extrapolation and/or summation, but also specific techniques. The main interest of the corresponding chapter is the study of the speed of convergence of "nice" continued fractions, as well as the *Quotient-Difference algorithm*, which is essential in the computation of inverse Mellin transforms using the ideas of T. Dokchitser.

The last two subjects, the computation of inverse Mellin transforms and of *L*-functions are more sophisticated and less basic than the previous subjects, but are also essential in number theory. Inverse Mellin transforms are essential for computing *L*-functions, and although they are expressible as integrals, in addition to the numerical integration methods studied in the previous chapters, there exist specific and highly efficient methods for computing them. The main focus of the final chapter is the computation of *L*-functions in reasonable ranges. We then explain the completely different methods that are available for computing them at large height within the critical strip, and end by a presentation of the so-called "explicit formulas" together with some applications.

1.2. Experimental protocols

Since the subject is relatively new, we have devised a large number of tests for many different methods. These tests have allowed us to decide on the relative efficiency of these methods, and to decide which are the best ones, for given accuracies between 38 and 1000 decimal digits. In all cases, except for very small accuracies such as 38 digits, the result of the test is uniform, in that the same methods are the best independently of the accuracy (but sometimes strongly dependent on the test functions). Thus, we give comparative tables for the largish accuracy of 500 decimal digits, the corresponding tables for other accuracies being similar.

Note that the most useful methods, such as the two best numerical integration and the two best numerical summation methods, are built-in functions in the Pari/GP system, but we still give the corresponding GP programs as they explain in complete detail the workings of the methods. The timings that we will give correspond to the GP script implementations. In fact, since some function names may clash with existing GP names, we have decided to systematically capitalize the names of the functions presented in this book, since almost no built-in GP function starts with a capital letter. A typical example is the Sumalt function for summing alternating series, which will not clash with the built-in sumalt function in GP.

1.3. Multiprecision algorithms and working accuracy

When we talk about multiprecision in this book, we mean a *reasonable* number of decimal digits, in practice between 38 and 1000 as mentioned above (this strange number 38 comes from the fact that now most computers are 64 bit machines, and $2^{64} \approx 10^{19.266}$, so most decimal accuracies that we use are multiple of 19). This implies in particular that we make no effort to use FFT-type algorithms, which are only useful with numbers having thousands of decimal digits.

Second, several algorithms which are very standard in numerical analysis are quite inapplicable in multiprecision, for instance Romberg integration. As a consequence of this, it will be necessary to select known algorithms that can be adapted to multiprecision computations, and also to create new algorithms which have no equivalent (or no interest) in standard numerical analysis.

A big advantage of working with a multiprecision-enabled software is that we can *cheat* in many ways, and compel the algorithm to give correct results in situations where it may otherwise not do so. A first simple example is to increase the working accuracy (the requested bit accuracy will always be denoted by B in this book, meaning that we expect to approximate a quantity with absolute error bounded by 2^{-B}): to avoid the loss of several bits in the computation of a sum of millions of terms, we change the accuracy to $B+32$, and restore the default accuracy at the end of the computation. (The GP language supports this automagically.) Another example is when a known fixed quantity will multiply the computed estimate, thereby increasing the absolute error by a fixed amount: we can and will increase B to compensate.

Much "worse", when we heuristically notice that a fixed fraction of decimals are computed incorrectly by an algorithm, we change B to $3B/2$ or $2B$. Of course this slows down the algorithm by a constant factor which (depending) can be of the order of 4 for instance, but at least we obtain a perfect result. A more rigorous solution would be to use interval or ball arithmetic as in the **Arb** system [**Joh17**] and iterate the computation until the automatic error bounds fall under a tolerance threshold.

The worst possible cheating is when the algorithm is inherently unstable: although it is not always a miracle solution, we can change B into for instance $B \log(B)$. This is a huge loss of speed, but which in many cases can compensate for the instability.

Thus, in most programs the reader will see that we change the working accuracy. The modifications which are given are heuristically reasonable, but nothing rigorous is proved about them, and the reader is welcome to either analyze the problem, or find better heuristics for compensating instability. These accuracy changes in programs will usually not be commented.

1.4. Comments on the GP language

This book is *not* intended as a tutorial for the **Pari/GP** system[1], but can serve as an introduction to the practical use of the language in nontrivial problems. We feel, however, that it is necessary to emphasize a few important points.

1.4.1. Functions and closures.

A basic question is how to implement in a computer language a function or a sequence f, so as to be able to write (for instance) `intnum(t=0, 1, f(t))` for integrating f between 0 and 1. In the GP language, a function is a first class object: typically, `(t->expr(t))`, where `expr` is some expression in t, defines an anonymous function, and it can be assigned to a variable `f` in the obvious way `f = (t->expr(t))`. This is actually equivalent to setting `f(t) = expr(t)`. Of course, one can have several variables, such as `f = (x,y,z)->x+2*y+3*z`.

If `f` is such a function, nothing prevents us from defining a new function in terms of it, such as `g = x -> f(1 / x)` or `N -> vector(N, x, f(x))` (on input N, the latter function returns the N-components vector $[f(1), f(2), \ldots, f(N)]$). What we just used in these last two examples, and this is immensely powerful and by no means obvious, is that such a function definition is able to make use of variables existing in the scope of its definition (in both cases, the existing definition of f): it "captures" everything it needs from its context. For instance:

---

[1]Tutorials are included in the package itself and more can be found on the website `http://pari.math.u-bordeaux.fr/tutorials.html`

```
? add(a) = x -> x + a; \\ remembers the 'a' used as argument
? f = add(2);
? f(3)
% = 5
? g = add(4);
? g(3)
% = 7
```

We will make heavy use of this powerful *closure* mechanism in this book: binding a function and (part of) its environment to easily define new functions.

By default, function arguments are passed by value: modifying arguments in a function call does not affect variables in the calling function. But it is possible to pass containers such as vectors or lists by reference: a function may change the content of an argument whose name is prefixed by a tilde ~, both in the function call and prototype:

```
? L = List([1, 2]);
? listput(~L, 3); \\ modify list L, by appending an element
? L
% = List([1, 2, 3])
? f(~v) = v[1]++ \\ increment first component of v
? f(~L);
? L
% = List([2, 2, 3])
```

We will use this sparingly.

1.4.2. Variable scope. (We advise to skip this subsection on first reading).

Another issue in the GP language is the behavior of "local" variables. As in any other computer language, a function will want to define its own variables so as not to collide with names used in other functions or in the calling program. GP is quite tolerant: you may very well forget to declare a variable (at your own risk), GP will not complain, but of course we strongly encourage you to declare all variables. Let us make this more precise.

A *scope* is the region of a program where a given name (or user variable) refers to a given value. In the GP language, a user's function body, the body of a loop, an individual command line, all define scopes; the whole program defines the *global* scope. It is good programming practice to limit the scope of user variables: it reduces the likelihood of mistakes (rogue subprograms may not modify variables inadvertently), it allows simple recursive programs, and it helps the GP optimizer to decide whether a given variable will not be modified by a given code section thereby avoiding unnecessary copies in that range. The last reason is critical by itself.

This can be implemented in two ways in GP: lexical and dynamic scoping, both of which have their uses. In this book, we will only use lexical scopes and the corresponding keyword my. The my declaration claims ownership of a given list of variables in a given block of program text, for instance

```
my(a, b, c);
my(a = sqrt(2), b = Pi + a); /* with initial values */
my([X, W] = vec);   /* set X = vec[1], W = vec[2] */
```

The value of a my variable is then private to the block and may not be accessed or modified from the outside. For instance

```
my(x); /* declare local variable x in this scope */
x = 1; /* from now on x is 1 */
for (...
   my(x = 2); /* NEW local variable in this scope, still named x */
   ... /* x is 2, unrelated to the 'x' (= 1) outside
);
/* we left the loop body scope: x refers to 1 again */
```

Of course it is good practice to not actually "shadow" variables by recycling the same variable name to refer to different quantities in successive scopes. But this shows that you may use the same variable name in different contexts without worrying about unwanted interactions.

Function arguments are lexically scoped to the function body. The loop index in a for loop is automatically declared and lexically scoped to the loop body, and analogously for loop-like constructs like sum, prod or vector and *must not* be declared outside of the loop (well, you can always declare them, but they will be considered as different variables, as our x above). Consider the following trivial example:

```
? my(a = 10, s = 0); for (a = 1, 5, s += a); [s, a]
% = [15, 10]
```

Note how the initial a = 10 was effectively ignored in the loop and how the final value of a is *not* 5 or 6 as could have been anticipated, but the original value 10: the loop index is lexically scoped to the loop body and does not survive it.

1.4.3. Inexact objects, precision issues. A floating point number is a rational $m\beta^e$ where β is a fixed base, m is an integer of fixed length (or precision, or accuracy) when written in base β and e is an integer in a fixed range. A real number can be rounded to the nearest floating point number and we think of a floating point number x as representing the range of real numbers that round to x. In this book, a mathematical object such as a complex number, a polynomial or a matrix will be called *exact* if all its components are rational or algebraic numbers; and *inexact* if one of its components is a floating point number[2].

GP can work with bit ($\beta = 2$) or decimal ($\beta = 10$) accuracies and we will always use the former, always denoted by B, which is thus the bit length of m. The basic GP command for setting a default bit accuracy B is \pb B, or in developed form

```
default(realbitprecision, B);
```

The command bitprecision(S, B) compels the object S to have binary accuracy B, which means that if S is exact (for instance if S is an integer), then nothing changes; and if not, the internal representation of those floating point components are set to the given bit precision. A warning concerning this: if S is the result of some already approximate computation, setting S to a larger precision will *not* improve the result: it just completes the binary expansion with the appropriate number of trailing zeros. The reason this is useful is that the result of an operation on the original S has a priori lower precision than S (it can increase in rare cases, such as $1 + S$ for a tiny S which will have roughly $-\text{exponent}(S)$ more significant

---

[2]We shall briefly encounter more general inexact objects, namely the field $\mathbb{C}_\ell$ of ℓ-adic numbers in Section 6.3.

bits that the original S). Artificially increasing the accuracy of S allows further computations involving S to be done at the new accuracy. This is critical in self-correcting schemes such as Newton iteration, see the function `Newtonrefine` in Chapter 2 for an example.

In the scripts given in this book we have always preferred to work with bit instead of decimal accuracy. First, because it is slightly more elegant, but also because it is more accurate: modifying the decimal accuracy by one unit is tantamount to modifying the bit accuracy by $\log_2(10) \approx 3.3$ bits, so operating with bits offers a finer granularity. This is useful in cases where a tiny increase in accuracy has a large influence on the running time, for instance in the computation of L-functions.

Furthermore, the `localbitprec(B)` command sets the precision to the given value only *locally* in the program which calls it, without modifying the default global accuracy. This uses a dynamic scoping semantic: all computations performed until the control flow leaves the scope occur at this accuracy; this includes subroutines called from within the scope! This corresponds to the typical use cases where we temporarily increase the accuracy to make up for cancellation or decrease it because at some point we only need a rough (and cheap) estimate, independently of the global accuracy.

Also `getlocalbitprec()` with no argument returns the current bit accuracy: almost all functions in this book will use it to determine at which (absolute) accuracy the returned result is to be computed.

1.4.4. Infinity. The symbols ∞ and $-\infty$ are recognized by the GP language and denoted `oo` and `-oo`. They *cannot* be operated on (for instance `oo + 5` outputs an error), but should be used for instance to specify limits of integration or as loop limits:

```
intnum(t = -oo, oo, 1 / (1 + t^2))
forprime(p = 2, oo, ...)
```

Note that the 0 polynomial has `poldegree` equal to `-oo`, 0 has `valuation` equal to `oo` at all primes and finally that the `precision` or `bitprecision` of an exact object is `oo` (meaning: infinite).

1.4.5. Parallelism. Pari/GP supports parallel computation (with MPI or POSIX threads). Although most algorithms in this book are easy to split into independent tasks, we made the choice of avoiding any reference to parallel computing, in order to better highlight the mathematical ideas and present more readable programs. In particular, parallel GP programs must be free of side effects, and suppressing access to global data is hard on the beginner, especially global functions. We leave the adaptation of our programs as exercises.

In many cases, it is as simple as

(1) replacing control flow instructions by their parallel equivalent.
(2) adding `export` or `exportall` statements at the end of the module defining the functions you want to use in parallel constructs.

Here is an example:

```
/* weighted sum, compute sum_i w[i] * f(x[i]) */
Sum(f, x, w) = sum(i = 1, #x, w[i] * f(x[i]));
```

```
/* parallel version */
ParSum(f, x, w) = parsum(i = 1, #x, w[i] * f(x[i]));

/* make both functions usable by the parallel engine */
export(Sum, ParSum); /* or exportall() */
```

Random timing example on a laptop with 8 threads: no parallelism 460 ms, with parallelism: cpu time 893 ms (almost twice), but real time 129 ms.

1.5. Warnings

1.5.1. Mathematical rigor. Even though we try to give the best mathematical justification for all the methods that we present, we emphasize the fact that almost all the (numerical) algorithms that we give are heuristic and not guaranteed. As mentioned, probably the best way to be sure of the correctness of the result is to redo the computation to a higher accuracy and check that the answers match. This is what numerical analysts call *adaptive methods*, but is nothing else than an acceptable cheat.

An additional warning which we may not repeat all the time: we will almost always assume that the functions that we consider in numerical work have an extremely regular behavior. This not only excludes clearly pathological functions, but even rather nice functions that seem to be regular but in fact are not quite so. A case in point is the extrapolation of continued fractions: since usually the difference $p_{n+1}/q_{n+1} - p_n/q_n$ between two partial quotients alternates in sign, it is almost always a good idea to consider p_n/q_n for n of a given parity, for instance p_{2n}/q_{2n}. The difference between two such even partial quotients will now usually be of constant sign and apparently quite regular, but in fact the extrapolation methods that we will study will most often fail, because of a hidden irregularity (see Chapter 7 for details).

1.5.2. Endpoints of basic methods.

(1) **Extrapolation**: a priori, computing $\lim_{n\to\infty} u(n)$ does not require any starting value, so there are no "endpoints". Nonetheless, the values of $u(n)$ for small n have usually no relation with its asymptotic behavior. Thus, it will often be useful to begin the sequence at $n = b$, i.e., consider instead $\lim_{n\to\infty} u(n + b)$, or to increase the speed of convergence of the sequence by using a multiplier, i.e., $\lim_{n\to\infty} u(an)$, or both together as $\lim_{n\to\infty} u(an + b)$. In most of the extrapolation algorithms that we give, we leave these modifications to the user.

(2) **Integration**: integrals over a compact interval pose no problem, so the only question is over semi-infinite intervals such as $\int_a^\infty f(x)\,dx$. In all cases, we have chosen to give the algorithms with $a = 0$, i.e., for computing $\int_0^\infty f(x)\,dx$. To compute $\int_a^\infty f(x)\,dx$, there are at least two possible formulas:

$$\int_a^\infty f(x)\,dx = \int_0^\infty f(x + a)\,dx = a\left(\int_0^\infty - \int_0^1\right) f(ax)\,dx\ ,$$

and it is up to the user to see what suits him best (when a is large, the second formula is usually better).

(3) **Summation**: for the computation of $\sum_{n \geq a} f(n)$, in almost all cases we have chosen $a = 1$ (*not* $a = 0$) so as to be able to write for instance $\sum_{n \geq 1} 1/n^2$. Here, if $a \neq 1$ there is no choice, we must write

$$\sum_{n \geq a} f(n) = \begin{cases} \sum_{n \geq 1} f(n) - \sum_{1 \leq n < a} f(n) & \text{when } a > 1 , \\ \sum_{n \geq 1} f(n) + \sum_{a \leq n < 1} f(n) & \text{when } a < 1 . \end{cases}$$

(4) **Alternating Sums**: here the problem is not the endpoint, which is still $a = 1$, but the *summand*. Several programs need the summand $f(n)$ to be defined over the positive reals, and not only over the integers. Thus, to say that $\sum_{n \geq 1} f(n)$ is an alternating sum means that $(-1)^{n-1} f(n)$ has constant sign, but this does not make much sense if n is not integral. We have thus decided that all of our `Sumaltxxx` programs have the following semantic meaning: `Sumaltxxx(n -> f(n))` (possibly with additional parameters) computes $\sum_{n \geq 1} (-1)^{n-1} f(n)$. Thus, the $(-1)^{n-1}$ is *not* included in the summand. Note that this is in contradiction with the built-in `sumalt` program, where the alternating sign must be included, which is possible in that case since it only uses integer values of n. For instance `sumalt(n = 1, (-1)^(n-1) / n)` is equivalent to `Sumalt(n -> 1/n)`.

1.6. Examples

To give a taste of what awaits the reader in the the next three chapters, we give some examples with timings of limits, integrals, and sums at 115 decimal digits, using preprogrammed GP functions which are analogous to, but faster, than the scripts that we give in this book. The same examples at 38 digits would take negligible time, and at 500 or 1000 digits would still be very fast: for information, we give between parentheses the times for 500 digits.

```
                            introex.txt
1   \p115
2   ? limitnum(n -> sum(j=0,n-1,1./(1+(j/n)^2))/n)
3   time = 4 ms. (148 ms.)
4   % = 0.78539816339744830961... \\ = Pi/4
5   ? asympnum(n -> sum(j=0,n-1,1./(1+(j/n)^2))/n-Pi/4)
6   time = 4 ms. (248 ms.)
7   % = [0,1/4,-1/24,0,0,0,1/2016,0,0,0,-1/4224,0,0,0,1/1536,...]
8   ? limitnum(n -> sum(j=1,n,j^2*log(j)))\
9                   -(n^3/3+n^2/2+n/6)*log(n)+n^3/9-n/12)
10  time = 88 ms. (10420 ms.)
11  % = 0.030448457058393270788... \\ = zeta(3) / (4*Pi^2)
12  ? asympnum(n -> sum(j=1, n, j^2 * log(j)) \
13                  - (n^3/3+n^2/2+n/6)*log(n) \
14                  + n^3/9 - n/12 - zeta(3)/(4*Pi^2))
15  time = 96 ms. (10777 ms.)
16  % = [0,-1/360,0,1/7560,0,-1/25200,0,1/33264,0,-691/16216200,\
17              0,1/10296,0,-3617/11138400,...]
18  ? intnum(x=0,1,sin(x)/x)
19  time = 8 ms. (200 ms.)
```

```
20   % = 0.94608307036718301494...
21   ? intnum(x=0,1,lngamma(1+x))
22   time = 44 ms. (1583 ms.)
23   % = -0.0810614667953272582... \\ = log(2*Pi)/2 - 1
24   ? intnumgauss(x=0,1,lngamma(1+x))
25   time = 12 ms. (491 ms.)
26   % = -0.0810614667953272582...
27   ? intnum(x=2,oo,1/(x^2*log(x)))
28   time = 12 ms. (344 ms.)
29   % = 0.37867104306108797672...
30   ? intnum(x=3,oo,1/(x*lngamma(x)))
31   time = 92 ms. (3895 ms.)
32   % = 0.51875414563535767897...
33   ? intnum(x=0,[oo,1],1/gamma(x))
34   time = 44 ms. (1781 ms.)
35   % = 2.80777024202851936522...
36   ? intnum(x=0,[oo,1],exp(-x)/sqrt(x^2+1))
37   time = 12 ms. (352 ms.)
38   % = 0.75461002577097216866...
39   ? sumnum(n=1,(exp(1/n)-1)/n)
40   time = 16 ms. (576 ms.)
41   % = 2.47966052573232990761...
42   ? sumnum(n=1,lngamma(1+1/n)/n)
43   time = 118 ms. (5896 ms.)
44   % = -0.2316676276361361542...
45   ? sumnummonien(n=1,lngamma(1+1/n)/n)
46   time = 22 ms. (942 ms.)
47   % = -0.2316676276361361542...
48   ? sumalt(n=1,(-1)^n*lngamma(1+1/n))
49   time = 11 ms. (395 ms.)
50   % = -0.0602960083664145846...
51   ? sumalt(n=1,(-1)^n*lngamma(n))
52   time = 1 ms. (27 ms.)
53   % = -0.1128956763223637161... \\ = -log(Pi/2) / 4
```

REMARKS 1.6.1.

(1) Note that this last example is a divergent series, for which we can still give a meaningful sum.

(2) The timings of some of the programs (especially intnumgauss and sumnummonien) can be reduced by a large factor (2 to 10 times) by a suitable *initialization* step done once and for all independently of the function. In the timings that we will give, this initialization step will always be given separately. For instance:

```
\p500
? sumnummonien(n = 1, lngamma(1 + 1/n) / n);
time = 942 ms.
? tab = sumnummonieninit();
```

```
time = 819 ms.
? sumnummonien(n = 1, lngamma(1 + 1/n) / n, tab);
time = 125 ms.
```

(3) Some functions such as `lngamma` use a large table of Bernoulli numbers which it stores once and for all. Thus, repeating the command a second time will often give faster results.

CHAPTER 2

Numerical extrapolation

2.1. Introduction

2.1.1. Generalities on basic numerical methods. In this book we will study several types of numerical methods, in particular what we will call the *basic types* which are extrapolation, differentiation, integration, and summation, which are dealt with in this chapter and the two that follow, and more specialized types which are continued fractions, Mellin transforms, and L-functions. We would like to make some general comments on the basic numerical methods.

Since all the corresponding problems are *linear* in nature, we will always proceed in the following way: let f be some function and $L(f)$ the quantity that we need to compute depending linearly on f (limit, derivative, integral, sum for instance). Depending on the desired bit accuracy B we will choose some parameter N, *nodes* $x_1, \ldots, x_N$, and *weights* $w_1, \ldots, w_N$, and approximate $L(f)$ by $\sum_{1 \le j \le N} w_j f(x_j)$. It will also usually be necessary to increase the working accuracy if the weights w_j are large (we will see a typical example in Lagrange extrapolation below).

Note in passing that even for such linear problems there does exist nonlinear methods, such as Sidi m-W extrapolation (the `LimitSidi` and analogous programs), but although they are quite sturdy and efficient in very low accuracy, they are not competitive in large accuracy.

The crucial problem is the choice of the x_j, since once chosen, there is usually a best possible choice for the w_j, and a minimal choice for N to obtain a given accuracy. The important point is that there is no unique solution to this problem, it entirely depends on the class of functions to which we want to apply our methods.

A case in point is numerical summation, which we will study in Chapter 4. Two of the most important methods that we will study are on the one hand *Euler–Maclaurin* type methods, and on the other hand *Gaussian* methods such as *Monien summation*. A look at the table in Section 4.11 shows that Monien summation is far superior in all parameters (number of function evaluations, size of x_j, necessary bit accuracy), but fails miserably on a number of relatively straightforward sums, while Euler–Maclaurin has no problem and is still quite fast. However, the most interesting fact is that the *nodes* x_j (and hence also the weights w_j) are completely different from one method to the next, even when the methods all give perfect results. As an example, consider the computation of $\sum_{n \ge 1} 1/n^2$ to 19 decimal digits, knowing that an antiderivative of the summand is $-1/n$. All methods give perfect results and instantly, but with the following nodes:

- Sumpos: 831 nodes from 1 to $1.1 \cdot 10^{21}$.
- SumLagrange: 22 nodes $1, 2, 3, \ldots, 22$.
- SumEulerMaclaurin: 128 nodes $1, 2, 3, \ldots, 128$.

2.4. Extrapolating by interpolation: Lagrange

Even though we want to *extrapolate* a sequence, we can use *interpolation* methods for this: simply set $f(1/n) = u(n)$, and assume that $f(x)$ is a well-behaved function on $[0, 1]$. We want to compute $f(0)$. The standard method for doing this is to use Lagrange interpolation. We will see that although in principle rather unstable, it is easy to modify it so as to make it quite robust.

2.4.1. The basic method. We choose some integer N, and set

$$P_N(X) = \sum_{1 \le n \le N} \prod_{\substack{1 \le j \le N \\ j \ne n}} \frac{X - 1/j}{1/n - 1/j} u(n) \,.$$

It is clear that $P_N(1/n) = u(n)$ for all $n \le N$. Thus, we can hope that $P_N(0)$ will be a reasonable approximation to $u(\infty)$, i.e., to the limit of $u(n)$.

It is easy to compute $P_N(0)$: we have

$$\prod_{j \ne n} (-1/j)/(1/n - 1/j) = \prod_{j \ne n} n/(n - j) = (-1)^{N-n} \binom{N}{n} \frac{n^N}{N!} \,,$$

so that

$$S := \lim_{n \to \infty} u(n) \approx P_N(0) = \sum_{1 \le n \le N} (-1)^{N-n} \binom{N}{n} \frac{n^N}{N!} u(n) \,.$$

It is in fact quite easy to give an upper bound for the error, but first it will be far too pessimistic, and second in any case there is catastrophic cancellation which we can try to compensate in two ways: first by increasing the accuracy of the computation, and second by replacing the sequence $u(n)$ by $u(n + n_0)$ for some fixed n_0 or by $u(m \cdot n)$ for some multiplier m. By Stirling's formula we see that the largest coefficient in the above formula is obtained for $n \approx aN$ where a is the unique solution in $]0, 1]$ of $e^{1/a}(1 - a) = a$, i.e., $a = 0.782188 \cdots$, and for that value of n the coefficient is approximately $e^{N/a}$, so we need to increase the bit accuracy by $N/(a \log(2)) \approx 1.8444N$ (in fact this constant 1.8444 is equal to $(W(\exp(-1)) + 1)/\log(2)$, where W is Lambert's function).

The main problem is the choice of N as a function of the bit accuracy B. We can reason as follows: if $u_k(n) = 1/n^k$, the method gives as limit

$$S = \frac{1}{N!} \sum_{1 \le n \le N} (-1)^{N-n} \binom{N}{n} n^{N-k} \,,$$

which by expanding $(1 - x)^N$ and applying $N - k$ times the operator $x \, d/dx$ is trivially seen to be equal to 1 for $k = 0$ and to vanish for $1 \le k \le N - 1$, so the result is exact for those sequences. We want to minimize the error for larger values of k: for $k = N$ we have $S = (-1)^{N-1}/N!$, and for $k > N$ it is not difficult to see that $|S|$ will still be of the same order of magnitude, more precisely that

$$S \sim \frac{(-1)^{N-1}}{N!} \frac{\log(N)^{k-N}}{(k - N)!} \,.$$

Thus, to have an error less than 2^{-B} we want $N! > 2^B$, and the corresponding value of N can easily be obtained by a low accuracy estimate, or simply by trial and error, see the program `Limitaccuracy` given below.

This estimate applies when the difference $u(n) - S$ is a polynomial of small degree in $1/n$. On the other hand, if $u(n) - S$ has only an *asymptotic expansion* in powers of $1/n$ this estimate does not apply. We thus consider the next simplest case $u_k(n) = \sum_{1 \le j \le n} 1/j^k$ for $k \ge 2$ with limit $S = \zeta(k)$. We need to estimate

$$f_k(N) = \frac{1}{N!} \sum_{1 \le n \le N} (-1)^{N-n} \binom{N}{n} n^N \sum_{1 \le j \le n} \frac{1}{j^k} - \zeta(k) .$$

A slightly more general result is as follows, and we thank D. Zagier and V. Petrov for the idea of its proof:

PROPOSITION 2.4.1.

(1) *For $k \ge 2$ and $a \ge 1$ integral, set*

$$f_{a,k}(N) = \frac{1}{N!} \sum_{1 \le n \le N} (-1)^{N-n} \binom{N}{n} n^N \sum_{1 \le j \le an} \frac{1}{j^k} - \zeta(k) .$$

Then $\log(|f_{a,k}(N)|) \sim -c_a \cdot N$ with $c_a = \log(|C_a|)$, where C_a is the a-th complex root of the equation $e^{z-1} = z$ ordered by increasing positive imaginary part.

(2) *In particular, to obtain B bits of accuracy using the multiplier a, we need $N \ge r_a B$ with $r_a = \log(2)/c_a$.*

PROOF. We give a sketch of the proof sent to us by D. Zagier, by first proving the following lemma.

LEMMA 2.4.2. *Let $A(t)$ be a continuous function on $[0, \infty[$ which grows at most polynomially as $t \to \infty$, let $a > 0$, and let*

$$L_{a,A}(n) = \int_0^\infty A(t) e^{-ant} \, dt \quad and \quad f_{a,A}(N) = \frac{1}{N!} \sum_{n=1}^N (-1)^{N-n} \binom{N}{n} n^N L_{a,A}(n) .$$

We have the formal identity

$$F_{a,A}(x) := \sum_{N \ge 1} f_{a,A}(N) x^{N-1} = \frac{1}{a} \int_0^1 A(-(\log(1-v) + xv)/a) \, dv .$$

PROOF. Setting $D = d/dt$, it is clear that $D^k(1 - e^{-at})^N$ vanishes for $t = 0$ and $t = \infty$ for $1 \le k \le N - 1$, so integrating $N - 1$ times by parts, we have

$$f_{a,A}(N) = \frac{1}{a^N N!} \int_0^\infty A(t) D^N \big((1 - e^{-at})^N\big) \, dt$$

$$= \frac{(-1)^{N-1}}{a^N N!} \int_0^\infty D^{N-1}\big(A(t)\big) \cdot D\big((1 - e^{-at})^N\big) \, dt$$

$$= \frac{(-1)^{N-1}}{a^{N-1}(N-1)!} \int_0^\infty D^{N-1}\big(A(t)\big) \cdot e^{-at}(1 - e^{-at})^{N-1} \, dt .$$

It follows that

$$F_{a,A}(x) = \int_0^\infty A(t - x(1 - e^{-at})/a) e^{-at} \, dt = \frac{1}{a} \int_0^1 A(-(\log(1-v) + xv)/a) \, dv$$

after making the change of variable $v = 1 - e^{-at}$. □

Now from the trivial relation $m^{-k} = (1/(k-1)!) \int_0^\infty e^{-mt} t^{k-1} \, dt$ we deduce that

$$\sum_{1 \le m \le an} \frac{1}{m^k} - \zeta(k) = -\sum_{m > an} \frac{1}{m^k} = \int_0^\infty A(t) e^{-ant} \, dt$$

with $A(t) = -(1/(k-1)!) t^{k-1}/(e^t - 1)$, so applying the lemma we deduce that if we let $F_{a,k}(x) = \sum_{N \ge 1} f_{a,k}(N) x^{N-1}$ be the generating function of the $f_{a,k}(N)$ we have

$$F_{a,k}(x) = \frac{(-1)^k}{(k-1)! \cdot a^k} \int_0^1 \frac{\left(\log(1-v) + vx \right)^{k-1}}{e^{-\left(\log(1-v) + vx \right)/a} - 1} \, dv \ .$$

Although the formula of the lemma was formal, we check immediately that this last expression converges for all $|x| < 1$, and that by deforming the path of integration we can analytically continue $F_{a,k}(x)$ into a meromorphic function on $\mathbb{C}$.

To estimate the radius of convergence of the function $F_{a,k}(x)$, we must find the closest singularity to the origin. Those of the numerator of the integrand are easily seen to be irrelevant (so the parameter k will not enter in the result), and those of the denominator correspond to $x = \phi_m(v) = (2\pi a m i - \log(1-v))/v$; since $m = 0$ does not correspond to a singularity, we look at the next largest $m = 1$. The function $\phi_1'(v)$ vanishes for $\log(1 - v_0) + v_0/(1 - v_0) = 2\pi a i$, and $\phi_1(v_0) = 1/(1 - v_0) := x_0$ which satisfies $e^{x_0 - 1} = x_0$, so the radius of convergence is $|x_0|$. A more precise analysis gives the result. $\qquad\square$

Since the equation $e^{z-1} = z$ is equivalent to $(-z)e^{-z} = -1/e$, note also that the $-C_a$ are the values of the *complex* branches of the Lambert function at $-1/e$. The function $a \mapsto r_a$ is monotonous and decreases slowly to 0. The constants C_a and r_a can be computed by a simple Newton iteration:

──────────────────── LimitmulNZagier.gp ────────────────────

```
1   /* Low accuracy approximation of the constants C_a = -W_a(-1/e)
2    * and r_a = log(2) / log(abs(C_a)), where W_a is the a-th
3    * complex branch of the Lambert function. */
4   LimitmulNZagier(a = 1) =
5   { my(F, C = I);
6     localbitprec(64);
7     (F(x) = log(x) - x + 1 - 2 * Pi * I * a);
8
9     for (i = 1, 6, C -= C * F(C) / (1 - C));
10    log(2) / log(abs(C)); /* C ~ C_a */
11  }
12
```

EXAMPLE.

```
? LimitmulNZagier(1)
% = 0.33183306518442008061
? LimitmulNZagier(2)
% = 0.26018372785612779930
? LimitmulNZagier(10)
% = 0.16633700432935816368
```

```
? LimitmulNZagier(100)
% = 0.10753811807764592797
```

We can now write a basic function to estimate the limit of a sequence:

────────────────────────────── Limit1.gp ──────────────────────────────
```
1    /* Compute lim_{n -> oo} u(n) using Lagrange interpolation,
2     * assuming u(n) expands in integral powers of 1/n. */
3    Limit1(u) =
4    { my(B = getlocalbitprec(), N, C, S);
5
6      N = ceil(LimitmulNZagier(1) * B);
7      localbitprec(B + Limitaccuracy(1) * N);
8      C = binomial(N);
9      S = sum(n = 1, N, (-1)^(N-n) * C[n+1] * n^N * u(n), 0.);
10     return (bitprecision(S / N!, B));
11   }
```

We have already mentioned that it is often better to apply this program not to the sequence $u(n)$ (or $u(n + n_0)$) itself, but to the sequence $u(a \cdot n)$ for some small integer a. In that case, the constant $r_1 = $ LimitmulNZagier$(1) \approx 0.3318$ should be replaced by r_a given by proposition 2.4.1. We do not include this extra parameter a in the programs that we give, but any serious implementation should use it.

EXAMPLES.
```
? \p 500
  realprecision = 500 significant digits
? u(n) = log(n!) - ((n + 1/2) * log(n) - n);
? Limit1(u) - log(2*Pi)/2
time = 81 ms.
% = 0.E-500 /* Perfect */
? u(n) = sum(j = 1, n - 1, 1. / (1 + (j / n)^2)) / n;
? Limit1(u) - Pi/4
time = 146 ms.
% = 4.37... E-366 /* 134 decimal digits lost. */
? Limit1(n -> u(2*n)) - Pi/4
time = 297 ms.
% = 0.E-500 /* Twice as slow but perfect. */
```
A more sophisticated example:
```
? \p38
? f(N) = (N + 1/2) * acos(-polrootsreal(pollegendre(N))[1]);
? L = Limit1(f)
time = 34 ms.
% = 2.4048255576957727686216318793264546431
? besselj(0, L)
% = 7.19... E-41
```
The limit L is a zero (in fact the smallest) of the Bessel function $J_0(x)$.

2.4.2. Variant 1: Behavior in $1/n^{\alpha m}$. From the rough analysis that we have made it is clear that the basic method will work when the asymptotic expansion of $u(n)$ of the type $u(n) = a_0 + \sum_{m \geq 1} a_m/n^m$ with the a_m behaving reasonably regularly and not growing too fast. We can generalize this, as follows.

Assume that $u(n) = a_0 + \sum_{m \geq 1} a_m/n^{\alpha m}$ for some $\alpha > 0$. It is now natural to set $u(n) = f(1/n^\alpha)$, and performing the same Lagrange interpolation on f, we find that

$$S = \lim_{n \to \infty} u(n) \approx \sum_{1 \leq n \leq N} w_{N,n} u(n) \,,$$

where

$$w_{N,n} = \prod_{1 \leq j \leq N, \ j \neq n} \frac{1}{1 - (j/n)^\alpha} \,.$$

As before, we will both need to determine a working accuracy and a suitable value for N.

For this, we first write an auxiliary program `Limitaccuracy` which computes the necessary accuracy for a given α: by computing the asymptotic behavior of the constants $w_{N,n}$, we find that to compensate for their growth we must increase the working accuracy by $-\log_2(a)N$ bits, where a is the unique positive solution of

$$\log(a) + \int_0^{a+1} \frac{x^{1/\alpha} - 1}{x - 1} \, dx = 0 \,.$$

Second, for a few specific values of α we compute a reasonable value of N as a function of the bit accuracy, as we did above for $\alpha = 1$. For this, we would need to estimate the asymptotic behavior of the error. The standard error estimates given in the literature are not sufficient for this. For the moment, we have only used an experimental approach, taking as test sequence

$$u(n) = \sum_{1 \leq j \leq n} 1/j^{\alpha+1} - 1/(2n^{\alpha+1})$$

and we use the experimentally found value in the programs. The values that we use are given in the following programs:

——————————————— | Limitconstants.gp | ———————————————

```
/* Find the required accuracy constant c for extrapolation,
 * so that bitprec = B + c*N: */
Limitaccuracy(al)=
{ my(a, T, A = 1 / al);

  localbitprec(32); T = intnuminit(0, 1);
  a = solve(b = 1e-5, 1, log(b) +
      intnum(x = 0, b + 1, if (x == 1, A, (x^A-1)/(x-1)), T));
  return (-log(a) / log(2));
}

/* Give magic constant corresponding to multiplier mul = 1,
 * obtained using the Limitmagic program. Only given for
 * al >= 1/4. */
LimitmulN(al) =
{ my(c);
```

```
17
18    localbitprec(32);
19    c = if (al >= 2,    0.2270,
20             al >= 1,    0.3318,
21             al >= 1/2, 0.6218,
22             al >= 1/3, 1.2,
23             al >= 1/4, 3, error("unsupported alpha"));
24    return (c);
25  }
```

EXAMPLES.

```
? Limitaccuracy(1)
% = 1.8444...
? (lambertw(exp(-1)) + 1) / log(2)
% = 1.8444... /* same */
? Limitaccuracy(2)
% = 1.1869...
? Limitaccuracy(1/2)
% = 2.6440...
```

The test programs are as follows, where the Limitweights program will be given below:

─────────────────────────── │ Limitmagic.gp │ ───────────────────────────

```
1   /* [v1, ..., vN] -> [v1/2+...+vN, v2/2+...+vN, ..., vN/2]. */
2   Vecsumrev(v) =
3   { my(N = #v, S = 0.);
4     forstep (n = N, 1, -1, S += v[n]; v[n] = S - v[n]/2);
5     return (v);
6   }
7   /* Test on sum_{0 < j <= n} 1/j^(al+1) - 1/(2n^(al+1)). */
8   Tst(N, al) =
9   { my(v = Vecsumrev( Limitweights(N, al) ));
10    return (sum(n = 1, #v, v[n] / n^(al + 1)));
11  }
12  /* Magic constant when asymp is ~ sum a_i n^(-al*i) */
13  Limitmagic(N, al) = localbitprec(10 * N);\
14    -log(2) * N / log(abs(Tst(N, al) - zeta(al + 1)));
```

EXAMPLES.

```
? Limitmagic(20000, 1)
% = 0.331813...   /* Recall: LimitmulNZagier(1) = 0.331833... */
? Limitmagic(20000, 2)
% = 0.226987...
? Limitmagic(5000, 1/2)
% = 0.621823...
```

Returning to the problem of extrapolation when the asymptotic behavior is in integral powers of $1/n^\alpha$, a first important case is $\alpha = 2$, which may occur for instance in the Romberg integration method that we will see in Chapter 3. In that case it is easy to compute that

$$w_{N,n} = \frac{2}{(2N)!}(-1)^{N-n}\binom{2N}{N-n}n^{2N}$$

for $1 \leq n \leq N$. We can thus write the following program:

──────────────── Limit2.gp ────────────────

```
1   /* Compute lim_{n -> oo} u(n) using Lagrange interpolation,
2    * assuming u(n) expands in powers of 1/n^2. */
3   Limit2(u) =
4   { my(B = getlocalbitprec(), N, C, S);
5
6     N = ceil(LimitmulN(2) * B);
7     localbitprec(B + Limitaccuracy(2) * N);
8     C = binomial(2*N);
9     S = sum(n = 1, N, (-1)^(N - n) * C[N - n + 1]
10                          * n^(2 * N) * u(n), 0.);
11    return (bitprecision(2 * S / (2 * N)!, B));
12  }
```

EXAMPLES.

```
? \p 2000
? u(n) = sum(j = 1, n - 1, 1. / j) - log(n) + 1 / (2*n);
? ga = Euler;
? Limit1(u) - ga
time = 5,812 ms.
% = 0.E-2003 /* Perfect. */
? Limit2(u) - ga
time = 2,360 ms.
% = 0.E-2003 /* Still perfect, more than twice as fast. */
```

In general, however, $w_{N,n}$ does not have an explicit expression and is computed using Lagrange polynomials. This finally leads to the following program which precomputes the $w_{N,n}$ for general α while keeping the more efficient formula for the two special cases $\alpha = 1$ and $\alpha = 2$ seen above:

──────────────── Limitweights.gp ────────────────

```
1   /* [L_i(0), i=1..N], where L_i is i-th Lagrange interpolation
2    * polynomial at the n^al, n <= N (such that L_i(i^al) = 1) */
3   Limitweights(N, al = 1) =
4   { my(V, W);
5     if (al == 1, W = vector(N);
6                  W[1] = (-1)^(N-1)/factorial(N-1);
7                  for (n=2, N, W[n] = W[n-1] * (n-1-N) / n);
8                  for (n=2, N, W[n] *= n^N),
9           al == 2, W = vector(N);
```

```
10              W[1] = (-1)^(N-1)/(factorial(N-1)^2*(N^2+N)>>1);
11              for (n=2, N, W[n] = W[n-1] * (n-1-N) / (N+n));
12              for (n=2, N, W[n] *= n^(2*N)),
13        V = dirpowers(N, al);  /* V[n] = n^al */
14        W = vector(N, n, V[n]^(N-1)
15                     / vecprod([ V[n]-V[j] | j <- [1..N][^n] ])));
16    return (W);
17 }
```

In the above program, the notation `[1..N][^n]` constructs the vector of consecutive integers between 1 and N with the n-th entry removed.

2.4.3. Variant 2: Behavior in $f(n)/n^{\alpha m}$.

The Euler–Maclaurin expansions of expressions such as $\sum_{j \le n} 1/n^{\beta+1} - \zeta(\beta+1)$ for $\beta > 0$ are of the form $a_1/n^\beta + a_2/n^{1+\beta} + \cdots$, so we need to extrapolate sequences with asymptotic behavior $u(n) = a_0 + f(n) \sum_{k \ge 1} a_k/n^{k\alpha}$ where $f(n)n^{-\alpha} \to 0$ as $n \to \infty$. We cannot use directly the previous method, but it is immediate to modify it: we write $u(n)/f(n) = a_0/f(n) + v(n)$, where $v(n) = a_1/n^\alpha + a_2/n^{2\alpha} + \cdots$ is of the form considered above with limit 0. Thus, using again the weights $w_{N,n}$ defined above, the quantity $\sum_{0 \le n \le N} w_{N,n} v(n)$ will be close to 0, hence $\sum_{0 \le n \le N} w_{N,n} u(n)/f(n)$ will be very close to $T_N \cdot a_0$, with $T_N = \sum_{0 \le n \le N} w_{N,n}/f(n)$. Assuming T_N is not close to 0, we obtain a good approximation to a_0.

─────────────────────── Limit.gp ───────────────────────

```
1  Limitalf(alf) =
2  { if (type(alf) == "t_VEC",        alf,
3         type(alf) == "t_CLOSURE", [1, alf],
4         [alf, 0]); }
5
6  /* Compute lim_{n -> oo} u(n) using Lagrange interpolation,
7   * assuming u(n) = a_0 + f(n) * sum_{m>=1} a_m / n^{al m},
8   * alf is the scalar al or a closure f (al = 1) or the pair
9   * [al, f] */
10 Limitinit(alf = 1) =
11 { my(B = getlocalbitprec(), N, B1, W, T);
12   my([al, f] = Limitalf(alf));
13
14   N = ceil(LimitmulN(al) * B);
15   B1 = ceil(B + Limitaccuracy(al) * N);
16   localbitprec(B1); al = bitprecision(al, B1);
17   W = Limitweights(N, al);
18   if (!f, return ([B1, W, 1]));
19   W = vector(#W, n, W[n] / f(n));
20   T = vecsum(W); if (exponent(T) < -B/2, error("T ~ 0"));
21   return ([B1, W, vecsum(W)]);
22 }
23
24 /* vu(N) = [u_1, ..., u_N] */
```

```
25   Limitvec(vu, alf = 1, tab = 0) =
26   {
27     if (!tab, tab = Limitinit(alf));
28     my(U, B = getlocalbitprec(), [B1, W, T] = tab, S);
29     localbitprec(B1); U = vu(#W);
30     S = sum(n = 1, #W, W[n] * U[n], 0.);
31     return (bitprecision(S / T, B));
32   }
33
34   /* u(n) = u_n */
35   Limit(u, alf = 1, tab = 0) =
36     Limitvec(N -> vector(N, n, u(n)), alf, tab);
```

REMARKS 2.4.3.

(1) In the above programs the parameter alf encodes the α and f parameters: a scalar is understood as α (and we take $f(n) = 1$), a function is understood as giving $f(n)$ (and we take $\alpha = 1$) and finally a pair $[\alpha, f]$ specifies both parameters.

(2) The main program is Limitvec in which the sequence is this time given by a function $N \mapsto (u(1), \ldots u(N))$ giving N consecutive terms. It may be faster to compute all such terms at once than independently, for instance if the terms satisfy a nice recursion. There is no loss in flexibility: given a sequence $n \mapsto u(n)$ as previously, the Limit program simply calls Limitvec with argument N -> vector(N,n,u(n)).

EXAMPLE. Use of the alf parameter:

```
? \p 308
  realprecision = 308 significant digits
? u(n) = sum(j = 1, n, j^(-1/3)) - (3/2) * n^(2/3);
? z = zeta(1/3);
? Limit(u) - z
time = 531 ms.
% = 0.00954... \\ catastrophic
? Limit(u, 1/3) - z
time = 24,726 ms.
% = -5.56... E-309 /* Perfect but very slow. */
? Limit(u, n -> n^(2/3)) - zeta(1/3)
time = 536 ms.
% = -1.66... E-308 /* Perfect and very fast. */
? Limit(n -> u(n) - 1 / (2 * n^(1/3)),\
                 [2, n -> n^(2/3)]) - zeta(1/3)
time = 228 ms.
% = -5.56... E-309 /* Perfect and even faster. */
```

The asymptotic expansion of $u(n)$ is of the form

$$u(n) = \zeta(1/3) + n^{2/3}((1/2)/n + a_2/n^2 + a_4/n^4 + \cdots) .$$

Thus the default alf=1 is totally inappropriate and gives only 2 correct decimals. Choosing alf=1/3 gives a perfect answer, but since this assumes only that the expansion is of the form $\zeta(1/3) + c_1/n^{1/3} + c_2/n^{2/3} + c_3/n^{3/3} + \cdots$, there is an enormous loss of time in ignoring the fact that $c_i = 0$ unless $i \equiv 1 \pmod 3$. Choosing `alf=n->n^(2/3)` is orders of magnitude faster since it uses $\alpha = 1$, although the program ignores that $a_i = 0$ if $i \geq 3$ is odd. Choosing `alf=[2,n->n^(2/3)]` (and removing $1/(2n^{1/3})$ from $u(n)$) illustrates the use of both parameters, does not ignore the vanishing of the a_i, and is the fastest of all.

EXAMPLE. Use of the `vec` version:

```
? \p 2003
   realprecision = 2003 significant digits
? u(n) = lngamma(n+1) - ((n+1/2)*log(n)-n);
? Limit(u) - log(2*Pi)/2
time = 3,004 ms.
% = 0.E-2003 /* Perfect. */
? U(N) = my(s = 0.);\
         vector(N, n, my(L = log(n)); s += L;\
                                 s - ((n+1/2)*L-n));
? Limitvec(U) - log(2*Pi)/2
time = 1,508 ms.
% = 0.E-2003 /* Perfect and twice as fast. */
? Limit(u, [2, n -> n]) - log(2*Pi)/2
time = 1,446 ms.
% = 2.209... E-2004 /* Pefect and also twice as fast. */
? Limitvec(U, [2, n -> n]) - log(2*Pi)/2
time = 734 ms.
% = 2.209... E-2004 /* Pefect and fastest. */
```

This example illustrates that in some cases it is faster to use the `vec` version, and shows once again how the knowledge of the vanishing coefficients of the asymptotic expansion can increase the speed even more.

2.4.4. Variant 3: Recursions and sums. When $u(n)$ is given by a recursion, we can often obtain the main term in its asymptotic expansion, but *only* up to a multiplicative or additive constant which must therefore be determined (a typical case being Euler–Maclaurin expansions). The previous `Limitvec` function allows this: we only need to write the function producing the vector $(u(1), \ldots u(N))$ from the recursion. To improve accuracy, it may be necessary to compute $(u(m), \ldots u(mN))$ for some small multiplier m.

A first important case of recursions is the determination of the asymptotic behavior of the partial quotients of continued fractions (see Chapter 7). If $p(n)/q(n)$ is the nth partial quotient, the techniques of that chapter allows us to find a function $f(n)$ such that $q(n) \sim C \cdot f(n)$ for a suitable constant C. More generally, we can write the following program for a second order linear recurrence (we include a multiplier `mul`):

───────────────── | Limitrec.gp | ─────────────────

```
1  /* Let u be a sequence defined by u(0)=u0, u(1)=u1, and the
2   * recursion u(n+1)=a(n+1)u(n)+b(n)v(n-1). Assume known that
3   * u(n)/f(n) tends to a constant C with asymptotic expansion
4   * in powers of 1/n^{al}. Compute C. */
```

```
5
6   Limitrec2(a, b, u0, u1, f, al = 1, mul = 10) =
7   {
8     Limitvec(N ->
9       my(V = vector(mul * N));
10      V[1] = u1 * 1.;
11      V[2] = a(2) * V[1] + b(1) * u0;
12      for (n = 2, mul * N - 1, V[n+1] = a(n+1)*V[n]+b(n)*V[n-1]);
13      vector(N, n, V[mul * n] / f(mul * n)));
14  }
15
16  /* Example of the Motzkin sequence. */
17  LimitMotzkin(mul = 10) =
18  { my(a = n -> (2*n - 1) / (n + 1));
19    my(b = n -> 3*(n - 1) / (n + 2));
20    my(f = n -> 3^n / n^(3/2));
21    Limitrec2(a, b, 0, 1, f, 1, mul);
22  }
```

Note that because of the convention used for continued fractions the coefficient of V[n] is a(n+1) and not a(n). For $q(n)$, set u0=1 and u1=a(1), and for $p(n)$ set u0=a(0) and u1=a(0)a(1)+b(0).

Note also that this program should be considered as a template for more general recursions. We have also included the program for the next example:

EXAMPLE. Consider the sequence u defined by $u(0) = 0$, $u(1) = 1$, and the recursion

$$u(n+1) = \frac{2n+1}{n+2}u(n) + \frac{3(n-1)}{n+2}u(n-1)$$

for $n \geq 1$ (this sequence enumerates *Dyck words* in combinatorics, and $u(n)$ is called the nth *Motzkin number*).

This sequence begins by $u(2) = 1$, $u(3) = 2$, $u(4) = 4$, $u(5) = 9$, $u(6) = 21$, ... Using the methods that we will use below in the context of continued fractions (Theorem 7.3.7), it is not difficult to prove that as $n \to \infty$ we have $u(n) \sim C \cdot 3^n/n^{3/2}$ for a certain constant C. We then use the LimitMotzkin program given above with multiplier mul = 10, and we immediately find that $C = 0.488602511902919921586\cdots$, which can be "recognized" as $1/(4\pi/3)^{1/2}$, so that

$$u(n) \sim \frac{3^n}{((4/3)\pi n^3)^{1/2}} \cdot$$

Exercise: prove this (amusing coincidence: $(4/3)\pi n^3$ is the volume of a sphere of radius n).

REMARK 2.4.4. The reader can check that if we had not used a multiplier (i.e., set mul = 1), the program would have returned nonsense.

Changing the multiplier mul from 2 to 10, say, would marginally change the speed, and will increase the accuracy of the result: at the default 38 decimals of accuracy, with mul = 2 we would obtain 14 correct decimals, and with mul = 9 we would obtain perfect accuracy. This is a general phenomenon: at 1000 decimals of accuracy, we obtain nonsense with `mul` $= 1$, 318 correct decimals with `mul` $= 2, \ldots,$ 933 decimals with `mul` $= 9$ and a perfect result for `mul` $= 10$.

Another important special case of recursion is of the type $u(n) = u(n-1)+c(n)$. This corresponds to summing infinite *series*, and in this case it is not necessary to store the sequence u. We refer to Chapter 4 for the corresponding (trivial) programs.

2.4.5. Zagier's interpretation. An alternative exposition of the Lagrange extrapolation method given above was communicated to us by D. Zagier (and was independently proposed by Levin [**Lev73**]). It proceeds as follows. Choose some reasonable integer k, depending on the desired accuracy, for instance $10 \leq k \leq 13$ for 38 decimals, set $v(n) = n^k u(n)$, and compute the kth forward difference $\Delta^k(v(n))$ of this sequence, where Δ is such that $\Delta(w)(n) = w(n+1) - w(n)$. Note that

$$v(n) = a_0 n^k + \sum_{1 \leq i \leq k} a_i n^{k-i} + O(1/n) \,.$$

The two crucial points are the following:

- The kth forward difference of a polynomial of degree less than or equal to $k-1$ vanishes, and that of n^k is equal to $k!$.
- Assuming reasonable regularity conditions, the kth forward difference of an asymptotic expansion beginning at $1/n$ will begin at $1/n^{k+1}$.

Thus, under reasonable assumptions we have

$$a_0 = \Delta^k(v)(n)/k! + O(1/n^{k+1}) \,,$$

so choosing n large enough can give a good estimate for a_0.

Because of the standard formula for Δ^k this method is equivalent to the basic Lagrange interpolation. As usual, one should use multipliers and/or translations to minimize the effect of small values of n on the sequence u.

Zagier's method can be generalized to cases where $\alpha \neq 1$. For instance, assume that $u(n) = \sum_{0 \leq i \leq p} a_i/n^{i/2} + O(1/n^{(p+1)/2})$. We can modify the above method as follows. First write $u_n = v_n + w_n/n^{1/2}$, where $v_n = \sum_{0 \leq i \leq q} a_{2i}/n^i + O(n^{-q-1})$ and $w_n = \sum_{0 \leq i \leq q} a_{2i+1}/n^i + O(n^{-q-1})$ are two sequences as above. Once again we choose some reasonable integer k such as $k = 10$, and we now multiply the sequence u_n by $n^{k-1/2}$, so we set $u'_n = n^{k-1/2}u_n = n^{k-1/2}v_n + n^{k-1}w_n$. Thus, when we compute the kth forward difference we will have

$$\Delta^k(n^{k-1/2}v_n) = \frac{(k-1/2)(k-3/2)\cdots 1/2}{n^{1/2}} \left(a_0 + \sum_{0 \leq i \leq q+k} b_{k,i}/n^i \right)$$

for certain coefficients $b_{k,i}$, while as above since $n^{k-1}w_n = P_{k-1}(n) + O(1/n)$ for some polynomial $P_{k-1}(n)$ of degree $k-1$, we have $\Delta^k(n^{k-1}w_n) = O(1/n^k)$. Thus we have essentially eliminated the sequence w_n, so we now apply the usual method to $v'_n = n^{1/2}\Delta^k(n^{k-1/2}v_n)$, which has an expansion in integral powers of $1/n$: we will thus have

$$\Delta^k(v'_n)/k! = ((k-1/2)(k-3/2)\cdots(1/2))a(0) + O(1/n^k)$$

(in fact we do not even have to take the same k for this last step).

We have mentioned that for $\alpha = 1$ Zagier's method is in fact a reinterpretation of Lagrange's. On the contrary, the above generalization is completely different from Lagrange with $\alpha = 1/2$, and experimentation shows that it is considerably slower, so we do not give the corresponding program.

2.5. Extrapolation using Sidi's mW algorithm

In a series of papers [**Sid79, Sid82a, Sid82b**], A. Sidi introduced several algorithms generalizing both Richardson's and Lagrange's extrapolation, mainly for the computation of difficult integrals (see Section 3.9.3). These algorithms can also be used for summation.

Let $u(n) = f(1/n)$ and assume that $f(x) = \ell - \phi(x) \sum_{i \geq 0} a_i x^i$ as $x \to 0$, where ϕ is now a *chosen* function tending to 0 at infinity; ℓ and the a_i are unknown. Sidi actually considers a superposition of m such expansions in $x^{\alpha_1 i}, \ldots, x^{\alpha_m i}$ and chooses $\phi_1, \ldots, \phi_m$, but we shall stick to the simplest case $m = 1$ and $\alpha_1 = 1$ (see [**Sid79, FS87**] for the general case).

Let y_j, $j \geq 0$ be a given sequence of positive real numbers such that $\phi(y_j)$ never vanishes, fix an integer n and consider the following linear system in $n + 2$ unknowns $\hat{\ell}, \widehat{a_0}, \ldots, \widehat{a_n}$:

$$\hat{\ell} = f(y_j) + \phi(y_j) \sum_{i=0}^{n} \widehat{a_i} y_j^i, \quad j = 0, \ldots, n + 1.$$

Dividing each line by $\phi(y_j)$, this is almost a polynomial interpolation problem, with an extra term $\hat{\ell}/\phi(y_j)$. Nevertheless, we can either solve the linear system by Gaussian elimination or exploit the system's special shape and apply Sidi's W algorithm [**Sid82a**], which is both more efficient and stable under suitable assumptions. We give the result for $\hat{\ell}$ only, similar formulae give the successive $\widehat{a_i}$ by induction:

THEOREM 2.5.1. *Let $D_k\{g\} = g[y_0, \ldots, y_k]$ be the divided differences operator defined by $g[y_j] = g(y_j)$ for $v \in \{0, \ldots, k\}$ and the recursion*

$$g[y_\nu, \ldots, y_{\nu+j}] = \frac{g[y_{\nu+1}, \ldots, y_{\nu+j}] - g[y_\nu, \ldots, y_{\nu+j-1}]}{y_{\nu+j} - y_j}$$

for $j \in \{1, \ldots, k\}$ and $\nu \in \{0, \ldots, k - j\}$. We have

$$\hat{\ell} \cdot D_n\{1/\phi\} = D_n\{f/\phi\} .$$

PROOF. The equality $\hat{\ell}/\phi(y) = (f/\phi)(y) + \sum_{i=0}^{n} \widehat{a_i} y^i$ holds at all the y_j. The operator $g \mapsto D_n\{g\}$ is linear and vanishes on polynomials of degree $\leq n$. □

General properties of divided differences show that $D_n\{g\}$ is a linear combination of the $g(y_i)$. The weights are positive if and only if the $\phi(y_j)$ alternate in sign, and Sidi [**Sid82a**] proves that the algorithm has good stability properties in this case; these good properties subsist if $\phi(y_j)\phi(y_{j+1}) < 0$ for large j only [**Sid88**].

When ϕ is constant, we recover Richardson's algorithm. We now make the natural choice $y_j = 1/(j+1)$ for all j; for $\phi(x) = x$ we recover Lagrange extrapolation. The following script implements the algorithm with $\phi(1/n) = n\big(u(n) - u(n-1)\big)$, a choice adapted from [**Sid88**] which guarantees oscillations of ϕ around the limiting value 0 for a wide class of sequences, for instance alternating ones. It uses the standard Neville algorithm to compute the two divided differences $D_n(\{u/\phi\})$ and $D_n(\{1/\phi\})$ for increasing $n = 1, 2, \ldots$: the key feature of the algorithm is that, although computing $D_n\{g\}$ uses $O(n^2)$ operations, adding a new data point $(y_{n+1}, g(y_{n+1}))$ to the list requires only $O(n)$ operations. This allows a dynamic scheme, which could have been used for Lagrange interpolation as well: we simply try $n = 1, 2, \ldots$ and stop when the difference between the $\hat{\ell}$ for the current and previous n no longer decreases significantly for 10 consecutive values of n. The program prints a warning if that difference is larger than the expected absolute error 2^{-B}. The downside of this strategy is that the corresponding weights can no longer be precomputed, but lots of sequences benefit from this early abort.

──────────────────────────────── | LimitSidi.gp | ────────────────────────────────

```
1   /* SUM = 1: Sum, SUM = 0: Limit */
2   LimitSidicommon(f, SUM, mu) =
3   { my(B = getlocalbitprec(), eps = 2.^(-B), fail = 0);
4     my(M = List(), N = List(), E = oo, Ekeep = 0, Wkeep = 0);
5     my(S, t, Wp, W, fn);
6
7     localbitprec(mu * B + 32);
8     S = 0.; fn = t = f(1);
9     for (n = 1, oo, \\ fn = f(n)
10      my(e = oo, c);
11      S += t; t = f(n + 1);
12      if (!SUM,
13        my(fN = t); t -= fn; fn = fN); \\ t = f(n + 1) - f(n)
14      if (!t, t = eps);
15      /* Sidi's W algorithm */
16      c = 1. / (n * t); listput(M, S * c); listput(N, c);
17      if (n == 1, Wp = S; next);
18      forstep (s = n - 1, 1, -1,
19        my(d = s * n / (n - s));
20        M[s] = d * (M[s] - M[s + 1]);
21        N[s] = d * (N[s] - N[s + 1]));
22      if (N[1], /* if N[1] = 0, count as failure */
23        W = M[1] / N[1];
24        e = exponent(W - Wp); if (e < -B, break));
25      if (fail++ >= 10,
26        warning("reached accuracy of ", -Ekeep, " bits.");
27        B = -Ekeep; W = Wkeep; break);
28      if (e < E, fail = 0; E = e; Ekeep = E; Wkeep = W);
29      Wp = W);
30    return (bitprecision(W, B));
31  }
32
33  LimitSidi(f, mu = 1.56) = LimitSidicommon(f, 0, mu);
34
35  /* sum_{n >= 1} f(n) using LimitSidi */
36  SumSidi(f, safe = 1) =
37    LimitSidicommon(f, 1, if (safe, 1.56, 1));
38
```

```
39   SumaltSidi(f) =
40     LimitSidicommon(n -> (-1)^(n - 1) * f(n), 1, 1);
```

REMARKS 2.5.2.

(1) The program is as described above when the flag SUM is zero: in this case, the parameter t at the start of Sidi's W algorithm (l. 15) is equal to $f(n+1) - f(n)$; line 22, we have $M[1] = D_n\{f/\phi\}$, $N[1] = D_n\{1/\phi\}$ and if $N[1]$ is non-zero we can compute $W = \hat{\ell}$ on the next line.

The flag SUM = 1 is used in SumSidi, which implicitly applies the extrapolation algorithm to the partial sums $S(n) = \sum_{j \le n} f(j)$. Instead of performing this very computation, we use $S(n+1) - S(n) = f(n+1)$ to set t directly to this value without actually computing $S(n)$.

(2) The computation of $f(n+1) - f(n)$ introduces cancellation as n increases, so the interface includes a multiplier mu to increase the working accuracy. The default value 1.56, a little above 50% extra accuracy, was determined experimentally. This is not always a miracle solution, in that even if we increase the desired accuracy and/or mu, the algorithm may fail to give more than a certain number of digits (see an example below).

EXAMPLE.

```
? \p500 /* Corresponds to 1664 bits. */
? LimitSidi(n -> n * lngamma(1 + 1 / n), 1) + Euler
reached accuracy of 1305 bits. /* mu too small */
time = 212 ms.
% = 1.333... E-393
? LimitSidi(n -> n * lngamma(1 + 1 / n), 1.3) + Euler
time = 397 ms.
% = 0.E-500 /* perfect. */
? u(n) = (Pi/(2*n)) * sum(j=1, n-1, 1 / sin(Pi*j/n)) - log(n);
? LimitSidi(u, 1.5) - Euler + log(Pi / 2) /* with mu = 1.5. */
reached accuracy of 1595 bits. /* mu too small */
time = 13,349 ms.
% = -1.32... E-480
? LimitSidi(u) - Euler + log(Pi / 2)
time = 15,312 ms.
% = -6.096... E-502 /* Now perfect with default mu = 1.56. */
```

As mentioned above, some limits are simply impossible to compute using this algorithm, even after increasing mu. For example:

```
? \p115
? LimitSidi(n -> sum(j = 2, n, (-1)^(j-1) * log(j) / j^2));
time = 216 ms. /* Perfect */
? LimitSidi(n -> sum(j = 2, n, log(j) / j^2))
reached accuracy of 18 bits. /* Completely wrong */
time = 12 ms.   /* Increasing mu would not improve anything */
```

This example shows the essential difference between sequences which *alternate* around their limit, and those which tend *monotonically*. In this latter case, the algorithm is extremely fragile, and it would need to be modified to handle expressions as above involving logs. On the contrary, the alternating case is very robust, and the expressions can be modified essentially at will.

2.6. Computing asymptotic expansions

2.6.1. The rational case. By iteration of extrapolation algorithms we may hope more generally to obtain *asymptotic expansions*, but we have to be careful about successive loss of accuracy. In nice cases the asymptotic expansion is regular, one has "recognized" the limit, and the coefficients of the asymptotic expansion may be rational numbers, which helps with keeping the loss of accuracy to a minimum. The following heuristic program assumes that the expansion is a *rational* combination of integral powers of $1/n^\alpha$:

──────────────── Asymp.gp ────────────────

```
1   /* Given a sequence u(n) [closure] with asymptotic expansion
2    * u_n = a0 + a1/n^al + a2/n^{2al} + ... where the a_i are
3    * rational numbers, find as many a_i as possible. */
4
5   /* vu(N) = [u_1, ..., u_N] */
6   Asympvec(vu, al = 1, LIM = 100) =
7   { my(B = getlocalbitprec(), v = [], W, LB, U, R, B1);
8
9     LB = al * log(B)/log(2);
10    [B1, W] = Limitinit(al);
11    localbitprec(B1); al = bitprecision(al, B1);
12    U = vu(#W); R = dirpowers(#W, al);
13    for (k = 0, oo,
14      my(p, q, a, z = sum(i = 1, #W, W[i] * U[i]));
15      z = bitprecision(z, max(B - floor(k*LB/2), 32));
16      [p, q] = lindep([1, z]); a = -p/q;
17      if (exponent((z-a) * q^2) > -16 || #v >= LIM, return (v));
18      v = concat(v, a); /* found a new a_i */
19      U = vectorv(#U, n, (U[n] - a) * R[n]));
20  }
21  /* u(n) = u_n */
22  Asymp(u, al = 1, LIM = 100) =
23    Asympvec(N -> vector(N, n, u(n)), al, LIM);
```

The program uses Limitinit but not the Limit function, so as to compute the sequence coefficients only once instead of recomputing them each time a new a_k is approximated. Note that we have included a limiting factor LIM, otherwise the program could run indefinitely on simple asymptotic expansions, whenever $a_k = 0$ for all $k > k_0$. Note that the crucial instructions are [p, q] = lindep([1, z]); a = -p/q; which are specific to the rational case.

The program does not support directly more general asymptotic expansions of the form $u(n) = a_0 + f(n)\sum_{k\geq 1} a_k n^{-k\alpha}$: just estimate a_0 to high precision using Limit then apply Asymp to $v(n) = (u(n) - a_0)/f(n)$. As usual, the Asympvec variant is more suitable for sequences defined by recursions rather than by a direct formula.

EXAMPLES. At 38 decimal digits of accuracy.

```
? \p38
? u(n) = n! / (n^n / exp(n)*sqrt(2*Pi*n));
? V = Asymp(u); [#V, V]
% = [9, [1, 1/12, 1/288, -139/51840, -571/2488320, ...]]
? u(n) = lngamma(n+1) - ((n+1/2)*log(n)-n+log(2*Pi)/2);
? V = Asymp(u); [#V, V]
% = [23, [0, 1/12, 0, -1/360, 0, 1/1260, 0, -1/1680, 0, ...]]
? V = Asymp(n -> n*u(n), 2); [#V, V] /* Expansion in 1/n^2. */
% = [11, [1/12, -1/360, 1/1260, -1/1680, 1/1188, -691/360360, ...]]
? u(n) = sum(j = 0, n - 1, 1 / (1 + j/n), 0.) / n - log(2);
? V = Asymp(u); [#V, V]
% = [20, [0, 1/4, 1/16, 0, -1/128, 0, 1/256, 0, -17/4096, ...]]
? u(n) = sum(j = 1, n, 1/sqrt(j)) - 2*sqrt(n) - zeta(1/2);
? V = Asymp(u, 1/2); [#V, V]
% = [35, [0, 1/2, 0, -1/24, 0, 0, 0, 1/384, 0, 0, 0, -1/1024, ...]]
```

2.6.2. The general case. We have assumed that the a_i are rational numbers (at least for $i \geq 1$ since we can subtract a_0 if necessary). It often happens however that a_i is of the form $\omega^i \cdot b_i$ for some known "period" ω and rational numbers b_i, in which case the program is easily adapted (note, however, that since the internal accuracy is increased to compute the coefficients, it is essential to give the period to a much greater accuracy than the desired one or, better, as a *closure* such as x -> Pi).

It may happen that the a_i are not even of the above form, i.e., are some arbitrary real or complex numbers. In that case we must suppress the `lindep` command. However, for each a_i found we will have a loss of accuracy, which must be compensated. More precisely, assume we have computed $a_0, \ldots, a_{L-1}$; we want to compute a_L up to an absolute error 2^{-B} as the limit of $v(n) := n^{\alpha L}\big(u(n) - \sum_{0 \leq i < L} a_i n^{-\alpha i}\big)$ using the `Limit` program. We choose a parameter $N = cB$, where $c = $ `LimitmulN`(α), and then compute $a_L \approx \sum_{1 \leq n \leq N} w_{N,n} v(n)$ at bit accuracy $B_1 = B + dN$, where $d = $ `Limitaccuracy`(α).

Because of the factor $n^{\alpha L}$ we must have computed the previous a_i with absolute error bounded by $2^{-B} N^{-\alpha(L-i)}$. Since we evaluate the $u(n)$ only once, our previous analysis shows that we must have computed them at bit accuracy $B_1 = B + \alpha L \log_2 N + dN$ and chosen $N = c(B + \alpha L \log_2 N)$. As mentioned above, this can be solved explicitly using Corollary 2.3.4, but we will use instead the simple-minded `Solvedivlog` program given above:

────────────────────────────── Asympraw.gp ──────────────────────────────

```
1   /* Given a sequence u(n) as a closure with asymptotic expansion
2    * u(n) = a0 + a1/n^al + a2/n^{2al} + ... Find a0, ..., a_LIM
3    * where the a_i are left as floating point numbers. */
4
5   /* vu(N) = [u_1, ..., u_N] */
6   Asymprawvec(vu, al = 1, LIM = 10) =
```

```
7    { my(B = getlocalbitprec(), v = vector(LIM+1));
8      my(c = LimitmulN(al), d = Limitaccuracy(al));
9      my(N, B1, W, U, R);
10
11     localbitprec(32);
12     N = ceil(Solvedivlog(c * al * LIM / log(2), c * B));
13     B1 = ceil(B + N / c + d * N);
14     localbitprec(B1); al = bitprecision(al, B1);
15     W = Limitweights(N, al);
16     U = vu(N);
17     R = vectorv(N, n, n^al);
18     for (k = 0, LIM,
19       my(z = sum(i = 1, #W, W[i] * U[i]));
20       U = vectorv(#U, n, bitprecision((U[n] - z) * R[n], B1));
21       v[k+1] = bitprecision(z, B));
22     return (v);
23   }
24   /* u(n) = u_n */
25   Asympraw(u, al = 1, LIM = 10) =
26     Asymprawvec(N -> vector(N, n, u(n)), al, LIM);
```

Setting the precision of $(U[n] - z) \times R[n]$ to B_1 is essential otherwise the routine quickly loses accuracy: we expect cancellation when subtracting $U[n] - z$ and this causes the next $z = \sum_i W[i]U[i]$ to be computed at a lower bit accuracy than expected.

EXAMPLE.

```
? V = Asympraw(n -> n * lngamma(1 + 1 / n), , 50)
% = [-0.577..., 0.822..., ...]]
? W = vector(#V, k, if(k == 1, -Euler, (-1)^k * zeta(k) / k));
? vector(10, k, exponent(V[5*k] - W[5*k]))
% = [-130, -131, -131, -132, -132, -132, -133, -103, -64, -26]
```

Since the bit accuracy is 128, we see that the first 35 coefficients are perfect, and then the accuracy of the result decreases.

EXAMPLE. The above example gave in fact the Taylor expansion of $\log(\Gamma(1+x))$ around $x = 0$. For a more complicated example, consider the function $f(x) = \sum_{k \geq 1} 1/(e^{k^2 x} - 1)$ studied by Ramanujan. It can be shown that as $x \to 0$, $f(x)$ has an expansion of the form $f(x) = a/x + b/x^{1/2} + c + o(1)$ for certain constants a, b, and c. We thus set $u(n) = f(1/n)/n = a + b/n^{1/2} + c/n + o(1/n)$. We first try to find the limit a using the built-in suminf summation function since the series defining f converges exponentially fast:

```
? f(x) = suminf(k = 1, 1 / (exp(k^2*x) - 1));
? u(n) = f(1 / n) / n;
? Limit(u, 1/2) /* must use alf = 1/2 */
% = -103249.19... /* nonsense */
? Limit(u -> u(10 * n), 1/2)
```

```
% = 1.6449340668482264364724151666460251891
? Limit(u -> u(12 * n), 1/2)
% = 1.6449340668482264364724151666460251892
```

This tells us that it is necessary to use a multiplier of at least 10. We can now write:

```
? V = Asympraw(n -> u(12 * n), 1/2);
? V = vector(#V, j, 12^((j - 1) / 2) * V[j]);
% = [1.6449..., -1.2942..., 0.24999..., 2.3...E-30,...,-3.1... E-12]
```

The second command transforms the asymptotic expansion of $u(12n)$ into that of $u(n)$.

We immediately recognize the limit $V[1]$ as being $\zeta(2) = \pi^2/6$, and apparently $V[3] = 1/4$, and $V[j] = 0$ for $j \geq 4$. $V[2]$ is harder to recognize, but one can show that $V[2] = \sqrt{\pi}\zeta(1/2)/2$. Thus, we have the experimental expansion around $x = 0$:

$$\sum_{k \geq 1} \frac{1}{e^{k^2 x} - 1} = \frac{\pi^2}{6x} + \frac{\sqrt{\pi}\zeta(1/2)}{2x^{1/2}} + \frac{1}{4} + O(x^8) .$$

In fact, one can show that the error term $O(x^8)$ is exponentially small, and in particular is $O(x^N)$ for any N.

EXAMPLE. Consider the *Franel numbers* defined by

$$F_n = \sum_{0 \leq j \leq n} \binom{n}{j}^3 .$$

(**Exercise:** evidently $\sum_{0 \leq j \leq n} \binom{n}{j} = 2^n$; compute explicitly $\sum_{0 \leq j \leq n} \binom{n}{j}^2$.)

It is clear that $F_n < 8^n$, and that $F_n/8^n \to 0$ when $n \to \infty$. We want to compute the asymptotic expansion of $u_n = F_n/8^n$. Consider first the problem of computing its limit (known to be equal to 0):

```
? F(n) = sum(j = 0, n, binomial(n, j)^3);
? u(n) = F(n) / 8^n;
? Limit(u)
% = -6.63... E-5 /* Bad. */
? Limit(n -> u(2 * n))
% = 5.02... E-26 /* Better. */
? Limit(n -> u(4 * n))
% = -4.15... E-38 /* Perfect. */
```

It thus seems that the sequence $u(4n)$ is better behaved with respect to our extrapolation process. We thus write:

```
? V = Asympraw(n -> u(4 * n))
% = [4.80... E-54, 0.09188..., -0.0076..., ...]
? C = V[2]; W = V / C
% = [5.23... E-53, 1.0000..., -0.08333..., ...]
? bestappr(W, 10^12)
% = [0, 1, -1/12, 1/432, 1/5184, 1/62208, 11/2239488,
        49/80621568. -317/322486272, ...]
```

In addition, writing `algdep(C * Pi, 2)` shows that $C = \sqrt{3}/(6\pi)$. Changing back $4n$ into n we thus have the asymptotic expansion

$$u(n) = \frac{2\sqrt{3}}{3\pi} \left(\frac{1}{n} - \frac{1}{3n^2} + \frac{1}{3^3 n^3} + \frac{1}{3^4 n^4} + \frac{1}{3^5 n^5} + \frac{11}{3^7 n^6} + \frac{49}{3^9 n^7} - \frac{317}{3^9 n^8} + \cdots \right).$$

Exercise: do the same for $u(n) = \sum_{0 \le j \le n} \binom{n}{j}^4 / 16^n$; warning: either use $\alpha = 1/2$ or study $n^{1/2} u(n)$. Which is the most efficient ?

2.6.3. Example: Zeros of J and Y-Bessel functions. As another example of the `Asympraw` program, we now show how to compute the asymptotic expansion of zeros of the J and Y-Bessel functions (see [**WW96**] for definitions). Before doing so, we give a general program for Newton refinement. Recall that if f is a C^2 function in some domain, the Newton iteration $x_{n+1} = x_n - f(x_n)/f'(x_n)$ is a quadratically convergent method to find a root z of $f(x) = 0$ under certain conditions: first, we must be in the "attraction basin" of the iteration (if we begin the iteration at x_0 too far away from a zero, the Newton iteration may wildly diverge), and second, the derivative $f'(x)$ must not be too small around the zero (in particular the method usually fails around a multiple zero). More precisely, using Taylor's theorem to order 2 we have $x_{n+1} - z = k_n(x_n - z)^2$, with $k_n = f''(c)/(2f'(c))$ for some c close to z. Thus, under our assumptions we may assume that k_n is close to a constant during the iteration, hence if $|x_n - z| < 2^{-b_n}$, we should have $|x_{n+1} - z| < 2^{-b_{n+1}}$, with $b_{n+1} \approx 2b_n - e$, where $e = \log_2(|k_n|)$. In addition, because of this it is unnecessary to perform the iterations at full precision, we begin at rather low accuracy, and double it (more precisely use $b_{n+1} = 2b_n - e$) at each step, doing only the last iteration at full precision, so that the whole process takes approximately twice the time of the last iteration.

We can thus write the following generic program for doing this:

──────────────── `Newtonrefine.gp` ────────────────

```
1   /* Assuming we are in the basin of attraction and
2    * |f'| is not too small, refine x using Newton.
3    * T is the function f / f'. */
4   Newtonrefine(T, x) =
5   { my(B = getlocalbitprec(), e, n, c, x0, e0, e2);
6
7     x0 = bitprecision(x, 96); e0 = exponent(T(x0));
8     e = max(exponent(1 - T'(x0)) - e0 - 1, 0);
9     e2 = min(-e0, 64) - e;
10    if (e2 <= 0, error("Insuficcient accuracy in Newtonrefine"));
11    n = 1 + exponent((B + 32 - e) \ e2);
12    c = 1 + e + (B - e) >> n;
13    for (j = 1, n,
14      c = 2 * c - e; x = bitprecision(x, c); x -= T(x));
15    return (bitprecision(x, B));
16  }
```

──

Note that in most practical cases, the derivative of the function f is known explicitly, so instead of computing it using numerical derivation techniques that we will study in the next chapter, it is better to give it explicitly. We have preferred instead to give explicitly the function $T = f/f'$ since, as is the case of Bessel functions, it will be 50% faster to compute f/f' than to compute f and f' separately. Note also that $f''/f' = (1 - T')/T$.

We now apply this program to Bessel functions: the following programs compute a vector of the first N zeros using the known fact that the nth zero of J is approximately equal to $(n + \nu/2 - 1/4)\pi$, and that of Y to $(n + \nu/2 - 3/4)\pi$.

Besselzeros.gp

```
/* Zeros for Bessel functions J_nu or Y_nu. */

/* f = besselj or bessely. */
Besselrefine(f, nu, x) =
{ my(T = (z -> 1 / ((nu / z) - f(nu + 1, z) / f(nu, z))));
  Newtonrefine(T, x);
}

Besseljzeros(nu, N) =
{ my(z = (nu/2 - 1/4)*Pi);
  vector(N, n, Besselrefine(besselj, nu, z += Pi));
}

Besselyzeros(nu, N) =
{ my(z = (nu/2 - 3/4)*Pi);
  vector(N, n, Besselrefine(bessely, nu, z += Pi));
}
```

EXAMPLE.

```
? Besseljzeros(0, 5)
% = [2.4048..., 5.5200..., 8.6537..., 11.7915..., 14.9309...]
? Besseljzeros(1/2, 5) / Pi
% = [1.0000..., 2.0000..., 3.0000..., 4.0000..., 5.0000...]
? Besselyzeros(1/2, 5) / Pi
% = [0.5000..., 1.5000..., 2.5000..., 3.5000..., 4.5000...]
```

We now want to find the asymptotic expansion of the zeros of J_0, so we simply do as follows:

```
? V = Asymprawvec(N -> Besseljzeros(0, N) \
                    - Pi * vector(N, n, n - 1/4))
% = [2.2030... E-59, 0.0397..., 0.0099..., -0.00011..., ...]
```

This leads us to believe that these zeros indeed have an asymptotic expansion in integral powers of $1/n$, but with coefficients that do not seem easy to recognize. However, if we do the following

```
? bestappr(subst(Ser(V), x, Pi * x / (1 + Pi * x / 4)), 10^20)
% = 1/8*x - 31/384*x^3 + 3779/15360*x^5 - 6277237/3440640*x^7 \
         + 2092163573/82575360*x^9 + O(x^11)
```

we find an expansion involving only reasonable rational numbers. Note that the strange change of variable $\pi x/(1 + \pi x/4)$ corresponds to the fact that the natural expansion is in odd powers of $(n + \nu/2 - 1/4)\pi$ for J and $(n + \nu/2 - 3/4)\pi$ for Y. As an exercise, the reader can try to guess the formula for the first few terms of the corresponding expansion for arbitrary ν.

2.7. Sample timings for Limit programs

The reader can find on the book's website complete comparative timings for the methods given in this chapter, and others that we may not even have mentioned (this is of course true for all the chapters of this book). In this section, we only give some samples.

We will assume that our sequence u tends to a limit a_0 as $n \to \infty$, that u is monotonic, and that it has an asymptotic expansion in $1/n$ or more generally in $1/n^{\alpha m + \beta}$ as above. If u is given directly as a function (i.e., not by a recursion), we may use the following commands or variants of those:

```
a0 = LimitRichardson(u);
a0 = Limit(u);
a0 = Limitvec(u);
a0 = LimitSidi(u);
```

We will see in Chapter 4 summation methods that could be used here as well (by summing $u(n) - u(n-1)$); however the two best methods, Sumdelta and SumMonien, cannot be used because they use an integral $\int_N^\infty u(t)\,dt$ which in principle does not make sense since u is only defined on the integers.

If u monotonic is given by a sufficiently long precomputed vector V, we cannot use LimitRichardson since that function uses values of n which will usually be much too large compared to the size of the vector. Thus, we can only use a0=Limit(u) and its variants, where u=(n->V[n]), and care must be taken that the vector is long enough.

If u is given by a recursion, once again LimitRichardson is essentially useless, but the method explained above for recursions using Lagrange extrapolation is tailor-made for this.

We consider the following examples. We only consider the case of monotonous $u(n)$, since in the alternating case we use Sumalt or variants. The first examples are cases where one can give $u(n)$ as a function:

$$u_1(n) = (1 + 1/n)^n, \quad u_2(n) = (1 + 1.0/n)^n : \quad a_0 = \exp(1),$$

$$u_3(n) = \log(n!) - \big((n + 1/2)\log(n) - n\big) : \quad a_0 = \log(2\pi)/2,$$

$$u_4(n) = \log(\Gamma(n+1)) - \big((n + 1/2)\log(n) - n\big) : \quad a_0 = \log(2\pi)/2,$$

$$u_5(n) = n\sin(1/n) : \quad a_0 = 1, \quad u_6(n) = n^2\big(1 - \cos(1/n)\big) : \quad a_0 = 1/2,$$

$$u_7(n) = \sum_{1 \le j \le n} 1/j^2 : \quad a_0 = \pi^2/6,$$

$$u_8(n) = \sum_{1 \le j \le n} 1/j - \log(n) - 1/(2n) : \quad a_0 = \gamma,$$

$$u_9(n) = \sum_{2 \le j \le n} \log(j)/j - \log^2(n)/2 : \quad a_0 = \gamma_1,$$

$$u_{10}(n) = \sum_{2 \le j \le n} j \log(j) - (6n^2 + 6n + 1)\frac{\log(n)}{12} + \frac{n^2}{4} : \quad a_0 = \frac{1}{2} - \zeta'(-1) ,$$

$$u_{11}(n) = \sum_{2 \le j \le n} j^2 \log(j) - \left(\frac{n^3}{3} + \frac{n^2}{2} + \frac{n}{6}\right) \log(n) + \frac{n^3}{9} - \frac{n}{12} : \quad a_0 = \frac{\zeta(3)}{4\pi^2} ,$$

$$u_{12}(n) = \frac{\pi}{2n} \sum_{1 \le j \le n-1} \frac{1}{\sin(\pi j/n)} - \log(n) : \quad a_0 = \gamma - \log(\pi/2) ,$$

$$u_{13}(n) = \frac{\pi}{2n} \sum_{1 \le j \le n-1} \frac{1}{\Im(e(\pi j/(2n)))} - \log(n) : \quad a_0 = \gamma - \log(\pi/2) ,$$

$$u_{14}(n) = \frac{3}{4n} + \frac{1}{n} \sum_{1 \le j \le n-1} \frac{1}{1 + j/n} : \quad a_0 = \log(2) ,$$

$$u_{15}(n) = \frac{3}{4n} + n \sum_{1 \le j \le n-1} \frac{1}{j^2 + n^2} : \quad a_0 = \pi/4 ,$$

$$u_{16}(n) = \frac{1}{(n-1)!} \sum_{1 \le j \le n} (-1)^{n-j} \binom{n}{j} j^{n-1/2} : \quad a_0 = \sqrt{2/\pi} ,$$

$$u_{17}(n) = \frac{n! e^n}{2n^n} - \sum_{0 \le j \le n} \frac{n!}{(n-j)! n^j} : \quad a_0 = -2/3 ,$$

$$u_{18}(n) = \sum_{1 \le j \le n} 1/j^{1/2} - 2n^{1/2} : \quad a_0 = \zeta(1/2) ,$$

$$u_{19}(n) = \frac{1}{n! \sqrt{n}} \sum_{0 \le j \le n-1} (n-j)^j (n-j)! : \quad a_0 = \sqrt{\pi/2} ,$$

$$u_{20}(n) = \frac{n!}{\sqrt{n}} \sum_{1 \le j \le n} \frac{1}{(n-j)! n^j} : \quad a_0 = \sqrt{\pi/2} ,$$

$$u_{21}(n) = \frac{1}{n!} \sum_{1 \le j \le n} e^{-j} j^n : \quad a_0 = 1/2 ,$$

$$u_{22}(n) = \sum_{1 \le j \le n} 1/j^{3/5} - n^{2/5}/(2/5) : \quad a_0 = \zeta(3/5) ,$$

$$u_{23}(n) = \sum_{1 \le j \le n} 1/j^{\pi/2} : \quad a_0 = \zeta(\pi/2) .$$

REMARKS 2.7.1.

(1) Of course some of these sequences are simply sums of infinite series, so are better treated using one of the summation programs that we shall study in Chapter 4.

(2) γ is Euler's constant, and γ_1 is the first Stieltjes constant that we will see as S_{29} in the summation examples.

(3) u_1 and u_2 are of course mathematically identical, but will not have the same running time depending on how the function is defined. Similarly for u_3 and u_4, for u_{13} and u_{14} (where each successive exponential is computed with a single multiplication from the previous one).

The next examples correspond to the case where u is more naturally defined by a recursion.

(1) Once again $r_1(n) = \sum_{1 \le j \le n} 1/n^2$, but now given by $r_1(0) = 0$, $r_1(1) = 1$, and the recursion $r_1(n+1) = r_1(n) + 1/(n+1)^2$.

(2) The constant C occurring in the enumeration of Dyck words, as explained in an example above, and the corresponding sequence will be called $r_2(n)$.

(3) A large source of sequences defined by recursions are continued fractions, which we will study in great detail in the following chapters. Consider for instance the following recursion, due to Apéry:

$$u(n+1) = ((34n^3 + 51n^2 + 27n + 5)u(n) - n^3 u(n-1))/(n+1)^3 \ ,$$

with on the one hand the initial terms $q(0) = 1$, $q(1) = 5$, and on the other hand $p(0) = 0$, $p(1) = 6$. Apéry proves that $\lim_{n \to \infty} p(n)/q(n) = \zeta(3)$. On the other hand, the study of recurrences of this type that we make below shows that $u(n) \sim C_u \cdot n^{-3/2} \cdot (1+\sqrt{2})^{4n}$ for constants C_u (of course different for the sequences p and q) which are difficult to compute. We find that for the sequence $q(n)$ we have $C_q = 0.22004 \cdots$, and with a little habit, it is easy to recognize that

$$C_q = \frac{(1+\sqrt{2})^2}{\pi^{3/2} 2^{9/4}} \ ,$$

and of course $C_p = \zeta(3) C_q$. We choose $r_3(n) = q(n)$.

(4) We will see below the beautiful continued fraction due to Ramanujan (Proposition 7.5.5):

$$\frac{\Gamma^2(1/4)}{\Gamma^2(3/4)} = \cfrac{8}{1 - \cfrac{1^2}{12 - \cfrac{3^2}{24 - \cfrac{5^2}{36 - \ddots}}}} \ .$$

If as usual we call $p(n)/q(n)$ the nth convergent, one can prove that $q(n) \sim C' \cdot 2^n (1+\sqrt{2})^{2n} n!/n^2$ for some constant C'. Using Lagrange extrapolation, we find $C' = 0.0068779 \cdots$ which we "recognize" as being equal to $1/((1+\sqrt{2})^2 \sqrt{\pi} 2^{9/4} \Gamma(1/4)/\Gamma(3/4))$. We denote by $r_4(n)$ the sequence above.

Among the examples given above, we distinguish four types of sequences.

(1) Sequences given as reasonably fast expressions (i.e., not requiring time say proportional to n to compute, but more like a power of $\log(n)$). These are the u_j for $1 \le j \le 6$. These are the only sequences for which it is plausible to use Richardson-type or Sumpos methods, which require the computation of $u(n)$ for exponentially large n.

(2) Sequences whose nth term needs time proportional to n (or at least exponential in $\log(n)$) to compute and asymptotic expansion in integral powers of $1/n$, so u_j with $7 \le j \le 17$ with the exception of u_9.

(3) Sequences with an asymptotic expansion in powers of $1/n^\alpha$ with $\alpha \ne 1$, for u_j with $18 \le j \le 22$ (treated with the general Limit program), or in $1/n^\beta$ times integral powers of $1/n$, for u_{22} and u_{23} (treated with the general Limit program).

(4) Sequences defined naturally by recursions r_j for $1 \leq j \leq 4$ (treated with `Limitrec`).

Detailed tables for many accuracies are available from the authors, but we only give a typical one for $D = 500$, in the form of several distinct tables corresponding to the four types of sequences mentioned above and different values of α and β. The times are in seconds; ∞ means that the program takes an unreasonable amount of time (more than 5 minutes), $\otimes$ means either that the program is currently unable to compute the result, or that the result is terribly wrong (fewer than 25% of digits are correct); fastest timings are in boldface, and the numbers in parentheses indicate the number of lost decimal digits (out of 500) in our current implementation.

u	LimitRichardson	LimitSidi	Limit
Init	0.00	0.00	0.00
u_1	∞	0.12	**0.00**
u_2	**0.00**	0.12	0.01
u_3	∞	0.52	**0.08**
u_4	0.02 (5)	0.51	**0.08**
u_5	**0.00**	0.12	0.05
u_6	0.00 (28)	0.11	**0.05**
u_7	∞	0.44	**0.00**
u_8	∞	0.48	**0.04**
u_9	∞	$\otimes$	$\otimes$
u_{10}	∞	0.51	**0.08**
u_{11}	∞	0.52	**0.08**
u_{12}	∞	10.5	**0.41**
u_{13}	∞	1.81	**1.29**
u_{14}	∞	0.62	**0.16**
u_{15}	∞	0.99 (48)	0.15 (135)
$u_{15} \circ 2$	∞	0.67	**0.32**
u_{16}	∞	4.93 (129)	0.21 (253)
$u_{16} \circ 4$	∞	11.4	**1.95**
u_{17}	∞	0.71	**0.25**

Timings for 500 Decimals ($\alpha = 1$).

In these examples, using a multiplier regains perfect accuracy in the bad cases u_{15} and u_{16} instead of losing hundreds of decimal digits. The loss of accuracy of `LimitRichardson` on u_6 is entirely due to the catastrophic cancellation occurring in the computation of $1 - \cos(1/n)$ for huge values of n. The result would be perfect if we replaced $1 - \cos(1/n)$ by $2\sin^2(1/(2n))$.

The `Limit` column corresponds to either `Limit` or `Limitvec`, where we used the latter whenever there was a significant advantage in computing all consecutive terms $u(1), \ldots, u(N)$ at once, rather than independently. For instance, all sums were computed using `Limitvec`. `LimitSidi` on the other hand uses a dynamic scheme introducing consecutive terms one at a time until the requested accuracy is reached, which prevents computing a vector of terms since the value of N is unknown. The simple solution to make a fair comparison with `Limitvec` is to use the recursive formula with memoization, storing all sequence terms as they are computed: should `LimitSidi` need to compute another term in the sequence it can proceed from the already computed ones. For instance, $u_7(n) = \sum_{1 \le j \le n} 1/j^2$ was implemented as follows:

```
u7(n, ~M) =
{ my(z);
  if (!mapisdefined(M, n, &z),
    z = u7(n - 1, ~M) + 1. / n^2;
    mapput(~M, n, z));
  return (z);
}
M = Map(Mat([0,0])); LimitSidi(n -> u7(n, ~M))
```

u	LimitSidi	Limit
Init	0.00	0.00
u_6	0.13	**0.02**
u_8	0.51	**0.02**
u_{10}	0.50	**0.04**
u_{13}	1.80	**0.44**
u_{14}	0.62	**0.06**
u_{15}	0.98 (48)	0.06 (76)
$u_{15} \circ 2$	0.68	**0.14**

Timings for 500 Decimals ($\alpha = 2$).

Comparing to the table with $\alpha = 1$, we see that if applicable one should use $\alpha = 2$ in the `Limit` program since the program is at least twice as fast.

u	LimitSidi	Limit
Init	0.00	3.62
u_{18}	0.47	**0.01**
u_{19}	$\otimes$	$\otimes$
$u_{19} \circ 8$	$\otimes$	**8.23**
u_{20}	$\otimes$	**0.86**
u_{21}	$\otimes$	$\otimes$
$u_{21} \circ 8$	**1.35**	5.24

Timings for 500 Decimals ($\alpha = 1/2$).

u	LimitSidi	Limit
Init	0.00	0.02
u_{22}	0.48	**0.02**

Timings for 500 Decimals ($\beta = 3/5$).

u	LimitSidi	Limit
Init	0.00	0.17
u_{23}	0.57	**0.11**

Timings for 500 Decimals ($\beta = \pi/2$).

u	Limitrec
r_1	**0.00**
r_2	$\otimes$
$r_2 \circ 10$	**0.01**
r_3	0.01 (103)
$r_3 \circ 2$	**0.01**
r_4	0.01 (260)
$r_4 \circ 3$	**0.02**
InitMax	0.00

Timings for 500 Decimals (recursions).

As noticed above, in the case of the Motzkin sequence r_2 we obtain nonsense without a multiplier, but obtain perfect accuracy with a multiplier of 10. For r_4, the multiplier 3 also gives perfect accuracy with essentially no speed loss.

2.8. Conclusion

What we find is the following:

- When the sequence is very easy to compute, in our samples $u_2(n) = (1 + 1.0/n)^n$, $u_4(n) = \log(\Gamma(n+1)) - ((n+1/2)\log(n) - n)$, $u_5(n) = n\sin(1/n)$, and $u_6(n) = n^2(1 - \cos(1/n))$, use the basic `LimitRichardson` program (note that even though they are given by "closed" expressions, $u_3(n) = \log(n!) - ((n+1/2)\log(n) - n)$ and $u_1(n) = (1 + 1/n)^n$ take a huge amount of time to compute when n is extremely large, so `LimitRichardson` cannot be used on such sequences). Of course, `Limit` is also very fast.
- When the sequence takes a long time to compute, for instance because it involves a number of terms proportional to n, use `Limit` or one of its variants. It is also important to choose the multiplier when one suspects that accuracy may be lost, such as in u_{15} and u_{16}, and in the recursions.
- If `Limit` does not work, for instance if the asymptotic behavior is of the form $f(n)/n^{\alpha m}$ for some *unknown* function f or even some unknown α, then its generalization `LimitSidi` is a user-friendly choice because of its robustness: it requires little knowledge about the sequence while remaining reasonably fast. It is not competitive with `Limit` in terms of speed, with the single exception of sequence u_{21}.

The reader will have noticed that we have not treated the sequence $u_9(n) = \sum_{2 \le j \le n} \log(j)/j - \log^2(n)/2$. This is because its asymptotic expansion involves *both* terms in $\log(n)/n^m$ *and* terms in $1/n^m$ and it should be treated by Sidi's $W^{(2)}$ algorithm [**FS87**]. However, the summation methods explained in Chapter 4 already allow to compute the limit of $u_9(n)$: we write

$$u_9(n) - u_9(n-1) = \log(n)/n - (\log^2(n) - \log^2(n-1))/2$$
$$= \log(n)/n + \log(1 - 1/n)\log(n(n-1))/2 .$$

Using the `Sumdelta` program given in Chapter 4 (or the built-in `sumnum` program), we can compute the limit of $u_9(n)$, but we must be careful. Consider the following three functions:

```
f1(n) = log(n)/n - (log(n)^2-log(n-1)^2)/2;
f2(n) =
{ my(B=getlocalbitprec());

  localbitprec(3*B); n = bitprecision(n, 3*B);
  log(n)/n - (log(n)^2-log(n-1)^2)/2;
}
f3(n) = log(n)/n + log(1-1/n)*log(n*(n-1))/2;
```

Writing `Sumdelta(f1)` gives a reasonable (but slightly wrong) result at 38 decimal digits, but which becomes completely wrong at 500 D. This is because of catastrophic cancellation in the computation of the difference of the two log squared. The function `f2` corrects this by tripling the internal accuracy. This now gives a perfect result, at the expense of speed. But the best solution is to write `Sumdelta(f3)` which minimizes the cancellation, and gives a perfect result quite fast:

$$\lim_{n \to \infty} u_9(n) = -0.0728158454836767248605863758749013191377 \cdots$$

Note that there are specific methods to compute this type of limit, see for instance Exercises 55 to 57 of Chapter 10 of [**Coh07b**].

CHAPTER 3

Numerical integration

Let f be a well-behaved function on some interval $[a, b]$, possibly noncompact. In this book, well-behaved may mean several things, the weakest being continuous (or sometimes piecewise continuous), the strongest being C^∞, or even holomorphic. The goal of numerical integration is of course to compute $I = \int_a^b f(x)\, dx$, assuming that this makes sense, in particular when the interval is noncompact.

Before embarking on this, we devote a long section to the inverse problem of *numerical differentiation*.

3.1. Numerical differentiation

The problem is as follows: given a function f, say defined and C^∞ on a real interval, compute $f'(x_0)$ for a given value of x_0. To be able to analyze the problem, we will assume that $f'(x_0)$ is not too close to 0, and that we want to compute it to a given *relative accuracy*, which is what is usually required in numerical analysis.

The naïve, although reasonable, approach, is to choose a small $h > 0$ and compute $(f(x_0 + h) - f(x_0))/h$. However, it is clear that (using the same number of function evaluations) the formula $(f(x_0 + h) - f(x_0 - h))/(2h)$ will be better. Let us analyze this in detail. For simplicity we will assume that all the derivatives of f around x_0 that we consider are neither too small nor too large in absolute value. It is easy to modify the analysis to treat the general case.

Assume f computed to a relative accuracy of ε, in other words that for any given x we know values $\tilde{f}(x)$ such that $\left|f(x) - \tilde{f}(x)\right| < \varepsilon\left|\tilde{f}(x)\right|$. The absolute error in computing $(f(x_0 + h) - f(x_0 - h))/(2h)$ is thus bounded by $\varepsilon|f(x_0)|/h$. On the other hand, by Taylor's theorem we have

$$\bigl(f(x_0 + h) - f(x_0 - h)\bigr)/(2h) = f'(x_0) + (h^2/6)f'''(x)$$

for some x close to x_0, so the absolute error made in computing $f'(x_0)$ as $(f(x_0+h) - f(x_0 - h))/(2h)$ is close to $\varepsilon|f(x_0)|/h + (h^2/6)|f'''(x_0)|$. For a given value of ε (i.e., the accuracy to which we compute f) the optimal value of h is $(3\varepsilon|f(x_0)/f'''(x_0)|)^{1/3}$ for an absolute error bound of $(1/2)(3\varepsilon|f(x_0)f'''(x_0)|)^{2/3}$ hence a relative error of

$$(3\varepsilon|f(x_0)f'''(x_0)|)^{2/3}/(2|f'(x_0)|) \ .$$

Since we have assumed that the derivatives have reasonable size, the relative error is roughly $C\varepsilon^{2/3}$, so if we want this error to be less than η, say, we need ε of the order of $\eta^{3/2}$, and h will be of the order of $\eta^{1/2}$.

Note that this result is not completely intuitive. For instance, assume that we want to compute derivatives to 38 decimal digits. With our assumptions, we choose h around 10^{-19}, and perform the computations with 57 decimals of relative accuracy. If for some reason or other we are limited to 38 decimals in the computation of f, the "intuitive" way would be also to choose $h = 10^{-19}$, and the above analysis shows that we would obtain only approximately 19 decimals. On the other hand, if we chose $h = 10^{-13}$ for instance, close to $10^{-38/3}$, we would obtain 25 decimals.

EXAMPLE. (At large accuracy):

```
? \p 1001
? F(h) = (log(2 + h) - log(2 - h)) / (2 * h);
? F(10^(-334)) - 1 / 2
% = 6.6... E-669 /* 2/3 decimals using cube root of accuracy */
? F(10^(-500)) - 1 / 2
% = 4.0... E-503 /* 1/2 decimals using square root of accuracy */
```

A simple-minded program to do this is as follows:

─────────────────────────── Diff.gp ───────────────────────────

```
1   /* Compute f'(a) numerically. */
2   Diff(f, a) =
3   { my(B = getlocalbitprec(), B1 = (3*B) \ 2, h, S);
4
5     localbitprec(B1);
6     h = 2^(-B\2);
7     a = bitprecision(a, B1);
8     S = (f(a + h) - f(a - h)) / (2*h);
9     return (bitprecision(S, B));
10  }
```

REMARKS 3.1.1.
  (1) Here, the change in accuracy to $B_1 = 3B/2$ is part of the algorithm, and not a "cheat" as mentioned above.
  (2) We could have written h=sqrt(2^(-B)) instead of h=2^(-B\2), but this would be much slower: first because of the extra square root replacing a trivial shift, but more importantly because whenever a is an exact number of the order of 1, the expression $a + h$ would have bit accuracy $B_1 + B/2$ instead of the intended B_1. The two function evaluations would then take place at a higher accuracy than needed.
  (3) On the other hand, in case a is an inexact number, we make sure its accuracy is at least B_1 before adding $\pm h$. Otherwise, the following computations would take place at the accuracy of a, lower than the needed B_1.
  (4) This is in fact preprogrammed in GP as the following two commands: either derivnum(x=a,f(x)), or even simpler f'(a), but it is instructive to give the program explicitly because of the above caveats.

There are of course many other formulas for computing $f'(x_0)$, or for computing higher derivatives, which can all easily be analyzed as above. To simplify notations, we will assume from now on that $x_0 = 0$: this causes no loss of generality since one may always apply the resulting formulas to $f(x + x_0)$.

For instance (exercise), one can look for approximations to $f'(0)$ of the form $S = (\sum_{1 \leq i \leq 3} \lambda_i f(h/a_i))/h$, for any nonzero and pairwise distinct a_i, and we find that this is possible as soon as $\sum_{1 \leq i \leq 3} a_i = 0$ (for instance, if $(a_1, a_2, a_3) = (-3, 1, 2)$

we have $(\lambda_1, \lambda_2, \lambda_3) = (-27, -5, 32)/20$, and the absolute error is then of the form $C_1/h + C2h^3$, so the same analysis shows that we should work with accuracy $\varepsilon^{4/3}$ instead of $\varepsilon^{3/2}$. Even though we have $3/2$ times more evaluations of f, we require less accuracy: for instance, if f requires time $O(B^\omega)$ to be computed to B bits, as soon as $(3/2) \cdot ((4/3)B)^\omega < ((3/2)B)^\omega$, i.e., $3/2 < (9/8)^\omega$, hence $\omega \geq 3.45$, this new method will be faster.

Perhaps the best known method with more function evaluations is the approximation

$$f'(0) \approx (f(-2h) - 8f(-h) + 8f(h) - f(2h))/(12h) ,$$

which requires accuracy $\varepsilon^{5/4}$, and since this requires 4 evaluations of f, this is faster than the first method as soon as $2 \cdot (5/4)^\omega < (3/2)^\omega$, in other words $\omega > 3.81$, and faster than the second method as soon as $(4/3) \cdot (5/4)^\omega < (4/3)^\omega$, in other words $\omega > 4.46$. To summarize, use the first method if $\omega < 3.45$, the second method if $3.45 \leq \omega < 4.46$, and the third if $\omega > 4.46$. Of course this game can be continued at will, but there is not much point in doing so. In practice the first method is sufficient, since it is extremely fast.

Finally, let us mention the "complex step" algorithm which applies to an *analytic* function f whose restriction to $\mathbb{R}$ takes real values. For a real parameter h, we write $f(ih) = f(0) + ihf'(0) - h^2/2!f''(0) + O(h^3)$ and approximate $f'(0) = \Im f(ih)/h + O(h^2)$. This involves a single function evaluation, avoids cancellation altogether, and crucially does not need to increase the working accuracy by 50%!

———————————————— DiffC.gp ————————————————

```
1  /* f: R->R, restriction of an analytic function in the
2   * neighborhood of a; compute f'(a). */
3  DiffC(f, a) =
4  { my(B = getlocalbitprec(), h = 2^(-B\2));
5
6    return (imag(f(a + I*h)) / h);
7  }
```

EXAMPLE.

```
? \p 5009
   realprecision = 5009 significant digits
? z = zeta'(3);
time = 6,494 ms. /* Warning: only after Bernoulli initialization. */
? Diff(zeta, 3) - z
time = 6,498 ms.
% = 1.81... E-5010 /* Same time. */
? DiffC(zeta, 3) - z
time = 1,787 ms.
% = 0.E-5009 /* Much faster. */
```

For higher derivatives, we can use the formula recursively, for instance

$$f''(0) \approx \frac{f(h) - 2f(0) + f(-h)}{h^2} .$$

In this case the optimal h is $C \cdot \varepsilon^{1/4}$, and the error is of the form $C' \cdot \varepsilon^{1/2}$, so we must work to accuracy ε^2, which is more expensive. If we want to work with accuracy $\varepsilon^{3/2}$ as above, we can use for instance the formula

$$f''(0) \approx \frac{-f(2h) + 16f(h) - 30f(0) + 16f(h) - f(2h)}{12h^2}$$

with $h = C \cdot \varepsilon^{1/6}$. However, if f is *holomorphic*, we can use the simpler formula

$$f''(0) \approx \frac{(f(h) + f(-h)) - (f(ih) + f(-ih))}{2h^2}$$

also with $h = C \cdot \varepsilon^{1/6}$ and accuracy $\varepsilon^{3/2}$.

Of course it gets worse for derivatives of higher order: let $M \geq 0$ be the order of the highest derivative we wish to approximate, we need to evaluate f in at least $M+1$ points, at an accuracy that grows with M. There is no reason to restrict to equally spaced grids, so let $(\alpha_1, \ldots, \alpha_N)$ be N complex numbers. We want to find universal weights $w_{m,n}$ such that

$$(3.1) \qquad f^{(m)}(0) \approx h^{-m} \sum_{k=1}^{N} w_{m,k} f(h\alpha_k), \quad 0 \leq m < N ,$$

for any sufficiently regular function f, where h is a real parameter tending to 0. It turns out that there is a unique solution if we require an approximation error of $O(h^{N-m})$ or better in the above approximations !

LEMMA 3.1.2. *Assume that f is of class C^N. The absolute error in the above is $O(h^{N-m})$ if and only if*

$$\sum_{k=1}^{N} w_{m,k} \alpha_k^m = m! \quad and \quad \sum_{k=1}^{N} w_{m,k} \alpha_k^n = 0$$

for all $n < N$, $n \neq m$.

PROOF. Expand $f(z)$ into a Taylor-Maclaurin series with remainder $O(z^N)$ and substitute in the right hand side of

$$h^m f^{(m)}(0) + O(h^N) = \sum_{k=1}^{N} w_{m,k} f(h\alpha_k) .$$

Equating powers of h, we obtain the desired equalities.   □

We now prove that there *is* a solution and at the same time give a more appealing formulation in terms of Lagrange interpolation: for $k = 1, \ldots, N$, let $L_k(X)$ be the Lagrange basis polynomial with value 1 at α_k and 0 at α_i, $i \neq k$, namely

$$L_k(X) = \frac{Q(X)}{Q'(\alpha_k)(X - \alpha_k)} , \quad \text{where} \quad Q(X) = \prod_{k=1}^{N}(X - \alpha_k) .$$

Applying (3.1) to $f(X) = L_k(X/h)$, we find that we must have $L_k^{(m)}(0) = w_{m,k}$. And indeed, since L_k has degree $N-1$ we have $L_k(X) = \sum_{0 \leq m < N} L_k^{(m)}(0) X^m / m!$, which gives the equations of the lemma.

REMARKS 3.1.3.

(1) This yields an efficient way to compute the $w_{m,k}$, as the coefficients of $L_k(X)$ multiplied by $m!$. Once the polynomials L_k are precomputed, we get the $L_k^{(m)}(0)$ for free for all m: this yields formulas to compute $f^{(m)}(0)$ for all $m \le M$ at once.

(2) If we fix h and let $L_f(X) := \sum_k L_k(hX)f(h\alpha_k)$ be the Lagrange polynomial interpolating the values of the function f at the $(h\alpha_k)$, then (3.1) becomes $f^{(m)}(0) \approx L_f^{(m)}(0)$.

(3) If the grid points α_k are symmetric about 0 (for every k, there is a k' such that $\alpha_{k'} = -\alpha_k$), then $L_k(-X) = L_{k'}(X)$ hence $w_{m,k} = (-1)^m w_{m,k'}$.

It is possible to have higher order of accuracy $O(h^{N-m+b})$, if and only if $\sum_{k=1}^{N} w_{m,k}\alpha_k^n = 0$ also holds for all $N \le n < N + b$. In particular, if the grid points α_k are symmetric about 0 then the above remark implies that

$$\sum_{k=1}^{N} w_{m,k}\alpha_k^N = \sum_{k=1}^{N} w_{m,k'}\alpha_{k'}^N (-1)^{N-m} \sum_{k=1}^{N} w_{m,k}\alpha_k^N .$$

COROLLARY 3.1.4. *If the grid points α_k are symmetric about 0 and $N - m$ is odd, then the absolute error in (3.1) is $O(h^{N-m+1})$.*

It can be proved [**SV14**] that we cannot have $b > m$; if the grid points are real numbers, we even have $b \le 1$ so the above corollary is optimal in this case (and always applies at the expense of possibly increasing N by 1 to make $N - m$ odd).

Given a largest order of derivation $M \ge 1$ and desired absolute accuracy 2^{-B}, we now make choices for the α_k and the parameters N and h. As in the above, we assume that f requires time $O(B^\omega)$ to be computed to B bits. We consider that all relevant $w_{m,k}$ are precomputed and only take into account the time needed to evaluate f.

We choose the (α_k) symmetric about 0 and N odd for simplicity. The choice $h \approx 2^{-B/(N-M)}$ is suitable to have a final absolute error bounded by 2^{-B} and we finally choose N such that $N \approx (\omega + 1)M$. The working accuracy is then $B(1 + M/(N - M)) \approx B(1 + 1/\omega)$ bits and the evaluation cost is dominated by $MB^\omega(\omega + 1)^{\omega+1}/\omega^\omega$.

In the corresponding program, we choose a symmetric equally spaced grid, i.e., $\{0, \pm 1, \ldots, \pm A\}$ where $N = 2A + 1$ is odd. With this choice, the polynomials $M! \cdot L_k$ have small integer coefficients (which can be explicitly expressed in terms of Stirling numbers but the resulting formula does not seem to be useful to speed up the computation).

––––––––––––––––––––– DiffM.gp –––––––––––––––––––––

```
/* Compute the M!/m! * w_{m,k} for m <= M; interpolating at
 * 2*N+1 points: (0, +-1, +-2, ..., +- N). */
DiffMinit(M, om) =
{ my(C, h, L, W, Q, N2, N, X, x = 'x);

  N2 = ceil((om+1) * M); N2 += (N2 + 1) % 2; \\ make N2 odd
  N = N2 \ 2;
  Q = x * vecprod(vector(N, k, x^2 - k^2));
  C = binomial(2*N);
```

```
10      \\ (2N)! / Q'(X[k+1]) = (-1)^(N-k) binomial(2*N, N-k)
11      X = vector(N2); L = vector(N2);
12      X[1] = 0;
13      L[1] = ((Q \ x) + O(x^(M+1))) * (C[N+1] * (-1)^N);
14      for (i = 1, N, my(p = 2*i, m = p+1);
15        X[p] =  i;
16        L[p] = (Q \ (x-i) + O(x^(M+1))) * (C[N-i+1] * (-1)^(N-i));
17        X[m] = -i;
18        L[m] = subst(L[p], x, -x));
19      /* L[k] = (2N)! L_k(x), W[m+1][k] = (2N)!/m! * w_{m,k} */
20      W = vector(M+1, m, vector(N2, k, polcoef(L[k], m-1)));
21      my(eW = [exponent(w) | w <- W[M+1], w != 0]);
22      my(E = vecmax(eW) - vecmin(eW));
23      my(B1, B = getlocalbitprec());
24      B1 = ceil(B * (1 + 1 / om) + E + 32);
25      localbitprec(B1);
26      h = 1.0 >> (ceil(B / (N2 - M)));
27      return ([X, W, h]);
28    }
29
30    /* [f(a), f'(a), ..., f^(M)(a)/M!]. Optimize assuming that
31     * f is computed to B bits in time O(B^om) */
32    DiffM(f, a = 0, M = 2, om = 2) =
33    { my(B = getlocalbitprec(), B1, X, W, Q, h, F, vF);
34      [X, W, h] = if (type(om) == "t_VEC", om
35                                          , DiffMinit(M, om));
36      B1 = bitprecision(h); localbitprec(B1);
37      a = bitprecision(a, B1);
38      F = [ f(a + h*v) | v <- X ]~;
39      Q = powers(1 / h, M, 1 / (#X-1)!);
40      \\ Q[m+1] = 1 / (h^m * (2*N)!), vF[m+1] ~ f^(m)(a)/m!
41      vF = vector(M+1, m, (W[m] * F) * Q[m]);
42      return (bitprecision(vF, B));
43    }
44
45    /* Compute Taylor expansion of f around x = a
46     * up to O((x-a))^M */
47    Taylor(f, a = 0, M = 3, om = 2) = Ser(DiffM(f, a, M - 1, om));
```

This is in fact preprogrammed in GP using derivnum(x = a, f(x), [0..M]). Of course, it is then trivial to obtain the Taylor expansion of f around $x = a$, as we have done in the Taylor program.

EXAMPLE.

```
? \p 5009
   realprecision = 5009 significant digits
? V = DiffM(lngamma, 2, 10);
time = 6,772 ms.
```

```
? W = vector(#V, n, if (n == 1, 0, \
                  n == 2, 1 - Euler, \
                  (-1)^(n-1) * (zeta(n-1) - 1) / (n-1)));
time = 80 ms.
? exponent(V - W)
% = -16641 /* Perfect accuracy since bitprecision is 16640. */
```

In the case where f is *holomorphic*, a final possibility to compute derivatives is by using *numerical integration* methods and *Cauchy's formula*:

$$\frac{f^{(k)}(a)}{k!} = \frac{1}{2\pi i} \int_{C_a} \frac{f(x)}{(x-a)^{k+1}}\, dx \ ,$$

where C_a is a circle centered at a not containing any poles of f. Experiment shows that most numerical integration methods that we will see in the rest of this chapter give good results with respect to accuracy but are not competitive with the above method, which requires fewer function evaluations. Note that $(1/(2\pi i)) \int_{C_a}$ is preprogrammed in GP under the name `intcirc`.

3.2. Integration of rational functions

Before coming to the question of integration of general functions, we begin with the simpler problem of integrating rational functions. In principle this is easy and taught in first-year calculus courses. In practice, however, there are a number of difficulties to overcome. Let $F \in \mathbb{C}(x)$ be a rational function. We want to compute $J = \int_a^b F(x)\, dx$, where $[a, b]$ is an interval of $\mathbb{R}$ with $b > a$: note that we may always easily reduce to that case by an affine change of variable. Also, a and b may possibly be equal to $\pm\infty$ (when the integral converges), it will not change the present discussion.

The standard method for computing J is first to expand F into partial fractions, say

$$F(x) = P(x) + \sum_{\alpha \text{ pole of } F} \sum_{1 \le j \le v(\alpha)} \frac{c_{\alpha,j}}{(x-\alpha)^j} \ ,$$

where $P(x) \in \mathbb{C}[x]$ is a polynomial, $v(\alpha)$ is the order of the pole α, and $c_{\alpha,j} \in \mathbb{C}$ are certain complex numbers. Each term can then be integrated individually: assuming for instance that $[a, b]$ is a compact interval of $\mathbb{R}$ (the modifications for a or b equal to $\pm\infty$ are immediate), we have

$$\int_a^b \frac{dx}{(x-\alpha)^j} = \frac{1}{j-1}\left(\frac{1}{(a-\alpha)^{j-1}} - \frac{1}{(b-\alpha)^{j-1}}\right) \quad \text{for } j \ge 2 \ ,$$

while for $j = 1$ we have

$$\int_a^b \frac{dx}{x-\alpha} = \log((b-\alpha)/(a-\alpha)) + k(a, b, \alpha)2\pi i$$

for some integer $k(a, b, \alpha)$, and log denotes the principal determination of the complex logarithm, such that $-\pi < \Im(\log(z)) \le \pi$. An easy but not totally trivial exercise shows that in fact we always have $k(a, b, \alpha) = 0$.

Thus, the remaining problem is to compute the expansion into partial fractions. Once again, this is completely standard: writing $F = N/D$ with N and D coprime polynomials in $\mathbb{C}[x]$, we first perform the Euclidean division of N by D as $N = DP + R$, giving the polynomial P. We then compute the roots α of D and their multiplicities $v(\alpha)$ with a polynomial root finder (`polroots` in GP^{1})[1], and for each such root α the coefficients $c_{\alpha,j}$ are the first $v(\alpha) - 1$ terms of the Taylor expansion around $x = 0$ of $R(x + \alpha)/(D(x + \alpha)/x^{v(\alpha)})$.

And this is where the difficulties begin. If the coefficients of F are in some *exact* field, such as $\mathbb{Q}$ or $\mathbb{Q}(i)$, all is well. On the other hand, if the coefficients are inexact, typically floating point numbers to some (possibly large) accuracy, we must think about every step of the above procedure.

(1) First, we say that N and D are coprime polynomials in $\mathbb{C}[X]$: since the coefficients are inexact, this does not make any sense. However, this is unimportant: if N and D are not coprime, in the procedure above we will have too many poles α or poles with too large multiplicity, but the coefficients $c_{\alpha,j}$ that we will find will be negligible, so that these extra poles or multiplicities will not really modify the result of the computation of the integral. In addition, we can test whether inexact polynomials are coprime simply by checking that their *resultant* is "far" from 0.

(2) For the same reason, the fact that the coefficients of the numerator N (or equivalently, of the remainder R) may be inexact does not really matter.

(3) Another case in which there is nothing to worry about is when the coefficients of the *denominator* D are exact: in that case we can compute correctly the multiplicities and approximate the roots.

(4) A final case where we can compute the roots of D is when D is *squarefree*: of course, if D is inexact this does not seem to make much sense, but we can say that D is squarefree again if the *resultant* of D and D' is not too small.

Thus, the only remaining case where there is a problem is when D has inexact coefficients and is possibly not squarefree. In that case, the computation of the roots of D may be incorrect (at least when they may be multiple roots), so we will usually obtain a result with many wrong decimals.

Here are a number of useful programs for working with rational functions:

―――――――――――――――――― Ratdec.gp ――――――――――――――――――

```
1   /* Squarefree factorization [[P1,c1],...,[Pk,ck]] of exact
2    * polynomial P = P1^c1 ... Pk^ck; characteristic is 0 */
3   PolSQF(P) =
4   { my(SQF = List(), T, V, c);
5
6     T = gcd(P, P'); V = P \ T; c = 0;
7     while (poldegree(V),
8       my(W = gcd(T, V), P = V \ W);
9       c++; if (poldegree(P), listput(SQF, [P,c]));
10      V = W; T \= W);
11    return (Vec(SQF));
12  }
13
14  /* Format of complex roots of polynomial P: [[z, c, rr], ...],
15   * where z are the roots (up to conjugation if P is real),
```

―――

[1]This root finder is a sophisticated program written by X. Gourdon based on the splitting circle method of A. Schönhage, which will not be discussed in this book; see [**Gou93**].

```
16      * c the multiplicity, and rr = 1 unless P is real and z
17      * nonreal, in which case rr = 2. */
18     Polroots(P) =
19     {
20       if (precision(P) == oo, return (Polrootsexact(P)));
21       my(B = getlocalbitprec());
22       /* P inexact. Error if we can't rule out multiple roots */
23       if (exponent(polresultant(P, P')) < -B / poldegree(P),
24         error("inexact and possibly nonsquarefree polynomial"));
25       return ([[z, 1, 1] | z <- polroots(P)]);
26     }
27
28     /* P is an exact polynomial */
29     Polrootsexact(P) =
30     { my(B = getlocalbitprec(), SQF = PolSQF(P), L = List());
31
32       for (i = 1, #SQF,
33         my([pol, c] = SQF[i], ro = polroots(pol));
34         my(flre = (imag(pol) == 0));
35         for (j = 1, #ro,
36           my(z = ro[j], rr);
37           rr = if (flre && exponent(imag(z)) > -B, 2, 1);
38           listput(L, [z, c, rr]);
39           /* conjugate roots adjacent in ro, skip conj(z) */
40           if (rr == 2, j++)));
41       return (Vec(L));
42     }
43
44     /* Partial fraction decomposition of rational function F with
45      * denominator either exact or squarefree. The answer is a
46      * vector of 4-component vectors [z, c, rr, sz], where z
47      * ranges through the distinct poles of F, and only one among
48      * two conjugates if the denominator DF of F is real, c is the
49      * order of the pole, rr = 2 if DF is real and z nonreal, and
50      * sz is the power series giving the polar decomposition
51      * around z: if sz = sum(k = 0, v-1, ck * X^k) + O(x^c) then
52      *    F = sum(k = 0, v-1, ck/(X-z)^(v-k)) + O(1).
53      * The integer part is ignored. */
54     Ratdec(F) =
55     { my(DF = denominator(F), NF = numerator(F), X = variable(F));
56       my(L, Lconj, Poles);
57
58       Poles = Polroots(DF);
59       L = vector(#Poles); Lconj = List();
60       for (i = 1, #Poles,
61         my([z, c, rr] = Poles[i]);
62         my(DFz = DF \ (X - z)^c, Qz = NF / DFz);
63         my(sz = subst(Qz, X, X + z) + O(X^c));
```

```
64      if (rr == 2 && imag(Qz),
65        rr = 1; /* split into z and conj(z) */
66        listput(Lconj, [conj(z), c, 1, conj(sz)]));
67      L[i] = [z, c, rr, sz]);
68    return (concat(L, Vec(Lconj)));
69  }
```

EXAMPLE.

```
? bestappr(Ratdec(1 / ((x^2 - 4) * (x + 1)^2 * (x^2 + 1))))
% = [[-2, 1, 1, -1/20 + O(x)], [2, 1, 1, 1/180 + O(x)],\
     [-I, 1, 1, 1/20 + O(x)], [-1, 2, 1, -1/6 - 1/18*x + O(x^2)],\
     [I, 1, 1, 1/20 + O(x)]]
```

In many cases we can completely avoid the above problems by using a completely *different* method to perform the computation instead of expanding into partial fractions.

Denote by $M = \max |\alpha|$ the maximum modulus of all the poles of F. Recall that $b > a$ are real, and assume that $a > M$. If we compute the Taylor series expansion of $x^d F(1/x)$ around $x = 0$ with $d = \deg(N) - \deg(D)$, we will obtain a power series with nonzero constant term and radius of convergence R equal to the modulus of the closest pole of $F(1/x)$ to the origin, i.e., $R = 1/M$. If we write $x^d F(1/x) = \sum_{k\geq 0} u_k x^k$, it follows that u_k is of the order of M^k. Thus for $|x| > M$ we have $F(x) = \sum_{k\geq 0} u_k x^{d-k}$, so

$$\int_a^b F(x)\,dx = \sum_{k\geq 0} \frac{u_k}{d+1-k}(b^{d+1-k} - a^{d+1-k}).$$

Since we assume that $b > a > M$ and u_k is of the order of M^k, it follows that the series converges geometrically, so can be computed directly.

More generally, assume that we can find some complex number c such that $\max(|a - c|, |b - c|) < \min |\alpha - c| = R$, where as usual α runs over the poles of F. Once again we can compute the Taylor expansion around $x = c$, say $F(x) = \sum_{k\geq 0} u_k(x - c)^k$, and u_k will be of order $1/R$, so $\int_a^b F(x)\,dx = \sum_{k\geq 0}(u_k/(k + 1))((b - c)^{k+1} - (a - c)^{k+1})$, which will be a geometrically convergent series.

To summarize the above discussion: we can compute numerically the integral on $[a, b]$ of a rational function with denominator D in the following cases.

(1) D is an exact polynomial.
(2) D is possibly nonexact but is squarefree, in the sense that the resultant of D and D' is "sufficiently far" from 0.
(3) If $b > a > \max |\alpha|$, where α runs over all the zeros of D. In practice, to avoid too long computations, we will ask that $b > a \geq 1.1 \max |\alpha|$.
(4) If one can find $c \in \mathbb{C}$ such that $\max(|a - c|, |b - c|) < \min |\alpha - c|$, where again α runs over all the zeros of D. Once again in practice, we will ask that $1.1 \max(|a - c|, |b - c|) \leq \min |\alpha - c|$,

We now write the corresponding programs. Note that in GP, ∞ is represented by oo or +oo, and $-\infty$ by -oo, so we will use this. We first give the programs using partial fractions; for simplicity, we assume $-\infty < a < b \leq \infty$ or $(a, b) = (-\infty, \infty)$:

──────────────────────── | Intratdec.gp | ────────────────────────

```
 1  /* Integration of rational function, assuming that denominator
 2   * is exact or squarefree, using Ratdec. */
 3
 4  Badpole(z, a, b) =
 5  { my(B = getlocalbitprec(), s = real(z), t = imag(z));
 6    return (exponent(t) <= -B && s >= a && s <= b);
 7  }
 8
 9  /* Integral from a to b, finite. Assume a < b */
10  Intratdeca_b(F, a, b) =
11  { my(X = variable(F), DF = denominator(F), NF = numerator(F));
12    my(q, S, Poles, co);
13
14    [q, NF] = divrem(NF, DF);
15    my([Qa, Qb] = subst(intformal(q), X, [a,b]));
16    S = Qb - Qa;
17    Poles = Ratdec(NF / DF);
18    for (j = 1, #Poles,
19      my([z, v, rr, sz] = Poles[j]);
20      my(T, P = powers(X - z, v - 1)); \\ P[i+1] = (X - z)^i
21      if (Badpole(z, a, b), error("diverging integral"));
22      T = sum(k = 0, v-2, polcoef(sz, k) * P[k+1] / (k+1-v));
23      if (rr == 2, T = 2 * real(T * conj(P[v])) / norm(P[v])
24                 , T /= P[v]);
25      [Qa, Qb] = subst(T, X, [a,b]);
26      S += Qb - Qa;
27      co = polcoef(sz, v - 1); if (!co, next);
28      co *= log((b - z) / (a - z));
29      S += if (rr == 2, 2*real(co), co));
30    return (S);
31  }
32
33  /* Integral from a to +oo. */
34  Intratdeca_oo(F, a) =
35  { my(S = 0, Poles = Ratdec(F), X = variable(F), co);
36
37    if (poldegree(F) >= -1, error("diverging integral"));
38    for (j = 1, #Poles,
39      my([z, v, rr, sz] = Poles[j]);
40      my(T, P = powers(X - z, v-1)); \\ P[i+1] = (X - z)^i
41      if (Badpole(z, a, oo), error("diverging integral"));
42      T = sum(k = 0, v-2, polcoef(sz,k) * P[k+1] / (k+1-v));
43      if (rr == 2, T = 2 * real(T * conj(P[v])) / norm(P[v])
44                 , T /= P[v]);
45      S -= subst(T, X, a);
46      co = polcoef(sz, v-1); if (!co, next);
47      co *= log(a - z);
```

```
48       S -= if (rr == 2, 2 * real(co), co));
49     return (S);
50   }
51
52   /* Integral from -oo to +oo. */
53   Intratdecoo_oo(F) =
54   { my(S = 0, Poles = Ratdec(F), co);
55
56     if (poldegree(F) >= -1, error("diverging integral"));
57     for (j = 1, #Poles,
58       my([z, v, rr, sz] = Poles[j]);
59       if (Badpole(z, -oo, oo), error("diverging integral"));
60       co = polcoef(sz, v - 1); if (!co, next);
61       co *= I * sign(imag(z));
62       S += if (rr == 2, 2 * real(co), co));
63     return (S * Pi);
64   }
65
66   /* Driver routine when one can use partial fractions.
67   Here -oo <= a < b <= oo and a == -oo only if b == oo. */
68   Intratdec(F, a, b) =
69   {
70     if (a == -oo, return (Intratdecoo_oo(F)));
71     if (b ==  oo, return (Intratdeca_oo(F, a)));
72     return (Intratdeca_b(F, a, b));
73   }
```

We now give the programs when one can use the Taylor expansion at infinity, still assuming $a > b$.

──────────────────────────── IntratTaylor.gp ────────────────────────────

```
1    /* Use Taylor expansion at infinity instead of partial
2     * fractions. We assume b > a > R. */
3    IntratTaylor_oo(F, a, b, R) =
4    { my(B = getlocalbitprec(), X = variable(F));
5      my(L, ser, v, Sa, Sb);
6
7      if ((b == oo || a == -oo) && poldegree(F) >= -1,
8        error("diverging integral"));
9      localbitprec(32); L = ceil(log(2)*B / log(a/R)) + 1;
10     localbitprec(B);
11     ser = 1. * subst(F, X, 1/X) + O(X^(L+2));
12     v = vector(L, k, polcoef(ser, k+1) / k);
13     Sa = Sb = 0;
14     forstep (k = L, 1, -1, Sa = (v[k] + Sa) / a);
15     if (b != oo, forstep (k = L, 1, -1, Sb = (v[k] + Sb) / b));
16     return (Sa - Sb);
17   }
```

```
18
19   /* Use c such that max(|a-c|, |b-c|) < min(|al-c|). */
20   IntratTaylorc(F, a, b, c = "unset") =
21   { my(B = getlocalbitprec(), X = variable(F));
22     my(L, ser, Sa, Sb, m, r, D = denominator(F)); \\ deg D > 0
23
24     if (c == "unset", c = (a + b) / 2);
25     localbitprec(32);
26     m = vecmin([norm(z - c) | z <- polroots(D)]);
27     a -= c;
28     b -= c; r = sqrt(m / max(norm(a), norm(b)));
29     if (r <= 1.1, error("Intrat fails"));
30     L = ceil(log(2) * B / log(r));
31     localbitprec(B);
32     ser = intformal(1. * subst(F, X, X + c) + O(X^L));
33     [Sa, Sb] = subst(truncate(ser), X, [a, b]);
34     return (Sb - Sa);
35   }
```

Finally, here is a general driver program for integrating rational functions.

──────────────────────────── │ Intrat.gp │ ────────────────────────────

```
1    /* Returns either 0 if a = b, or [s, a, b, u] such that the
2     * integral is equal to s int_a^b f(ux) dx with s,u = 1 or -1
3     * and a < b. */
4    Intchangevar(a, b) =
5    {
6      if (a == b, return ([0, 0, 0, 0]));
7      if (a == oo, return ([-1, b, oo, 1]),
8          a ==-oo, return ([1, -b, oo, if (b == oo, 1,-1)]),
9          /* now a is finite */
10         b == oo, return ([1, a, oo, 1]),
11         b ==-oo, return ([-1, -a, oo, -1]),
12         return ([1, a, b, 1]));
13   }
14
15   /* Driver routine. c is set to a complex number in last resort
16    * (see text). */
17   Intrat(F, a, b, c = "unset") =
18   { my(B = getlocalbitprec(), s, u, R, S);
19     my(X = variable(F), D = denominator(F));
20
21     [s, a, b, u] = Intchangevar(a, b);
22     if (!s, return (0));
23     if (u == -1, F = subst(F, X, -X));
24     localbitprec(B + 32);
25     if (precision(D) == oo ||
26         exponent(polresultant(D, D')) >= -B / poldegree(D),
```

```
27        S = Intratdec(F, a, b)
28      ,
29        R = polrootsbound(D);
30        if (a > 1.1 * R,
31          S = IntratTaylor_oo(F, a, b, R)
32        ,
33          if (a == -oo || b == oo, error("Intrat fails"));
34          S = IntratTaylorc(F, a, b, c)));
35      if (!imag(F), S = real(S));
36      return (bitprecision(s * S, B));
37    }
```

Note that if all else fails and if the user does not give a specific value of c, as a last resort we try $c = (a + b)/2$.

EXAMPLES. (All essentially instantaneous):

```
? \p 38
? F = 1 / ((x^2 - 4) * (x + 1)^2 * (x^2 + 1));
/* Exact denominator. No problem apart from the poles. */
? Intrat(F, 3, 5)
% = 0.00064063379579153750082403536321269008579
? Intrat(F, -3, 5)
  ***    user error: diverging integral
? Intrat(F, 3, oo)
% = 0.00069232770088554668660862964466634582684
? Intrat(F,-3,-oo)
% = -0.0022290663543392135366131178377983323732
? F = 1 / ((x^2 + 1.1) * (x + 3.9));
/* Squarefree denominator. Still no problem. */
? Intrat(F, 0, 1)
% = 0.16780392718729633365957560255514096110
? Intrat(F, 0, oo)
% = 0.27760215298390801679006468138968821376
? F = 1 / (x^2 + Pi / 3)^2;
/* Nonsquarefree nonexact denominator, problems begin. */
? Intrat(F, 2, 3)
% = 0.020648757479902711377582458624640362207
? Intrat(F, -1, 2)
  ***    user error: Intrat fails
```

In this last example the simplest solution is to split the interval of integration from -1 to 0 and from 0 to 2.

Note also that immediate modifications of the IntratTaylor program can be used if F is an integrable function on $[N, \infty[$ such that $F(1/X)$ has a power series with nonzero radius of convergence at 0. As an example which we will use below, we can let $F(x) = \log(f(x))$ where f is a rational function such that the degree of $f(x) - 1$ is smaller than or equal to -2. The radius of convergence will then be $1/\max|\alpha|$, where now α ranges over all the poles *and* the zeros of f. We can also use integration by parts:

$$\int_N^\infty \log(f(x))\,dx = x\log(f(x))\Big|_N^\infty - \int_N^\infty xf'(x)/f(x)\,dx$$

$$= -N\log(f(N)) - \int_N^\infty xf'(x)/f(x)\,dx\ .$$

Since $xf'(x)/f(x)$ is a rational function of degree less than or equal to -2, the above program is able to compute the last integral.

3.3. Generalities on numerical integration

3.3.1. Introduction. The subject matter of numerical integration is vast, and numerous books are devoted partly or entirely to the subject. It is of course out of the question to be in any way exhaustive, but our main purpose is to explain a number of useful methods in the context of multiprecision computations. We do not claim that our methods are the best, but they are very efficient and implemented in Pari/GP.

We will deal almost exclusively with the computation of integrals $\int_a^b f(x)\,dx$, where $[a,b]$ is a not necessarily compact interval and f some suitable regular function, in fact usually assumed to be C^∞ or even holomorphic in a domain containing the interval of integration.

Note that by making a trivial affine change of variable we may always assume that a and b are real (possibly $\pm\infty$) with $a < b$. We will make this assumption from now on.

Some other important types of integration, which we will only briefly mention, are as follows:

(1) Computation of the antiderivative $\int_a^x f(t)\,dt$ for a number of values of x: it is possible to compute this faster than simply by using some integration method for each desired value of x.

(2) More generally, computation of solutions of ordinary differential equations for several values of the variable.

(3) Computation of contour integrals: after a suitable change of variable corresponding to a parametrization of the contour, this reduces to our standard type of integral.

(4) Computation of multiple integrals: one can of course use Fubini's theorem, and reduce to the computation of simple integrals. However, in general there are more efficient methods for the computation: first, one should try whenever possible to use some variant of Stokes's theorem to reduce to integrals of lower dimension. And second, there are usually specific integration methods which are more efficient than the straightforward use of Fubini's theorem.

Almost all of the methods that we will consider (with the unique exception of IntoscSidi, see Section 3.9.3 below) are of the following form. For a suitable class of functions, and for each integer $n \geq 1$ we compute *nodes* (or abscissas) x_i and *weights* w_i for $1 \leq i \leq n$ such that

$$\int_a^b f(x)\,dx = \sum_{1 \leq i \leq n} w_i f(x_i) + E_n(f)\ ,$$

where $E_n(f)$ is some "small" error depending on n and f, and tending to 0 (if possible at least geometrically) when $n \to \infty$.

The corresponding basic summation program is simple but so ubiquitous that we write it here once and for all:

────────────────────────────── │ SumXW.gp │ ──────────────────────────────

```
1  /* XW = [X, W] contains nodes and weights */
2  SumXW(f, XW) =
3  { my(B = getlocalbitprec(), [X, W] = XW);
4    localbitprec(B + 32);
5    return (sum(i = 1, #X, W[i] * f(X[i])));
6  }
```

Even when the method is not directly given in this way, inasmuch as possible we will also give the formulas in this form. More precisely, if $[a, b]$ is a compact interval and the formula for the interval $[-1, 1]$ is $\int_{-1}^{1} f(x)\,dx = \sum_{1 \le i \le n} w_i f(x_i) + E_n(f)$, we have

$$\int_a^b f(x)\,dx = \frac{b-a}{2}\left(\sum_{1 \le i \le n} w_i f\left(\frac{a+b}{2} + \frac{b-a}{2}x_i\right) + E_n(f_{a,b})\right),$$

where $f_{a,b}$ is the function on $[-1, 1]$ defined by $f_{a,b}(t) = f((a+b)/2 + ((b-a)/2)t)$.

If instead we are given $\int_0^1 f(x)\,dx = \sum_{1 \le i \le n} w_i f(x_i) + E_n(f)$, then

$$\int_a^b f(x)\,dx = (b-a)\left(\sum_{1 \le i \le n} w_i f(a + (b-a)x_i) + E_n(f_{a,b})\right),$$

where here $f_{a,b}(t) = f(a + (b-a)t)$.

Note that GP closures allow a simple and elegant implementation of both formulas in terms of SumXW:

```
my(c = (a + b) / 2; d = (b - a) / 2);
d * SumXW(t -> f(c + d * t), XW);
```

```
my(d = b - a);
d * SumXW(t -> f(a + d * t), XW);
```

The nodes x_i *usually* (but not always) belong to the interval $[a, b]$, but in certain cases, when the functions considered are meromorphic, they can even be complex.

An example of integration methods of a different type is

$$\int_a^b f(x)\,dx = \sum_{1 \le i \le n} w_i f(x_i) + \sum_{1 \le j \le n} v_i f'(y_i) + E_n(f),$$

where the values of the *derivative* are also used, but we will not consider this type of method.

3.3.2. Dealing with singularities.

Thus, let us focus our attention to integrals $\int_a^b f(x)\,dx$, and for the moment to the case where $[a, b]$ is compact. A number of things may complicate matters. We begin by looking at the case where f has *singularities* on the interval $[a, b]$, always assumed to be finite in number. Possibly after splitting the interval into several sub-intervals and making an affine change of variable, we may assume that $[a, b] = [0, 1]$ and that there is a unique possible singularity, at $x = 0$.

Without any attempt to be exhaustive, from best to worst the singularity may be of the following types:

(1) A *removable* singularity, which is therefore not a singularity at all. For instance $\int_0^1 x/(e^x - 1)\,dx$ or $\int_0^1 (1/(e^x - 1) - 1/x)\,dx$. A reasonable method to deal with this kind of behavior is to compute sufficiently many terms of the Taylor expansion of the integrand around $x = 0$: for instance, if $f(x) = 1/(e^x - 1) - 1/x$, we integrate instead from 0 to 1 the function $\widetilde{f}(x)$ defined by $\widetilde{f}(x) = -1/2 + x/12 - x^3/720$ if $|x| < \varepsilon^{1/5}$ and $\widetilde{f}(x) = 1/(e^x - 1) - 1/x$ if $|x| \geq \varepsilon^{1/5}$, where ε is the desired accuracy. Note that CAS such as Pari/GP can automatically compute for you the Taylor expansion of most functions to any desired number of terms.

(2) A *logarithmic type* singularity. An integrand such as $f(x) = 1/\log(x)$ has a singularity at $x = 0$, which cannot be "removed" although it is reasonable to set $f(0) = 0$. Here, the solution depends very much on the type of integration method which is used. For instance, for the doubly-exponential type methods that we will study, this singularity can be ignored. On the other hand, the Newton–Cotes and Gaussian methods give very bad results (although of course one can devise a Gaussian method which is *specifically* adapted to integrands of the form $g(x)/\log(x)$). Exactly the same comments are true for integrands which have an infinite logarithmic singularity at $x = 0$ such as $f(x) = \log(x)$.

(3) A *power* singularity, i.e., if $f(x) \sim C \cdot x^{-\alpha}$ for some $\alpha < 1$ as $x \to 0$. In that case, the change of variable $y = x^{1-\alpha}$ makes the singularity disappear:

$$\int_0^1 f(x)\,dx = \frac{1}{1-\alpha} \int_0^1 f(y^{1/(1-\alpha)})y^{\alpha/(1-\alpha)}\,dy \,,$$

and $f(y^{1/(1-\alpha)})y^{\alpha/(1-\alpha)} \sim C$ as $y \to 0$.

Of course, we could also use changes of variable to make $\log(x)$ type singularities disappear, but this is almost always a bad idea since such singularities are so mild.

(4) For a *known essential singularity*, once again a suitable change of variable will do the same: for instance, if $f(x) \sim Ce^{-1/x}$ on $[0, 1]$, the change of variable $y = 1/x$ gives

$$\int_0^1 f(x)\,dx = \int_1^\infty f(1/y)/y^2\,dy \,,$$

which is a nicely convergent integral since $f(1/y) \sim Ce^{-y}$ as $y \to \infty$.

3.3.3. Dealing with poles and large intervals. If f is a meromorphic function, it may have poles *near* the interval $[a, b]$. Equivalently, the radius of convergence of the Taylor expansion of f around some point $c \in [a, b]$ is limited by the presence of these poles. This has important consequences on the error of the integration method. More precisely, assume that we have some integration method on $[0, 1]$ depending on some parameter n, whose error is $f^{(n)}(c)/(d^n n!)$ for some constant $d > 1$ independent of the function f and some $c \in [0, 1]$, and assume that the interval $[a, b]$ is compact. We have $\int_a^b f(x)\,dx = (b - a)\int_0^1 f(a + (b - a)t)\,dt$, so the error is now $(b-a)((b-a)/d)^n f^{(n)}(c)/n!$ for some $c \in [a,b]$. If we denote by r the minimum distance of a pole of f to the interval $[a,b]$, the radius of convergence of the Taylor expansion of $f(x)$ around $x = c$ is at least equal to r, in other words $f^{(n)}(c)/n!$ is roughly of order $1/r^n$, and usually not smaller. Thus the error is roughly of order $((b-a)/rd)^n$.

This has the following important consequence. If r is "small", i.e., if there exist a pole near the interval of integration, or if the length $|b-a|$ of the interval of integration is "large", we may have $|(b-a)/rd| \geq 1$ so the integration method will not work at all. It is then necessary to *split* the interval of integration into sub-intervals $[a_{j-1}, a_j]$, so that for instance $|(a_j - a_{j-1})/rd| \leq 1/2$, and then choose n large enough so that $|((a_j - a_{j-1})/rd)^n|$ is less than the desired accuracy. Note that choosing n very large has two disadvantages: first, the computation of the nodes and weights for the integration method may become very expensive. And second, the irregularities of the function due to the computation of higher derivatives will be considerably enhanced, which is of course bad. Thus instead, we will consider n as *fixed* (depending on the accuracy), and choose the splitting of the interval as a function of f and its poles.

As we have just mentioned, a suitable splitting of the interval depends in large part (but not entirely) on a, b, and n, and on the position of the poles (hence on the radius of convergence at different points). Let us consider a number of typical cases. For this, note that we may reduce to the case $a = 1$ by the trivial change of variable $x \mapsto x + a - 1$, so we will only consider integrals of the form $\int_1^N f(x)\,dx$ with $N > 1$.

(1) Assume first that f has no poles at all, i.e., that f is an entire function, or possibly that the poles of f are so far from the interval of integration $[1, N]$ as to have little influence: more precisely, whose distance to the interval is at least n/e (with $e = 2.718\cdots$). Even in that case, it is usually necessary to split the interval. Assume for instance that $f(x) = \sin(x)$, which is the simplest case. Here the bound $|f^{(n)}(c)| \leq 1$ is essentially optimal, so the error will be roughly $((N-1)/d)^n/n!$. Since n is fixed, the quasi optimal strategy is here to cut the interval $[1, N]$ into k equal pieces, with k chosen so that $((N-1)/(kd))^n/n! < 2^{-B}$ (where B is the desired bit accuracy), i.e., $k > e(N-1)2^{B/n}/(nd)$ (with $e = 2.718\cdots$), where we approximate $n!^{1/n}$ with n/e by Stirling's formula.

For other entire functions, the above strategy may not be optimal, but is in general quite sufficient.

(2) Assume on the contrary that f has poles, all to the left of the interval $[1, N]$, and assume first that the poles of f are all in $\Re(z) \leq 0$ with a pole at 0, and let us split the interval as $[a,b] = \bigcup_{1 \leq j \leq k}[a_{j-1}, a_j]$. If $c \in [a_{j-1}, a_j]$, the closest distance from c to a pole is equal to a_{j-1}, so the radius of convergence of f around c is at least a_{j-1}, in other words $f^{(n)}(c)/n!$ is roughly of order $1/a_{j-1}^n$. It follows that the error is roughly $\sum_{1 \leq j \leq k}((a_j - a_{j-1})/(da_{j-1}))^n$. It is not difficult to see that a close to optimal choice is to choose the a_j in geometric progression, i.e., $a_j = N^{j/k}$, so that the error is roughly of order $k((N^{1/k} - 1)/d)^n$, and since k will be small, we want approximately $(N^{1/k} - 1)/d < 2^{-B/n}$, in other words $k > \log(N)/\log(d\,2^{-B/n} + 1)$.

Up to small constants, this argument is still valid if f has poles z having real part less than or equal to 1, but such that $|z - 1| \geq 1$.

(3) Assume now that all the poles of f have real part less than or equal to 1, but that at least one pole z_0 is close (but not too close) to 1, i.e., is such that $|z - 1| < |1|$. Denote by $r > 0$ the minimal distance of a pole of f to 1. If $N \geq 2$ we write

$$\int_1^N f(x)\,dx = \int_1^2 f(x)\,dx + \int_2^N f(x)\,dx$$
$$= r\int_1^{1/r+1} f(r(t-1)+1) + \int_1^{N-1} f(x+1)\,dx\ ,$$

and it is clear that for both integrals we are in the case where the distance from 1 to the nearest pole is at least equal to 1, so the preceding case is applicable. If $1 < N \leq 2$ the formula is even simpler since we write $\int_1^N f(x)\,dx = r\int_1^{1+(N-1)/r} f(r(t-1)+1)$. Note that of course this formula is valid even if $N > 2$, but it is more efficient to use the cut-off at $x = 2$ as we have done above.

(4) If all the poles have real part greater than or equal to N, we can apply both of the preceding arguments to the function $f(N + 1 - x)$.

(5) If f has poles near the interval $[1, N]$ but some having real part larger than 1 and less than N, at least two methods may be applicable.

    (a) If one knows that the distance of *all* poles to $[1, N]$ is at least equal to $r > 0$, one can split the interval into k equal pieces, where we must have $((N - 1)/(krd))^n < 2^{-B}$, in other words $k > 2^{B/n}(N - 1)/(rd)$.

    (b) Otherwise, if they are known, the simplest is to *subtract* from f the sum F of the *polar parts* corresponding to poles close to $[1, N]$ in the above sense; F will be a rational function and $f - F$ a function whose poles are away from the interval, so we can integrate F using integration of rational functions, and $f - F$ using (1), (2), or (3).

We now write a *driver program* incorporating these ideas. Assume that we have computed a quadruple $\mathtt{vv} = [n, \mathtt{X}, \mathtt{W}, d]$ depending only on the desired accuracy 2^{-B}, where d is as above, such that for the nicest functions f, we have

$$\int_0^1 f(x)\,dx = \sum_{1 \leq i \leq n} \mathtt{W}[\mathtt{i}]f(\mathtt{X}[\mathtt{i}]) + E \quad \text{with} \quad |E| < 2^{-B}\ .$$

We want to write a program using this data which takes into account the interval $[a, b]$ (in particular if it is large) and the possible existence of poles, as explained above. Note that we assume that f has no singularities, removable or otherwise: they should be taken into account beforehand as explained above.

The syntax of the driver program is $\mathtt{Intgen(f, a, b, XW = 0, poles = [])}$: this means that $\mathtt{f}$ is a function, $\mathtt{XW}$ gives nodes and weights as above and may be omitted, in which case it must be recomputed internally, and $\mathtt{poles = []}$ means that by default we ignore the poles, but they can be included as fifth parameter. We allow the following formats for $\mathtt{poles}$: either it is a vector whose entries are either some of the poles, or some of the polar parts of f (possibly summed together), or it is a strictly positive real number r, in the case where one only knows that the poles are at a distance r from the interval. We choose as initialization example

Gauss–Legendre integration, which is one of the most useful methods as we will see below, but it is of course immediate to adapt it to all other methods.

───────────────────────────── Intgen.gp ─────────────────────────────

```
1   /* Generic Programs. */
2
3   /* Driver programs for numerical integration. Reduce to an
4    * interval [1, N]. Here, [a, b] can be any compact complex
5    * interval. Poles can be given either as complex numbers,
6    * as polar parts, or as a minimal distance r. */
7
8   /* Sample initialization: Gauss-Legendre (d = 4), which reduces
9    * to [0, 1] for integration on compact intervals. */
10  Intgeninit() =
11  { my([X, W] = IntGaussLegendreinit(), n = 2 * #X, d = 4);
12    /* Note: X contains only the positive nodes in ]-1, 1[ */
13
14    X = vector(n, i, 1 + (-1)^i * X[(i + 1) \ 2]) / 2;
15    W = vector(n, i, W[(i + 1) \ 2]) / 2;
16    return ([X, W, d]);
17  }
18
19  /* Basic formula. */
20  Intgen0(f, a, b, XW) =
21    my(c = b - a); c * SumXW(x -> f(a + c * x), XW);
22
23  /* Sum basic formula over a partition A of an interval. */
24  Intgensum(f, A, XW) =
25    sum(j = 1, #A-1, Intgen0(f, A[j], A[j+1], XW));
26
27  /* Poles of have distance at least r to [1, N].
28   * Subdivide [1, N] in equal pieces */
29  Intgenpolesmin(f, r, N, XW) =
30  { my(B = getlocalbitprec(), n = #XW[1], d = XW[3], k, h);
31
32    k = ceil((N - 1) * 2^(B/n) / (r * d));
33    h = (N - 1) / k; /* equal pieces */
34    return (Intgensum(f, vector(k+1, j, 1+(j-1)*h), XW));
35  }
36
37  /* Poles of f satisfy real(z) <= 1 and |z-1| >= 1.
38   * Subdivide [1, N] using a geometric progression */
39  Intgenpoles1(f, N, XW) =
40  { my(B = getlocalbitprec(), n = #XW[1], d = XW[3], k);
41
42    k = ceil(log(N) / log(d * 2^(-B/n) + 1));
43    return (Intgensum(f, powers(N^(1/k), k), XW));
44  }
```

```
45
46    /* Approximate (n!)^(1/n) */
47    Intgennsure(XW) = my(n = #XW[1]); return (n * exp(-1));
48
49    /* Case where poles can be ignored. */
50    Intgennopoles(f,N,XW) =
51      Intgenpolesmin(f, Intgennsure(XW), N, XW);
52
53    /* Case where the polar part is known. */
54    Intgenpolar(f, F, N, XW) =
55    { my(g = t -> f(t) - subst(F, variable(F), t));
56      return (Intgennopoles(g, N, XW) + Intrat(F, 1, N));
57    }
58
59    /* Main driver routine on [a, b]. */
60    Intgen(f, a, b, XW = 0, poles = []) =
61    { my(eith, c = b - a, N = abs(c) + 1);
62
63      if (c == 0, return (0.));
64      if (!XW, XW = Intgeninit());
65      eith = if (imag(c), exp(I * arg(c))
66                          , sign(real(c)));
67      (f = x -> f(eith * (x-1) + a)); /* now on [1, N] */
68      if (type(poles) != "t_VEC", /* minimal distance r > 0 */
69        return (eith * Intgenpolesmin(f, poles, N, XW)));
70      if (#poles == 0,
71        return (eith * Intgennopoles(f, N, XW)));
72      my(polar = 1);
73      poles = vector(#poles, j,
74          my(p = poles[j]);
75          if (type(p) == "t_RFRAC",
76            my(vx = variable(p));
77            subst(p, vx, eith * (vx - 1) + a)
78          , /* else */
79            polar = 0;
80            1 + conj(eith) * (p - a)));
81      if (polar, /* all polar parts are known */
82        return (eith * Intgenpolar(f, vecsum(poles), N, XW)));
83      /* convert poles to list of complex numbers */
84      poles = concat(vector(#poles, j,
85        my(a = poles[j]);
86        if (type(a) == "t_RFRAC", polroots(denominator(a)), [a])));
87      return (eith * Intgenpoles(f, N, XW, poles));
88    }
89
90    /* Distance of complex z to [1, N]. */
91    Distpole(N,z) =
92    { my(x = real(z));
```

```
93      if (x < 1, return (abs(z - 1)));
94      if (x > N, return (abs(z - N)));
95      return (abs(imag(z)));
96    }
97
98    /* General case, on [1, N]; poles are complex numbers */
99    Intgenpoles(f, N, XW, poles) =
100   { my(S, r, p1, pN, nsure = Intgennsure(XW));
101
102     if (vecmin([Distpole(N,z) | z <- poles]) >= nsure,
103       return (Intgenpolesmin(f, nsure, N, XW)));
104     p1 = pN = List([nsure]);
105     foreach (poles, a,
106       my(x = real(a));
107       if (x <= 1 && abs(a - 1) >= 0.1, listput(~p1, abs(a - 1)),
108           x >= N && abs(a - N) >= 0.1, listput(~pN, abs(a - N)),
109           error("please give polar parts in Intgen")));
110     if (#pN == 1, r = vecmin(Vec(p1)),
111         #p1 == 1, r = vecmin(Vec(pN)); f = x -> f(N+1-x),
112       error("please give polar parts in Intgen"));
113     /* Poles of f satisfy real(z) <= 1 and |z-1| >= r >= 0.1 */
114     if (r >= 1, return (Intgenpoles1(f, N, XW)));
115     /* A pole is close to 1 */
116     S = r * Intgenpoles1(x -> f(r * (x - 1) + 1),
117                           1 + min(N - 1, 1) / r, XW);
118     if (N > 2, S += Intgenpoles1(x -> f(x + 1), N - 1, XW));
119     return (S);
120   }
```

EXAMPLE. Let us compute $\int_a^b f(x)\,dx$ for $f(x) = e^{\pi x}/(x^2 + 1)^2$ and $[a, b] = [0, 10]$ or $[a, b] = [-10, 10]$ at 115 decimal digits. For $[a, b] = [0, 10]$ we first simply write Intgen(f,0,10). The result is only correct to a relative accuracy of 10^{-62}, i.e., half of the desired accuracy. If we add the position of the poles and write Intgen(f,0,10,,[-I,I]), the result is perfect. This works because the poles have real part less than or equal to 0.

For $[a, b] = [-10, 10]$, the command Intgen(f,-10,10) only gives a relative accuracy of 10^{-30}. Writing Intgen(f,-10,10,,[-I,I]) gives an error message since the real parts of the poles are inside the interval of integration. To obtain the exact result, we must include the polar part G=(-1+Pi/2*x*(1+x^2))/(1+x^2)^2 and write Intgen(f,-10,10,,[G]), which gives a perfect result. An alternative method is of course to split the interval of integration at $x = 0$.

Note that most integration methods will also lose decimals for $\int_{-10}^{10} f(x)\,dx$: for instance, the doubly-exponential method Inta_b that we will study below also only gives a relative accuracy of 10^{-18}, so it must be corrected to take into account the polar parts. This simply means changing the Intgeninit program given above from the use of Gauss–Legendre to the use of doubly-exponential methods.

EXAMPLE. Let $N(a, b)$ be the number of zeros of the Riemann zeta function $\zeta(s)$ from height a to height b, where we assume $b \geq a > 1$. We can compute

$$N(a, b) = \frac{1}{2\pi i} \int_{C(a,b)} \frac{\zeta'(s)}{\zeta(s)} \, ds \,,$$

where $C(a, b)$ is the rectangle with vertices ai, $ai + 1$, $bi + 1$, bi oriented positively, which is a sum of four easily written integrals. Note that this is far from being an efficient way to compute this quantity, but is only given as an example.

For instance with $a = 5$ and $b = 100$, if the integral is written without taking care of the poles of the integrand, then all of the methods that we will describe give nonsense results, with the exception of Newton–Cotes (with or without Chebyshev nodes), but those take an inordinate amount of time. On the other hand, if we include artificial values of r as minimal distance to the poles, such as $r = 1$ or 2, whose effect is only to split the intervals of integration into much smaller pieces, then the `Intgen` program that we wrote, using Gauss–Legendre initialization, gives perfect results, and reasonably fast (if we initialize with the doubly-exponential method, the results will also be perfect, but much slower). Incidentally, we find in this way that there are 29 zeros up to height 100. Note once again that this can be obtained several orders of magnitude faster by using other methods specific to L-functions with functional equation similar to that of $\zeta(s)$, see for instance Booker's nice exposition of Turing's method [**Boo06**].

3.4. Newton–Cotes type methods

3.4.1. Trapezes, Simpson, etc. These are probably the oldest numerical integration methods. The idea is as follows. Assume that f is a C^∞ function on some compact interval $[a, b]$. By definition of the Riemann integral, we have $I = \lim_{n \to \infty} u(n)$, where

$$u(n) = \frac{b-a}{n} \left(\frac{f(a)}{2} + \sum_{1 \leq j \leq n-1} f\left(a + \frac{(b-a)j}{n}\right) + \frac{f(b)}{2} \right) .$$

This corresponds to summing vertical trapezes, i.e., approximating the function f piecewise by polynomials of degree 1. It is easy to show, and in fact is an immediate consequence of the Euler–Maclaurin formula (see Proposition 4.2.40), that $u(n) = I + c_2/n^2 + c_4/n^4 + \cdots$, with $c_2 = ((b-a)^2/12)(f'(b) - f'(a))$ and similar formulas for c_{2j}. Note that it is *essential* to include the terms $f(a)/2 + f(b)/2$ (i.e., to sum on trapezes instead of rectangles), otherwise there will be a term c_1/n in the expansion of $u(n)$.

If we compute $u(n)$ for $n = 10^6$, say, the result will already be pretty good since we will have approximately 12 decimals of accuracy. But clearly this is totally unsuitable if we want hundreds of decimals.

The idea of Newton–Cotes methods such as Simpson's rule is to approximate f by polynomials of degree larger than 1. Let $\delta = (b-a)/n$ be the "step size". We will need the values of f at each $a + j\delta$, and $N - 1$ *equally spaced* interpolation points inside each interval $]a + j\delta, a + (j+1)\delta[$ for $0 \leq j < n$, so we will need to compute Nn values of f. Setting $g(t) = f(a + \delta t)$, we have

$$I = \int_a^b f(x) \, dx = \delta \int_0^n g(t) \, dt = \delta \sum_{0 \leq j < n} \int_0^1 h_j(u) \, du \,, \quad \text{with} \quad h_j(u) = g(j + u) \,.$$

Thus $h = h_j$ is a function on $[0, 1]$ that we want to interpolate by a polynomial of degree N at the $N+1$ equally spaced points i/N for $0 \le i \le N$. There is evidently a single such polynomial, given by the Lagrange interpolation formula:

$$P_N(X) = \sum_{0 \le i \le N} \prod_{0 \le j \le N,\ j \ne i} \frac{NX - j}{i - j} h(i/N) \ .$$

It follows that

$$\int_0^1 h(x)\, dx = \sum_{0 \le i \le N} c_{N,i} h(i/N) + E_N \quad \text{with} \quad c_{N,i} = \int_0^1 \prod_{0 \le j \le N,\ j \ne i} \frac{Nx - j}{i - j}\, dx \ ,$$

and an error term E_N for which one can prove that

$$E_N = \begin{cases} \dfrac{h^{(N+2)}(c)}{N^{N+3}(N+2)!} r(N) & \text{if } N \text{ is even} , \\[3mm] \dfrac{h^{(N+1)}(c)}{N^{N+2}(N+1)!} r(N) & \text{if } N \text{ is odd} , \end{cases}$$

for some c in $]0, 1[$, where

$$r(N) = \int_0^N t^2 \prod_{1 \le j \le N} (t - j)\, dt \ .$$

One can also prove that $r(N) < 0$ and that for N large enough $N! < |r(N)| < (N+1)!$. It follows that, up to small powers of N, if we choose N even then E_N is at most of order $h^{(N+2)}(c)/N^{N+4}$, so for $h(u) = h_j(u) = f(a + \delta(j + u))$, of order at most $f^{(N+2)}(c)\delta^{N+2}/N^{N+4}$. Thus if N is even

$$I = \delta \sum_{0 \le i \le N} c_{N,i} \sum_{0 \le j < n} f\big(a + \delta(j + i/N)\big) + E_{N,n}(f) \ ,$$

where the error $E_{N,n}(f)$ is again at most of order $f^{(N+2)}(c)\delta^{N+2}/N^{N+4}$.

In particular, if we want to put this into canonical form as explained above, this method can be written

$$\int_0^1 f(x)\, dx = \sum_{0 \le i \le Nn} w_i f(x_i) + E_{N,n}(f)$$

where $x_i = i/(nN)$, and if we write $i = Nj + r$ with $0 \le r < N$ then $w_i = c_{N,r}/n$ unless $r = 0$ and $i \ne 0$ and $i \ne nN$, in which case $w_i = 2c_{N,0}/n$. Up to small powers of N, the error $E_{N,n}(f)$ is thus of order $\dfrac{f^{(N+2)}(c)}{(nN)^{N+2}}$. Using a very rough approximation and neglecting small powers of N and n we thus have

$$E_{N,n}(f) \approx \left(\frac{b-a}{en}\right)^{N+2} \frac{f^{(N+2)}(c)}{(N+2)!} \ ,$$

where $e = \exp(1)$.

For instance, Simpson's rule (corresponding to $N = 2$) gives

$$6w_i = 1, 4, 2, 4, 2, \ldots, 4, 1 \ ,$$

while for $N = 4$ we have

$$90w_i = 7, 32, 12, 32, 14, 32, 12, 32, \ldots, 32, 7 \ .$$

Now we have decided that the initialization time is of no importance since it can be done once and for all, and that we only want to minimize the number nN of function evaluations. To analyze what this entails, we make the assumption that f is meromorphic and that the distance of the interval $[a, b]$ to the closest pole of f is at least r, so that very roughly $f^{(N)}(c)/N!$ behaves like r^{-N} (of course if $r = \infty$ this doesn't make sense). Thus, again approximately, if we want the error to be less than 2^{-B} we need $(enr/(b-a))^N > 2^B$, in other words $N > B \log(2)/(\log(enr/(b-a)))$. We must therefore minimize $n/\log(enr/(b-a))$, under the restriction that $n > (b-a)/(er)$. An immediate computation shows that the optimal value of n is $(b-a)/r$, so we choose $n = \lceil (b-a)/r \rceil$, so that if we ignore rounding we will have $\log(enr/(b-a)) = 1$, so $N = B\log(2)$, which is independent both of the interval $[a, b]$ and the radius of convergence r. Of course we have ignored rounding on n, so this value of N can be decreased in general, but since it is preferable to have an initialization independent of everything apart from the needed accuracy, we will keep this value of N.

Now note that an annoying phenomenon occurs here, called *Runge's phenomenon*, similar to the Gibbs phenomenon in Fourier analysis: even though the interpolating polynomials pass through the given points, between these points their behavior becomes usually wildly different from that of f, so one should not use too large a value of N. Nonetheless, the analysis that we have made should compensate for this.

-- IntNewton.gp --

```
1   /* Compute int_a^b f(x) dx by Newton-Cotes methods
2    * with equally spaced abscissas. */
3
4   IntNewtoninit(N = 0) =
5   { my(B = getlocalbitprec(), C, D, T, Q, X, W);
6
7     if (!N, N = 2 * ceil(B * log(2) / 2)); /* We want N even. */
8     T = vector(N + 1, j, N * 'x - (j - 1));
9     Q = vecprod(T); C = binomial(N);
10    W = vector(N + 1, j,
11      my(B = subst(intformal(Q \ T[j]), 'x, 1));
12      B * (-1)^(N - j + 1) * C[j]);
13    D = content(W); W /= D; D /= factorial(N);
14    X = vector(#W, i, i);
15    return ([X, W, D, N, B + exponent(W) + exponent(D)]);
16  }
17
18  IntNewton(f, a, b, r = 1, XW = 0) =
19  { my(c = b - a, n, S, h, A);
20
21    if (type(XW) == "t_INT", XW = IntNewtoninit(XW));
22    my(D = XW[3], N = XW[4], Bnew = XW[5]);
23    localbitprec(32); n = max(ceil(abs(c) / r), 1);
24    localbitprec(Bnew);
25    h = c / (N*n);
```

```
26    A = vector(n, j, a + h * (N * (j-1) - 1));
27    S = SumXW(x -> my(y = h * x);
28                   sum(j = 1, n, f(A[j] + y), 0.), XW);
29    return (c * (D / n) * S);
30  }
```

The main remark concerning this program is that the weights XW are large, so we must compensate this by working at a much larger accuracy depending on their size, given by the command exponent(D) + exponent(W).

EXAMPLES.

```
? \p 500
? XW = IntNewtoninit();
time = 7,365 ms.
? IntNewton(x -> 1 / (1 + x), 0, 1, 1, XW) - log(2)
time = 4 ms.
% = 0.E-500
? A = log((1 + I / 4) / (-1 + I / 4));
? f = x -> 1 / (x + I / 4);
? IntNewton(f, -1, 1, 1, XW) - A
time = 28 ms.
% = -1.23... E-511 - 1.27... E-187*I /* Very wrong. */
? IntNewton(f, -1, 1, 1 / 4, XW) - A
time = 106 ms.
% = 2.45... E-509 - 4.87... E-501*I /* Perfect. */
```

These examples illustrate several things. First, even though this program should in principle not be used for high precision calculations, in particular because of the Runge phenomenon, it nonetheless gives excellent answers for reasonable functions. For instance, after initialization $\log(2)$ is computed in 5 milliseconds. Second, we really must pay attention to the poles: the computation of $\int_{-1}^{1} dx/(x + i/4)$ has only 187 correct decimal digits with the default $r = 1$, and gives a perfect result with $r = 1/4$, with a longer computation time.

As in the rest of this book, other detailed examples and timings are given at the end of this chapter.

As mentioned above, the choice $n \approx (b-a)/r$ (and even that of N) is tailored for functions having a small radius of convergence. It is complete overkill for functions having infinite radius of convergence such as $f(x) = e^x$.

3.4.2. Chebyshev nodes. [2] In the above program, the Runge effect forced us to increase accuracy and choose a relatively large value of n. Another method is to avoid taking equally spaced nodes, but to take nodes which are more concentrated towards the endpoints: one popular choice is *Chebyshev* nodes, i.e., to choose the roots of Chebyshev polynomials, which are cosines of simple rational numbers.

---

[2] Do not confuse this notion with Gauss–Chebyshev integration which we will study in Section 3.6.3.

More precisely, with the same notation as above, we have

$$I = \delta \sum_{0 \le j < n} \int_j^{j+1} g(t)\,dt = \frac{\delta}{2} \sum_{0 \le j < n} \int_{-1}^1 h_j(u)\,du \;,$$

where now $h_j(u) = g(j + (u+1)/2)$. As above we interpolate the function $h = h_j$ on $[-1,1]$ by a polynomial of degree $N-1$, but this time at the N points $x_{N,i} = \cos((2i-1)\pi/(2N))$ for $1 \le i \le N$. The unique such polynomial is given by Lagrange as

$$P_{N-1}(X) = \sum_{1 \le i \le N} \prod_{1 \le j \le N,\ j \ne i} \frac{X - x_{N,j}}{x_{N,i} - x_{N,j}} h(x_{N,i}) \;,$$

so that

$$\int_{-1}^1 h(x)\,dx \approx \sum_{1 \le i \le N} c_{N,i} h(x_{N,i}) \quad \text{with} \quad c_{N,i} = \int_{-1}^1 \prod_{1 \le j \le N,\ j \ne i} \frac{x - x_{N,j}}{x_{N,i} - x_{N,j}}\,dx \;,$$

hence

$$I \approx \frac{\delta}{2} \sum_{1 \le i \le N} c_{N,i} \sum_{0 \le j < n} f\big(a + \delta(j + (1 + x_{N,i})/2)\big) \;.$$

In canonical form, this method can be written

$$\int_0^1 f(x)\,dx \approx \sum_{1 \le i \le nN} w_i f(x_i) \;,$$

where if we write $i = Nj + r$ with $1 \le r \le N$ (note: not $0 \le r < N$), we have $x_i = (2j + 1 + \cos((2r-1)\pi/(2N)))/(2n)$ and $w_i = c_{N,i}/(2n)$.

One can show that (again using a very rough approximation) the error $E_{N,n}(f)$ is given by

$$E_{N,n}(f) \approx \left(\frac{b-a}{4n}\right)^{N+2} \frac{f^{(N+2)}(c)}{(N+2)!} \;.$$

We see that the essential difference with the case of equally spaced nodes is that $(b-a)/(en)$ is replaced with $(b-a)/(4n)$. The same analysis as above now gives $n = \lceil (b-a)e/(4r) \rceil$ (which for large $(b-a)/r$ is approximately $0.68(b-a)/r$), and once again $N = B\log(2)$. We thus have approximately $2/3$ of function evaluations compared with the preceding method.

This gives the following program, essentially identical to the previous one apart from the lack of accuracy increase and the initialization of the abscissas and weights: we use the parity of the Chebyshev polynomial to compute half as many values.

———————————— IntChebyshev.gp ————————————

```
/* Compute int_a^b f(x) dx using Chebyshev nodes.
 * If given, N must be odd. */
IntChebyshevinit(N = 0) =
{ my(B = getlocalbitprec(), Q, e1, XY, X, Y, W, N2);

  if (!N, N = 2 * ceil(B*log(2)/2) - 1); /* We want N odd. */
  Q = polchebyshev(N);
  localbitprec(B + exponent(Q) + 32);
  e1 = exp(I*Pi / (2*N)); N2 = (N - 1) / 2;
  XY = powers(e1^2, N2, e1); X = real(XY); Y = imag(XY);
```

```
11      X[N2 + 1] = 0;
12      W = vector(N2 + 1, i,
13        my(B = Q \ ('x - X[i]));
14        (-1)^i * Y[i] * sum(j = 0, N2, my(k = 2*j);
15                                      polcoef(B, k) / (k + 1)));
16      return (bitprecision([X, W, N], B + 32));
17    }
18
19    IntChebyshev(f, a, b, r = 1, XW = 0) =
20    { my(B = getlocalbitprec());
21      if (type(XW) == "t_INT", XW = IntChebyshevinit(XW));
22      my([X, W, N] = XW, c = b - a, n, S, h, A);
23      localbitprec(32); n = ceil(abs(c) * exp(1) / (4 * r));
24      localbitprec(B + 32);
25      h = c / (2 * n);
26      A = vector(n, j, a + h * (2 * j - 1));
27      S = SumXW(x -> if (x, my(y = h * x);
28                         sum(j = 1, n, f(A[j]+y) + f(A[j]-y))
29                         , sum(j = 1, n, f(A[j]), 0.)), XW);
30      return (bitprecision(- h * S * (2 / N), B));
31    }
```

We expect this program to be more stable with respect to the Runge effect and the poles.

EXAMPLES. We choose exactly the same as for IntNewton so we do not repeat the definitions of f and A:

```
? \p 500
? XW = IntChebyshevinit();
time = 1,726 ms.
? IntChebyshev(x -> 1 / (1 + x), 0, 1, 1, XW) - log(2)
time = 2 ms.
% = 0.E-500
? IntChebyshev(f, -1, 1, 1, XW) - A
time = 10 ms.
% = 0 E-516 - 4.66... E-373*I /* 128 decimals lost. */
? IntChebyshev(f, -1, 1, 1 / 4, XW) - A
time = 29 ms.
% = 0.E-516 + 0.E-500*I /* Perfect. */
```

Note that all the computations are faster than with IntNewton, and also that even when r is given incorrectly, the loss of accuracy is much smaller.

It is easy to improve on the above program when the integer n is far from abs(b - a) * exp(1) / (4 * r). For instance, assume that $b - a = 1$ and $r = 1$, which will be the case for many of our test integrals. This expression is equal to $\exp(1)/4 = 0.679\cdots$, so we choose $n = 1$ but it is then sufficient to chose N close to $\exp(1)/4$ times the previous value of N, and therefore this will make the number of function evaluations, hence essentially the whole program, faster by a factor of approximately 0.68. We leave the corresponding improvements to the reader.

3.4.3. Romberg integration. Once again, assume that $[a, b]$ is a compact interval. By definition of the Riemann integral, we have $I = \lim_{n \to \infty} u(n)$, where as above

$$u(n, 1) := u(n) = \frac{b - a}{n} \left(f(a)/2 + \sum_{1 \le j \le n-1} f(a + (b - a)j/n) + f(b)/2 \right).$$

As already mentioned, by the Euler–Maclaurin formula (see Proposition 4.2.40), we have

$$u(n) = I + c_2/n^2 + c_4/n^4 + O(1/n^6)$$

for some computable constants c_{2m} depending on the derivatives of f.

The idea of Romberg integration (which is a special case of Richardson extrapolation which we studied in Section 2.2) is to use both $u(n)$ and $u(2n)$: since $u(n) = I + c_2/n^2 + c_4/n^4 + O(1/n^6)$ we have $u(2n) = I + c_2/(4n^2) + c_4/(16n^4) + O(1/n^6)$, so $(4u(2n) - u(n))/3 = I - c_4/(4n^4) + O(1/n^6)$. At this stage, three things should be noted:

- First, since the convergence is in $1/n^4$, with $n = 10^6$ we will now have 24 decimals.
- Second, note that we only need to compute $u(2n)$, since the terms occurring in the sum for $u(n)$ are exactly those of $u(2n)$ with j even. More precisely, setting $S(n) = \sum_{1 \le j \le n-1} f(a + (b - a)j/n)$, we have $S(2n) = S_0(2n) + S_1(2n)$ and $S(n) = S_1(2n)$, the index of S representing the 2-adic valuation of the corresponding j's. Thus since $u(n) = (f(a) + f(b))/(2n) + S(n)/n$ and $u(2n) = (f(a) + f(b))/(4n) + S(2n)/(2n)$, we have

$$u(n, 2) := \frac{4u(2n) - u(n)}{3} = \frac{f(a) + f(b)}{6n} + \frac{2S_0(2n) + S_1(2n)}{3n}.$$

  The reader will recognize this as Simpson's rule.
- Finally, there is of course no reason not to continue: we have $u(n, 2) = (4u(4n) - u(2n))/3 = I - c_4/(64n^4) + O(1/n^6)$, so in the same way we have $u(n, 3) := (64u(4n) - 20u(2n) + u(n))/45 = I + O(1/n^6)$, and once again the computation of $u(4n)$ "includes" that of $u(2n)$ and $u(n)$, more precisely we have $S(4n) = S_0(4n) + S_1(4n) + S_2(4n)$, $S(2n) = S_1(4n) + S_2(4n)$, and $S(n) = S_2(4n)$, so

$$u(n, 3) = \frac{64u(4n) - 20u(2n) + u(n)}{45}$$
$$= \frac{1}{45n} \left(7 \frac{f(a) + f(b)}{2} + 16S_0(4n) + 6S_1(4n) + 7S_2(4n) \right).$$

Since by induction we have $u(n, k) = I + O(1/n^{2k})$, if we iterate k times, we obtain a formula of the type

$$I = d_k \frac{f(a) + f(b)}{2n} + \frac{1}{n} \sum_{0 \le v \le k} c_{k,v} S_v(2^k n) + O \left(\frac{1}{n^{2k}} \right).$$

for some constants $c_{k,v}$ and d_k.

Before computing these constants, note that we will need to compute $2^k n$ values of f, while on the other hand we need $n^{2k} > 2^B$, where as usual B is the desired number of binary digits. Thus we will choose $n > 2^{B/2k}$ and we must minimize $k \log(2) + (B/(2k)) \log(2)$, which gives $k = (B/2)^{1/2} \approx 0.71 B^{1/2}$ and

$$n > \exp(\log(2)B/(2k)) = \exp(\log(2)(B/2)^{1/2}) \approx \exp(0.49 B^{1/2}) \;,$$

so the number of function evaluations is approximately $\exp(0.98 B^{1/2})$. Since this is already larger than $40 \cdot 10^6$ for $B = 320$ (96 decimal digits), it is unreasonable to use this method for higher accuracies, but reasonable in low accuracy.

We can, however, do better. In the chapter on extrapolation we have mentioned that there exists a variant of the above where the number $2^k n$ of function evaluations is replaced by something of the order of $2^{k/2}n$. This gives $k \approx B^{1/2}$ and a number of function evaluations of the order of $\exp(0.69 B^{1/2})$. Although still exponential, this is of the order of $40 \cdot 10^6$ for $B = 640$ ($D = 192$), a notable improvement. Since it is much slower than other methods, we do not give any program.

A much better Romberg-type method is as follows: in the above program, for simplicity we forget that $S(n)$ is included in the computation of $S(2n)$, which may waste a time factor of up to 2. But what if we forget entirely that there is any link between $S(n)$ and $S(2n)$, and simply consider the sequence $u(n)$ that we need to extrapolate ? Looking at the different extrapolation methods in Chapter 2, the Limit2 method seems the most efficient, used on the sequence $u(n) = (f(a) + f(b))/2 + \sum_{1 \le j \le n-1} f(a + (b-a)j/n)$, which usually has an asymptotic expansion in integral powers of $1/n^2$. We can slightly improve on this by noting that the rational number j/n may occur more than once, and by combining this idea with the formula used in the Limit2 method and exchanging summations we gain a factor of approximately $\pi^2/6$ as follows:

────────────────────────────── | IntLagrange.gp | ──────────────────────────────

```
1   /* Romberg integration using Lagrange interpolation. */
2
3   IntLagrangeinit() =
4   { my(B = getlocalbitprec(), N, C, D, V);
5
6     N = ceil(LimitmulN(2) * B); if (N%2, N++);
7     C = binomial(2*N); \\ C[i+1] = binomial(2N,i)
8     D = dirpowers(N, 2*N-1); \\ D[i] = i^(2N-1)
9     V = vector(N, m, D[m] * sum(d = 1, N \ m,
10                            (-1)^(d*m) * C[N-d*m+1] * D[d]));
11    return (V);
12  }
13
14  IntLagrange(f, a, b, V = 0) =
15  {
16    if (type(V) != "t_VEC", V = IntLagrangeinit());
17    my(B = getlocalbitprec(), N = #V, D = factorial(2 * N));
18    localbitprec(B + Limitaccuracy(2) * N);
19    S = sum(m = 2, N, my(h = (b - a) / m);
20            V[m] * sum(j = 1, m - 1,
21                   if (gcd(j, m) == 1, f(a + j * h))), 0.);
```

```
22    return ((2*S + V[1] * (f(a) + f(b))) / D);
23  }
```

Note that, as in many other methods, we have to be careful with integrands f having removable singularities at the endpoints, such as $f(x) = (e^x - 1)/x$. This method is much faster than all preceding Romberg-type methods, and in particular does not require huge numbers of evaluations of f, see below for timings. As we will see, however, it is not competitive with respect to other methods such as Gaussian or Doubly-Exponential integration.

3.5. Orthogonal polynomials

The next type of numerical integration algorithms that we will examine are Gaussian integration methods. They are based in a fundamental way on the theory of orthogonal polynomials. Even though this is very classical and can be found in numerous textbooks such as [**Sze75**], we believe it necessary to review the formulas that we need, with proof, since some of them are not so easy to find, and since we will also need this theory in Gaussian *summation* methods (see Section 4.8). Among other references, we have borrowed several results and algorithms from [**PTVF07**] and [**KZ98**].

3.5.1. Definition and basic properties.
We consider the following setting. Let W be an $\mathbb{R}$-algebra containing the ring of polynomials $\mathbb{R}[X]$, let $\phi\colon W \to \mathbb{R}$ be a linear form, and assume that $\langle F, G \rangle := \phi(FG)$ defines a positive definite Euclidean scalar product on W, in other words that $\phi(F^2) \geq 0$ and $\phi(F^2) = 0$ if and only if $F = 0$ in W.

The basic example of this in the context of numerical integration is to consider a (possibly infinite) interval $[a, b] \subset \mathbb{R}$, a positive function $w(x)$ from $]a, b[$ to $\mathbb{R}^+$, and to let W be the space of continuous functions F on $]a, b[$ such that both $\int_a^b w(x) F(x)\, dx$ and $\int_a^b w(x) F^2(x)\, dx$ converge. Then $\phi(F) = \int_a^b w(x) F(x)\, dx$ satisfies the desired assumptions, the scalar product being $\langle F, G \rangle = \int_a^b w(x) F(x) G(x)\, dx$.

Clearly this can be generalized to an L^2 setting, and to complex valued functions, but the above will be sufficient.

PROPOSITION 3.5.1.

(1) *There exists a* unique *sequence of* monic *polynomials $P_n \in \mathbb{R}[X]$ of degree exactly equal to n which are orthogonal with respect to the scalar product.*

(2) *The $(P_m)_{0 \leq m \leq n}$ form a basis of the space of polynomials of degree less than or equal to n, and P_n is orthogonal to all polynomials of degree less than or equal to $n - 1$.*

(3) *We have $P_0(X) = 1$, $P_1(X) = X - a_0$, and a second order linear recurrence for $n \geq 1$:*

$$P_{n+1}(X) = (X - a_n) P_n(X) - b_n P_{n-1}(X)$$

for suitable constants a_n for $n \geq 0$ and b_n for $n \geq 1$ given by

$$a_n = \frac{\langle X P_n, P_n \rangle}{\langle P_n, P_n \rangle} \quad and \quad b_n = \frac{\langle P_n, P_n \rangle}{\langle P_{n-1}, P_{n-1} \rangle} .$$

PROOF. (1). We must clearly set $P_0(X) = 1$. Assume proved existence and uniqueness up to degree $n-1$. Note that any set of polynomials of exact degree k for $0 \le k \le n$ forms a basis of polynomials of degree less than or equal to n (which already proves part of point (2)). Thus if P_n exists it must be of the form $P_n(X) = X^n + \sum_{0 \le i \le n-1} \lambda_i P_i(X)$. Taking the scalar product with $P_i(X)$ and using orthogonality gives immediately the standard Gram–Schmidt formula $\lambda_i = -\langle P_i, X^n \rangle / \langle P_i, P_i \rangle$ for $0 \le i \le n-1$, proving both uniqueness and existence by induction.

(2). The first part has just been seen, and the second is immediate since any polynomial of degree less than or equal to $n-1$ is a linear combination of the P_i for $i \le n-1$, which are all orthogonal to P_n.

(3). The formulas for P_0 and P_1 are clear. The polynomial $P_{n+1}(X) - XP_n(X)$ has degree less than or equal to n since P_n and P_{n+1} are monic, so as before we can write $P_{n+1}(X) - XP_n(X) = \sum_{0 \le i \le n} \lambda_i P_i(X)$ with

$$\lambda_i = \langle P_{n+1} - XP_n, P_i \rangle / \langle P_i, P_i \rangle = -\langle XP_n, P_i \rangle / \langle P_i, P_i \rangle$$

since P_{n+1} is orthogonal to all the P_i for $i \le n$. Now we use the essential fact that $\langle XP_n, P_i \rangle = \phi(XP_nP_i) = \langle P_n, XP_i \rangle$, and since P_n is orthogonal to all polynomials of degree less than or equal to $n-1$ we have $\lambda_i = 0$ for $i \le n-2$, and $\lambda_n = -\langle P_n, XP_n \rangle / \langle P_n, P_n \rangle$,

$$\lambda_{n-1} = -\langle P_n, XP_{n-1} \rangle / \langle P_{n-1}, P_{n-1} \rangle = -\langle P_n, P_n \rangle / \langle P_{n-1}, P_{n-1} \rangle$$

since $XP_{n-1} = P_n + Q_{n-1}$ with Q_{n-1} of degree at most $n-1$, proving the recursion. $\qquad\square$

REMARKS 3.5.2.
  (1) One does not always (in fact quite rarely) normalize the family of orthogonal polynomials by asking that they are monic. The formulas and the recursion must then be trivially modified.
  (2) Except in rather simple cases where everything is explicit, such as those that we will see below, the above recursion, while explicit, is not the best way to compute the $P_n(X)$, and one of the goals of the sequel is to find a better way.

PROPOSITION 3.5.3. *Let ϕ be any linear form from W to $\mathbb{R}$, with no positiveness condition. Assume that the recursion of (3) above is well-defined, and more precisely that the polynomials P_n defined by the induction of (3) satisfy $\langle P_n, P_n \rangle > 0$. Then ϕ does indeed define a positive definite scalar product on $\mathbb{R}[X]$.*

PROOF. Let $F \in \mathbb{R}[X]$. We can write $F = \sum_n \lambda_n P_n$ for suitable coefficients $\lambda_n = \langle F, P_n \rangle / \langle P_n, P_n \rangle$, so by orthogonality $\langle F, F \rangle = \sum_n \lambda_n^2 \langle P_n, P_n \rangle$, and by assumption this is nonnegative, and can vanish only if $\lambda_n = 0$ for all n. $\qquad\square$

3.5.2. Using moments to compute $P_n(X)$.

DEFINITION 3.5.4. We define the *nth moment* m_n by $m_n = \langle X^n, 1 \rangle = \phi(X^n)$, and we let $\Phi(T)$ be the formal power series in T^{-1} defined by

$$\Phi(T) = \sum_{n \ge 0} m_n T^{-n-1} = \frac{m_0}{T} + \frac{m_1}{T^2} + \frac{m_2}{T^3} + \cdots .$$

THEOREM 3.5.5.

(1) *If F and G are two polynomials in $\mathbb{R}[X]$, the scalar product $\langle F, G \rangle$ is equal to the coefficient of X^{-1} in the Laurent expansion of $F(X)G(X)\Phi(X)$.*

(2) *Define the polynomial Q_n of degree exactly equal to $n-1$ by $P_n(X)\Phi(X) = Q_n(X) + O(X^{-1})$. Then the coefficients of X^{-m} in $P_n(X)\Phi(X)$ vanish for $1 \le m \le n$. Equivalently, we have the stronger identity*

$$P_n(X)\Phi(X) = Q_n(X) + O(X^{-n-1}) \,,$$

or, equivalently

$$\frac{Q_n(X)}{P_n(X)} = \Phi(X) + O(X^{-2n-1})$$

$$= \frac{m_0}{X} + \frac{m_1}{X^2} + \cdots + \frac{m_{2n-1}}{X^{2n}} + O(X^{-2n-1}) \,.$$

(3) *The polynomials Q_n satisfy the same recursion as P_n with initial terms $Q_0(X) = 0$ and $Q_1(X) = m_0 = \langle 1, 1 \rangle$.*

(4) *We have $Q_n(Y) = \langle (P_n(X) - P_n(Y))/(X - Y), 1 \rangle$.*

(5) *We have the continued fraction identity*

$$\frac{Q_n(X)}{P_n(X)} = \cfrac{m_0}{X - a_0 - \cfrac{b_1}{X - a_1 - \cfrac{b_2}{X - a_2 - \cfrac{\cdots}{\ddots - \cfrac{b_{n-1}}{X - a_{n-1}}}}}} \,.$$

(6) *We have the identity*

$$P_n(X)Q_{n+1}(X) - P_{n+1}(X)Q_n(X) = \langle P_n, P_n \rangle \,.$$

In particular, P_n and Q_n are coprime for all n.

PROOF. (1). Since the coefficient of X^{-1} in $F(X)G(X)\Phi(X)$ is bilinear in F and G, it is sufficient to prove this for $F(X) = X^m$ and $G(X) = X^n$. In that case $\langle X^m, X^n \rangle = \phi(X^{m+n}) = m_{m+n}$ by definition, which is indeed the coefficient of X^{-1} in $X^{m+n}\Phi(X)$.

(2). Since $m_0 = \langle 1, 1 \rangle$ is nonzero, it is clear that Q_n has degree exactly one less than that of P_n, i.e., $n - 1$. Write

$$P_n(X)\Phi(X) = Q_n(X) + \sum_{m \ge 0} a(m)X^{-m-1} \,.$$

For any $m < n$ we know that P_n is orthogonal to X^m, so by (1) the coefficient of X^{-1} in $X^m P_n(X)\Phi(X)$ vanishes. We thus have $a(m) = 0$ for $m < n$, hence $P_n(X)\Phi(X) = Q_n(X) + O(X^{-n-1})$ as claimed. The equivalence with the second formula is clear since P_n has degree n.

(3). Multiplying the recursion for P_n by $\Phi(X)$, we deduce that $Q_{n+1}(X) = (X - a_n)Q_n(X) - b_n Q_{n-1}(X) + O(X^{-n})$, and since the $Q_k(X)$ are polynomials, this implies that the $O(X^{-n})$ is in fact equal to 0, so the Q_n satisfy the same recursion. In addition, since $P_0 = 1$ we have $Q_0 = 0$, and since $P_1 = X - a$ for some constant a and $\Phi(X) = m_0/X + O(1/X^2)$ we have $Q_1 = m_0$.

(4). By induction, this is true for $n = 0$ since $P_0(X) = 1$ and $Q_0(X) = 0$, and for $n = 1$ since $P_1(X) = X - a$ and $Q_1(X) = \langle 1, 1 \rangle$. Assuming it true up to n, if we call $R_n(X)$ the right-hand side, for $n \geq 1$ we have

$$R_{n+1}(Y) - \big((Y - a_n)R_n(Y) - b_n R_{n-1}(Y)\big) = \langle S_n(X,Y)/(X - Y), 1 \rangle$$

with

$$
\begin{aligned}
S_n(X,Y) &= P_{n+1}(X) - P_{n+1}(Y) - (Y - a_n)\big(P_n(X) - P_n(Y)\big) \\
&\quad + b_n\big(P_{n-1}(X) - P_{n-1}(Y)\big) \\
&= P_{n+1}(X) - (Y - a_n)P_n(X) + b_n P_{n-1}(X) = (X - Y)P_n(X)
\end{aligned}
$$

using the recursion for P_n, proving (4) since $\langle P_n(X), 1 \rangle = 0$.

(5). and (6). By the basic results on continued fractions (5) follows from (3) and the initial values for P_n and Q_n. As usual in this context the matrix identity

$$\begin{pmatrix} P_{n+1} & P_n \\ Q_{n+1} & Q_n \end{pmatrix} = \begin{pmatrix} P_n & P_{n-1} \\ Q_n & Q_{n-1} \end{pmatrix} \begin{pmatrix} X - a_n & 1 \\ -b_n & 0 \end{pmatrix}$$

implies by taking determinants and induction that

$$P_{n+1}Q_n - Q_n P_{n+1} = b_n \cdots b_1 \cdot (P_1 Q_0 - Q_1 P_0) = b_n \cdots b_1 \cdot (-m_0) \ .$$

Since $b_n = \langle P_n, P_n \rangle / \langle P_{n-1}, P_{n-1} \rangle$ the product telescopes, proving (6). $\square$

Note that (2) means that (Q_n, P_n) is the $(n-1, n)$th *Padé approximant* of $\Phi(X)$.

COROLLARY 3.5.6. *Assume that the odd moments vanish, i.e., that $m_{2n+1} = 0$ for all $n \geq 0$.*

(1) *We have $a_n = 0$ for all n and the polynomials P_n and Q_n all have a parity, more precisely $P_n(-X) = (-1)^n P_n(X)$ and $Q_n(-X) = (-1)^{n-1} Q_n(X)$.*

(2) *For $n \geq 2$ we have the recursion*

$$P_{n+2} = \big(X^2 - (b_n + b_{n+1})\big)P_n - b_n b_{n-1} P_{n-2}$$

(and similarly for Q_n) with the initial terms $P_0 = 1$, $P_1 = X$, $P_2 = X^2 - b_1$, $Q_0 = 0$, $Q_1 = m_0$, $Q_2 = m_0 X$.

(3) *Equivalently, we have the continued fraction*

$$\frac{Q_{2n}(X)}{P_{2n}(X)} = \cfrac{m_0 X}{X^2 - A_0 - \cfrac{B_1}{X^2 - A_1 - \cfrac{B_2}{X^2 - A_2 - \cfrac{\ddots}{\quad} - \cfrac{B_{n-1}}{X^2 - A_{n-1}}}}} \ ,$$

where $A_0 = b_1$, $A_k = b_{2k} + b_{2k+1}$ for $k \geq 1$, and $B_k = b_{2k-1}b_{2k}$ (and a similar result for $Q_{2n+1}(X)/P_{2n+1}(X)$).

PROOF. (1). By (2) of the theorem we have $(Q_n/P_n)(-X) = -(Q_n/P_n)(X) + O(X^{-2n-1})$, so multiplying by $P_n(-X)P_n(X)$ which has degree $2n$ we deduce that we have the *identity* $(Q_n/P_n)(-X) = -(Q_n/P_n)(X)$, so that Q_n/P_n is an odd rational function, and since P_n and Q_n are coprime we deduce that both P_n and Q_n have a parity, given by their respective degree. In addition, $\langle XP_n, P_n \rangle = \langle XP_n^2, 1 \rangle$, and $XP_n^2(X)$ is thus a polynomial with only odd monomials, and since all the odd moments vanish we have $a_n = \langle XP_n, P_n \rangle / \langle P_n, P_n \rangle = 0$.

(2) and (3). The immediate computation is left to the reader. $\square$

Now let $\phi^* \colon W \to \mathbb{R}$ be the linear form on W defined by $\phi^*(X^{2n+1}) = 0$, $\phi^*(X^{2n}) = \phi(X^n) = m_n$ and extended by linearity; equivalently, if $F(X) + F(-X) = 2F^*(X^2)$ we define $\phi^*(F) = \phi(F^*)$. Note for future reference that this implies that $\phi^*(F(X)) = \phi^*(F(-X))$. We make the fundamental assumption (which is not necessarily satisfied) that ϕ^* defines a positive definite scalar product $\langle \ , \ \rangle^*$, we let P_n^* be the monic orthogonal polynomials attached to ϕ^* and denote by Q_n^* the corresponding Q-polynomials given by the previous theorem. They are easily related to the previous P_n and Q_n as follows:

THEOREM 3.5.7. *Assume that ϕ^* defines a positive definite scalar product.*

(1) *We have $P_n^*(-X) = (-1)^n P_n^*(X)$ and the two-term recursion for P_n^* is of the simpler form*

$$P_{n+1}^*(X) = X P_n^*(X) - c_n P_{n-1}^*(X)$$

with $c_n = \langle P_n^*, P_n^* \rangle^* / \langle P_{n-1}^*, P_{n-1}^* \rangle^*$.

(2) *We have $P_{2n}^*(X) = P_n(X^2)$, $a_0 = c_1$ (where we recall that $P_1(X) = X - a_0$), and for $n \geq 1$ the coefficients a_n and b_n of the recursion for P_n are given in terms of the c_n by*

$$a_n = c_{2n} + c_{2n+1} \quad and \quad b_n = c_{2n-1} c_{2n} \ .$$

(3) *We have the continued fraction expansion*

$$\frac{Q_{n+1}^*(X)}{P_{n+1}^*(X)} = \cfrac{m_0}{X - \cfrac{c_1}{X - \cfrac{c_2}{X - \cdots - \cfrac{c_n}{X}}}} = \cfrac{m_0 X^{-1}}{1 - \cfrac{c_1 X^{-2}}{1 - \cfrac{c_2 X^{-2}}{1 - \cdots - c_n X^{-2}}}} \ .$$

(4) *We have $Q_{2n}^*(X) = X Q_n(X^2)$ and the following continued fractions:*

$$\frac{Q_n(T^{-1})}{P_n(T^{-1})} = \cfrac{m_0 T}{1 - \cfrac{c_1 T}{1 - \cfrac{c_2 T}{1 - \cdots - c_{2n-1} T}}} \ , \quad and$$

$$\Phi(T^{-1}) = m_0 T + m_1 T^2 + m_2 T^3 + \cdots = \cfrac{m_0 T}{1 - \cfrac{c_1 T}{1 - \cfrac{c_2 T}{1 - \cdots}}} \ .$$

Together with the formulas obtained above expressing a_n and b_n in terms of the c_n, note that this allows to compute these quantities directly from the m_n. Note also that the c_i should not be computed by heavy formal power series manipulations, but by the much simpler and faster *Quotient-Difference algorithm*, see Sections 3.5.7 and 7.6.

PROOF. (1). (a). Assume by induction that for $k < n$ we have $P_k^*(-X) = (-1)^k P_k^*(X)$. Since P_n^* is monic of degree n the polynomial $(-1)^n P_n^*(-X)$ is monic. In addition, by our induction hypothesis if $k < n$ we have

$$\langle (-1)^n P_n^*(-X), P_k^*(X) \rangle^* = (-1)^{n+k} \langle P_n^*(-X), P_k^*(-X) \rangle^*$$
$$= (-1)^{n+k} \langle P_n^*(X), P_k^*(X) \rangle^* = 0 \ ,$$

where we have used $\phi^*(F(X)) = \phi^*(F(-X))$ for $F = P_n^* P_k^*$. So $(-1)^n P_n^*(-X)$ is orthogonal to all the $P_k^*(X)$ for $k < n$ and is monic, so by uniqueness must be equal to $P_n^*(X)$.

(b). The recursion of Proposition 3.5.1 is of the form

$$P_{n+1}^*(X) = (X - d_n)P_n^*(X) - c_n P_{n-1}^*(X) \ .$$

We change X into $-X$ and multiply by $(-1)^{n+1}$; using what we have just proved shows that $P_{n+1}^*(X) = (X + d_n)P_n^*(X) - c_n P_{n-1}^*(X)$, so that $d_n = 0$ as claimed. As in the proposition we have $c_n = \langle P_n^*, P_n^* \rangle^* / \langle P_{n-1}^*, P_{n-1}^* \rangle^*$.

(2). (a). Once again, assume by induction that for $k < n$ we have $P_{2k}^*(X) = P_k(X^2)$. By definition of ϕ^* any polynomial in X^2 is orthogonal to any odd power of X. On the other hand, let X^{2k} be any even power of X with $k < n$. By parity proved in (1), it is a linear combination of the $P_{2j}^*(X)$ for $j \leq k$ (the $P_{2j+1}^*(X)$ cannot occur), so to prove that $P_n(X^2)$ is orthogonal to all the X^{2k} it is sufficient to prove that it is orthogonal to all the $P_{2j}^*(X)$. However, by our induction hypothesis we have $P_{2j}^*(X) = P_j(X^2)$, and $\phi^*(P_n(X^2)P_j(X^2)) = \phi(P_n(X)P_j(X)) = 0$ by assumption.

(b). We have

$$\begin{aligned}
P_{2n+2}^*(X) &= X P_{2n+1}^*(X) - c_{2n+1} P_{2n}^*(X) \\
&= X \big(X P_{2n}^*(X) - c_{2n} P_{2n-1}^*(X) \big) - c_{2n+1} P_{2n}^*(X) \\
&= (X^2 - c_{2n+1}) P_{2n}^*(X) - c_{2n} X P_{2n-1}^*(X) \\
&= (X^2 - c_{2n+1}) P_{2n}^*(X) - c_{2n} \big(P_{2n}^*(X) + c_{2n-1} P_{2n-2}^*(X) \big) \\
&= \big(X^2 - (c_{2n} + c_{2n+1}) \big) P_{2n}^*(X) - c_{2n-1} c_{2n} P_{2n-2}^*(X) \ .
\end{aligned}$$

Using $P_{2n}^*(X) = P_n(X^2)$ and identifying with the recursion for P_n gives $a_n = c_{2n} + c_{2n+1}$ and $b_n = c_{2n-1} c_{2n}$.

(3). This is a direct applications of Theorem 3.5.5 (4) to the new scalar product.

(4). With evident notation, by definition we have $m_{2n}^* = m_n$, $m_{2n+1}^* = 0$, hence

$$\Phi^*(T) = \sum_{n \geq 0} m_n^* T^{-n-1} = \sum_{n \geq 0} m_n T^{-2n-1} = T \Phi(T^2) \ .$$

By definition $P_{2n}^*(X)\Phi^*(X) = Q_{2n}^*(X) + O(X^{-1})$. Since we proved that $P_{2n}^*(X) = P_n(X^2)$, we have $P_n(X^2) X \Phi(X^2) = Q_{2n}^*(X) + O(X^{-1})$, but on the other hand $P_n(X^2)\Phi(X^2) = Q_n(X^2) + O(X^{-2})$ so we deduce that $Q_{2n}^*(X) = X Q_n(X^2)$.

Replacing n by $2n - 1$ in the second continued fraction obtained above we thus obtain

$$X Q_n(X^2)/P_n(X^2) = m_0 X^{-1}/(1 - c_1 X^{-2}/(1 - c_2 X^{-2}/(1 - \cdots c_{2n-1} X^{-2}))) \ ,$$

so multiplying by X^{-1} and replacing X^2 by T^{-1} gives

$$Q_n(T^{-1})/P_n(T^{-1}) = m_0 T/(1 - c_1 T/(1 - c_2 T/(1 - \cdots c_{2n-1} T))) \ ,$$

which is the first identity. By Theorem 3.5.5 (2) we know that $Q_n(T^{-1})/P_n(T^{-1}) = \Phi(T^{-1}) + O(T^{2n-1})$ and we obtain the second identity

$$m_0 T + m_1 T^2 + m_2 T^3 + \cdots = m_0 T/(1 - c_1 T/(1 - c_2 T/(1 - c_3 T/(1 - \cdots)))) \ .$$

$\square$

COROLLARY 3.5.8. *The map ϕ^* defines a positive definite scalar product (so that the above theorem is valid) if and only if $\Phi(T^{-1})$ has a continued expansion of the above form with $c_i > 0$ for all $i \geq 1$.*

PROOF. Clear from the theorem and Proposition 3.5.3. $\qquad\square$

3.5.3. The Christoffel–Darboux formula. We will say that a polynomial $P \in \mathbb{R}[X]$ is *positive* if P is nonzero and takes only nonnegative values on $\mathbb{R}$. We keep all the previous notation; in particular ϕ is a linear form which defines a positive definite scalar product on $\mathbb{R}[X]$, and P_n the corresponding family of monic orthogonal polynomials. We denote by $(x_j)_{1 \leq j \leq n}$ the complex roots of P_n.

THEOREM 3.5.9.
1. *For any positive polynomial P we have $\phi(P) > 0$.*
2. *All the roots x_j of the orthogonal polynomials P_n are real and simple.*
3. *Let F be a polynomial of degree less than or equal to $2n$, and denote by f_{2n} its coefficient of X^{2n}. We have the identity*

$$\phi(F) = \langle F, 1 \rangle = \sum_{1 \leq j \leq n} F(x_j) \frac{Q_n(x_j)}{P_n'(x_j)} + f_{2n} \langle P_n, P_n \rangle ,$$

   where Q_n is as in Theorem 3.5.5 (2). In particular, if F has degree less than or equal to $2n - 1$ we have

$$\phi(F) = \sum_{1 \leq j \leq n} w_j F(x_j) \quad \text{with} \quad w_j = Q_n(x_j)/P_n'(x_j) .$$

4. *We have $Q_n(x_j)/P_n'(x_j) > 0$ for all j.*
5. *For any root x_j of P_n we have*

$$Q_n(x_j) = \langle P_n/(X - x_j), 1 \rangle = \frac{\langle P_{n-1}, P_{n-1} \rangle}{P_{n-1}(x_j)} = -\frac{\langle P_n, P_n \rangle}{P_{n+1}(x_j)} .$$

PROOF. (1). Let P be a positive polynomial with leading coefficient λ. We first claim that λ is positive (otherwise $\lim_{x \to \infty} P(x) = -\infty$) and that P/λ is a product of monic positive polynomials of degree 2. Indeed, let

$$P(X) = \lambda \prod_j (X - x_j)^{m_j} \prod_k \big((X - z_k)(X - \overline{z_k})\big)^{n_k} ,$$

where the x_j and the z_k are the distinct real roots and complex roots of P respectively. For each j, the multiplicity m_j of the real root x_j must be even, otherwise P changes sign at x_j. Hence P/λ is a product of positive polynomials of degree 2 of the form $(X - x_j)^2$ or $(X - z_k)(X - \overline{z_k})$.

Let $Q(X) = X^2 + bX + c$ be a monic positive polynomial of degree 2, so that $4c - b^2 \geq 0$. It follows that $Q = Q_1^2 + Q_2^2$ with $Q_1 = (X + b/2)$ and $Q_2 = \sqrt{4c - b^2}/2$. Now a product of sums of two squares is still a sum of two squares (multiplicativity of the norm of complex numbers), so we deduce from our claim that P/λ is a sum of two squares, hence so is $P = A^2 + B^2$. Since $\phi(A^2 + B^2) = \langle A, A \rangle + \langle B, B \rangle \geq 0$ with equality if and only if $A = B = 0$, which is forbidden since $P \neq 0$, we have proved (1).

(2). Let $(x_j)_{1 \leq j \leq m}$ be the roots of P_n which are both real and of odd multiplicity, and set $R(X) = \prod_{1 \leq j \leq m}(X - x_j)$. By construction, the polynomial $P_n R$ is real and all its real roots have even multiplicity, so it is a positive polynomial. By

(1) we have $\langle P_n, R \rangle = \phi(P_n R) > 0$. Since P_n is orthogonal to all polynomials of degree $m < n$ and is not orthogonal to R, it follows that $m \geq n$, hence that $m = n$ so that all the roots of P_n are real and of odd order, necessarily of order 1 otherwise $m < n$.

(3). By Theorem 3.5.5 (1), we know that $\phi(F) = \langle F, 1 \rangle$ is the coefficient of X^{-1} in $F(X)\Phi(X)$, and by (4) of that theorem we have $\Phi(X) = Q_{n+1}(X)/P_{n+1}(X) + O(X^{-2n-3})$. Thus, if F has degree less than or equal to $2n$ (in fact even $2n+1$, but we do not need this), we have $F(X)\Phi(X) = F(X)Q_{n+1}(X)/P_{n+1}(X) + O(X^{-2})$. Now by Theorem 3.5.5 (6) we have

$$\frac{Q_{n+1}(X)}{P_{n+1}(X)} = \frac{Q_n(X)}{P_n(X)} + \frac{\langle P_n, P_n \rangle}{P_n(X)P_{n+1}(X)} \, ,$$

and since $P_n P_{n+1}$ has degree $2n + 1$ and is monic it follows that

$$F(X)\Phi(X) = F(X)Q_n(X)/P_n(X) + f_{2n} \langle P_n, P_n \rangle X^{-1} + O(X^{-2}) \, ,$$

so the coefficient of X^{-1} in $F(X)\Phi(X)$ is equal to $f_{2n} \langle P_n, P_n \rangle$ plus the coefficient of X^{-1} in $F(X)Q_n(X)/P_n(X)$. Now since the roots of P_n are (real and) simple, by expanding into partial fractions we have

$$F(X)\frac{Q_n(X)}{P_n(X)} = G(X) + \sum_{1 \leq j \leq n} \frac{u_j}{X - x_j} \, ,$$

where $G(X)$ is a polynomial (of degree less than or equal to $2n - 1$), and the u_j are obtained by multiplying by $X - x_j$ and making X tend to x_j, hence $u_j = F(x_j)Q_n(x_j)/P_n'(x_j)$. The result follows since the coefficient of X^{-1} in the right-hand side is equal to $\sum_{1 \leq j \leq n} u_j$.

(4). Consider the polynomial $R_j = \prod_{i \neq j}(X - x_i)^2$. It is the square of a real polynomial, so $\phi(R_j) = \langle R_j, 1 \rangle > 0$. Since it has degree $2n - 2$, by (3) we have $\langle R_j, 1 \rangle = \sum_{1 \leq i \leq n} R_j(x_i)Q_n(x_i)/P_n'(x_i)$, and since by construction $R_j(x_i) = 0$ for $i \neq j$ and $R_j(x_j) > 0$ since the zeros of P_n are simple, it follows that $Q_n(x_j)/P_n'(x_j) = \langle R_j, 1 \rangle / R_j(x_j) > 0$ as claimed.

(5). For the first formula we simply set $Y = x_j$ in Theorem 3.5.5 (4). For the second, we use Theorem 3.5.5 (6) with n replaced by $n - 1$, which implies after taking $X = x_j$ that $P_{n-1}(x_j)Q_n(x_j) = \langle P_{n-1}, P_{n-1} \rangle$, and the last is proved similarly. $\square$

COROLLARY 3.5.10. *If $\phi(F) = \int_a^b w(x)F(x)\,dx$ with $w \geq 0$ as in the basic example at the start of Section 3.5.1, then all the roots x_j of P_n belong in fact to the open interval $]a, b[$.*

PROOF. In the above proofs, we replace the notion of positive polynomial by positive on the interval $]a, b[$. The proof of (1) shows that if P is such a polynomial then P is equal to the product of a sum of two squares times a polynomial Q having all its roots real and outside the interval $]a, b[$, and in particular Q has a constant sign in $]a, b[$. Thus once again we deduce that $\phi(P) \neq 0$, and the proof of (2) goes as before. $\square$

THEOREM 3.5.11 (Christoffel–Darboux).

(1) *We have the formulas*

$$\sum_{0 \leq k \leq n} \frac{P_k(X)P_k(Y)}{\langle P_k, P_k \rangle} = \frac{1}{\langle P_n, P_n \rangle} \frac{P_{n+1}(X)P_n(Y) - P_{n+1}(Y)P_n(X)}{X - Y} ,$$

$$\sum_{0 \leq k \leq n} \frac{P_k(X)^2}{\langle P_k, P_k \rangle} = \frac{1}{\langle P_n, P_n \rangle} (P'_{n+1}(X)P_n(X) - P_{n+1}(X)P'_n(X)) .$$

(2) *The zeros of P_n and P_{n+1} interlace.*

PROOF. (1). Set $\Delta_n(X, Y) = P_{n+1}(X)P_n(Y) - P_{n+1}(Y)P_n(X)$. By the recurrence relation we have

$$P_{n+1}(X)P_n(Y) = (X - a_n)P_n(X)P_n(Y) - b_n P_{n-1}(X)P_n(Y) \quad \text{and}$$
$$P_{n+1}(Y)P_n(X) = (Y - a_n)P_n(X)P_n(Y) - b_n P_{n-1}(Y)P_n(X) ,$$

so subtracting we obtain $\Delta_n(X, Y) = (X - Y)P_n(X)P_n(Y) + b_n \Delta_{n-1}(X, Y)$.
Recalling that $b_n = \langle P_n, P_n \rangle / \langle P_{n-1}, P_{n-1} \rangle$, we thus have

$$\frac{\Delta_n(X, Y)}{\langle P_n, P_n \rangle} = (X - Y)\frac{P_n(X)P_n(Y)}{\langle P_n, P_n \rangle} + \frac{\Delta_{n-1}(X, Y)}{\langle P_{n-1}, P_{n-1} \rangle} ,$$

from which we deduce immediately that

$$\frac{\Delta_n(X, Y)}{\langle P_n, P_n \rangle} = (X - Y)\sum_{k=1}^{n} \frac{P_k(X)P_k(Y)}{\langle P_k, P_k \rangle} + \frac{\Delta_0(X, Y)}{\langle P_0, P_0 \rangle} ,$$

and the first formula follows since $\Delta_0(X, Y) = P_1(X) - P_1(Y) = X - Y$ and $P_0(X) = 1$. To obtain the second formula we make $Y \to X$ and apply l'Hospital's rule.

(2). By Theorem 3.5.9 we know that the zeros of P_{n+1} are real and simple, so let $z_k < z_{k+1}$ be two consecutive zeros of P_{n+1}. The derivatives $P'_{n+1}(z_k)$ and $P'_{n+1}(z_{k+1})$ must have opposite signs (and are nonzero since the zeros of P_{n+1} are simple). The second formula of (1) implies that $P'_{n+1}(X)P_n(X) - P_{n+1}(X)P'_n(X) > 0$, so we have $P'_{n+1}(z_k)P_n(z_k) > 0$ and $P'_{n+1}(z_{k+1})P_n(z_{k+1}) > 0$. It follows that $P_n(z_k)P_n(z_{k+1}) < 0$; by continuity, the polynomial P_n has at least one zero x such that $z_k < x < z_{k+1}$. In addition, this zero x must be unique since P_n has only n zeros. Thus, if we call x_j the zeros of P_n, we must have

$$z_1 < x_1 < z_2 < x_2 < \cdots < z_n < x_n < z_{n+1} ,$$

in other words the zeros interlace. $\square$

3.5.4. Using moment matrices. It is easy (although not very useful from a computational standpoint) to give the polynomials $P_n(X)$ and $Q_n(X)$ explicitly.

DEFINITION 3.5.12.

(1) We define

$$M_n = \begin{pmatrix} m_0 & m_1 & \cdots & m_n \\ m_1 & m_2 & \cdots & m_{n+1} \\ \vdots & \vdots & \vdots & \vdots \\ m_n & m_{n+1} & \cdots & m_{2n} \end{pmatrix}$$

and

$$N_n(X) = \begin{pmatrix} m_0 & m_1 & \dots & m_n \\ m_1 & m_2 & \dots & m_{n+1} \\ \vdots & \vdots & \vdots & \vdots \\ m_{n-1} & m_n & \dots & m_{2n-1} \\ 1 & X & \dots & X^n \end{pmatrix}.$$

(2) For any polynomial $A(X) = \sum_{0 \le i \le n} a_i X^i$ we denote by $\widetilde{A}$ the column vector $(a_0, a_1, \dots, a_n)^t$, where ^{t} denotes transpose.

PROPOSITION 3.5.13.

(1) *For any polynomial A of degree less than or equal to n we have*

$$M_n \widetilde{A} = (\langle A, 1 \rangle, \langle A, X \rangle, \dots, \langle A, X^{n-1} \rangle, \langle A, X^n \rangle)^t \quad \text{and}$$
$$N_n(X) \widetilde{A} = (\langle A, 1 \rangle, \langle A, X \rangle, \dots, \langle A, X^{n-1} \rangle, A(X))^t.$$

(2) *For any two polynomials A and B of degree less than or equal to n we have*

$$\widetilde{B}^t M_n \widetilde{A} = \langle A, B \rangle.$$

(3) *We have the explicit formulas*

$$P_n(X) = \frac{\det(N_n(X))}{\displaystyle\prod_{0 \le j < n} \langle P_j, P_j \rangle} = \frac{\det(N_n(X))}{\det(M_{n-1})} \quad \text{and}$$

$$Q_n(X) = \sum_{0 \le j < n} m_j \left\lfloor P_n(X)/X^{j+1} \right\rfloor,$$

where $\left\lfloor P_n(X)/X^{j+1} \right\rfloor$ denotes the quotient of the Euclidean division of $P_n(X)$ by X^{j+1}.

(4) *In particular we have $\det(M_n) = \prod_{0 \le j \le n} \langle P_j, P_j \rangle$, or equivalently $\langle P_n, P_n \rangle = \det(M_n)/\det(M_{n-1})$.*

PROOF. (1) is trivial by using the definition of the m_i and standard matrix multiplication, and (2) follows by linearity. This is of course the standard way of expressing a scalar product in terms of its Gram matrix. For (3), by (1) we have

$$N_n(X) \widetilde{P_j} = (\langle P_j, 1 \rangle, \langle P_j, X \rangle, \dots, \langle P_j, X^{n-1} \rangle, P_j(X))^t$$
$$= (0, \dots, \langle P_j, X^j \rangle, \dots, P_j(X))^t,$$

and since $X^j - P_j$ is a polynomial of degree strictly less than j it is orthogonal to P_j, so $\langle P_j, X^j \rangle = \langle P_j, P_j \rangle$. We thus have a matrix identity of the form $N_n(X)O_n = U_n$, where O_n is the matrix whose columns are the $\widetilde{P_j}$, hence upper triangular with 1 on the diagonal since the P_j are monic, and U_n is a lower triangular matrix with the $\langle P_j, P_j \rangle$ on the diagonal for $0 \le j \le n - 1$, and $P_n(X)$ for $j = n$. Taking determinants proves the first formula of (3). Since P_n is monic, this shows that the $n \times n$ upper-left minor of $N_n(X)$ is equal to $\prod_{0 \le j \le n-1} \langle P_j, P_j \rangle$, proving (4) and the second formula of (3). The formula for Q_n follows immediately from the first formulas of (3) and Theorem 3.5.5 (4), and its proof is left to the reader. $\square$

The matrix M_n is the nth Gram matrix of the positive definite scalar product, hence using the *Cholesky decomposition* for which we will give an algorithm below in Section 3.5.6, there exists an *upper triangular* matrix $R_n = (r_{i,j})_{0 \le i,j \le n}$ such that $R_n^t R_n = M_n$.

LEMMA 3.5.14.

(1) *For* $0 \le j \le n$ *we have*
$$R_n \widetilde{P_j} = (0, 0, \dots, r_{j,j}, 0, \dots, 0)^t ,$$
with $r_{j,j}$ *as* jth *coordinate (starting at 0).*

(2) *We have* $\langle P_j, P_j \rangle = r_{j,j}^2$.

(3) *If we write* $P_j(X) = \sum_{0 \le i \le j} p_{j,i} X^i$ *with* $p_{j,j} = 1$, *we have* $p_{j,j-1} = -r_{j-1,j}/r_{j-1,j-1}$.

PROOF. By (2) of the proposition we have $\widetilde{B}^t R_n^t R_n \widetilde{A} = \langle A, B \rangle$, so that the $R_n \widetilde{P_j}$ form an orthogonal basis for the ordinary scalar product of vectors in $\mathbb{R}^{n+1}$, and since P_j has degree j and is monic, we have $R_n \widetilde{P_j} = (*, *, \dots, r_{j,j}, 0, \dots, 0)^t$, so by orthogonality the $*$ vanish, proving (1). (2) immediately follows, again by (2) of the proposition, and (3) by identifying the $(j-1)$st coordinate of the matrix product. $\square$

COROLLARY 3.5.15. *Recall that we denoted by* a_n *and* b_n *the coefficients of the recursion* $P_{n+1}(X) = (X - a_n)P_n(X) - b_n P_{n-1}(X)$. *For* $1 \le j \le n-1$ *we have the formulas*
$$a_j = \frac{r_{j,j+1}}{r_{j,j}} - \frac{r_{j-1,j}}{r_{j-1,j-1}} \quad and \quad b_j = \frac{r_{j,j}^2}{r_{j-1,j-1}^2} .$$

PROOF. By Proposition 3.5.1 we have $b_j = \langle P_j, P_j \rangle / \langle P_{j-1}, P_{j-1} \rangle$, so the formula for b_j follows from the lemma. For a_j, we multiply the recursion at j by R_n and obtain with evident notation e_i for the ith basis vector:
$$r_{j+1,j+1} e_{j+1} = R_n \widetilde{XP_j} - a_j r_{j,j} e_j - b_j e_{j-1}$$
from which identification of the coefficients of e_j gives
$$r_{j,j} p_{j,j-1} + r_{j,j+1} = a_j r_{j,j} ,$$
and the formula for a_j follows from the third statement of the lemma. $\square$

3.5.5. Further properties of orthogonal polynomials. For completeness we include here without proof an additional result which is not needed in this book.

DEFINITION 3.5.16. We define the nth orthogonal polynomial matrix O_n as follows: if $P_j(X) = \sum_{0 \le i \le n} a_{i,j} X^i$ (with $a_{i,j} = 0$ for $i > j$ and $a_{i,i} = 1$), then $O_n = (a_{i,j})_{0 \le i,j \le n}$, in other words

$$O_n = \begin{pmatrix} 1 & a_{0,1} & a_{0,2} & \dots & a_{0,n} \\ 0 & 1 & a_{1,2} & \dots & a_{1,n} \\ \vdots & \vdots & \ddots & \vdots & \vdots \\ 0 & 0 & \dots & 1 & a_{n-1,n} \\ 0 & 0 & \dots & 0 & 1 \end{pmatrix} .$$

Note that we already used O_n above in the proof of Proposition 3.5.13.

PROPOSITION 3.5.17. *By abuse of notation, let $O_{n,m}$ be the $(n+1-m) \times (n+1-m)$ matrix extracted from O_n by taking the first $n+1-m$ rows and the last $n+1-m$ columns, in other words $O_{n,m} = (a_{i,j})_{\substack{0 \le i \le n-m \\ m \le j \le n+1}}$.*

On the other hand, let $T_{n,m} = (m_{i+j})_{0 \le i,j \le m-1}$ be the upper-left principal minor and $B_{n,m} = (m_{i+j})_{\substack{n-m+1 \le i \le n \\ 0 \le j \le m-1}}$ be the bottom-left minor.

(1) *We have the identity*

$$\det(O_{n,m}) = (-1)^{nm} \frac{\det(B_{n,m})}{\det(T_{n,m})} \ .$$

(2) *In particular we can recover the moments m_n from the polynomials P_n thanks to the formula*

$$m_n = (-1)^n m_0 \det(O_{n,1}) = (-1)^n m_0 \begin{vmatrix} a_{0,1} & a_{0,2} & a_{0,3} & \cdots & a_{0,n} \\ 1 & a_{1,2} & a_{1,3} & \cdots & a_{1,n} \\ 0 & 1 & a_{2,3} & \cdots & a_{2,n} \\ \vdots & \vdots & \ddots & \vdots & \vdots \\ 0 & 0 & \cdots & 1 & a_{n-1,n} \end{vmatrix} \ .$$

For instance, we also have

$$\det(O_{n,2}) = \frac{m_{n-1} m_{n+1} - m_n^2}{m_0 m_2 - m_1^2} \ .$$

3.5.6. Computing orthogonal polynomials: Cholesky. We now consider the *practical* problem of computing the polynomials $P_n(X)$ and $Q_n(X)$, since these will be needed in both Gaussian integration and summation methods.

Because of the numerous formulas that we proved above (and several more that we have not mentioned) involving the *moments* m_n, we will base our methods on the assumption that we have computed these moments in some way (see the Gaussian integration methods below for examples of this). Note, however, that this is not the only way, nor the best, to give our scalar product. For instance, we may prefer to give a "better" polynomial basis than $(1, X, \ldots, X^n)$.

There are at least two ways to obtain the orthogonal polynomials from the moments: the first is using the Cholesky decomposition of the moment matrix M_n, and the second is to use the continued fractions for $\Phi(X)$ obtained above. We begin by the Cholesky decomposition. A standard procedure for this is as follows:

───────────────────── MatCholesky.gp ─────────────────────

```
1   /* Given a symmetric positive definite matrix A, finds an
2    * upper triangular matrix R such that R~ * R = A. */
3   MatCholesky(A) =
4   { my([m, n] = matsize(A), R = matrix(n, n));
5
6     if (n != m, error("A not square"));
7     for (i = 1, n,
8       R[i,i] = sqrt(A[i,i] - sum(k = 1, i - 1, R[k,i]^2));
9       my(c = 1 / R[i,i]);
10      for (j = i + 1, n,
11        R[i,j] = c * (A[i,j]
12                  - sum(k = 1, i - 1, R[k,i] * R[k,j]))));
```

```
13    return (R);
14  }
```

REMARKS 3.5.18.

(1) Note that because all GP vector indices begin at 1, here the indices vary from 1 to n instead of from 0 to n. This will of course have to be taken into account in the next program.

(2) The algorithm is stable as long as the $r_{i,i}$ do not become too small. If A is positive definite, the expression under the square root is always positive in *exact* arithmetic; if it becomes negative because of round-off errors, the program will fail.

(3) A technical remark is that the value of A[n,n] is only used in the computation of R[n,n]. Therefore, if we do not need R[n,n] later, we can set A[n,n] to any value we like.

Obtaining the coefficients a_n and b_n of the recursion and hence the orthogonal polynomials P_n and the Q_n is now immediate from the above Corollary 3.5.15:

`OrthopolyCholesky.gp`

```
1   /* Given 2n moments m_k for 0 <= k < 2n stored in
2    * M[1]...M[2n], compute the coefficients
3    * [a_0, ..., a_{n-1}] and [b_1, ..., b_{n-1}] of the 3-term
4    * recursion P_{j+1}(X) = (X - a_j)P_j(X) - b_jP_{j-1}(X);
5    * P_1 = X - a_0, P_0 = 1. */
6
7   Threeterm(M) =
8   { my(n = #M \ 2, N = n + 1, A, R, p, v, a, b);
9
10    /* M[2n+1] is outside of range: replace by 0 */
11    A = matrix(N, N, i, j, if (i != j || i != N, M[i+j-1]));
12    R = MatCholesky(A);
13    p = vector(n, j, 1 / R[j, j]);
14    v = vector(n, j, R[j, j+1] * p[j]);
15    a = vector(n - 1, j, if (j == 1, M[2]/M[1], v[j] - v[j-1]));
16    b = vector(n - 1, j, (R[j+1, j+1] * p[j])^2);
17    return ([a, b]);
18  }
19
20  /* Q1 = q1, ab = [a, b], compute [P_n, Q_n]. */
21  Orthopolyrec(ab, q1) =
22  {
23    my([a, b] = ab, P0 = 1, P1 = 'x - a[1], Q0 = 0, Q1 = q1);
24    for (j = 1, #a - 1,
25      my(A = 'x - a[j+1]);
26      [P0, P1] = [P1, A * P1 - b[j] * P0];
27      [Q0, Q1] = [Q1, A * Q1 - b[j] * Q0]);
28    return ([P1, Q1]);
29  }
```

```
30
31   /* Given 2n moments m_k for 0 <= k < 2n, compute [P_n, Q_n]. */
32   OrthopolyCholesky(M) = Orthopolyrec(Threeterm(M), M[1]);
```

Note that we use the above technical remark to avoid using `mom[2*n+1]` which is outside the given range. Note also that we could use the `CFback` program from Chapter 7 to compute (P_n, Q_n), but in the present case it is simpler to directly use the recursions.

3.5.7. Computing orthogonal polynomials: Continued fractions.
As we have seen, a second method, which is usually preferable, to obtain the orthogonal polynomials is to use continued fractions. We have at our disposal (at least) two continued fractions: the one given by Theorem 3.5.5 (5), and the one given by Theorem 3.5.7 (4). The latter has a simpler form and is generally preferable, but we must be careful about vanishing coefficients. Indeed, as we will see in Chapter 7 the use of the *Quotient-Difference algorithm* (QD for short) is an efficient method for computing the coefficients of the continued fraction without heavy handling of power series. However, it may (and does) happen that some coefficients c_i vanish, which makes impossible the direct application of QD (it can be suitably modified). It can be shown that the c_i are quotients of products of suitable *Hankel determinants*, so this phenomenon corresponds to the vanishing of such determinants.

We will not dwell on this problem, although we will come back to it in Chapter 8, and make the following simplifying assumption: with the notation of Theorem 3.5.7, we assume that one of the following two conditions hold:

- the continued fraction is of the form given by Theorem 3.5.7 (4) with the c_i not too small;
- $m_{2j+1} = 0$ for all j, in which case the continued fraction is of the form

$$\Phi(T^{-1}) = m_0 T + m_2 T^3 + \cdots = m_0 T / (1 - d_1 T^2 / (1 - d_2 T^2 / (1 - \cdots)))$$

or, equivalently, $\Phi(T) = m_0 / (T - d_1 / (T - d_2 / (T - \cdots)))$, and we assume that the d_i are not too small.

Under these assumptions, we can apply the QD algorithm of Chapter 7 to obtain the d_i or the c_i. Using Theorem 3.5.7, it is now immediate to compute the P_n and Q_n:

———————————————— | OrthopolyCF.gp | ————————————————

```
1    /* MOM(n) returns the 2n moments m_k for 0 <= k < 2n,
2     * compute [P_n, Q_n]. */
3    OrthopolyCF(MOM, n) =
4    { my(B = getlocalbitprec(), N = 2*n, v, M, PQ, P, Q);
5
6      localbitprec(3*B / 2); M = MOM(n);
7      if (exponent(M[2]) >= -B, /* m_1 != 0 */
8        for (j = 1, N,
9          if (exponent(M[j]) < -B, error("untreatable case")));
10       /* normal Quodif, hope for the best */
11       v = Quodif(M);
12       PQ = CFback(N, j -> v[j+1] / 'x, j -> (j != 0))
```

```
13   , /* else */
14      forstep (j = 4, N, 2, /* m_{j-1} ~ 0 for all j */
15         if (exponent(M[j]) >= -B, error("untreatable case")));
16      /* Quodif with only even terms, hope for the best */
17      v = Quodif(vector(n, j, M[2*j - 1]));
18      my(R = 'x * CFback(n - 1, j -> v[j + 2] / 'x^2));
19      PQ = v[1] / R);
20   P = denominator(PQ);
21   Q = numerator(PQ);
22   return (bitprecision([P, Q] / pollead(P), B + 32));
23 }
```

We refer to Chapter 7 for the continued fraction algorithms used in this program.

We could of course also have used the relations $a_n = c_{2n} + c_{2n+1}$ and $b_n = c_{2n-1}c_{2n}$, where $-c_n$ are the coefficients output by the Quodif algorithm, together with the usual recursions for P_n and Q_n, but only in the case $c_n > 0$ for all n, thanks to Corollary 3.5.8.

3.5.8. Computing the roots of P_n. In application to Gaussian integration and to Monien summation, it is essential to be able to compute the roots of P_n, and to large accuracy since we desire multiprecision answers. This is an extremely important problem, since this part of the computation can represent 99% of the total computation time, especially if not properly done.

Pari/GP has a built-in general purpose root finder polroots which is *guaranteed* to find the complex roots to the desired accuracy, using Schönhage's splitting circle method. Note that this is highly nontrivial, but we will not discuss it here (see [**Gou93**]). Finding *real* roots is much easier, and the simpler and more efficient polrootsreal function can be used in this case. This function uses a combination of Descartes's rule of sign to isolate roots and Newton's iterations to compute them to the desired accuracy. When the polynomial has only real roots, as in the case of orthogonal polynomials, we will of course use polrootsreal.

These general purpose programs run in polynomial time, but they nonetheless become very slow when the degree increases, and it is essential to find better methods whenever possible.

In some cases, this is trivial: for instance, in the case of the Chebyshev polynomials of the first or second kind that we will mention below, there are explicit formulas for the roots in terms of trigonometric functions. In other cases, such as many of the other classical orthogonal polynomials, there exist excellent starting values that isolate roots well enough to perform Newton iterations, see for instance the Legendreroots program below. This is also the case for one-half of the roots occurring in the Monien summation method.

However, in general such specialized methods are either not available or have not been considered, so we must proceed differently. A method suggested in [**PTVF07**] and due to Wilf is as follows. Consider the basic recursion for the polynomials $P_n(X)$ given by Proposition 3.5.1. This can be written in matrix form as follows:

$$
X \begin{pmatrix} P_0(X) \\ P_1(X) \\ \vdots \\ P_{n-2}(X) \\ P_{n-1}(X) \end{pmatrix} = \begin{pmatrix} a_0 & 1 & 0 & \dots & 0 & 0 \\ b_1 & a_1 & 1 & 0 & \dots & 0 \\ \vdots & \vdots & \vdots & \vdots & \vdots & \vdots \\ 0 & 0 & \dots & b_{n-2} & a_{n-2} & 1 \\ 0 & 0 & 0 & \dots & b_{n-1} & a_{n-1} \end{pmatrix} \begin{pmatrix} P_0(X) \\ P_1(X) \\ \vdots \\ P_{n-2}(X) \\ P_{n-1}(X) \end{pmatrix} + \begin{pmatrix} 0 \\ 0 \\ \vdots \\ 0 \\ P_n(X) \end{pmatrix}
$$

which we can abbreviate as $XC_n(X) = TC_n(X) + P_n(X)e_{n-1}$, where $C_n(X)$ is the column of the $P_i(X)$ for $0 \le i \le n-1$, T is an $n \times n$ tridiagonal matrix, and $e_{n-1} = (0,0,\dots,1)^t$. It follows that $P_n(x_j) = 0$ if and only if $TC_n(x_j) = x_j C_n(x_j)$, and since $P_m(x_j) \ne 0$ for all $m < n$ we have $C_n(x_j) \ne 0$, so x_j is an *eigenvalue* of the matrix T. Since the roots of P_n are all simple, it follows that they are exactly the eigenvalues of T, so instead of a root finding problem, we now have an eigenvalue computation, and for this there exist much more efficient methods. Without entering into details (see again [**PTVF07**] for instance), it is immediate to show that the above matrix T is similar (hence has the same eigenvalues) to the matrix

$$
\begin{pmatrix} a_0 & b_1^{1/2} & 0 & \dots & 0 & 0 \\ b_1^{1/2} & a_1 & b_2^{1/2} & 0 & \dots & 0 \\ \vdots & \vdots & \vdots & \vdots & \vdots & \vdots \\ 0 & 0 & \dots & b_{n-2}^{1/2} & a_{n-2} & b_{n-1}^{1/2} \\ 0 & 0 & 0 & \dots & b_{n-1}^{1/2} & a_{n-1} \end{pmatrix}
$$

which is now a *symmetric* tridiagonal matrix, and there exist efficient algorithms to find the eigenvalues of such matrices. We give a general program `Tqli` (QL algorithm for Triangular matrices, with Implicit shifts) directly copied from [**PTVF07**] and adapted to multiprecision and the **GP** language, which computes the eigenvalues of such a matrix:

————————————— Tqli.gp —————————————

```
1   /* Tridiagonal QL Implicit algorithm (TQLI) */
2
3   /* a << b to accuracy 2^(-B) ? */
4   Negligible(a, b, B) = b && exponent(b) - exponent(a) > B;
5
6   Getm(d, e, N, j, B) =
7   {
8     for (m = j, N - 1,
9       if (!e[m] || Negligible(e[m], d[m], B)
10               || Negligible(e[m], d[m+1], B), return (m)));
11    return (N);
12  }
13
14  Tqli(d, e) =
15  { my(B = getlocalbitprec(), N = #d, n = if (!d, N >> 1, N), m);
16
17    localbitprec(B + 32);
```

```
18    e = concat(e, 0);
19    for (j = 1, n,
20      while ((m = Getm(d, e, N, j, B)) != j,
21        my(p = 0, s = 1, c = 1, g, r);
22        g = (d[j+1] - d[j]) / (2 * e[j]);
23        r = sqrt(1 + g^2);
24        g = d[m] - d[j] + e[j] / if (g < 0, g - r, g + r);
25        forstep (i = m - 1, j, -1,
26          my(f = s * e[i], b = c * e[i]);
27          e[i+1] = r = sqrt(f^2 + g^2);
28          if (exponent(r) < -B, d[i+1] -= p; e[m] = 0; next(2));
29          s = f / r; c = g / r; g = d[i+1] - p;
30          r = (d[i] - g) * s + 2 * c * b;
31          p = s * r; d[i+1] = g + p; g = c * r - b);
32        d[j] -= p; e[j] = g; e[m] = 0));
33    return (bitprecision(vecsort(d[1..n]), B));
34  }
```

This program is quite efficient in low accuracy. However, in multiprecision computations linked to orthogonal polynomials, it is preferable to proceed as follows. Using `Tqli`, we first compute low accuracy (64 bits) values for the roots of the orthogonal polynomial P_N. We then want to apply Newton's method to refine the roots up to the desired accuracy. For this, we need to be careful about quite a number of things:

(1) Since we need to compute $(b(n))^{1/2}$ as input to the `Tqli` program, to avoid accuracy loss this should be done at a slightly larger accuracy than the desired one (for instance 96 bits instead of 64).

(2) The basic Newton iteration formula is $r \leftarrow r - P(r)/P'(r)$. However, computing the value of $P(r)$ simply by replacing r in the explicit formula for P entails a complete loss of accuracy. On the contrary, if we use the recursion giving P_{n+1} in terms of P_n and P_{n-1} this is completely stable. By differentiating the recursion, it is immediate to generalize this to the computation of $P'(r)$, $P''(r)/2$, etc., and this is what will be done in the `Orthopolyeval` program below

(3) For the Newton iteration itself, we use the `Newtonrefine` program given in Chapter 2 adapted to our recursion. Simply note that if $e \geq 64$, our initial approximation using the `Tqli` program will need to be done at a larger accuracy, and we leave to the reader the necessary modifications. On the contrary, if e is small, we can remove a few initial iterations since we know that `Tqli` will already have given us approximately 64 bits of accuracy.

―――――――――――――――――― | Orthopolyroots.gp | ――――――――――――――――――

```
1   /* 3-term recursion coefficients a, b attached to orthogonal
2    * polynomial P_N are given as vectors
3    *   a = [a(0), ..., a(N-1)], b = [b(1), ..., b(N-1)]
4    * (they may contain further coefficients) */
5
```

```
 6    /* Let r be a (complex or formal) value and a, b be the 3-term
 7     * recursion coefficients vectors. Let f be the attached MONIC
 8     * orthogonal polynomial f = P_N / pollead(P_N).
 9     * If k > 0, return
10     *   [f(r), f'(r), f''(r) / 2, ..., f^(k-1)(r) / (k-1)!]
11     * If k <= 0, only return f^(-k)(r) / (-k)! */
12    Orthopolyeval(a, b, N, r, k = 0) =
13    { my(K = if (k > 0, k, 1 - k), u0 = vector(K), u1 = vector(K));
14
15      u0[1] = 1; u1[1] = r - a[1]; if (K > 1, u1[2] = 1);
16      for (m = 1, N - 1,
17        my(u2 = (r - a[m+1]) * u1 - b[m] * u0);
18        u0 = u1; u1 = u2 + concat(0, u1[1..K-1]));
19      return (if (k <= 0, u1[K], u1));
20    }
21
22    /* Adaptation of Newtonrefine. */
23    OrthoNewtonrefine(a, b, N, r, B) =
24    { my(c, p, p1, p2, n, e);
25
26      [p, p1, p2] = Orthopolyeval(a, b, N, r, 3);
27      e = max(exponent(p2 / p1) - 1, 0);
28      if (e >= 63, error("Insuficcient accuracy in Newtonrefine"));
29      n = 1 + exponent((B + 32 - e) \ (64 - e));
30      c = 1 + e + (B - e) >> n;
31      for (j = 1, n,
32        c = 2*c - e; r = bitprecision(r, c);
33        [p, p1] = Orthopolyeval(a, b, N, r, 2);
34        r -= p / p1);
35      return (bitprecision(r, B));
36    }
37
38    /* Given the 3-term recursion coefficients a, b as closures
39     * or the corresponding vectors with N and N-1 components,
40     * compute the roots of the orthogonal polynomial P_N.
41     * If a = 0, only half the roots are computed */
42    Orthopolyroots(a, b, N) =
43    { my(B = getlocalbitprec(), r, R, sym, vsb);
44
45      if (type(a) == "t_CLOSURE", a = vector(N, n, a(n)));
46      if (type(b) == "t_CLOSURE", b = vector(N - 1, n, b(n)));
47      localbitprec(96); vsb = sqrt(b); localbitprec(64);
48      r = Tqli(a, vsb); /* roots with 64 bit accuracy */
49      sym = (#r < N); /* complete by symmetry ? */
50      R = vector(N);
51      for (i = 1, #r,
52        R[i] = OrthoNewtonrefine(a, b, N, r[i], B + 32);
53        if (sym, R[N + 1 - i] = -R[i]));
```

54   return (R);
55   }

EXAMPLES.
? a(n) = 0; /* Recursion for Legendre. */
? b(n) = n^2 / (4 * n^2 - 1);
? \p 1001   /* or 3328 bits */
? P = pollegendre(1000);
? A = polrootsreal(P);
time = 39,725 ms.
? B = Orthopolyroots(a, b, 1000);
time = 11,239 ms.
? exponent(A - B~)
% = -3328   /* Perfect. */
? C = Legendreroots(1000)[1]; /* Gives only the positive roots */
time = 1,824 ms.
? exponent(abs(B[1..500]) - C)
% = -3328   /* Perfect. */

Thus in this example Orthopolyroots is almost four times faster than polrootsreal, and it also uses much less memory. Note that a pure Tqli at 1000 decimal digits (instead of using Newton refinement of a 64 bit first approximation) would take 46.5s.

Note that when, as here, $a(n) = 0$, we can speedup the Orthopolyeval program by a factor of almost 2 by computing two steps at a time, using the formula of Corollary 3.5.6 (2). This is left to the reader, but in the special case of Legendre polynomials this is essentially what the Legendreroots program that we will see below does.

EXAMPLES. (Continued):
? a(n) = 2 * n - 1;
? b(n) = n^2;
? \p 115 /* or 384 bits */
? P = pollaguerre(1000);
? A = polrootsreal(P);
time = 5min, 40,110 ms.
? B = Orthopolyroots(a, b, 1000);
time = 8,452 ms.
? exponent(A - B~) - exponent(A)
% = -383 /* Perfect. */

In this case, polrootsreal is absolutely not competitive. Note that a pure Tqli would be a little faster (7,837 ms.), but this is because the required accuracy is quite low, at larger accuracy it becomes slower.

3.6. Gaussian integration methods

3.6.1. General Gaussian integration. We can now apply our algorithms for orthogonal polynomials to Gaussian integration. Note that we will see in the next chapter (Section 4.8) a similar idea for numerical *summation*.

The idea is this: in the case of the Newton–Cotes formulas or Chebyshev nodes (or accelerated methods such as Romberg), the abscissas (called *nodes*) where the values of f are taken are given in advance, for instance equally spaced in the case of Newton–Cotes. Thus, if we use n nodes, we can ask that the integral be exact for polynomials of degree up to $n-1$ since this gives a linear system of n equations in n unknowns, which does have a unique solution given by Newton–Cotes. As we have mentioned, if n is odd, then one can prove that the integral is in fact exact also for polynomials of degree n (for instance Simpson's rule, which is exact for quadratic polynomials by definition, is in fact also exact for cubic polynomials).

But if we do *not* fix the abscissas nor the weights, we now have $2n$ unknowns, so we can hope that the integral be exact for polynomials of degree up to $2n-1$, and this is indeed the case: this is called Gaussian integration. Note that here we have a highly nonlinear system to solve, so it is not clear how to go about it.

Luckily, the theory of orthogonal polynomials comes to our aid, and once again we refer to the numerous books and papers on this subject and on Gaussian integration methods.

The general setting of Gaussian integration is as follows: let $[a, b]$ be some interval, bounded or not, and let $w(x)$ be a *weight function* on $[a, b]$, i.e., a piecewise continuous nonzero and nonnegative function. We want an integration rule of the form

$$\int_a^b w(x) f(x) \, dx = \sum_{1 \le j \le n} w_j f(x_j) + E_n \,,$$

with nodes x_j, weights w_j, and absolute error E_n. Since we have $2n$ indeterminates x_j and w_j, we can hope to be able to have $E_n = 0$ for all polynomials of degree less than or equal to $2n-1$, and this is indeed possible thanks to Theorem 3.5.9 (3). More precisely, we have the following basic theorem of Gaussian integration, which we state with not necessarily monic polynomials since this will be the case in general:

THEOREM 3.6.1. *Let $(x_j)_{1 \le j \le n}$ and $(w_j)_{1 \le j \le n}$ be the nodes and weights to use for an optimal integration rule as above, let $(P_n)_{n \ge 0}$ be a family of (not necessarily monic) orthogonal polynomials corresponding to $\phi(F) = \int_a^b w(x) F(x) \, dx$, and denote by k_n the leading term of P_n.*

(1) *The x_j are the roots of the polynomial $P_n(X)$, and they are simple, real, and lie in the open interval $]a, b[$.*

(2) *The weights w_j are given by the three formulas*

$$w_j = \frac{k_n}{k_{n-1}} \frac{\langle P_{n-1}, P_{n-1} \rangle}{P_{n-1}(x_j) P_n'(x_j)} = \frac{1}{P_n'(x_j)} \int_a^b \frac{P_n(x)}{x - x_j} w(x) \, dx = \frac{Q_n(x_j)}{P_n'(x_j)} \,,$$

where the Q_n are defined in Theorem 3.5.5 (2).

(3) *If as above we write*

$$\int_a^b w(x) f(x) \, dx = \sum_{1 \le j \le n} w_j f(x_j) + E_n \,,$$

the error E_n is given by

$$E_n = \frac{\langle P_n, P_n \rangle}{k_n^2} \frac{f^{(2n)}(c)}{(2n)!}$$

for some $c \in]a, b[$.

Note that if f is a polynomial of degree less than equal to $2n$ we recover Theorem 3.5.9 (3).

PROOF. Note that since the formulas do not change if we multiply P_n by a nonzero constant we may assume that the P_n are the canonical *monic* orthogonal polynomials, hence we can apply all the formulas proved above.

(1) and (2). Theorem 3.5.9 (3) tells us that if we choose the x_j as the roots of P_n and $w_j = Q_n(x_j)/P_n'(x_j)$ then for any polynomial F of degree at most equal to $2n - 1$ we will have $\int_a^b w(x) F(x)\, dx = \sum_{1 \le j \le n} w_j F(x_j)$, so this indeed achieves our first goal, and the roots are simple and in the interval $]a, b[$ by Theorem 3.5.9 (2) and Corollary 3.5.10. The other formulas for w_j follow from Theorem 3.5.9 (4) and (5).

(3). We give a sketch of the (very classical) proof. First note that there exists a unique polynomial H_{2n-1} of degree $2n - 1$ such that $H_{2n-1}(x_j) = f(x_j)$ and $H_{2n-1}'(x_j) = f'(x_j)$ for all j (it can be constructed explicitly, analogously to the Lagrange interpolation polynomial, and is called the *Hermite interpolation polynomial*, see below). We first claim that for any $x \in]a, b[$ there exists $d = d(x) \in]a, b[$ such that

$$f(x) = H_{2n-1}(x) + \frac{f^{(2n)}(d(x))}{(2n)!} P_n(x)^2 \ .$$

This is trivial if x is one of the x_j, so we assume not and consider the auxiliary function

$$\phi(t) = f(t) - H_{2n-1}(t) - \frac{f(x) - H_{2n-1}(x)}{P_n(x)^2} P_n(t)^2 \ ,$$

and note that $\phi(x_j) = \phi(x) = 0$, so by Rolle's theorem there exist n points y_j distinct from the x_j and x such that $\phi'(y_j) = 0$, and since $\phi'(x_j) = 0$ we thus have $2n$ distinct points where ϕ' vanishes, so applying Rolle $2n - 1$ times we deduce that there exists $d \in]a, b[$ such that $\phi^{(2n)}(d) = 0$, and a trivial computation using the fact that H_{2n-1} has degree $2n - 1$ proves our claim.

To prove (3), we now multiply our formula by $w(x)$ and integrate. Since Gaussian integration is exact up to degree $2n - 1$ we have $\int_a^b w(x) H_{2n-1}(x)\, dx = \sum_j w_j H_{2n-1}(x_j) = \sum_j w_j f(x_j)$. On the other hand by the mean value theorem, since $w(x) \ge 0$ we have

$$\int_a^b f^{(2n)}(d(x)) w(x) P_n(x)^2\, dx = f^{(2n)}(c) \int_a^b w(x) P_n(x)^2\, dx = f^{(2n)}(c) \langle P_n, P_n \rangle$$

for some $c \in]a, b[$, proving (3).   □

Note also the following additional result:

PROPOSITION 3.6.2. *For simplicity, set* $L_{n-1,j}(x) = P_n(x)/(P_n'(x_j)(x - x_j))$. *We have*

$$w_j = \int_a^b L_{n-1,j}(x) w(x)\, dx = \int_a^b L_{n-1,j}(x)^2 w(x)\, dx \ ,$$

and in particular the weights w_j are all positive.

PROOF. For $i \neq j$ we have $L_{n-1,j}(x_i) = 0$, while by L'Hospital's rule, for $i = j$ we have $L_{n-1,j}(x_j) = 1$. It follows that the polynomial $L_{n-1,j}(x)^2 - L_{n-1,j}(x)$ vanishes for all $x = x_i$, and since the x_i are distinct, it is thus divisible by $P_n(x)$, say $L_{n-1,j}(x)^2 - L_{n-1,j}(x) = P_n(x)R_{n-2}(x)$ for some polynomial R of degree $n-2$. Since P_n is orthogonal to all polynomials of degree up to $n-1$ the result follows from the second formula for w_j given in the theorem. $\qquad\square$

As an exercise, the reader can check that the Hermite interpolation polynomial H_{2n-1} used in the proof of the theorem is given by

$$H_{2n-1}(x) = \sum_j L_{n-1,j}(x)^2 \left(f(x_j) + (x - x_j)\left(f'(x_j) - 2L'_{n-1,j}(x_j)f(x_j)\right)\right) .$$

More generally, one can fix Taylor expansions: for given distinct nodes (x_j) and positive integers (r_j), the Chinese remainder theorem allows to write down the unique polynomial H of degree less than $\sum_j r_j$ such that

$$H \equiv \sum_{k < r_j} \frac{f^{(k)}(x_j)}{k!}(x - x_j)^k \pmod{(x - x_j)^{r_j}} ;$$

see for instance [**vzGG99**, §5.6].

We have seen above a number of practical methods for computing the P_n, the weights, etc. for a general weight function $w(x)$. We now consider a number of standard weight functions $w(x)$ for which everything is known concerning the corresponding orthogonal polynomials. Note that in each case, the standard choice of normalization makes these polynomials non-monic.

(1) $w(x) = 1$ and $[a, b] = [-1, 1]$: the $P_n(X)$ are the Legendre polynomials.
(2) $w(x) = 1/(1 - x^2)^{1/2}$ and $[a, b] = [-1, 1]$: $P_n(X) = T_n(X)$, the Chebyshev polynomials of the first kind.
(3) $w(x) = (1 - x^2)^{1/2}$ and $[a, b] = [-1, 1]$: $P_n(X) = U_n(X)$, the Chebyshev polynomials of the second kind.
(4) $w(x) = (1-x)^\alpha(1+x)^\beta$ and $[a, b] = [-1, 1]$: $P_n(X) = P_n^{\alpha,\beta}(X)$, the Jacobi polynomials. Note that this general case includes the three preceding ones. We leave this more general case to the reader, except for the case $\alpha = 0$ (or $\beta = 0$) which will be treated separately below.
(5) $w(x) = e^{-x}$ and $[a, b] = [0, \infty[$: $P_n(X) = L_n(X)$, the Laguerre polynomials.
(6) $w(x) = e^{-x^2}$ and $[a, b] =] - \infty, \infty[$: $P_n(X) = H_n(X)$, the Hermite polynomials.

3.6.2. Gaussian integration: Gauss–Legendre. Here $[a, b] = [-1, 1]$ and $w(x) = 1$. The corresponding orthogonal polynomials are the *Legendre polynomials* P_n, which can be defined by

$$P_n(X) = \frac{1}{2^n n!}\frac{d^n}{dX^n}((X^2 - 1)^n)$$

$$= \frac{1}{2^n}\sum_{0 \le k \le \lfloor n/2 \rfloor} (-1)^k \binom{n}{k}\binom{2n - 2k}{n}X^{n-2k} .$$

Indeed, it is immediate to check on this definition that they are orthogonal with respect to the scalar product defined by $w(x)$. From this definition it is easy to see that they satisfy the 3-term recursion

$$(n+1)P_{n+1}(X) = (2n+1)XP_n(X) - nP_{n-1}(X)$$

with $P_0(X) = 1$ and $P_1(X) = X$. In addition, we have

$$P_n'(X) = \frac{n}{X^2 - 1}(XP_n(X) - P_{n-1}(X)) \,.$$

Note for future use that the recursion coefficients a_n and b_n for the corresponding *monic* orthogonal polynomials are $a_n = 0$ and $b_n = n^2/(4n^2 - 1)$.

The nodes of Gauss–Legendre quadrature are the roots x_i of the nth Legendre polynomial $P_n(X)$, and the weights are given by $w_i = 2/((1 - x_i^2)P_n'(x_i)^2) = 2(1 - x_i^2)/(n^2 P_{n-1}(x_i)^2)$, so that

$$\int_a^b f(x)\, dx \approx \frac{b-a}{2} \sum_{1 \le i \le n} w_i f\left(\frac{a+b}{2} + \frac{b-a}{2} x_i\right) \,.$$

In canonical form

$$\int_{-1}^1 f(x)\, dx \approx \sum_{1 \le i \le n} w_i f(x_i) \,,$$

with x_i and w_i as above.

Since Legendre polynomials are so well known, there are extremely efficient methods to find their roots, see for instance [**JM18**]. It would carry us too far afield to describe them, so we will be content with the following programs:

——————————————— Legendreroots.gp ———————————————

```
1   /* If !last, Newton step z - P_N(z) / P'_N(z),
2    * else [z - P_N(z) / P'_N(z), (N-1)!P_{N-1}(z)]. */
3   Legendrenext(N, z, last) =
4   { my(q, Z, lz = #z, z2 = apply(x -> x^2, z), n0 = N % 2, a, b);
5
6     if (n0, b = vector(lz, i, 1); a = z2, a = z; b = 0);
7     forstep (n = n0, N - 1, 2,
8       b = (2 * n + 1) * a - n^2 * b;
9       a = (2 * n + 3) * vector(lz, i, z2[i] * b[i])
10                      - (n + 1)^2 * a);
11    q = vector(lz, i, a[i] / (a[i] - N * b[i]));
12    Z = vector(lz, i, z[i] - (z2[i] - 1) * q[i] / (N * z[i]));
13    if (!last, return (Z));
14    return ([Z, vector(lz, i, b[i] * (1 + q[i]))]);
15  }
16
17  /* initial approximation for positive roots of P_N */
18  Legendreroots0(N) =
19  { localbitprec(32);
20    my(z = exp(I * Pi / (4 * N + 2)), z3 = z^3);
21    return ((1 - (N-1) / (8 * N^3))
22            * real(powers(z3 * z, (N-2) \ 2, z3)));
23  }
24
25  /* positive roots z_i of P_N and (N-1)!P_{N-1}(z_i). */
```

```
26   Legendreroots(N) =
27   { my(z = Legendreroots0(N), e = -exponent(1 - z[1]^2));
28     my(B = getlocalbitprec(), n = exponent(B + 32 - e) - 2);
29     my(pr = 1 + e + (B - e) >> n);
30
31     for (j = 1, n,
32       pr = 2 * pr - e;
33       z = Legendrenext(N, bitprecision(z, pr), j == n));
34     return (bitprecision(z, B));
35   }
```

REMARKS 3.6.3.

(1) The Legendreroots program computes the positive roots (z_i) of P_N as well as the values $(N-1)!P_{N-1}(z_i)$ needed to compute the integration weights. The roots of P_N are thus $\pm z_i$, and in addition the root 0 when N is odd.

(2) The jth root of the Nth Legendre polynomial is close to
$$\cos\big(\pi(4j-1)/(4N+2)\big),$$
the extra factor $1 - (N-1)/(8N^3)$ being an approximation to the beginning of the asymptotic expansion. It can be shown that the above approximation is already in the basin of attraction of Newton iteration. The program performs this iteration handling the working accuracy with care, in particular when the root to refine is close to ± 1.

(3) The Legendrenext function performs a simple Newton step using the recursion for P_N, and when last is set it also returns $(N-1)!P_{N-1}(z)$. Note that $P_N(z)/P_N'(z)$, which is the essential ingredient for Newton iteration, is computed using 2 steps of the recursion at once. Also note that instead of this specific program, we could use the more general Orthopolyeval program, which would be more than twice slower.

(4) At the end of the Newton iteration (last = 1), the roots z_i are at full accuracy, and to obtain $(N-1)!P_{N-1}(z_i)$ also to full accuracy we simply use the Taylor expansion of P_{N-1} to order 1, which is the reason for the factor $1 + q[i]$.

The last thing that we need is an estimate of n as a function of the accuracy. One can prove that the error $E_n(f)$ is given by
$$E_n(f) = \frac{(b-a)^{2n+1}n!^4}{(2n)!(2n+1)!}\frac{f^{(2n)}(c)}{(2n)!}$$
for some $c \in]a,b[$. Note that $n!^4/((2n)!(2n+1)!) \sim (\pi/2)/2^{4n}$. As in Section 2.3, assume that f is analytic in a neighborhood of $[a,b]$ and denote by r the distance to the nearest pole. Then, neglecting lower order terms, $E_n(f)$ is of the order of $(4r/(b-a))^{-2n}$.

If $4r > (b-a)$ we can make this less than 2^{-B} by choosing
$$n > B \cdot \frac{\log(2)}{2\log\big(4r/(b-a)\big)}.$$

On the other hand, if $4r \leq (b - a)$ (or even if $4r/(b - a)$ is barely larger than 1) it will be necessary to *split* the interval of integration, as already mentioned in Section 3.3.3, and this is what we will do in the `IntGaussLegendresplit` program below. In the following simple approach, we assume that we do not have these problems and take as default $r = b - a = 1$, so that $n > B/4$.

———————————————— | IntGaussLegendre.gp | ————————————————

```
1   /* Assume r / (b - a) >= 1. See text. */
2   IntGaussLegendreinit(N = 0) =
3   { my(B = getlocalbitprec(), X, Y, W, c);
4
5     if (type(N) == "t_VEC", return (N)); /* already an init */
6     if (!N, N = ceil(B / 4));
7     if (N % 2, N++); /* We want N even, see text. */
8     localbitprec(B + 32);
9     [X, Y] = Legendreroots(N);
10    c = 2 * factorial(N - 1)^2 / N^2;
11    W = vector(#X, i, c * (1 - X[i]^2) / Y[i]^2);
12    return (bitprecision([X, W], B + 32));
13  }
14
15  /* Simple Gauss-Legendre integration. */
16  IntGaussLegendre(f, a, b, XW = 0) =
17  { my(B = getlocalbitprec(), S, m, p);
18    my([X, W] = IntGaussLegendreinit(XW));
19
20    localbitprec(B + 32);
21    m = (b - a) / 2;
22    p = (b + a) / 2;
23    S = sum(i = 1, #X, my(y = m * X[i]);
24                       W[i] * (f(p + y) + f(p - y)));
25    return (bitprecision(m * S, B));
26  }
```

REMARKS 3.6.4.

(1) As we will see, if the nth derivative of f grows much *slower* than $n!$, for instance $f(x) = e^x$ or $f(x) = \sin(x)$, or essentially equivalently, if the power series has infinite radius of convergence, our choice of $N = B/4$ is overkill. Thus, we leave N as an optional parameter in the program, which can be adjusted to increase speed and/or accuracy. We will see below in the `IntGaussLegendresplit` program a better way to handle this problem.

(2) To compute the roots we use the `Legendreroots` program seen above. We could also use the general `Orthopolyroots` program, which is approximately twice slower, or even the built-in `polrootsreal` program which would be even slower and use more memory. We assume N even only for simplicity, for N odd we would need to add the contribution of the

root 0, with weight $2/(N^2 P_{N-1}(0)^2)$, where $P_{N-1}(0)$ is obtained from the explicit formula given above.

(3) We used the specific Legendreroots program to compute the $P_{N-1}(x_i)$, which are necessary for the weights. We could also have used the more general but slower Orthopolyeval program mentioned above. Note that replacing X by x_i in the explicit expression for the polynomial P_{N-1} would cause an enormous accuracy loss.

(4) Equivalent programs are already available in GP under the names intnum-gaussinit and intnumgauss.

EXAMPLES.

```
? \p 1001
   realprecision = 1001 significant digits
? H = IntGaussLegendre;
? H(x -> 1 / (1 + x), 0, 1) - log(2)
time = 1,254 ms.
% = 0.E-1001 /* Perfect, slow because of initialization. */
? XW = IntGaussLegendreinit();
time = 1,253 ms.
? H(x -> 1 / (1 + x), 0, 1, XW) - log(2)
time = 5 ms.
% = 0.E-1001 /* Perfect and now instantaneous. */
? H(x -> 1 / (1 + x^2), 0, 1, XW) - Pi/4
time = 6 ms.
% = 0.E-1001 /* Perfect and instantaneous. */
? f(x) = 1 / (1 + 6 * x);
? S = log(7) / 6; \\ integral from 0 to 1
? H(f, 0, 1, XW) - S
time = 2 ms.
% = -7.78... E-576 /* 425 decimals lost because of the pole. */
? 2 * #XW[1] /* XW has 832 nodes (XW[1] = the positive ones). */
% = 832
? H(f, 0, 1, 1450) - S /* Use 1450 nodes. */
time = 3,784 ms.
% = -7.43... E-1003 /* Perfect but slow: not the way to go. */
? H(f, 0, 1/6, XW) + H(f, 1/6, 1/2, XW) +  H(f, 1/2, 1, XW) - S
time = 16 ms.
% = 0.E-1002 /* Perfect and instantaneous. */
```

In the last examples, the presence of the pole at $x = -1/6$ creates problems. One way out is to increase the default value of N: in the present example, $N = 1450$ (instead of the default 832) solves the accuracy problem, but at the expense of a very slow program. As already mentioned above, the correct solution is to split the interval, here at $1/6$ and $1/2$, and the result is now instantaneous and perfect (the IntGaussLegendresplit program below would in fact split the interval into 5 sub-intervals). We now generalize this remark: when the interval of integration is "large", or when the poles are close to it, it may be useful to split the interval of integration, depending both on its size, and on the distance r of the interval to the nearest pole, which gives an estimate on the growth of the nth derivative. This is of course not specific to the Gauss–Legendre method.

Let us briefly analyze the problem: we consider that even if the initialization takes an inordinate amount of time, it is done once and for all, so we only want to minimize the number of function evaluations. For simplicity, assume that we split the interval of integration into n equal steps (this is of course in general not the best strategy, for instance if there is a single pole to the left of the interval). The length h of the small intervals is thus $(b-a)/n$, so by the analysis made above we should choose

$$N \approx B \log(2)/(2 \log(4r/h)) = B \log(2)/(2 \log(4rn/(b-a))) \,,$$

so the total number of function evaluations is nN. We must therefore minimize this function of n, and a small computation shows that the optimal value of n is $n \approx e(b-a)/(4r)$ (with $e = \exp(1)$), in other words $h = 4r/e$. This means that the optimal value of N only depends on the accuracy and should be chosen close to $B \log(2)/2 = 0.34657B$.

──────────────────────── IntGaussLegendresplit.gp ────────────────────────

```
1   /* Compute int_a^b f(x) dx using Gauss-Legendre integration
2    * and interval splitting; assume a and b real. XW is either
3    * 0, N, or the output of an IntGaussLegendreinit */
4   IntGaussLegendresplit(f, a, b, r = 1, XW = 0) =
5   { my(B = getlocalbitprec(), bma = b - a, h, S, n, N, X, W);
6
7     if (!bma, return (0.));
8     localbitprec(32);
9     n = abs(bma) / (4 * r);
10    N = if (!XW, ceil(B * log(2) / (2 * max(1, -log(n)))),
11           type(XW) == "t_VEC", 2 * #XW[1],
12           XW);
13    n = ceil(n * 2^(B / (2 * N)));
14    localbitprec(B);
15    if (type(XW) != "t_VEC", XW = IntGaussLegendreinit(N));
16    [X, W] = XW;
17    localbitprec(B + 32);
18    h = bma / (2*n);
19    S = sum(i = 1, #X, W[i] *
20           sum(k = 1, n, f(a + (2*k-1 + X[i]) * h) +
21                         f(a + (2*k-1 - X[i]) * h)));
22    return (bitprecision(h * S, B));
23  }
```

REMARKS 3.6.5.

(1) In this program, if the user does not give explicitly the number N of nodes (XW = 0), we choose

$$N = \lceil B \log(2)/(2 \max(\log(4r/(b-a)), 1)) \rceil \,,$$

so as to minimize the initialization time.

(2) On the other hand, if the user wants us to use a given value of N, we may need to split the interval into sub-intervals of length h such that $N \geq B \log(2)/(2\log(4r/h))$, in other words $h \leq 4r2^{-B/(2N)}$, so we must choose
$$n \geq ((b-a)/(4r))2^{B/(2N)} \ .$$

Note that although the number of summations will be multiplied by some constant n depending on $b-a$ and r, the initialization is done only once, so at large accuracy the time will only slightly increase, and the results will be completely accurate. We have already seen this in the above example of $\int_0^1 dx/(1+6x)$. Another example due to the size of the interval is $\int_0^{10} e^{-x}/(1+x)\,dx$ which has only 449 correct decimals out of 1000 if computed by IntGaussLegendre, but gives a perfect result (a little slower) if computed by IntGaussLegendresplit.

EXAMPLES.
```
? \p 500 /* 1664 bits */
? f(x) = ct++; 1 / (x + I / 4);
  /* ct will count the number of evaluations. */
? A = log((1 + I/4) / (-1 + I/4));
? ct = 0; [exponent(IntGaussLegendre(f, -1, 1) - A), ct]
time = 177 ms.
% = [-295, 416] /* Only 295 correct bits, 416 evaluations. */
? XW = IntGaussLegendreinit(); 2 * #XW[1]
time = 178 ms.
% = 416 /* 416 nodes => 416 evaluations. */
? H = IntGaussLegendresplit;
? ct = 0; [exponent(H(f, -1, 1, , XW) - A), ct]
time = 4 ms.
% = [-880, 832] /* With default r = 1, 880 correct bits. */
? ct = 0; [exponent(H(f, -1, 1, 1/4 , XW) - A), ct]
time = 11 ms.
% = [-1662, 3328] /* Perfect result but 3328 evaluations. */
? XW2 = IntGaussLegendreinit(300); /* Impose only 300 nodes. */
time = 93 ms. /* Faster initialization. */
? ct = 0; [exponent(H(f, -1, 1, 1/4 , XW2) - A), ct]
time = 20 ms.
% = [-1662, 4200] /* Perfect, but more evaluations. */
```

EXAMPLES. (Continued):
```
? f(x) = ct++; exp(-x) / (1 + x);
? \p 600
? A = intnum(x = 0, 10, f(x));
? \p 500 /* or 1664 bits */
? ct = 0; [exponent(IntGaussLegendre(f, 0, 10) - A), ct]
time = 200 ms.
% = [-744, 416] /* Only 744 correct bits, 416 evaluations. */
? ct = 0; [exponent(H(f, 0, 10, 1, XW) - A), ct]
time = 236 ms.
% = [-1664, 4160] /* With r = 1 perfect, 4160 evaluations. */
```

```
? ct = 0; [exponent(H(f, 0, 2, 1, XW)
                + H(f, 2, 10, 3, XW) - A), ct]
time = 116 ms.
% = [-1664, 2080] /* Still perfect, faster (2080 evaluations). */
```

This last example illustrates the fact that `IntGaussLegendresplit`'s naïve splitting can be improved with knowledge of the poles' location.

3.6.3. Gaussian integration: Gauss–Chebyshev. Here $[a, b] = [-1, 1]$ and $w(x) = 1/\sqrt{1 - x^2}$ or $w(x) = \sqrt{1 - x^2}$, which we treat simultaneously. The corresponding orthogonal polynomials are the *Chebyshev* polynomials T_n and U_n of the first and second kind, respectively, which can be defined by

$$T_n(\cos(t)) = \cos(nt) \quad \text{and} \quad U_n(\cos(t)) = \frac{\sin(n + 1)t}{\sin(t)} .$$

They satisfy the same 3-term recursions

$$T_{n+1}(X) = 2XT_n(X) - T_{n-1}(X) \quad \text{and} \quad U_{n+1}(X) = 2XU_n(X) - U_{n-1}(X) ,$$

we have as usual $T_0(X) = U_0(X) = 1$, but $T_1(X) = X$ and $U_1(X) = 2X$ (note that if we had set $T_{-1}(X) = 0$ we would have $T_1(X) = 2X$, so we must be careful here).

As usual, the nodes of Gauss–Chebyshev quadrature are given by the roots of the Chebyshev polynomials, but here these roots are completely explicit, and in fact so are the weights. For $w(x) = 1/\sqrt{1 - x^2}$, corresponding to T_n, we have simply

$$x_j = \cos\left(\frac{(2j - 1)\pi}{2n}\right) \quad \text{and} \quad w_j = \frac{\pi}{n} ,$$

with error given by

$$E_n = \frac{(b - a)^{2n+1}\pi}{2^{4n}} \frac{f^{(2n)}(c)}{(2n)!}$$

while for $w(x) = \sqrt{1 - x^2}$ corresponding to U_n, we have simply

$$x_j = \cos\left(\frac{j\pi}{n + 1}\right) \quad \text{and} \quad w_j = \frac{\pi}{n + 1}(1 - x_j^2) ,$$

with error given by

$$E_n = \frac{(b - a)^{2n+1}\pi}{2^{4n+2}} \frac{f^{(2n)}(c)}{(2n)!} .$$

This error estimate is very similar to the one for Gauss–Legendre, and the same remarks can be made here, but with a different conclusion: if $4r/(b - a) \le 1$, it will be necessary to split the interval as we did above. However, in that case the algebraic singularities at -1 and 1 become separate. Thus the first and last sub-intervals must be treated with a Gauss quadrature involving only one algebraic singularity at an endpoint, and we will treat this below in the `IntGaussJacobi` program. All the intermediate sub-intervals must simply be treated using the simple Gauss–Legendre integration method seen above.

We therefore only write the "simple" version of the program assuming no splitting is necessary, and only for the interval $[-1, 1]$ since is can be trivially modified to treat an arbitrary compact interval:

$$\boxed{\text{IntGaussChebyshev.gp}}$$

```
1   /* int_{-1}^1 (1-x^2)^{e/2} f(x) dx with e = -1 or 1. */
2   IntGaussChebyshevinit(e = -1, N = 0) =
3   { my(B = getlocalbitprec(), X, W);
4
5     if (!N, N = ceil(B / 4));
6     /* This assumes r >= 2. See text. */
7     localbitprec(B + 32);
8     X = vector(N);
9     if (e == -1,
10      my(c = exp(I*Pi / (2*N)));
11      X = real(powers(c^2, N - 1, c));
12      W = vector(N, j, Pi / N)
13    , /* else */
14      my(c = exp(I*Pi / (N + 1)));
15      my(Z = powers(c, N - 1, c));
16      X = real(Z);
17      W = Pi / (N + 1) * imag(Z)^2;
18    return ([X, W]);
19  }
20
21  IntGaussChebyshev(f, e = -1, XW = 0) =
22  {
23    if (type(XW) == "t_INT", XW = IntGaussChebyshevinit(e, XW));
24    return (SumXW(f, XW));
25  }
```

EXAMPLE.

```
\p2000
? /* For comparison, first compute with built-in function */
? T = intnuminit([-1, -1/2], [1, -1/2]);
time = 4,286 ms. /* Slow initialization. */
? A = intnum(x = [-1, -1/2], [1, -1/2], \
             exp(-x) / sqrt(1 - x^2), T);
time = 14,929 ms. /* Correct but slow. */
? XW = IntGaussChebyshevinit(-1);
time = 41 ms. /* Very fast initialization. */
? IntGaussChebyshev(x -> exp(-x), -1, XW) - A
time = 887 ms.
% = 0.E-2003 /* Perfect and very fast. */
```

As mentioned above, a generalization both of Gauss–Legendre and of Gauss–Chebyshev is when $w(x) = (1 - x)^\alpha (1 + x)^\beta$, and the corresponding orthogonal polynomials are the *Jacobi polynomials*. We leave to the reader to write the corresponding programs (see Section 3.6.7 for the special case where α or β is zero).

3.6.4. Gaussian integration: Gauss–Laguerre. Here $[a, b] = [0, \infty[$ and $w(x) = x^\alpha e^{-x}$. The corresponding polynomials are the generalized *Laguerre* polynomials $L_n^{(\alpha)}$ which can be defined by

$$L_n^{(\alpha)}(X) = \frac{X^{-\alpha} e^X}{n!} \frac{d^n}{dX^n} (X^{n+\alpha} e^{-X})$$

$$= \sum_{0 \le k \le n} (-1)^k \binom{n+\alpha}{n-k} \frac{X^k}{k!} \;,$$

and satisfy the 3-term recurrence relation

$$(n+1) L_{n+1}^{(\alpha)}(X) = (-X + (2n + \alpha + 1)) L_n^{(\alpha)}(X) - (n + \alpha) L_{n-1}^{(\alpha)}(X)$$

with $L_0^{(\alpha)}(X) = 1$ and $L_1^{(\alpha)}(X) = (-X + \alpha + 1)$ (which here corresponds to $L_{-1}^{(\alpha)}(X) = 0$).

The recursion coefficients a_n and b_n for the corresponding monic polynomials are $a_n = 2n + 1 + \alpha$ and $b_n = n(n + \alpha)$.

As usual the nodes of Gauss–Laguerre quadrature are the roots x_i of the nth Laguerre polynomial, the weights are given by

$$w_i = \frac{\Gamma(n + \alpha + 1)}{(n+1)(n+1)!} \frac{x_i}{L_{n+1}^{(\alpha)}(x_i)^2} \;,$$

and the error is given by

$$E_n(f) = n! \Gamma(n + \alpha + 1) \frac{f^{(2n)}(c)}{(2n)!} \;.$$

We must now discuss this error. When performing Gaussian integration on *compact* intervals, for instance Gauss–Legendre, Gauss–Chebyshev, or more generally Gauss–Jacobi, the range of functions to which the methods apply is quite large, since continuous functions can be uniformly approximated by polynomials. This is also reflected in the error estimate $E_n(f)$ which has (essentially) $((b-a)/4)^{2n}$ in front of $f^{(2n)}/(2n)!$, and by sufficient splitting this can be made as small as we like, at least if $r > 0$, where r is the distance to the nearest pole.

On the contrary, on infinite intervals the range of functions to which Gaussian methods apply is much more restricted, which is again reflected in the above error estimate. In practice, apart from polynomials themselves, which by definition will give good if not perfect results, the functions to which this method can be applied should have infinite radius of convergence, such as $\sin(x)$, $J_0(x)$, etc. This will also be the case for the Gauss–Hermite method which we will study next. Thus the notion of "r" that we used in the compact case does not apply here. Nonetheless, we still need to estimate $f^{(n)}(c)$: consider the typical case of a trigonometric function such as $f(x) = \sin(ax)$. We have trivially $|f^{(n)}(x)| \le a^n$, and it is important to take this value of a into account. Note that splitting the interval will not help here. We will therefore make the assumption that we know a value of a such that $f^{(n)}(x) = O(a^n)$, and since $n! \Gamma(n + \alpha + 1)/(2n)!$ is of the order of 2^{-2n}, we need N such that $(a^2/4)^N < 2^{-B}$. This is not possible if $a \ge 2$, and otherwise N is proportional to B. Nonetheless we can hope that the error estimate is pessimistic, so we will simply include an optional parameter ct set by default to $1/2$, use N = ct * B, and hope for the best.

| IntGaussLaguerre.gp |

```
 1   /* int_0^oo x^al e^{-x} f(x) dx with al > - 1
 2    * assumed exact. */
 3
 4   IntGaussLaguerreinit(al = 0, ct = 1/2) =
 5   { my(B = getlocalbitprec(), N, X, W, lt, lu, va, vb);
 6
 7     if (type(ct) == "t_VEC", return (ct));
 8     N = ceil(ct * B);
 9     va = vector(N + 1, n, 2*n - 1 + al);
10     vb = vector(N, n, n * (n + al));
11     X = Orthopolyroots(va[1..N],
12                        vb[1..N-1], N); /* roots of L_N */
13     lu = 1 / factorial(N + 1);
14     lt = (-1)^(N + 1) * lu; /* leading term of L_{N+1} */
15     W = vector(N);
16     for (i = 1, N,
17       my(Y = Orthopolyeval(va, vb, N + 1, X[i]) * lt);
18       W[i] = X[i] / Y^2);
19     W *= lu * gamma(N + al + 1) / (N + 1);
20     return (bitprecision([X, W], B + 32));
21   }
22
23   IntGaussLaguerre(f, a = 0, al = 0, XW = 1/2) =
24   {
25     XW = IntGaussLaguerreinit(al, XW);
26     return (exp(-a) * SumXW(x -> f(x + a), XW));
27   }
```

Experimenting with this program, we see that for $f(x) = \sin(kx)$ and $f(x) = J_0(kx)$, for $k = 1/2, 1, 2, 3, 4$ we need $ct = 1/4, 1/2, 1, 2, 3$ (approximately) to obtain a perfect result, essentially independently of the needed accuracy:

EXAMPLE.
```
? \p 38
? H = IntGaussLaguerre;
? f(x) = sin(4*x);
? H(f, , , 1) - 4/17
time = 153 ms.
% = -2.52... E-12
? H(f, , , 2) - 4/17
time = 597 ms.
% = -3.83... E-25
? H(f, , , 3) - 4/17
time = 1,340 ms.
% = 3.91... E-37 /* Finally almost perfect, slower. */
```
Note that in this specific example we can cheat, since $\int_0^\infty e^{-x} \sin(ax)\,dx = a/(1 + a^2)$, which is invariant when changing a into $1/a$, so continuing:

```
? H(x -> sin(x/4)) - 4/17
time = 37 ms.
% = -6.91... E-40 /* Perfect and very fast. */
```

3.6.5. Gaussian integration: Gauss–Hermite. The last special case of Gaussian integration that we will consider is that of Gauss–Hermite. Here $[a, b] =]-\infty, \infty[$ and $w(x) = e^{-x^2}$. The corresponding polynomials are the *Hermite* polynomials H_n which can be defined by

$$H_n(X) = (-1)^n e^{X^2} \frac{d^n}{dX^n} (e^{-X^2}),$$

and satisfy the 3-term recurrence relation

$$H_{n+1}(X) = 2X H_n(X) - 2n H_{n-1}(X)$$

with $H_0(X) = 1$ and $H_1(X) = 2X$ (which corresponds to $H_{-1}(X) = 0$).

As usual the nodes of Gauss–Hermite quadrature are the roots x_i of the nth Hermite polynomial, the weights are given by

$$w_i = \frac{2^{n-1} n! \pi^{1/2}}{(n H_{n-1}(x_i))^2},$$

and the error is given by

$$E_n = \frac{n! \pi^{1/2}}{2^n} \frac{f^{(2n)}(c)}{(2n)!}.$$

However, the coefficients of H_n grow rapidly and a better normalization is to use the polynomials $\widetilde{H}_n = \pi^{-1/4} H_n / (2^n n!)^{1/2}$ which are now *orthonormal* with respect to the scalar product. The three term recursion is now

$$\widetilde{H}_{n+1}(X) = \sqrt{\frac{2}{n+1}} X \widetilde{H}_n(X) - \sqrt{\frac{n}{n+1}} \widetilde{H}_{n-1}(X)$$

with initial terms $\widetilde{H}_0(X) = \pi^{-1/4}$, $\widetilde{H}_1(X) = 2^{-1/2} \pi^{-1/4} X$, and the weights are given by $w_i = 1/(n \widetilde{H}_{n-1}(x_i)^2)$.

We leave to the reader the task of writing a program similar to the program for Gauss–Laguerre integration. In fact, the change of variable $x = y^{1/2}$ shows that the methods are closely related, and in particular $H_n(X)$ is closely related to $L_n^{\pm 1/2}(X)$.

3.6.6. General Gaussian integration. We now come back to the general case, and assume that we do not a priori know anything about the orthogonal polynomials associated to our weight function $w(x)$. Note that it is correctly stated in the literature that the procedure that we are about to describe is very ill-conditioned, but we are going to do it anyway for two reasons: first, it is quite simple to implement. But second and most importantly, numerical analysts usually deal with computations in double precision, *not* in multiprecision. So we are going to do something which is completely anathema for them, but which works in so many cases that it is worthwhile: we will work with a much higher accuracy than what is required, and hope for the best. Evidently, this may not always work.

The standard way to proceed (although usually not the best, but we will not expand on this) is to start from the moments $m_k = \langle 1, x^k \rangle$ for $k \geq 0$. Then using either the Cholesky method or the continued fraction method (the programs OrthopolyCholesky and OrthopolyCF) we compute the polynomials P_n and Q_n

for a suitable n depending on the desired accuracy. We then compute the roots x_j of P_n in the most efficient manner possible (see Section 3.5.8), then the weights w_j using one of the three formulas of Theorem 3.6.1 (2), the simplest being probably $w_j = Q_n(x_j)/P'_n(x_j)$, and finally the error is given by $E_n = \gamma_n f^{(2n)}(c)/(2n)!$ with $c \in]a, b[$ and

$$\gamma_n = \langle P_n, P_n \rangle = \int_a^b P_n(x)^2 w(x)\, dx \ .$$

In the following general program, it is important to decide how to give the moments. First, it may happen that they are known symbolically, in which case one can give the moments as a vector of values of sufficient length. If we do not have any specific method, we simply compute the moments using the robust doubly-exponential method **Inta_xxx** (or the corresponding built-in **intnum**). It may also happen that we do have a specific method to compute the moments, which will only give approximate values, as with **Inta_xxx**. In these cases, we must be careful: when the coefficients of the polynomial P are only approximate, the **polrootsreal(P)** command may *miss* some real roots. Therefore in the program that follows, we increase the working accuracy (by a factor $3/2$) until all the roots have been found.

IntGauss.gp

```
/* Given a weight function w and an interval ab = [a, b],
 * perform Gaussian integration. Omit ab if w is already
 * the closure of moments */
IntGaussinit(w, ab = 0, N = 0) =
{ my(B = getlocalbitprec(), mom, NEWB);

  if (type(w) != "t_CLOSURE", return (w));
  localbitprec(32);
  NEWB = 2.25*B; /* or 3*B if Cholesky */
  if (!N, N = ceil(Solvemullog(0, B * log(2))));
  if (ab === 0,
    mom = w
  ,
    my([a,b] = ab);
    mom = (N -> intnum(t = a, b, powers(t, 2 * N - 1, w(t)))));
  while (1,
    localbitprec(NEWB);
    my(P, Q, W, X);
    /* or OrthopolyCholesky(mom(N)) */
    [P, Q] = OrthopolyCF(mom, N);
    X = polrootsreal(P);
    if (poldegree(P) == #X,
      W = subst(Q / P', 'x, X);
      return (bitprecision([X, W], B + 32)));
    /* accuracy loss, increase bitprec by 50%. */
    NEWB = (3*NEWB) >> 1);
}

/* Compute int_a^b w(x)f(x)dx. w can be a closure or the
```

```
30    * result of IntGaussinit. */
31  IntGauss(f, a, b, w) =
32  { my(XW = IntGaussinit(w, [a, b]));
33    SumXW(f, XW);
34  }
```

REMARKS 3.6.6.

(1) We have used the continued fraction program OrthopolyCF to compute the orthogonal polynomials, but we can if desired simply substitute it by the OrthopolyCholesky program.

(2) In the initialization part we have set the accuracy to $2.25B$, and mentioned that we would need $3B$ if we use OrthopolyCholesky. These values seem reasonable experimentally in reasonable accuracies (say less than 500 decimal digits), but may need to be increased in larger accuracies.

(3) We have made absolutely no effort to minimize the *initialization* program IntGaussinit, which may take quite some time, especially since we increase considerably the working accuracy. For real-life applications, it should of course be improved. On the other hand, as we will see below, the IntGauss program itself, after initialization, is very competitive.

(4) Warning: the IntGauss program computes $\int_a^b w(x)f(x)\,dx$, so you must not give the integrand x->w(x)*f(x) to the function, but x->f(x) or simply f.

(5) Even though we have implicitly assumed that $[a, b]$ is a compact interval, this assumption is unnecessary. Indeed, the variables w, a, and b are *only* used in the computations of the moments, and not in the sequel. Thus, either we give the moments explicitly (for instance, if $w(x) = \exp(-x)$, $a = 0$, and $b = \infty$ we have $m_k = k!$), or we can compute them using the programs with no modification, using the standard GP syntax for ∞ such as oo or [oo,1].

Let us give an example of the use of this general Gaussian integration program. Assume that we want to compute numerically $\int_0^1 -x^k \log(x)/(1+x)\,dx$ for $k = 0, 1, 2$, say (with exact values $\pi^2/12$, $1 - \pi^2/12$, and $\pi^2/12 - 3/4$ respectively). Since $\log(x)$ has only a very mild singularity at 0, we could try to use directly Gauss–Legendre: we would then obtain 3, 7, 10 correct decimals respectively at the default 38 decimals. On the other hand, if we compute once and for all vv=IntGaussinit(x->-log(x),0,1), we now obtain (instantaneously) perfect accuracy when using IntGauss, and much faster than any other program.

3.6.7. General Gaussian integration: Examples. In this subsection, we use the general Gaussian method that we have just explained to treat three examples which will be needed later.

Example 1: Special Jacobi Polynomials

Consider the weight function $w(x) = x^\alpha$ on $]0, 1]$, where we assume that $\alpha > -1$. The corresponding orthogonal polynomials are a special case of (shifted) Jacobi polynomials (we will not consider general Jacobi polynomials). Since Gaussian integration deal with integrals of the form $\int w(x)f(x)\,dx$ with f well approximated by polynomials of degrees 0, 1, etc., we may assume if necessary that $\alpha \in]-1, 1]$, for instance by taking fractional parts, but we will only assume that $\alpha > -1$.

We have $m_k = \int_0^1 w(x)x^k\,dx = 1/(k+\alpha+1)$, so $\Phi(T) = \sum_{k\geq 1} T^{-k}/(k+\alpha)$ is a modified logarithm series. Now it is well-known and easy to give the corresponding continued fraction, so that we do not have to use the quotient-difference algorithm, nor the Cholesky method: we have $\Phi(T^{-1}) = c(0)z/(1-c(1)z/(1-c(2)z/(1-\cdots)))$, where $c(0) = 1/a$, $c(2k+1) = (a+k)^2/((a+2k)(a+2k+1))$ for $k \geq 0$, and $c(2k) = k^2/((a+2k-1)(a+2k))$ for $k \geq 1$, where for simplicity we have set $a = \alpha + 1 > 0$. Since $c_n > 0$ we can apply Theorem 3.5.7 (2), and we thus have $a_0 = c(1) = a/(a+1)$, and for $k \geq 1$

$$a_k = c_{2k} + c_{2k+1} = (a^2 + (2k-1)a + 2k^2)/((a+2k-1)(a+2k+1)) \quad \text{and}$$
$$b_k = c_{2k-1}c_{2k} = k^2(a+k-1)^2/((a+2k-2)(a+2k-1)^2(a+2k)) \ .$$

We can thus write the following specific programs:

──────────────────────── IntGaussJacobi.gp ────────────────────────

```
1   /* Gaussian integration with weight x^al on ]0, 1]
2    * with al > -1 assumed exact. */
3
4   IntGaussJacobiinit(al = 0, N = 0) =
5   { my(B = getlocalbitprec(), a = al+1, X, W, va, vb, P, Pp, Q);
6
7     if (!al, return (Intgeninit()[1..2]));
8     if (!N, N = ceil(B / 4));
9     /* This implicitly assumes r / (b - a) >= 1. See text. */
10    va = vector(N, k, (a^2+(2*k-3)*a+2*(k-1)^2)
11                       / ((a+2*k-2)^2-1)));
12    vb = vector(N-1, k, k^2*(a+k-1)^2
13                       / ((a+2*k-2)*(a+2*k-1)^2*(a+2*k)));
14    [P, Q] = Orthopolyrec([va, vb], 1 / a);
15    Pp = P';
16    localbitprec(B + exponent(Pp)
17                    - min(0, exponent(subst(Pp, x, 1))));
18    X = Orthopolyroots(va, vb, N);
19    W = subst(Q / Pp, 'x, X);
20    return (bitprecision([X, W], B + 32));
21  }
22
23  IntGaussJacobi(f, a, b, al = 0, XW = 0) =
24  { my(B = getlocalbitprec(), c = b - a, S);
25
26    if (type(XW) == "t_INT", XW = IntGaussJacobiinit(al, XW));
27    S = c^(al + 1) * SumXW(x -> f(a + c * x), XW);
28    return (bitprecision(S, B));
29  }
```

REMARKS 3.6.7.

(1) In this program, we have computed explicitly the polynomials P_N and Q_N so as to have the weights as $Q_N(x_i)/P'_N(x_i)$. In addition, we have evaluated this expression directly on the explicit expression of P_N and Q_N, and since this is rather unstable we have to double the working accuracy. To avoid this, we can do one of two things. The simplest is to generalize the formula in the Legendre case: the weights are simply given by $w_i = 1/(x_i(1-x_i)P'_N(x_i)^2)$. If we do not know this formula, we can still stably evaluate $Q_N(x_i)/P'_N(x_i)$ by generalizing the `Orthopolyeval` program to the polynomials Q_N, simply by changing the initialization `u0[1] = 1`, `u1[1] = r - a(0)` to `u0[1] = 0`, `u1[1] = m0`, where $m_0 = \langle 1, 1 \rangle$.

(2) See the discussion of the general `IntGaussLegendresplit` program for the reason of our choice `N = B / 4`.

EXAMPLES.

```
? \p 308 /* or 1024 bits */
? A = intnum(x = [0, 1/3], 1, x^(1/3) / (1 + x));
/* To compare with the correct value. */
? IntGaussJacobi(x -> 1 / (1 + x), 0, 1, 1/3) - A
time = 1,935 ms.
? 0.E-308 /* Perfect, slow, default N = B/4 = 256. */
? IntGaussJacobi(x -> 1 / (1 + x), 0, 1, 1/3, 205) - A
time = 1,144 ms.
? 0.E-308 /* Perfect, faster with a slightly lower N. */
? B = intnum(x = [0, 1/3], 1, x^(1/3) * exp(-x));
? IntGaussJacobi(x -> exp(-x), 0, 1, 1/3, 67) - B
time = 96 ms.
? 0.E-308 /* Perfect with a much smaller N. */
```

The above examples again illustrate the influence of the radius of convergence r, or equivalently the distance to the poles: in the first case where $r = 1$, the default $N = 256$ can be slightly reduced to 205 without losing accuracy, while in the second case where $r = \infty$, N can be reduced to 67 without losing accuracy.

Example 2: $1/\cosh(\pi x)$ **and** $x/\sinh(\pi x)$

Consider now the weight function $w(x) = 1/\cosh(\pi x)$ on the whole real line (we did not exclude this case in the above results). It is immediate to show essentially by definition that $m_k = 0$ if k is odd and that

$$m_{2k} = \int_{-\infty}^{+\infty} \frac{x^{2k}}{\cosh(\pi x)}\, dx = (-1)^k \frac{E_{2k}}{2^{2k}}\ ,$$

where $E_0 = 1$, $E_2 = -1$, $E_4 = 5$, $E_6 = -61$, etc. are the so-called *Euler numbers*. It follows that $\Phi(T^{-1}) = \sum_{k \geq 0} (-1)^k E_{2k} T^{2k+1}/2^{2k}$ is an odd function of T, so of the second form considered above. Now it can be shown that

$$\Phi(T) = 1/(T - (1^2/4)/(T - (2^2/4)/(T - (3^2/4)/(T - \cdots)))) ,$$

so that $a_n = 0$, $m_0 = 1$, and $b_n = n^2/4$ for $n \geq 1$. We are now going to apply Corollary 3.5.6 (3) in the following form: when N is even, we have

$$\frac{X^{-1/2}Q_N(X^{1/2})}{P_N(X^{1/2})} = \cfrac{m_0}{X - A_0 - \cfrac{B_1}{X - A_1 - \cfrac{B_2}{X - A_2 - \ddots - \cfrac{B_{N/2-1}}{X - A_{N/2-1}}}}}.$$

This has several advantages: first note that since P_N is an even and Q_N is an odd polynomial, the numerator and denominator of the left-hand side are polynomials. Second, the continued fraction has half the length of the standard one. Finally, and for the same reason, finding the roots of $P_N(X^{1/2})$, which is a polynomial of degree $N/2$, is considerably faster than finding the roots of $P_N(X)$ itself.

Evidently all of the above remarks are valid whenever the odd moments vanish.

In our case, we have $A_0 = b_1 = 1/4$, $A_k = b_{2k} + b_{2k+1} = 2k^2 + k + 1/4$ for $k \geq 1$, and $B_k = b_{2k-1}b_{2k} = k^2(2k-1)^2/4$. We can thus write the following specific program:

————————————————— IntGausscosh.gp —————————————————

```
1   /* int_{-oo}^{oo} f(x) / cosh(pi x) dx. */
2
3   IntGausscoshinit(n = 0) =
4   { my(B = getlocalbitprec(), X, W);
5
6     if (!n, n = ceil(B / 4));
7     if (n%2, n++);
8     my(N = 1 + n >> 1);
9     my(va = vector(N, k, (k - 1) * (2*k - 1) + 1/4));
10    my(vb = vector(N - 1, k, k^2 * (2*k - 1)^2 / 4));
11    my([P, Q] = Orthopolyrec([va, vb], 1));
12    localbitprec(5 * B / 4);
13    X = Orthopolyroots(va, vb, N);
14    W = subst(Q / (2 * P'), 'x, X);
15    return (bitprecision([sqrt(X), W], B + 32));
16  }
17
18  IntGausscosh(f, iseven = 0, XW = 0) =
19  { my(B = getlocalbitprec(), S);
20    if (type(XW) == "t_INT", XW = IntGausscoshinit(XW));
21    S = if (iseven, 2 * SumXW(f, XW)
22                  , SumXW(x -> f(x) + f(-x), XW));
23    return (bitprecision(S, B));
24  }
```

EXAMPLE.

```
? \p 1000
? A = intnum(x = [-oo, Pi], [oo, Pi], cos(x) / cosh(Pi * x));
```

```
? IntGausscosh(x -> cos(x)) - A
time = 7,067 ms.
% = -1.48... E-1002 /* Perfect. */
? XW = IntGausscoshinit();
time = 6,917 ms.
? IntGausscosh(x -> cos(x), 0, XW) - A
time = 131 ms.
% = -1.48... E-1002
? IntGausscosh(x -> cos(x), 1, XW) - A;
time = 66 ms.
% = -1.48... E-1002 /* Twice as fast: cos(x) is even. */
```

We will see in Chapter 4 that this allows us to compute alternating series using one of the Abel–Plana formulas.

A similar example is with the weight $w(x) = x/\sinh(\pi x)$. We leave to the reader to check that the corresponding moments are 0 for k odd and $m_{2k} = (-1)^k T_{2k+1}/2^{2k+1}$ for k even, where the T_k are the *tangent numbers* defined by $T_k = 2^{k+1}(2^{k+1} - 1)B_{k+1}/(k + 1)$ (which are integers: $T_1 = 1$, $T_3 = -2$, $T_5 = 16$, $T_7 = -272$, ...), and that the corresponding continued fraction is

$$\Phi(T) = (1/2)/(T - (1 \cdot 2/4)/(T - (2 \cdot 3/4)/(T - (3 \cdot 4/4)/T - \cdots)))$$

and to write the corresponding programs. This would also allow us to compute sums of alternating series using the other Abel–Plana formula.

Example 3: $x/(\exp(2\pi x) - 1)$

Our third example will similarly allow us to compute series with *positive* terms (which is more difficult) using another Abel–Plana formula, but the continued fraction will not be as simple. Here our interval is again the whole real axis, and $w(x) = x/(\exp(2\pi x) - 1)$ for $x > 0$, $w(x) = x/(1 - \exp(-2\pi x)) = w(-x)$ for $x < 0$. Since w is an even function, as in the previous example we have $m_k = 0$ if k is odd, and otherwise

$$m_{2k} = 2\int_0^\infty \frac{x^{2k+1}}{e^{2\pi x} - 1}\, dx = (-1)^k \frac{B_{2k+2}}{2k + 2},$$

where this time the B_n are the usual Bernoulli numbers. It follows that

$$\Phi(T^{-1}) = \sum_{k \geq 0}(-1)^k \frac{B_{2k+2}}{2k + 2} T^{2k+1}$$

which is again an odd function.

Unfortunately, the corresponding continued fraction does not seem to have any simple form: $\Phi(T) = c(0)/(T - c(1)/(T - c(2)/(T - \cdots)))$ with

$$c = [1/12, 1/10, 79/210, 1205/1659, 262445/209429, \dots].$$

Exercise: Show that the continued fraction for $\sum_{k \geq 0}(-1)^k B_{2k+2}T^{2k+1}$ *does* have a simple form, and find the corresponding weight function $w(x)$.

In our case it is therefore necessary to use the QD algorithm, and even though the coefficients $c(n)$ are rational numbers, they have huge denominators, so it is essential to use the floating point variant, leading to the following:

──────────────────── | IntGaussexp.gp | ────────────────────

```
1   /* int_{-oo}^oo f(x)w(x) dx with w(x)= x/(exp(2 pi x)-1)
2    * on [0, oo[ and w(-x) = w(x). */
3   IntGaussexpinit(N = 0) =
4   { my(B = getlocalbitprec(), v, QP, P, Q, X, W);
5
6     if (!N, N = ceil(B / 4));
7     if (N%2 == 0, N++);
8     localbitprec(3*B/2 + 32);
9     bernvec(N); /* initialize Bernoulli cache */
10    v = Quodif(vector(N+1, k, (-1)^(k-1)*bernfrac(2*k)/(2*k)));
11    QP = CFback(N, m -> v[m + 2] / 'x);
12    [Q, P] = [denominator(QP) / (12 * 'x), numerator(QP)];
13    X = polrootsreal(P);
14    if (#X != poldegree(P),
15        error("loss of accuracy in polrootsreal"));
16    W = subst(Q / (2 * P'), 'x, X);
17    return (bitprecision([sqrt(X), W], B + 32));
18  }
19
20  IntGaussexp(f, iseven = 0, XW = 0) =
21  { my(B = getlocalbitprec(), S);
22    if (type(XW) == "t_INT", XW = IntGaussexpinit(XW));
23    S = if (iseven, 2 * SumXW(f, XW)
24                  , SumXW(x -> f(x) + f(-x), XW));
25    return (bitprecision(S, B));
26  }
```

REMARKS 3.6.8.
  (1) Since we need to use Quodif we have used CFback to recover P and Q, but
      we could just as well have used Orthopolyrec together with the formulas
      of Theorem 3.5.7 (2).
  (2) For a similar reason, we used the polrootsreal program to compute the
      nodes instead of Orthopolyroots. However, since the output of Quodif
      is a priori inexact, we need to check that polrootsreal does find all the
      roots, whence the check made in the IntGaussexpinit program. For
      simplicity, we raise an exception if a root is missing; the interested reader
      can also adapt the more robust solution used in IntGaussinit (increase
      the accuracy until all roots are found).
  (3) As in most Gaussian integration programs, note that by far the most time
      consuming part of the initialization is the computation of the real roots
      of P.

EXAMPLE.
```
? \p 1000
? A = 2*intnum(x = 0, [oo, 2*Pi], x*cos(x) / (exp(2*Pi*x) - 1));
? XW = IntGaussexpinit();
time = 41,086 ms.
```

```
? IntGaussexp(x -> cos(x), 1, XW) - A /* Again cos(x) even */
time = 67 ms.
% = 0.E-1002
```

3.7. Gaussian Integration on $[a, \infty]$

In the preceding sections we have already seen several Gaussian methods for integrating on infinite intervals: the Gauss–Laguerre method, applicable to integrands of the form $x^\alpha e^{-x} f(x)$, where $f(x)$ can be well approximated by polynomials, and the Gauss–Hermite method for $e^{-x^2} f(x)$. Of course, the general Gaussian methods seen in Section 3.6.6, using either Gram–Schmidt, or better, continued fractions, are also applicable when the interval of integration is infinite.

However, almost by definition, the range of applicability of these methods is quite restricted. We give here two other quite naïve methods whose range of applicability is also restricted, but give a larger set of integrals which can be computed.

3.7.1. Using Gauss–Legendre.
In the case of functions f which are well-behaved at ∞ and in particular tend *slowly* (i.e., not exponentially) to 0, we can use Gauss–Legendre to compute integrals on $[a, \infty[$ for some a, essentially by changing x into $1/x$ or some other negative power of x.

Assume that as $x \to \infty$ the function f has an asymptotic expansion of the form $f(x) = \sum_{j \geq 0} a_j/x^{j\alpha+\beta}$ with $a_0 \neq 0$, so that for convergence we must assume that $\alpha > 0$ and $\beta > 1$ (this will for instance be the case for rational functions of degree less than or equal to -2). We want to make the change of variable $y = x^{-\gamma}$ for a suitable $\gamma > 0$ so that we can use generalized Gauss–Legendre (i.e., either IntGaussLegendre or IntGaussJacobi). Assuming $a \geq 1$ ($a > 0$ is sufficient, but it is preferable to assume $a \geq 1$), we have

$$\int_a^\infty f(x)\,dx = \frac{1}{\gamma} \int_0^{a^{-\gamma}} f(y^{-1/\gamma}) y^{-1/\gamma-1}\,dy \; ,$$

and as $y \to 0$ we have the asymptotic expansion

$$f(y^{-1/\gamma}) y^{-1/\gamma-1} = \sum_{j \geq 0} a_j y^{-1+(j\alpha+\beta-1)/\gamma} \; .$$

Since we want a good approximation by polynomials, we must have $\alpha/\gamma \in \mathbb{Z}_{>0}$, so the best is to choose $\gamma = \alpha$, hence we must use generalized Gauss–Legendre with weight $w(x) = x^{(\beta-1)/\alpha-1}$ (note that since $\alpha > 0$ and $\beta > 1$, the exponent of x is strictly larger than -1, as it should).

──────────────── IntGaussa_oo.gp ────────────────

```
1  /* Assume that f is a nice function on [a, oo[ with a > -1
2   * which has an asymptotic expansion at infinity of the type
3   * f(x) = sum_{j >= 0} a_j / x^{j*al+be}
4   * with al>0 and be>1. Compute int_a^oo f(x) dx using
5   * generalized Gauss-Legendre. */
6  IntGaussa_ooinit(alf = [1, 2], XW = 0) =
7  { my([al, be] = alf);
8    return (IntGaussJacobiinit((be - 1) / al - 1, XW));
9  }
10
```

```
11   IntGaussa_oo(f, a = 1, alf = [1, 2], XW = 0) =
12   {
13     if (a < -1, error("must have a > -1 in IntGaussa_oo"));
14     if (a < 1, return (IntGaussa_oo(x -> f(x+a-1), 1, alf, XW)));
15     my([al, be] = alf, c = 1 / al, d = c * be);
16     if (type(XW) != "t_VEC",  XW = IntGaussa_ooinit(alf, XW));
17     return (c * IntGaussJacobi(x -> f(x^(-c)) * x^(-d),
18                                0, a^(-al), d - c - 1, XW));
19   }
```

EXAMPLE. Compute $\int_1^\infty dx/(1+x^2)$:

```
? \p 500
? f(x) = 1 / (1 + x^2);
? IntGaussa_oo(f) - Pi/4
time = 174 ms.
% = 0.E-500 /* Perfect. */
```

The above program computes $\int_a^\infty f(x)\,dx$ for $a \geq 1$. If $a < 1$ (or even a complex) you have several choices:

(1) Use $\int_a^\infty f(x)\,dx = \int_1^\infty f(x+a-1)\,dx$.
(2) Use $\int_a^\infty f(x)\,dx = a\int_1^\infty f(ax)\,dx$.
(3) Use

$$\int_a^\infty f(x)\,dx = \int_a^N f(x)\,dx + \int_N^\infty f(x)\,dx$$

with $N \geq 1$ of your choosing.

The specific choice that you make evidently depends entirely on the function f and on the value of a. For instance, if a is large (say $a = 100$), we could use the program directly, but it is usually preferable to use the second formula above, changing x into ax. If $|a| < 1$ we may want to use the first, and if $a \leq -1$ the last, but all this is left to the judgment of the user.

EXAMPLES. (Continued): compute $\int_0^\infty dx/(1+x^2)$:

```
? IntGaussa_oo(x -> f(x - 1)) - Pi/2
time = 181 ms.
% = -8.82...E-319 /* 182 decimals lost. */
? IntGaussLegendre(f, 0, 1) + IntGaussa_oo(f) - Pi/2
time = 361 ms.
% = 0.E-500 /* Perfect but twice slower. */
```

Warning: here is another example (still at 500 decimals):

```
? f(x) = x^3 * log(1 + 1 / (x + 1)^5);
? A = intnum(x = 2, oo, f(x));
? IntGaussa_oo(f, 2) - A
% = 0.E-501 /* Perfect. */
? IntGaussa_oo(x -> f(x + 1), 1) - A
% = -1.289... E-398 /* 103 decimals lost. */
```

The reader can check that this is because, after the change of variable $x \mapsto 1/x$ done by IntGaussa_oo, there is a pole which is closer to the interval of integration.

For integrals from 0 to ∞, we can also use the change of variable $y = (1 - x)/(1 + x)$, which sends the interval $[0, \infty[$ to the interval $[-1, 1]$ on which we can apply Gauss–Legendre.

We give trivial specific examples of the use of the parameters α and β: if $f(x) = 1/x^{3/2}$, we must take $\beta = 3/2$, but the value of α is unimportant. If $f(x) = 1/(x^{3/2} + x^{5/2})$, we must again take $\beta = 3/2$, but now α must be a *divisor* of 1, in other words we could take $\alpha = 1$, but also $\alpha = 1/2$ or $\alpha = 1/3$, but not $\alpha = 2$. If $f(x) = 1/(x^{3/2} + x^2)$, we again take $\beta = 3/2$, but now α must be a divisor of $1/2$, so the simplest is to choose $\alpha = 1/2$.

As we will see below, this program works reasonably well with functions which tend to 0 slowly at infinity, although as we have just seen, it often loses a lot of accuracy. For functions which tend to 0 rapidly, for instance exponentially, as such it is totally useless. However if we combine it with the IntGaussLegendresplit program it again becomes useful. A primitive program can be as follows:

—————————————— | IntGaussa_oosplit.gp | ——————————————

```
1   /* Same assumptions as IntGaussa_oo. The parameter N
2    * is a cutoff between Gaussian integration on a compact
3    * interval and the use of IntGaussa_oo.*/
4   IntGaussa_oosplit(f, a, N, alf = [1, 2], v1 = 0, v2 = 0) =
5   {
6     if (!v1, v1 = IntGaussa_ooinit(alf, 0));
7     if (!v2, v2 = IntGaussLegendreinit());
8     return (IntGaussa_oo(f, N, alf, v1)
9             + IntGaussLegendresplit(f, a, N, 1, v2));
10  }
```

As a rule of thumb, choose $N = 2$ for functions tending to 0 slowly at infinity, and $N \approx 0.3B/u$ for functions tending to 0 like e^{-ux}, where as usual B is the number of binary digits.

3.7.2. Gaussian integration with polynomials in $1/x$.
When computing integrals on a compact interval $[a, b]$, using Gaussian integration based on orthogonal *polynomials* (possibly multiplied by functions such as $(x - a)^\alpha (b - x)^\beta$) is quite reasonable, in particular since polynomials are dense in continuous functions.

However the same is not true on noncompact intervals such as $[a, \infty[$: when the function to be integrated tends slowly to 0, such as a rational function, changes of variable such as $x \to 1/x$ transform the integral into an integral over a compact interval of a reasonable function, which can then sometimes be reasonably Gaussian integrated. This is the program IntGaussa_oo above. On the other hand, if the function f tends to 0 very fast, for instance if $f(x) = e^{-x}g(x)$ for some "reasonable" g, we cannot make this type of change of variable, nor in fact any other, since for instance the function $e^{-1/x}$ is much too singular at 0.

Of course, one solution is to cheat: assuming that the size of $g(x)$ is reasonably small compared to e^x, we can say that $\int_N^\infty e^{-x}g(x)\,dx$ will be of the order of e^{-N}, so if $N > B\log(2)$ (where as usual B is the desired bit accuracy), this integral can be neglected, so (if for instance $a = 0$) we are reduced to computing $\int_0^N e^{-x}g(x)\,dx$, which could be treated by Gauss–Legendre. Unfortunately, a little experimentation shows that this gives very bad results, so this idea must be scrapped.

The problem with using polynomials (as in Gauss–Laguerre, which is adapted to computing $\int_0^\infty e^{-x} g(x)\,dx$) is that the method works very well when $g(x)$ is well approximated by a polynomial in x, but in general not: for instance such a simple integral as $\int_0^\infty (e^{-x}/(1+x))\,dx$ cannot be computed with any reasonable accuracy. To generalize the (generalized) Gauss–Laguerre method, we would like to be able to compute exactly integrals of the form $\int_a^\infty x^{\alpha+\beta n} e^{-x}\,dx/x$, where $a > 0$, α, and $\beta > 0$ are fixed, and n is in some interval of $\mathbb{Z}$, not necessarily of nonnegative integers. Of course the only reason that we take integrals from a to ∞ instead of from 0 is to avoid problems with $x^{\alpha-1+\beta n}$ for $n < (1-\alpha)/\beta$, which have nothing to do with the question at hand.

As a first approach, let us assume that we want these integrals to be exactly computed for $n \in [-N, 0]$. This is exactly a problem of orthogonal polynomials, but now with a strange weight: setting $y = x^{-\beta}$, we have

$$\int_a^\infty x^{\alpha+\beta n} e^{-x}\,dx/x = \frac{1}{\beta} \int_0^{a^{-\beta}} y^{-\alpha/\beta-n} e^{-y^{-1/\beta}}\,dy/y \ ,$$

and since we want $-n \in [0, N]$, we are looking for Gaussian integration with weight function $x^{-\alpha/\beta-1} e^{-x^{-1/\beta}}$ on the interval $[0, a^{-\beta}]$, which we now have to study.

The scalar product (normalized by multiplying by e^a) is thus

$$\langle p, q\rangle = \frac{1}{\beta} \int_0^{a^{-\beta}} x^{-\alpha/\beta} e^{a-x^{-1/\beta}} p(x)q(x)\,dx/x = \int_a^\infty x^\alpha e^{a-x} p(x^{-\beta})q(x^{-\beta})\,dx/x \ .$$

To find the corresponding orthogonal polynomials, recall that we have two methods: one using Gram–Schmidt to find the three-term relation, and the second using continued fractions. As we have seen, this latter method is much more efficient, so we forget about the first.

We need first to compute the moments: we have

$$m_k = \langle 1, x^k\rangle = \int_a^\infty x^{\alpha-1-\beta k} e^{a-x}\,dx \ ,$$

so by integration by parts

$$m_k = \frac{x^{\alpha-\beta k}}{\alpha - \beta k} e^{a-x}\Big|_a^\infty - \frac{1}{\beta k - \alpha} \int_a^\infty x^{\alpha-\beta k} e^{a-x}\,dx$$

$$= \frac{a^{\alpha-\beta k}}{\beta k - \alpha} - \frac{m_{k-1/\beta}}{\beta k - \alpha} \ .$$

We will therefore assume that $1/\beta \in \mathbb{Z}_{>0}$ (otherwise we would simply compute the moments directly without using a recursion).

It is of course immediate to obtain from the above an explicit formula for m_k, but we do not need it for computation since the m_k will be computed using the above recursion. We thus simply initialize m_k for $0 \le k < 1/\beta$, then use the recursion except if $k = \alpha/\beta$ where we must again compute the value directly.

We can now write the programs for computing

$$\int_a^\infty x^{\alpha-1} e^{a-x} f(x)\,dx = b \int_0^{a^{-\beta}} x^{-1-b\alpha} e^{a-1/x^b} f(1/x^b)\,dx \ ,$$

where we have set $b = 1/\beta$ and where it is $f(1/x^b)$ which is well approximated by polynomials, and not $f(x)$ as in Gauss–Laguerre:

──────────────── IntGaussLaguerreinv.gp ────────────────

```
1   /* int_a^oo x^{al-1} e^{-x} f(x) dx where we assume that
2    * f(x) = a_0 + a_1/x^be + a_2/x^(2be) + ... using
3    * continued fractions. */
4   IntGaussLaguerreinvinit(aalbe = [1, 1, 1]) =
5   { my([a, al, be] = aalbe, B = getlocalbitprec(), X, W);
6
7     localbitprec(B + 32);
8     [X, W] = IntGaussinit(N -> Laguerreinvmoments(aalbe, 2*N));
9     X = vector(#X, j, X[j]^(-1 / be));
10    return (bitprecision([X, exp(-a)*W], B + 32));
11  }
12
13  IntGaussLaguerreinv(f, aalbe = [1, 1, 1], XW = 0) =
14  {
15    if (!XW, XW = IntGaussLaguerreinvinit(aalbe));
16    return (SumXW(f, XW));
17  }
18
19  /* Weight function w(x) = x^{-b al-1}e^{a-1/x^b}$
20   * on ]0, 1/a^be]. Given a, al, be, and n >= 0, this
21   * computes an (n+1)-component vector containing the
22   * moments m_k for 0 <= k <= n for the given scalar product. */
23  Laguerreinvmoments(aalbe, N) =
24  { my([a, al, be] = aalbe, b = 1 / be, ea = exp(a), v, w, ap);
25
26    if (type(b) != "t_INT" || b > N+1, error("incorrect beta"));
27    v = vector(N + 1);
28    for (k = 0, b - 1, v[k+1] = incgam(al - k * be, a) * ea);
29    w = a^(al - 1 + be); ap = a^(-be);
30    for (k = b, N,
31      w *= ap;
32      v[k+1] = if (k * be == al, eint1(a) * ea
33                        , (w - v[k+1-b]) / (k*be - al)));
34    return (v);
35  }
```

We will see in the timings below that, as usual with Gaussian methods, this is by far the most efficient program for numerical integration when it is applicable (in the examples that we give below this will be the case for $\int_0^\infty f(x)e^{-x}\,dx$ with $f(x) = 1/(1+x)$, $f(x) = e^{1/(x+1)}$, $f(x) = 1/\sqrt{1+x}$, and $f(x) = \tan(1/(x+1))$).

EXAMPLE.

```
? \p115
? A = intnum(x = 1, [oo, 1], exp(-x) / (1 + x));
```

```
? IntGaussLaguerreinv(x -> 1 / (1 + x)) - A
time = 72 ms.
% = 8.656... E-111 /* 4 decimals lost */
```

3.8. Doubly-exponential integration methods (DE)

3.8.1. Introduction. We now come to the very important *doubly-exponential* method, invented in the 1960's by M. Mori and H. Takahashi. For simplicity, we will use the acronym DE. (Some authors call it the *tanh-sinh* method, referring to a specific change of variable used for compact intervals, but this is too restrictive). Let us start with some comments:

- The method applies only to functions which are *holomorphic* (or reasonably meromorphic) on a domain around the path of integration. Also it gives poor results when the function has poles *near* the path of integration, although the method can be modified to take into account the presence of known poles, exactly as we did for Gaussian integration, but with different formulas.
- Although slower than Gaussian integration methods, it is usually much more *robust*, in that mild singularities of the integrand pose in general no problem.
- It sometimes gives the fastest known methods for computing certain *functions*, and it is in particular very useful in the computation of L-functions as we will see later. It was proved to be near optimal for suitable spaces of analytic functions [**Sug97**].

3.8.2. Cursory analysis of the DE. We begin by the following:

COROLLARY 3.8.1. *Assume that* $F \in C^\infty(\mathbb{R})$. *For any* $h > 0$, $N \in \mathbb{Z}_{\geq 1}$, *and* $k \in \mathbb{Z}_{\geq 1}$ *we have*

$$\int_{-Nh}^{Nh} F(t)\,dt = h \sum_{m=-N}^{N} F(mh) - h\frac{F(N) + F(-N)}{2}$$

$$- \sum_{1 \leq j \leq \lfloor k/2 \rfloor} \frac{B_{2j}}{(2j)!} h^{2j} \left(F^{(2j-1)}(N) - F^{(2j-1)}(-N) \right) + R_k(F, N)\,,$$

with

$$R_k(F, N) = \frac{(-1)^k}{k!} h^k \int_{-N}^{N} F^{(k)}(t) B_k(\{t\})\,dt\,.$$

PROOF. In Proposition 4.2.40 from the next chapter simply choose $a = -Nh$, $b = Nh$, and replace N by $2N$, so that we indeed have $h = (b - a)/(2N)$. □

COROLLARY 3.8.2. *Assume in addition that* $F(t)$ *and all its derivatives tend to* 0 *as* $t \to \pm\infty$ *at least as fast as* $|t|^{-\alpha}$ *for some* $\alpha > 1$. *For all* $N \in \mathbb{Z}_{\geq 1}$ *and* $k \in \mathbb{Z}_{\geq 1}$ *we have*

$$\left| \int_{-\infty}^{\infty} F(t)\,dt - h \sum_{m=-N}^{N} F(mh) \right| \leq h \sum_{|m|>N} |F(mh)| + h^{2k} C_{2k}\,,$$

where we set

$$C_{2k} = \frac{|B_{2k}|}{(2k)!} \int_{-\infty}^{\infty} |F^{(2k)}(t)|\,dt.$$

PROOF. Clear by making $N \to \infty$ in the preceding corollary.   □

Now assume that as $|t| \to \infty$ we have $|F(t)| = O\left(e^{-a \cdot e^{|t|} + o(e^{|t|})}\right)$ for some $a > 0$. We will say in that case (and similar ones) that F tends to 0 *doubly exponentially fast* at $\pm\infty$. Let us choose $h = \log(bN/\log(N))/N$ for some $b > 0$ (we will see below that this is close to optimal). Because of the exponential decrease of F, the sum for $|m| > N$ will be bounded by $c \cdot F(Nh)$ for a small multiplicative constant c, and

$$F(Nh) = F(\log(bN/\log(N))) = O\left(e^{-abN/\log(N) + o(N)}\right) .$$

Moreover, assume that F is holomorphic around $\mathbb{R}$, for instance in a horizontal strip $|\Im(z)| \le d$. This implies that the radius of convergence of the power series giving $F(z)$ around any $t \in \mathbb{R}$ is at least equal to d, in other words that $|F^{(2k)}|/(2k)! < 1/d_1^{2k}$ for any $d_1 < d$. Since $|B_{2k}| = O((2k)!/(2\pi)^{2k})$, we have $C_{2k} = O\left((2k)!/(2\pi d_1)^{2k}\right)$. Taking logarithms and using Stirling's formula, we have

$$\log(C_{2k}) \le (2k + 1/2)\log(2k) - 2k - 2k\log(2\pi d_1) + d_2$$

for some constant d_2. To choose k, we must therefore minimize $(2k+1/2)\log(k) - 2k(\log(\pi d_1) + 1 - \log(h))$. Neglecting the term $(1/2)\log(k)$, the derivative with respect to k is $2\log(k) - 2(\log(\pi d_1/h))$, so $k = \pi d_1/h$ is the quasi optimal choice, giving

$$\log(C_{2k}) + 2k\log(h) = -2\pi d_1/h + o(1/h) = -2\pi d_1 N/(\log(bN)) + o(N/\log(N)) .$$

Thus, the error made by approximating $\int_{-\infty}^{\infty} F(t)\,dt$ by $h\sum_{m=-N}^{N} F(mh)$ is of the order of $O(e^{-2\pi d_1 N/\log(bN)})$, an almost exponentially good approximation. This is remarkable, since it means that the number $2N+1$ of necessary function evaluations will be almost linear in the number of required decimals. In addition, since d_1 is in the exponent, the accuracy will increase as a function of the width of the strip of holomorphy.

REMARK 3.8.3. It is easy to see that the above choice of h is close to optimal: if we set $h = \log(\psi(N))/N$, the logarithm of the error made in truncation is essentially $-a\psi(N)$, and that due to C_{2k} is $-2\pi d_1 N/\log(\psi(N))$, so setting these two equal gives $\psi(N)\log(\psi(N)) = (2\pi d_1/a)N$, so $\psi(N)$ should be close to $(2\pi d_1/a)N/\log(N)$. Thus, the optimal choice of h should be $(\log(bN) - \log(\log(N)))/N$ with $b = 2\pi d_1/a$, close to what we have chosen.

3.8.3. DE over compact intervals $[a, b]$. Of course in actual practice very few functions tend to zero doubly exponentially fast, but it is easy to reduce to that case using a *change of variable*. Assume first that we want to compute numerically $I = \int_{-1}^{1} f(x)\,dx$ for some function f holomorphic in some region containing the real interval $[-1, 1]$. We make the change of variable

$$x = \phi(t) = \tanh(\lambda \sinh(t)) ,$$

where λ is a constant to be chosen. We have

$$\phi'(t) = \lambda \cosh(t)/\cosh^2(\lambda \sinh(t)) = 2\lambda e^{-\lambda e^{|t|} + |t| + o(1)}$$

when $|t| \to \infty$, so $F(t) = f(\phi(t))\phi'(t)$ indeed tends to zero doubly exponentially fast. Note that when t varies from $-\infty$ to ∞, $x = \phi(t)$ increases from -1 to 1.

Thus,

$$I = \int_{-1}^{1} f(x)\, dx = \int_{-\infty}^{\infty} f(\phi(t))\phi'(t)\, dt$$

is very well approximated by

$$h \sum_{m=-N}^{N} f(\phi(mh))\phi'(mh)$$

for suitable choices of N and h. Note in passing that as usual the *abscissas* (or *nodes*) $\phi(mh)$ where f must be computed, and the *weights* $\phi'(mh)$ can be computed once and for all for a given accuracy, independently of the function f.

Evidently, if instead f is defined on some compact interval $[a, b]$, we of course write

$$\int_{a}^{b} f(x)\, dx = \frac{b-a}{2} \int_{-1}^{1} f\left(\frac{a+b}{2} + \frac{b-a}{2}x\right)\, dx \ ,$$

and use the method explained above, i.e., the change of variable $x = \tanh(\lambda \sinh(t))$. A detailed analysis of the method (see for instance [**Mol10**]) shows that the optimal choice of λ for a wide class of functions is $\lambda = \pi/2$, so we will make this choice.

Although the programs are already implemented in GP, they are so simple that we give them anyhow. As usual we split the program in two, one which can once and for all perform the precomputations for a given accuracy, and the second which does the actual integration.

```
──────────────────────────  Inta_b.gp  ──────────────────────────
```

```
1   IntgetN(B, c) =
2   { localbitprec(32); B += 64; ceil(B * (log(B) - 5/4) / c); }
3
4   /* Precomputations for tanh-sinh numerical
5    * integration on a compact interval [a, b]. */
6   Inta_binit(mul = 1) =
7   { my(B = getlocalbitprec(), h, e, ei, X, W, q, r);
8
9     N = IntgetN(B, 6.18 / mul);
10    localbitprec(B + 32);
11    h = log(2*Pi * N/log(N)) / N;
12    X = vector(N); e = exp(h); q = Pi * e;
13    W = vector(N); ei = 1/e;   r = Pi * ei;
14    for (m = 1, N, \\ q = Pi exp(m h), r = Pi exp(-m h)
15      my(c = (q + r)/2, s = q - c, z = 2 / (exp(s) + 1));
16      X[m] = 1 - z;
17      W[m] = c * z * (2 - z);
18      if (exponent(z) < -B - 32,
19          X = X[1..m]; W = W[1..m]; break);
20      q *= e; r *= ei);
21    return ([X, W, h/2]);
22  }
23
24  /* Numerical integration of a closure f on compact [a, b],
25   * with precomputed XW = [X, W], otherwise recomputed. */
```

```
26   Inta_b(f, a, b, XW = 0) =
27   { my(B = getlocalbitprec(), BX, X, W, W0, c, d, S);
28
29     if (!XW, XW = Inta_binit());
30     localbitprec(B + 32);
31     [X, W, W0] = XW;
32     BX = bitprecision(X[#X]);
33     a = bitprecision(a, BX);
34     b = bitprecision(b, BX);
35     c = (b + a) / 2;
36     d = (b - a) / 2;
37     S = Pi * f(c) + sum(m = 1, #X,
38                         W[m] * (f(c+d*X[m]) + f(c-d*X[m])));
39     bitprecision(S * W0 * d, B);
40   }
```

REMARK 3.8.4. The main heuristic in this program (as in all the other DE programs that we will give) is the choice of the number N of nodes using the IntgetN program, which depends on the bit accuracy B and on an experimental constant c. In view of the analysis of the method, we need N to be $O(B \log(B))$, so we could choose $N = \lambda B \log(B)$ for suitable constants λ. Experiments show that a slightly better choice is $N = \lambda B \log(B) - \mu B$ for some positive constant μ, and we have decided to use systematically $\mu = 5\lambda/4$, so $N = \lambda(B \log(B) - 5B/4)$. This is certainly not the best choice, but it works well. In all DE programs, we will choose N in this way using a specific heuristic choice of λ. This choice can be modified by using the optional parameter mul in the initialization program, which multiplies the default N by mul:

EXAMPLE.

```
? \p115
? Inta_b(x -> 1 / (1 + x^2), -3, 3) - 2 * atan(3)
% = 2.50... E-33 /* 82 decimals lost due to the poles at +-i. */
? XW = Inta_binit(5); /* Use 5 times as many nodes. */
? Inta_b(x -> 1 / (1 + x^2), -3, 3, XW) - 2 * atan(3)
% = 0.E-114 /* Perfect. */
```

As already mentioned in the context of Gaussian integration, in this example the correct solution is not to increase the number of points as above, but either to split the interval, say at $x = 0$, or to take into account explicitly the poles, see below.

Even on a compact interval, this is of course far from being the whole story, and it is necessary to explain in detail the pitfalls of this program. In particular, when f has an integrable singularity, we must be careful. First, by splitting the interval, it is wise to put the singularity (or singularities) at one of the endpoints. Examples:

   (1) Inta_b(log,0,1): immediately gives the correct answer -1: the singularity is a mild logarithmic (and in particular integrable) singularity at the endpoint $x = 0$.

(2) `Inta_b(x->log(abs(x),-1,1))`: here we have an error message, because
the singularity is right in the middle of the interval, and the program tries
to compute $\log(0)$. To obtain the correct result, split at 0.
(3) `Inta_b(x->log(abs(x),-1,2))`: here we have no error message, but the
result has only one correct decimal ($-1.560\cdots$, while the correct answer
is $2\log(2) - 3 = -1.6137\cdots$). Once again, to obtain the correct result,
split at 0.

When f has a removable singularity, we must also help the program. For
instance, consider the integral

$$\int_0^1 \left(\frac{1}{e^x - 1} - \frac{1}{x} \right) dx \ .$$

If one applies directly the `Inta_b` function at $38D$, the result has only 20 correct
decimals. The reason is that the evaluation of the function $f(x) = 1/(e^x - 1) - 1/x$
near $x = 0$ induces bad cancellation. One method to correct this is to compute the
Taylor series expansion around 0, and to replace the definition of $f(x)$ by something
like
    `f(x)=if(exponent(x)>-B/5,1/(exp(x)-1)-1/x,-1/2+x/12-x^3/720)`
where as usual `B=getlocalbitprec()` is the current bit accuracy.

Exercise: In a similar manner, compute $\int_0^1 (e^x - 1 - x)/x^2 \, dx$.

From now on, we assume by splitting the interval that we have only one singularity, at the beginning of the interval, which by translation we may assume to be
$[0, 1]$. If the singularity is removable or of logarithmic type, we proceed as above. On
the other hand, if the singularity is algebraic, we must tell the program: for example,
working with the default accuracy of 38 decimal digits, `Inta_b(x->1/sqrt(x),0,1)`
has 25 correct digits, and `Inta_b(x->x^(-3/4),0,1)` only has 12. The solution is
of course simply to make an algebraic change of variable: if the singularity at 0 is
$1/x^a$ for some a with $0 < a < 1$, we set $x = y^{1/(1-a)}$ before using the integration
method, and since

$$\int_0^1 f(x) \, dx = \frac{1}{1-a} \int_0^1 f(y^{1/(1-a)}) y^{a/(1-a)} \, dy \ ,$$

the singularity of the integrand at 0 has disappeared. Note that this is built-in GP:
one writes `intnum(x=[0,-a],f(x))`. Compare
    `intnum(x=0,1,x^(-3/4))`, which gives $3.9999999999999987422295162236\cdots$,
with
    `intnum(x=[0,-3/4],1,x^(-3/4))`, which gives exactly 4 to 38 digits.
We could of course also incorporate this in the homemade `Inta_b` program if
desired.

If the singularity is a little wilder, we may be in trouble. The user must
then do a manual change of variable to remove it. As an example, consider
$\int_0^{1/2} 1/(x\log^2(x)) \, dx = 1/\log(2)$. Using `intnum` directly gives only 2 correct decimals, and using as initial endpoint $[0, -1]$ gives an error message since the change
of variable mentioned above in case of an algebraic singularity becomes itself singular. One can "cheat" and use as initial endpoint $[0, -0.99999]$, say, which gives
6 correct decimals, but one cannot go much closer to -1 without generating an
error message. Thus the user must himself perform the evident change of variable

$x = e^{-y}$, and

$$\int_0^{1/2} 1/(x \log^2(x)) \, dx = \int_{\log(2)}^\infty dy/y^2$$

which (notwithstanding the fact that the answer is now trivial) must be computed using doubly-exponential integration over a noncompact interval, which we will study below.

Another type of problem which may occur when integrating over a compact interval is the presence of *poles* near the interval of integration. Recall that to use the doubly-exponential integration we need holomorphic functions (although poles can easily be accounted for). A typical example is $\int_{-1}^1 \cosh(x) \, dx/(1 + x^2/a^2)$ when a is small. For $a = 0.1$ we already obtain only 5 correct decimals, and for $a = 0.01$ only 1. There are two solutions to this problem. The first is to split the interval so that the poles are near the endpoints: the integration routine will be less affected by them. For instance, in this example splitting at 0 gives 30 correct decimals for $a = 0.1$, and 20 decimals for $a = 0.01$.

The second solution is to use the doubly-exponential method on the function minus its polar part, and to compute the integral of the polar part directly since it is an elementary function, so for our example we can write:

```
Inta_b(x -> (cosh(x) - cos(a)) / (1 + x^2 / a^2), -1, 1) \
              + 2 * cos(a) * a * atan(1 / a)
```

which gives an essentially perfect answer even for very small values of a.

Note that if the polar part is not a rational function, i.e., if it is necessary to make a cut in the complex plane, the best solution is to split the interval: a typical example is $\int_{-1}^1 dx/\sqrt{1 + x^2/a^2}$.

3.8.4. DE over $]-\infty, \infty[$. We now consider integration over noncompact intervals. For the moment, we will assume that our function is non oscillating in some reasonable sense, and consider the case of oscillating functions below.

In view of the doubly-exponential philosophy, we should endeavor to make changes of variables which produce functions which tend to 0 doubly-exponentially, no more and no less.

We first consider integration over the whole of $\mathbb{R}$, i.e., on $]-\infty, \infty[$. Here the philosophy is clear: if the function tends to 0 doubly exponentially at both ends, we do not need to make a change of variable, we apply Riemann sums (e.g., Corollary 3.8.2) directly. If the function tends to 0 simply exponentially at both ends, we must make a change of variable which make it tend doubly-exponentially: the simplest is to set $x = \sinh(t)$. Finally, if the function tends to 0 polynomially, say, the simplest is to set $x = \sinh(\sinh(t))$.

Explicitly:

(1) In the case of functions tending to 0 doubly-exponentially at both ends we use Riemann sums directly:

$$\int_{-\infty}^\infty f(x) \, dx \approx h \sum_{m=-N}^N f(mh) \, .$$

(2) In the case of functions tending to 0 simply-exponentially at both ends we use $x = \phi(t) = \sinh(t)$:

$$\int_{-\infty}^{\infty} f(x)\,dx \approx h \sum_{m=-N}^{N} f(\sinh(mh)) \cosh(mh) \ .$$

(3) In case of functions tending to 0 polynomially at both ends we use $x = \phi(t) = \sinh(a\sinh(t))$ for some $a > 0$:

$$\int_{-\infty}^{\infty} f(x)\,dx \approx ah \sum_{m=-N}^{N} f(\sinh(a\sinh(mh))) \cosh(a\sinh(mh)) \cosh(mh) \ .$$

REMARK 3.8.5. In the case of a compact interval, say $[-1,1]$, the doubly-exponential decrease was due to that of $\phi'(t) = a\cosh(t)/\cosh^2(a\sinh(t))$, the behavior of the function f itself being unimportant. Here it is exactly the opposite: for instance in the case of polynomial decrease $\phi'(t) = a\cosh(a\sinh(t))\cosh(t)$ tends to ∞ doubly-exponentially, but since we need to assume convergence of the integral, this is more than compensated by the doubly-exponential decrease of $f(\phi(t)) = f(\sinh(a\sinh(t)))$ which is entirely due to the behavior of the function f.

Of course the above changes of variable cover only very special cases: for instance most functions do not tend to 0 doubly or simply exponentially or polynomially, and often not on both sides. These two problems are treated separately. If for instance the function does not have the same behavior at $-\infty$ and at $+\infty$, the simplest is to split the integral at 0 and compute the two integrals separately using the change of variable given in the next subsection. A typical (although perhaps slightly artificial) example is $\int_{-\infty}^{\infty} dx/(e^x + x^2)$, which tends to 0 simply exponentially when $x \to +\infty$ but polynomially when $x \to -\infty$.

The second problem, that of functions whose decrease is none of the above, is dealt with in a heuristic although reasonable manner, which can of course be made rigorous with an analysis analogous to the one we have done above. Typical examples are functions e^{-ax^2} (or more generally $e^{-ax^{2m}}$). Experimentation and theory shows the following: when a is not too small, say $a > 1/100$, one should always consider these functions as tending to 0 simply-exponentially, although when a is large, say $a > 10$, one may also consider them as tending to 0 doubly exponentially. On the other hand, when a is very small, say $a \leq 1/100$, one could consider the function as tending to 0 polynomially, except that we must modify the exponential function so that the program does not try to evaluate it at very negative arguments, by writing something like `myexp(x)=if(x<-0.694*B,0,exp(x))`, where as usual B is the current bit accuracy. The results will be correct but not completely accurate, so we advise to make a change of variable (in the present specific case $y = x\sqrt{a}$) to avoid the problem occurring for small a.

There is no specific GP command for integrating a function which tends to 0 doubly-exponentially: compute h and N and perform the sum. Perhaps better is to choose a reasonable N for which $|f(x)|$ is smaller than the desired accuracy for $|x| > N$, and simply compute $\int_{-N}^{N} f(x)\,dx$ using doubly-exponential integration on a compact interval as explained above.

For functions tending to 0 simply-exponentially at both ends, the GP command is `intnum(x=[-oo,1],[oo,1],f(x))`, where naturally -oo and oo code for $-\infty$ and

$+\infty$ respectively (note that oo is two lowercase o's), and the additional component ,1 in the endpoints says that the function tends to 0 simply-exponentially like $e^{-1\cdot|x|}$. Note that the constant 1 works even if the simply-exponential decrease is $e^{-c\cdot|x|}$ with $c \neq 1$, but if c is really far from 1 (say $c = 2\pi$ or $c = 0.1$), it is preferable to write explicitly [-oo,c] and [oo,c].

For functions tending to 0 polynomially at both ends, the GP command is simply intnum(x=-oo,oo,f(x)).

Nonetheless, we still give a homemade program:

| Intoo_oo.gp |

```
/* [x^a, x^(a+1), ..., x^(a+b)]; b >= 0 */
Powersab(x, a, b) = powers(x, b - 1, x^a);

/* Precomputations for numerical integration on ]-oo, oo[. */
Intoo_ooinit(fast = 0, mul = 1) =
{ my(B = getlocalbitprec(), N, h, e);
  my(vexp, vexpi, vsinh, vcosh, X, W);

  N = IntgetN(B, if (fast, 3.32, 2.213) / mul);
  localbitprec(B + 32);
  h = log(2*Pi*N / log(N)) / N;
  e = exp(h);
  vexp = Powersab(e, 1, N);
  vexpi = Powersab(1 / e, 1, N);
  vsinh = (vexp - vexpi) / 2;
  vcosh = (vexp + vexpi) / 2;
  if (fast,
    for (m = 1, N - 1,
      my(b = exponent(vcosh[m]) + B + 1);
      if (vsinh[m] > 0.347 * b + 1,
          vsinh = vsinh[1..m]; vcosh = vcosh[1..m]; break));
    return (bitprecision([h, vsinh, vcosh], B + 32)));
  vexp = vector(N, m, exp(vsinh[m]));
  vexpi = vector(N, m, 1 / vexp[m]);
  X = (vexp - vexpi) / 2;
  W = (vexp + vexpi) / 2;
  W = vector(N, m, W[m] * vcosh[m]);
  for (m = 1, N - 1,
    if (exponent(W[m]) - 2 * exponent(X[m]) < -B - 64,
        X = X[1..m]; W = W[1..m]; break));
  return (bitprecision([X, W, h], B + 32));
}

/* Numerical integration of a closure f on ]-oo, oo[, with
 * precomputed XW = [h, vabs, vwt], otherwise recomputed. */
Intoo_oo(f, fast = 0, XW = 0) =
{
  if (!XW, XW = Intoo_ooinit(fast));
```

```
39    my(B = getlocalbitprec(), h = XW[3], S);
40    S = f(0) + SumXW(x -> f(x) + f(-x), XW);
41    return (bitprecision(S * h, B));
42    }
```

3.8.5. DE over $[a, \infty[$ or $]-\infty, a]$. We now consider the problem of integrating over a half-line $[a, \infty[$ or $]-\infty, a]$. Possibly after changing x into $-x$ and/or x into $x + a$ or $a(x + 1)$, we may assume that the interval is $[0, \infty[$, so that we want to compute $\int_0^\infty f(x)\,dx$.

Once again, we distinguish three different behaviors of $f(x)$ as $x \to \infty$.

(1) In the case of functions tending to 0 doubly-exponentially, as above we choose a reasonable N such that $|f(x)|$ is less than the desired accuracy for $x > N$ and compute $\int_0^N f(x)\,dx$ using integration on a compact interval.

(2) In the case of functions tending to 0 simply-exponentially we use the change of variable $x = \phi(t) = \exp(t - \exp(-t))$:

$$\int_0^\infty f(x)\,dx \approx h \sum_{m=-N}^N f\left(e^{mh-e^{-mh}}\right)(1 + e^{-mh})e^{mh-e^{-mh}} \ .$$

(3) In the case of functions tending to 0 polynomially we use the change of variable $x = \phi(t) = \exp(a\sinh(t))$ for some $a > 0$ (typically $a = 1$ or 2):

$$\int_0^\infty f(x)\,dx \approx ah \sum_{m=-N}^N f\left(e^{a\sinh(mh)}\right)\cosh(mh)e^{a\sinh(mh)} \ .$$

Once again, some remarks concerning these changes of variable: in the case of functions tending to 0 simply exponentially, note that the summand tends to 0 doubly exponentially for $m \to +\infty$ because of the behavior of f (essentially $f(\exp(mh)))$, while for $m \to -\infty$ it tends to 0 doubly exponentially because of ϕ' (essentially $\exp(-\exp(-mh)))$. In the case of functions tending to 0 polynomially the same is true since, as in the case of integration on $]-\infty, +\infty[$, the convergence of the integral ensures that $f(\exp(a\sinh(mh)))\exp(a\sinh(mh))$ tends to 0 doubly exponentially for $m \to +\infty$, and for $m \to -\infty$ the factor $\exp(a\sinh(mh))$ ensures doubly-exponential behavior.

The GP syntax for the last two cases is naturally
    `intnum(x=0,[oo,1],f(x))` and `intnum(x=0,oo,f(x))`
respectively.

As a final warning, note that natural functions may be indirectly the sum of a function which tends exponentially to 0 and one which tends polynomially. A typical example is

$$J = \int_1^\infty \left(\frac{1 + e^{-x}}{x}\right)^2 dx \ .$$

Since the function tends polynomially to 0 at infinity, we should write
    `intnum(x=0,oo,((1+exp(-x))/x)^2)`.
However this will generate an error because the software will try to compute $\exp(-x)$ for large values of x. The solution is to split the integrand into its slow part $1/x^2$ and its exponential part $(2e^{-x} + e^{-2x})/x^2$, and write $J = J1 + J2$, with
    `J1=intnum(x=1,oo,1/x^2)` and

J2=intnum(x=1,[oo,1],(2*exp(-x)+exp(-2*x))/x^2).

Since we will need in an essential way integration over $[a, \infty[$ for numerical summation, we write explicitly the corresponding programs:

--------------------------------- Inta_oo.gp ---------------------------------

```
1   /* Precomputations for numerical integration on [a, oo[
2    * Behaviour of f at infinity:
3    *   fast > 0: exponential decrease as exp(-fast*x)
4    *   -2 < fast < -1: slow decrease, as x^fast
5    *   fast <= -2, fast = 0, or omitted: as x^(-2) or better */
6   Inta_ooinit(fast = 0, mul = 1) =
7   {
8     if (fast > 0, return (Inta_oofastinit(mul)));
9     if (fast < 0 && fast >= -1, error("impossible fast"));
10    if (fast <= -2, fast = 0);
11    return (Inta_ooslowinit(fast, mul));
12  }
13
14  Inta_ooslowinit(fast = 0, mul = 1) =
15  {
16    my(B, N, h, b, e1, q, Xp, Wp, Xm, Wm);
17    b = if (fast, -2 / (fast + 1), 2); \\ > 0
18    B = getlocalbitprec(); N = IntgetN(B, 3.49 / mul);
19    localbitprec(B + 32); h = log(2*Pi*N/log(N)) / N;
20    Xp = vector(N); Xm = vector(N);
21    Wp = vector(N); Wm = vector(N); e1 = exp(h); q = 1;
22    for (m = 1, N,
23      my(t);
24      q *= e1; t = (q + 1 / q)/2; \\ exp(mh), cosh(mh)
25      Xp[m] = exp(b * (q - t));
26      Xm[m] = 1 / Xp[m];
27      Wp[m] = Xp[m] * t;
28      Wm[m] = Xm[m] * t;
29      if (exponent(Xm[m]) < -B * b / 2 - 64,
30        Xp = Xp[1..m]; Wp = Wp[1..m];
31        Xm = Xm[1..m]; Wm = Wm[1..m]; break));
32    return (bitprecision([b * h, Xp, Wp, Xm, Wm], B + 32));
33  }
34
35  Inta_oofastinit(mul = 1) =
36  { my(B = getlocalbitprec(), N, h, e1, vexph, Xp, Xm, Wp, Wm);
37
38    N = IntgetN(B, 5.85 / mul);
39    localbitprec(B + 32); h = log(2*Pi*N/log(N)) / N;
40    e1 = exp(h); vexph = powers(e1, N-1, e1);
41    Xp = vector(N); Wp = vector(N);
42    Xm = vector(N); Wm = vector(N);
43    for (m = 1, N,
```

```
44      my(vm = vexph[m], vmi = 1 / vm);
45      Xp[m] = exp(m * h - vmi);
46      Xm[m] = exp(-(m * h + vmi));
47      Wp[m] = Xp[m] * (1 + vmi);
48      Wm[m] = Xm[m] * (1 + vm);
49      if (exponent(Xm[m]) < -B - 64,
50        my(b = exponent(Wp[m]) + B);
51        if (Xp[m] > 0.694 * b + 1,
52          Xp = Xp[1..m]; Wp = Wp[1..m];
53          Xm = Xm[1..m]; Wm = Wm[1..m]); break));
54    return (bitprecision([h, Xp, Wp, Xm, Wm], B + 32));
55  }
56
57  /* auxiliary sum */
58  Inta_oosum(f, a, Xp, Wp, Xm, Wm) =
59    sum(m = 1, #Xp, Wp[m] * f(a+Xp[m]) + Wm[m] * f(a+Xm[m]));
60
61  /* Numerical integration of a closure f on [a, oo[, with
62   * precomputed XW from Inta_ooinit, otherwise recomputed. */
63  Inta_oo(f, a = 1, fast = 0, XW = 0) =
64  {
65    if (fast > 0, return (Inta_oofast(f, a, fast, XW)));
66    if (!XW, XW = Inta_ooinit(fast));
67    my(S, [bh, Xp, Wp, Xm, Wm] = XW);
68    if (a <= 1, S = f(a + 1);
69            , S = f(2*a); Xp *= a; Xm *= a; bh *= a);
70    return (bh * (S + Inta_oosum(f, a, Xp, Wp, Xm, Wm)));
71  }
72
73  /* Assume f goes to 0 like exp(-fast * x). */
74  Inta_oofast(f, a = 1, fast = 1, XW = 0) =
75  {
76    if (!XW, XW = Inta_oofastinit());
77    my(S, [h, Xp, Wp, Xm, Wm] = XW, e1 = exp(-1) / fast);
78    if (fast != 1, Xp /= fast; Xm /= fast);
79    if (a <= 1, S = f(a + e1);
80            , S = f(a * (1 + e1)); Xp *= a; Xm *= a; h *= a);
81    S = 2 * e1 * S + Inta_oosum(f, a, Xp, Wp, Xm, Wm) / fast;
82    return (h * S);
83  }
84
85  /* Assume f goes to 0 like exp(-fast * x^{1/2}). */
86  Inta_oohalffast(f, a = 1, fast = 1, XW = 0) =
87  {
88    if (!XW, XW = Inta_oofastinit());
89    2 * Inta_oofast(x -> x * f(x^2), sqrt(a), fast, XW);
90  }
```

REMARKS 3.8.6.

(1) In the built-in function `intnum`, an integral such as $\int_a^\infty f(t)\,dt$ is simply treated as $\int_0^\infty f(a+t)\,dt$, which is mathematically correct, but induces errors when a is very large. Thus, as in the above program, when $a > 1$ (arbitrary but reasonable cut-off), we prefer to write the integral as the mathematically equal quantity $a\int_1^\infty f(at)\,dt$.

(2) The constants 3.49 and 5.85 in the two initialization procedure have been obtained experimentally: using test integrals, we have increased these constants so as to have the fastest possible program, while keeping almost perfect accuracy (we allow an error of three decimal digits).

(3) As for the built-in `intnum` program, `Inta_oofast` (or `Inta_oo` with `fast > 0`) handles integrands which behave like $\exp(-C \cdot x)$ as $x \to \infty$. We have added the program `Inta_oohalffast` which is a trivial modification and handles integrands behaving like $\exp(-C \cdot x^{1/2})$ since we will need them later in this book.

3.8.6. Contour integration and double integrals. It is important to note that doubly-exponential integration methods can be used in a natural manner on other domains than intervals of $\mathbb{R}$. An important case is that of *contour integration*. For instance, if we want to compute the contour integral of some complex function $f(z)$ on the circle C_R centered at the origin of radius R and divided by $2\pi i$, since

$$\frac{1}{2\pi i}\int_{C_R} f(z)\,dz = \frac{R}{2\pi}\int_{-\pi}^{\pi} f(Re^{it})e^{it}\,dt\ ,$$

we simply give as arguments a function f and positive real R:

```
cont(f,R) = R/(2*Pi) * intnum(t=-Pi,Pi, f(R*exp(I*t)) * exp(I*t));
```

For instance, setting `f = x -> x/(1+x^2)`, we see that `cont(f, 1/2)` is close to 0, while `cont(f, 3/2)` is close to 1, which gives a vivid numerical illustration of the residue theorem. Similarly, `cont(zeta, 1/2)` gives 0 while `cont(zeta, 3/2)` gives 1, because of the pole at $s = 1$ of $\zeta(s)$. As usual, it is preferable to give R as an exact rational number, in case our black box function computes at a higher accuracy than the current precision (for `intnum`, this is in fact the case).

Another possible use is for multiple integrals, although this is usually very slow. For instance, to integrate on the inside of a rectangle $[a,b] \times [c,d]$ one can write `Inta_b(x->Inta_b(y->f(x,y),c,d),a,b)`. However this requires the abscissas and weights of the inner integral to be recomputed each time, so instead we use `Inta_binit()` to precalculate this table, and write instead

```
tab = Inta_binit();
Inta_b(x -> Inta_b(y -> f(x, y), c, d, tab), a, b, tab)
```

For instance at 500 decimals, if $f(x,y) = xy$, without initializing the table the double integral requires more than 5 minutes, while it requires about 16 seconds (including the initialization) if we initialize the table (note that the Gaussian integration program `IntGaussLegendre` requires only 0.30 seconds after a 0.18 second initialization).

Of course, there is no reason to restrict to integration on a rectangle. For instance, to compute the Petersson square of the Ramanujan Δ function, we can write

```
? \p 115
? tab = Inta_oofastinit();
? f(x, y) = norm(eta(x + y * I, 1))^24 * y^10;
? g(x) = Inta_oofast(y -> f(x, y), sqrt(1 - x^2), 1, tab);
? Inta_b(x -> g(x), -1/2, 1/2)
```

which takes 12 seconds to give the result $1.035362\cdots 10^{-6}$ to 115 decimal digits. Note that there are infinitely faster methods to compute this quantity, we only give this as an example.

As a final remark, note that for double integrals one should try inasmuch as possible to use Stokes's theorem, to reduce them to curvilinear integrals which are much faster to compute. For instance, computing Petersson scalar products as above can always be reduced to a curvilinear integral, see [CS17], but this goes too far beyond the purpose of this book.

3.9. Integration of oscillatory functions

We now consider the case where the integrand f is an oscillating function, typically a trigonometric function, a Bessel function, a product of those, or a function involving fractional parts, all of these possibly multiplied by a regular nonoscillating factor. There is of course no specific problem when one integrates over a compact interval, and otherwise, to isolate the difficulty, we split an integral on $]-\infty, +\infty[$ as a sum of integrals on $]-\infty, 0]$ and $[0, \infty[$, and we may therefore always assume that we want to compute an integral on $[0, \infty[$.

Warning: some programs will require a period (or quasi-period in the case of Bessel functions): we will always denote it by T. Usually (but not always) T can also be a multiple of the period. For instance, in the case of $\sin(x)$, we have $T = 2\pi$, but usually one can also choose $T = 4\pi$, etc. The other programs require a half-period, which we will therefore always denote by H. Here it is essential that H be really a half-period, although in some cases H can be an *odd* multiple: thus for $\sin(x)$ we have $H = \pi$, but in some cases one can also take $H = 3\pi$ for instance.

3.9.1. The DE for periodic functions. We first note that if the function to integrate tends to 0 exponentially fast or even faster, the oscillating nature of the integrand will pose no problem to the DE method since the Euler–Maclaurin sum will be cut-off at some $m = \pm N$ where the value of the function is negligible. Thus in that case we can ignore the oscillating nature of the function and use the change of variable $\phi(t) = \exp(t - \exp(-t))$ given above. Typically:

```
? \p500
? Inta_oo(x -> exp(-x) * sin(x), 0, 1) - 1/2
% = -1.5 ... E-504   /* Perfect. */
```

On the other hand, if the function tends to 0 polynomially, or even worse, does not tend to 0 at all, the standard change of variable fails. For instance,

```
? \p500
? Inta_oo(x -> sin(x)^2 / x^2, 0) - Pi/2
% = -0.000348... /* Completely wrong */
```

It is therefore necessary to use a specific change of variable. We shall use the function $\varphi_K(t) = t/(1 - \exp(-K\sinh(t)))$ where $K > 0$ is a real parameter. It clearly satisfies the following:

$$\varphi_K(t) \sim \begin{cases} |t| \exp\left(-\frac{K}{2}\exp|t|\right), & t \to -\infty, \\ \frac{1}{K}, & t \to 0, \\ t + t\exp\left(-\frac{K}{2}\exp t\right), & t \to \infty, \end{cases}$$

and

$$\varphi'_K(t) \sim \begin{cases} \frac{K}{2}|t| \exp\left(|t| - \frac{K}{2}\exp|t|\right), & t \to -\infty, \\ \frac{1}{2}, & t \to 0, \\ 1, & t \to \infty. \end{cases}$$

Let us first consider the special case of a sine integral $\int_0^\infty f(x)\sin(x)\,dx$ for some regular function f. Let N and h be chosen as usual (i.e., N close to $B\log(B)$ and h close to $\log(bN)/N$, where B is the number of desired bits of accuracy), set $M = \pi/h$ and $x = \phi(t) := M\varphi_K(t)$. Then

$$\int_0^\infty f(x)\sin(x)\,dx = \int_{-\infty}^\infty f(\phi(t))\sin(\phi(t))\phi'(t)\,dt$$

$$\approx h\sum_{m=-N}^N f(\phi(mh))\sin(\phi(mh))\phi'(mh)\,.$$

Why can we truncate the sum? We first note that when $t \to -\infty$, the functions $\phi(t)$ and $\phi'(t)$ tend to 0 doubly-exponentially, so there is no problem for negative m. Second, when $t \to +\infty$ both $z(t) = \phi(t) - Mt$ and $\phi'(t) - M$ tend to 0 doubly exponentially. It follows that

$$\sin(\phi(mh)) = \sin(z(mh) + m\pi) = (-1)^m \sin(z(mh)) \sim (-1)^m z(mh)$$

will also tend to 0 doubly-exponentially.

REMARK 3.9.1.   (1) Note the remarkable fact that, contrary to all the preceding changes of variable, this one depends on the integration step h that we must choose beforehand.

(2) The computation of $\sin(\phi(mh))$ for large m suffers from cancellation because of the argument reduction modulo 2π (or rather modulo $\pi/2$) used to compute the sine function. A serious implementation should use

$$\sin(\phi(mh)) = (-1)^m \sin(z(mh))$$

and compute $z(t)/M = \varphi_K(t) - t$ without cancellation, as $tE/(1 - E)$ where $E = \exp(-K\sinh(t))$.

A more general algorithm was introduced by Ooura and Mori in [**OM91**]. Its advantage is that it is not restricted to functions of the form $f(x)\sin(x)$, but more generally to functions $f(x)g(x)$ where $g(x)$ vanishes periodically (it needs not be a periodic function) and which does not cross the real axis tangentially, forbidding functions like $\sin^k(x)$ or $\cos^k(x)$ with $k \geq 2$ (it is *not* applicable to *quasi-periodic* functions such as Bessel functions). The algorithm uses the same DE transformation and is based on the following result, which we give without proof:

LEMMA 3.9.2. *Let F be an analytic function in a fixed neighborhood of the real half-line $[0,\infty[$ such that there exist $H > 0$ and C two real constants such that*

$F(H(n+C)) = 0$ *for sufficiently large integers* n. *Let* $h > 0$ *be a mesh size and* $M := H/h$. *After the variable transformation* $x = \phi(t) = M\varphi_K(t)$, *we have*

$$\int_0^\infty F(x)\,dx = \int_{-\infty}^\infty F\big(\phi(t)\big)\phi'(t)\,dt$$

$$\approx h \sum_{m=-N}^{N} F\left(\phi\big((m+C)h\big)\right)\phi'\big((m+C)h\big).$$

For large positive m, *the summand is close to* $F(H(m+C)) = 0$; *for large negative* m, *the factor* $\phi'((m+C)h)$ *makes it tend to* 0. *For* $N \approx_K B\log B$ *and* $h = \log(2\pi N/\log(N))/N$, *the error in the above approximation is bounded by* 2^{-B}.

A possible program is as follows:

──────────────────── IntoscOouraMori.gp ────────────────────

```
 1   /* int_a^oo F(x) dx where F = fg, and g periodic such
 2    * that g(H * (n + C)) = 0 for n >> 1. */
 3   IntoscOouraMoriinit(H, C = 0) =
 4   { my(B = getlocalbitprec(), K = 28/10);
 5     my(N, h, M, X, W, ehpow, t);
 6
 7     localbitprec(32);
 8     N = ceil(B*log(B) / 1.5);
 9     h = log(2*N*Pi / log(N)) / N;
10     localbitprec(B + 32); h = bitprecision(h, B + 32);
11     M = H / h; t = (C - N) * h;
12     ehpow = powers(exp(h), 2 * N, exp(t));
13     X = vector(2 * N + 1); W = vector(2 * N + 1);
14     for (n = 1, 2 * N + 1,
15       my(D, E, phi, phip);
16       if (t, /* et = exp(t) */
17         my(et = ehpow[n], emt = 1 / et);
18         my(sht = (et - emt) / 2, cht = (et + emt) / 2);
19         E = exp(-K * sht); D = 1 - E;
20         phi  = t / D; phip = (1 - K * cht * phi * E) / D;
21       ,
22         phi  = 1 / K; phip = 1 / 2);
23       t += h;
24       X[n]  = M * phi; W[n] = H * phip);
25     return ([X, W]);
26   }
27
28   IntoscOouraMori(f, H, C = 0, XW = 0) =
29   {
30     if (!XW, XW = IntoscOouraMoriinit(H, C));
31     return (SumXW(f, XW));
32   }
```

REMARKS 3.9.3.

(1) We use the letter H (for Half) instead of T since H will almost always be *half* of the period of the oscillating part. Note that here we do not need to give it to a larger accuracy. Also, contrary to Intoscalt below, we must give H itself, and not an odd multiple.

(2) One can assume that $0 \leq C < 1$. The most important cases being $C = 0$ as in $F(x) = f(x) \cdot \sin(\pi x/H)$, and $C = 1/2$ as in $F(x) = f(x) \cdot \cos(\pi x/H)$.

(3) The "magic constants" 2.8 and 1.5 are heuristic, but seem to work quite well. They can of course be changed at will if necessary.

(4) The initialization program can be improved when $C = 0$ or $1/2$ because $\exp(-t) = 1/\exp(t)$ will already be in the ehpow table for most values of t hence needs not be computed. In fact, if $C = 0$, one can directly use $\cosh(-t) = \cosh(t)$ and $\sinh(-t) = -\sinh(t)$ and halve the number of nodes as in Inta_b.

(5) As in the sine integral, argument reduction and catastrophic cancellation may occur when computing $F(\phi(mh))$ for large m. This cannot be helped in full generality, but a more specific implementation would cater for this by exploiting periodicities and computing $\varphi_K(t) - t$ as in the above set of remarks, in particular for sine and cosine integrals.

EXAMPLES. Let us try this program on

$$\int_0^\infty \sin(x)/x \, dx = \pi/2, \qquad \int_0^\infty \cos(x)/(1+x^2) \, dx = \pi/(2e)$$

and the divergent integral

$$\int_0^\infty \sin(x) \log(x) \, dx := \lim_{\varepsilon \to 0} \int_0^\infty \exp(-\varepsilon x) \sin(x) \log(x) \, dx = -\gamma \ .$$

```
? \p200
? IntoscOouraMori(sinc, Pi) - Pi/2
time = 173 ms.
% = -3.80... E-211
? IntoscOouraMori(x -> cos(x)/(1+x^2), Pi, 1/2) - Pi/(2*exp(1))
time = 178 ms.
% = 3.56... E-212
? IntoscOouraMori(x -> sin(x)*log(x), Pi) + Euler
time = 244 ms.
% = 2.05... E-206
```

At larger accuracy, the two sine integrals remain accurately computed, but this is no longer the case for the cosine one:

```
? f(x) = cos(x) / (1 + x^2);
? \p500
? IntoscOouraMori(f, Pi, 1/2) - Pi/2 * exp(-1)
time = 1,451 ms.
% = -2.00... E-472
? \p1000
? IntoscOouraMori(f, Pi, 1/2) - Pi/2 * exp(-1)
time = 10,782 ms.
% = -3.77... E-882
```

```
? \p2000
? IntoscOouraMori(f, Pi, 1/2) - Pi/2 * exp(-1)
time = 1min, 7,797 ms.
% = -7.96... E-1654
```

Changing the magic constant K does not improve the situation. The problem is due to the pole at $\pm i$: the equation $M\varphi_K(t) = \pm i$ has a solution with $\Im(t) = O(1/\log M)$, which gets close to the real axis as the precision increases and h gets smaller, affecting the convergence rate. Ooura and Mori [**OM99**] later proposed a more robust change of variables of the form $M\psi_{a,b}(t)$ with

$$\psi_{a,b}(t) = \frac{t}{1 - \exp\left(-2t - a(1 - e^{-t}) - b(e^t - 1)\right)}, \quad 0 \le a \le b \le 1.$$

The function $\psi_{a,b}$ has the same asymptotic properties as φ_K, in particular $\psi(t) - t$ (resp. $\psi'(t)$) tends to 0 double-exponentially when $t \to \infty$ (resp. $t \to -\infty$). But complex singularities are no longer a problem: the closest solution to the real axis of the equation $M\psi_{a,b}(t) = z$ satisfies $\Im(t) = \frac{\arg z}{2} + O(1/\log M)$ provided $a = o(M \log M)^{-1/2}$. We found out experimentally that, with our heuristic choices for h and N, values of size $a \approx \log M/M^{1/2}$ perform better in practice. We obtain the following program:

_____ IntoscOouraMori2.gp _____

```
1   IntoscOouraMori2init(H, C = 0) =
2   { my(B = getlocalbitprec());
3     my(a, b, N, h, M, X, W, ehpow, t);
4
5     localbitprec(32);
6     N = ceil(B*log(B) / 1.5);
7     h = log(2*N*Pi / log(N)) / N;
8     localbitprec(B + 32); h = bitprecision(h, B + 32);
9     M = H / h; t = (C - N) * h;
10    b = 1 / 4;
11    a = bestappr(5 * log(M) / sqrt(M), 10^4);
12    ehpow = powers(exp(h), 2 * N, exp(t));
13    X = vector(2 * N + 1); W = vector(2 * N + 1);
14    for (n = 1, 2 * N + 1,
15      my(D, E, phi, phip);
16      if (t, /* et = exp(t), em = exp(-t) */
17        my(et = ehpow[n], emt = 1 / et);
18        E = exp(-2 * t - a * (1-emt) - b * (et-1)); D = 1 - E;
19        phi  = t / D;
20        phip = (1 + (-2 - a * emt - b * et) * E * phi) / D;
21      ,
22        phi  = 4 / (4*a + 9); /* specific to b = 1/4 */
23        phip = 1/2 + (8*a - 2) / (4*a + 9)^2);
24      t += h;
25      X[n] = M * phi; W[n] = H * phip);
26    return ([X, W]);
27  }
```

```
28
29   IntoscOouraMori2(f, H, C = 0, XW = 0) =
30   {
31     if (!XW, XW = IntoscOouraMori2init(H, C));
32     return (SumXW(f, XW));
33   }
```

The new program is marginally slower than the previous one, but indeed more robust in the presence of poles. On the other hand, neither function can handle $\int_0^\infty \sin^2(x)/x^2\, dx$:

```
? \p500
? IntoscOouraMori2(f, Pi, 1/2) - Pi/2 * exp(-1)
time = 1,480 ms.
% = -8.53... E-501
? \p1000
? IntoscOouraMori2(f, Pi, 1/2) - Pi/2 * exp(-1)
time = 11,006 ms.
% = -2.21... E-1000
? \p2000
? IntoscOouraMori2(f, Pi, 1/2) - Pi/2 * exp(-1)
time = 1min, 10,983 ms.
% = -2.89... E-2002
```

```
?\p200
? IntoscOouraMori(x->sinc(x)^2, Pi) - Pi/2
% = -0.0002... /* junk */
? IntoscOouraMori2(x->sinc(x)^2, Pi) - Pi/2
% = -0.0001... /* junk */
```

Of course, to estimate $\int_0^\infty \sin^2(x)/x^2\, dx$ we can use $\sin^2(x) = (1 - \cos(2x))/2$, hence after a few evident computations

$$\int_0^\infty \frac{\sin^2(x)}{x^2}\, dx = \frac{1}{2} \int_0^{\pi/4} \frac{1 - \cos(2x)}{x^2}\, dx + \frac{2}{\pi} + \int_0^\infty \frac{\sin(x)}{(x + \pi/2)^2}\, dx \ .$$

This is now handled directly for instance using

```
? S1 = Inta_b(x -> (1 - cos(2*x)) / x^2, 0, Pi / 4);
? S2 = IntoscOouraMori(x -> sin(x) / (x + Pi/2)^2, Pi);
? S1 / 2 + 2 / Pi + S2 - Pi / 2
% = -1.01... E-115 /* Perfect. */
```

This last example can be generalized: if the periodic part does not cross the real axis tangentially, cases we may be able to expand it into Fourier series and compute the sum of the corresponding integrals, assuming either that the Fourier series is finite or that the Fourier coefficients tend rapidly to 0 (this will for instance *not* be the case for the function $g(x) = \{x\}$, the fractional part of x). This corresponds to the following primitive program:

───────────────────────── Fourier.gp ─────────────────────────

```
1    /* Fourier expansion of a function g of period T to k terms.
2     * Returns [a0, C, S] with a0 the constant term, C the vector
```

```
3    * of cosine coefficients and S the sine coefficients. */
4
5    Fourier(g, T, k) =
6    { my(t = 2*Pi / T, T2 = T / 2);
7
8      return (intnum(x = -T2, T2,
9                    my(z = exp(I * t * x), e = powers(z, k-1, z));
10                   g(x) * [1, real(e), imag(e)]) / T2);
11   }
```

REMARKS 3.9.4.
(1) The **powers** call computes the vector of $\exp(ijtx)$, $1 \leq j \leq k$ using a single exponential and $k - 1$ multiplications, from which we derive the $\cos(jtx)$ and $\sin(jtx)$. This is more efficient than the direct formula involving k exponentials or $2k$ calls to the sine or cosine functions.
(2) We could of course have computed the Fourier coefficients individually instead of as a vector. The advantage of writing as we did above is that the function g is computed only once over the necessary abscissas, and not for each Fourier coefficients.

EXAMPLES.
```
? v = Fourier(t -> sin(t)^5, 2*Pi, 5); bestappr(v, 1000)
%= [0, [0, 0, 0, 0, 0], [5/8, 0, -5/16, 0, 1/16]]
? v = Fourier(t -> t/2, 2*Pi, 5); bestappr(v, 1000)
%= [0, [0, 0, 0, 0, 0], [1, -1/2, 1/3, -1/4, 1/5]]
```
The first command tells us that $\sin^5(t) = (\sin(5t) - 5\sin(3t) + 10\sin(t))/16$. The second tells us that when $t \in]-\pi, \pi[$ (which is the interval on which we compute our Fourier coefficients) the beginning of the Fourier expansion of the function equal to $t/2$ on that interval and of period 2π is

$$t/2 = \sin(t) - \sin(2t)/2 + \sin(3t)/3 - \sin(4t)/4 + \sin(5t)/5 - \cdots .$$

Using the above program, we can then write a program which can handle a few more functions than the initial IntoscOouraMori program:

_____ Intperiodic.gp _____

```
1    /* Integral from 0 to infinity of f(x)g(x) where f(x) tends
2     * to 0 smoothly and slowly at infinity and g is a periodic
3     * function of period T, given by its Fourier coefficients
4     * G in the format output by Fourier, with exact coefficients.
5     * Assume all partial Fourier integrals converge. */
6
7    Intperiodicsimple(f, G, T, XWC = 0, XWS = 0) =
8    { my(B = getlocalbitprec(), b = 15 - B, S, U, vA);
9
10     localbitprec(2 * B); U = 2 * Pi / T; localbitprec(B + 32);
11     my([a0, vC, vS] = G);
12     S = if (exponent(a0) > b, a0 * intnum(t = 0, oo, f(t)));
13     if (exponent(vC) > b,
```

```
14    my(fC = (x -> cos(U * x) * vector(#vC, n, f(x / n))));
15    if (!XWC, XWC = IntoscOouraMoriinit(T/2, 1/2));
16    vA = IntoscOouraMori(fC, T/2, 1/2, XWC);
17    S += sum(n = 1, #vC, vC[n] * vA[n] / n));
18  if (exponent(vS) > b,
19    my(fS = (x -> sin(U * x) * vector(#vS, n, f(x / n))));
20    if (!XWS, XWS = IntoscOouraMoriinit(T/2, 0));
21    vA = IntoscOouraMori(fS, T/2, 0, XWS);
22    S += sum(n = 1, #vS, vS[n] * vA[n] / n));
23  return (bitprecision(S, B));
24 }
25
26 /* Possible wrapper function. gG must be a pair [g, G] with
27  * G as above or the max nonzero Fourier coefficient. */
28 Intperiodic(f, gG, T, XWC = 0, XWS = 0) =
29 { my([g, G] = gG);
30
31   if (type(G) == "t_INT",
32     localbitprec(2 * getlocalbitprec());
33     G = Fourier(g, T, G));
34   return (Intperiodicsimple(f, G, T, XWC, XWS));
35 }
```

Some important remarks concerning these functions.

REMARKS 3.9.5.

(1) Since we perform the computation at an accuracy larger than the current accuracy, the period T must be given say at double the desired accuracy.

(2) The Intperiodicsimple program does the basic work using the Fourier expansion. However, it is limited by two things. First, the periodic function g must be given by a finite vector of Fourier coefficients (in the format explained in the Fourier program) which must be exact or computed at a larger accuracy, otherwise the program may completely fail. But more importantly, a simple integral such as the example of $\int_0^\infty \sin(x)^2 \, dx/x^2$ given above will give an error message, because $\sin(x)^2 = (1-\cos(2x))/2$ and the program will try to compute separately $\int_0^\infty dx/x^2$ and $\int_0^\infty \cos(2x) \, dx/x^2$. Thus, to compute this integral one should write it as

$$\int_0^\pi \sin(x)^2 \, dx/x^2 + \int_0^\infty \sin(x)^2 \, dx/(x+\pi)^2$$

and compute the first integral using Inta_b.

Note that for efficiency we explicitly test the possible vanishing of the Fourier coefficients, since this frequently happens, for instance if the function g is even or odd.

The main problem with this method is that it does not apply to almost periodic functions such as Bessel functions, or to functions whose Fourier expansion is infinite and converges too slowly, typically involving $\{x\}$, the fractional part of x. We will see below two methods which can handle this type of oscillating behavior.

EXAMPLE. Let us compute the integral

$$\int_0^\infty \frac{\sin^{k_1}(t)}{t^{k_2}}\, dt$$

for a few strictly positive integral values of k_1 and k_2. For the integral to converge at 0 we need $k_1 \geq k_2$, and at infinity we need either $k_2 \geq 2$ or $k_2 = 1$ and k_1 odd. Since we have a removable singularity at $t = 0$ we use the built-in function sinc and write $\sin^{k_1}(t)/t^{k_2} = \mathrm{sinc}^{k_1}(t) t^{k_1-k_2}$ to make the singularity disappear. Splitting the integral at $t = 2\pi$ for instance, we can then write:

_____ Intperiodicsine.gp _____

```
/* int_0^oo sin(t)^{k_1} / t^{k_2} dt, assuming convergence. */
Intperiodicsine(k1, k2) =
{ my(B = getlocalbitprec(), S);

  if (k2 >= k1 + 1, error("not integrable at 0"));
  if (k2 == 1 && (k1 % 2 == 0), error("not integrable at oo"));
  localbitprec(5/4 * B);
  S = Intperiodic(t -> 1 / t^k2,
                  [t -> sin(t)^k1, k1], 2*Pi, 2*Pi)
    + intnum(t = if (k1 >= k2, 0, [0, k1 - k2]),
             2*Pi, sinc(t)^k1 * t^(k1 - k2));
  return (bitprecision(S, B));
}
```

_____

We find to perfect accuracy the following results, where $*$ indicates that the integral does not converge:

$k_1 \backslash k_2$	1	2	3	4	5	6
1	$\dfrac{\pi}{2}$	$*$	$*$	$*$	$*$	$*$
2	$*$	$\dfrac{\pi}{2}$	$*$	$*$	$*$	$*$
3	$\dfrac{\pi}{4}$	$\dfrac{3}{4}\log(3)$	$\dfrac{3\pi}{8}$	$*$	$*$	$*$
4	$*$	$\dfrac{\pi}{4}$	$\log(2)$	$\dfrac{\pi}{3}$	$*$	$*$
5	$\dfrac{3\pi}{16}$	$\dfrac{5}{16}\log\left(\dfrac{3^3}{5}\right)$	$\dfrac{5\pi}{32}$	$\dfrac{5}{96}\log\left(\dfrac{5^{25}}{3^{27}}\right)$	$\dfrac{115\pi}{384}$	$*$
6	$*$	$\dfrac{3\pi}{16}$	$\dfrac{3}{16}\log\left(\dfrac{2^8}{3^3}\right)$	$\dfrac{\pi}{8}$	$\dfrac{1}{16}\log\left(\dfrac{3^{27}}{2^{32}}\right)$	$\dfrac{11\pi}{40}$

Of course there exists an explicit formula for these integrals which we leave to the reader to discover.

3.9.2. Integration of oscillatory functions: Sums. There is, however, a much more efficient way to handle this type of integral when the oscillation is very regular, which is of course the case for trigonometric functions, but also for Bessel functions. Assume that we can write the integrand $F(x)$ as $F(x) = f(x)g(x)$, where $f(x)$ is nonoscillating, and $g(x)$ is oscillating with a fixed quasi half-period, which up to homothety we may assume equal to π. We can write $\int_0^\infty F(x)\,dx = \sum_{n \geq 0} u(n)$ with $u(n) = \int_{n\pi}^{(n+1)\pi} F(x)\,dx$. If $u(n)$ alternates in sign and is regular, we may use the Sumalt program that we will study in the next chapter. The most efficient way to compute $u(n)$ is by using Gauss–Legendre integration. On the other hand, there may be a minor singularity at 0, which is better handled by DE. However, since we have not written any GP script for handling algebraic singularities, for once we use the built-in intnum program which has this capability. Also, once again, for this alternating method we use the letter H instead of T for the half-period.

──────────────────────── Intoscalt.gp ────────────────────────

```
1   /* int_a^oo F(x)dx where F = f g with f nonoscillating and g
2    * oscillating with quasi half-period H and int_{n H}^{(n+1) H}
3    * alternating. H must be given with a larger accuracy than the
4    * default. */
5
6   Intoscalt(F, H, XW = 0) =
7   { my(B = getlocalbitprec(), U);
8
9     localbitprec(B + 32);
10    if (!XW, XW = IntGaussLegendreinit());
11    U = (n -> (-1)^(n-1)*IntGaussLegendre(F, (n-1)*H, n*H, XW));
12    return (bitprecision(Sumalt(U), B));
13  }
```

Note that once again here we do not need to give the half-period H with larger accuracy than the current one. On the other hand, contrary to IntoscOouraMori, one can give an odd multiple of H instead of H itself, usually at the expense of speed and/or accuracy.

EXAMPLES.

```
? \p 38
? A(n) = Intoscalt(x -> sinc(x)^(2*n - 1), Pi);
? vector(5, n, bestappr(A(n) / Pi, 10^8))
time = 43 ms.
% = [1/2, 3/8, 115/384, 5887/23040, 259723/1146880] /* OK */
? Intoscalt(x -> besselj(0, x) / (x + 1), Pi)
time = 43 ms.
% = 0.75461002577097216866261271487098645062 /* Perfect */
? C = sqrt(3)/2*gamma(1/3);
? Intoscalt(x -> cos(x) / x^(2/3), Pi) - C
time = 21 ms.
% = -0.229... /* horrible, need to tell about singularity */
? intnum(x = [0, -2/3], Pi, cos(x) / x^(2/3)) \
    - Intoscalt(x -> cos(x) / (x + Pi)^(2/3), Pi) - C
time = 23 ms.
% =  3.52 ... E-38 /* Now perfect */
? Intoscalt(x -> besselj(0, x) * sin(4 * x), Pi) - 1 / sqrt(15)
time = 48 ms.
```

```
% = 1.46...E-38 /* perfect */
? Intoscalt(x -> besselj(0, x) * sin(3 * x), Pi) - 1 / sqrt(8)
time = 48 ms.
% = -0.0179... /* completely wrong */
? Intoscalt(x -> sinc(x)^4, Pi) - Pi / 3
time = 9 ms.
% = -4.72... E-8 /* only 8 correct decimals */
```

The reason that the last examples give wrong results is that $u(n)$ is not any more quasi-alternating, but mostly of constant sign. To handle this, we need a summation program for such series, which we will study in Chapter 4. Here we need to be excessively careful (we advise the reader to read this section only after going through all of the algorithms given in Chapter 4). The fastest programs are SumLagrange and SumMonien, but these programs are rather restrictive in their assumptions. However, they have the advantage of needing to evaluate the partial integrals $\int_n^{n+1}$ only for reasonable values of n. We thus give a program using SumLagrange, which we will take as default.

For other integrands, this will not work so we will need to use the more robust Euler–Maclaurin type summation method Sumdelta. However, in this program we will need to evaluate the partial integrals for huge values of n, typically $n = 10^{50}$ or more (this is also the case for other programs such as SumAbelPlana or Sumpos). For some integrands this is OK, for others not: for instance trigonometric functions such as $\sin(x)$ cannot be reasonably evaluated for x that large. Similarly, the fractional part $\{x\}$ will give an error message. Thus, we modify it to avoid this, but the unavoidable consequence is that it will be applicable only to *exactly* periodic functions, hence for instance not to integrals involving Bessel functions. For the latter, the use of Sumpos may work, see examples below.

────────────────────────── | Intosc.gp | ──────────────────────────

```
1   /* Integral from a to infinity of F(x) = f(x)g(x), where f(x)
2    * tends to 0 smoothly and slowly at infinity and g is a
3    * periodic function of period T, given to larger accuracy
4    * than default. */
5
6   IntoscLagrange(F, T, alf = [1, 0], XW = 0) =
7   { my(B = getlocalbitprec(), U);
8
9     localbitprec(2 * B);
10    if (!XW, XW = IntGaussLegendreinit());
11    U = (n -> IntGaussLegendre(F, (n-1) * T, n * T, XW));
12     return (bitprecision(SumLagrange(U, alf), B));
13  }
14  alias(Intosc, IntoscLagrange);
15
16  /* Use Sumpos instead */
17
18  Intoscpos(F, T, XW = 0) =
19  { my(B = getlocalbitprec(), U);
20
```

```
21    localbitprec(B + 32);
22    if (!XW, XW = IntGaussLegendreinit());
23    U = (n -> IntGaussLegendre(F, (n - 1) * T, n * T, XW));
24    return (Sumpos(U));
25  }
26
27  Intoscdelta(f, g, T, XW = 0) =
28  { my(B = getlocalbitprec(), U);
29
30    localbitprec(B + 32) ;
31    if (!XW, XW = IntGaussLegendreinit());
32    localbitprec(B);
33    U = (n -> IntGaussLegendre(x -> f(x + (n - 1) * T) * g(x),
34                               0, T, XW));
35    return (Sumdelta(U));
36  }
```

Remarks concerning the period T: here T is really the period itself of the oscillating part, *not* the half-period which would give nonsense. You may give an integral multiple of T at the possible expense of speed and/or accuracy. Note that in these programs we work in larger accuracy than the current one, so T should be given to at least twice the current accuracy.

EXAMPLES.

```
? \p115
? PI = Pi;
? \p38
? Intosc(x -> sinc(x)^4, PI) - PI / 3
time = 38 ms.
% = 0.E-38 /* Perfect */
? Intosc(x -> besselj(0, x) * sin(3 * x), PI) - 1 / sqrt(8)
time = 493 ms.
% = -0.0011... /* Still completely wrong */
? Intosc(x -> besselj(0, x) * sin(3 * x), PI, \
                          [1, n -> sqrt(n)]) - 1 / sqrt(8)
time = 493 ms.
% = 0.E-38 /* Now perfect. */
```

An additional advantage of the present method (which will also be the case for the IntoscSidi program that we give next) is that it can handle integrals involving the fractional part of x:

EXAMPLES.

```
? Intosc(t -> frac(t) / (t + 1)^2, 1) + Euler - 1
time = 9 ms.
% = 0.E-38
? Intosc(t -> (frac(t) - 1/2) / (t + 1), 1) - log(2*Pi) / 2 + 1
time = 10 ms.
% = 0.E-38
```

```
? Intosc(t -> frac(t)^2 / (t + 1)^2, 1) - (log(2*Pi) - Euler - 1)
time = 10 ms.
% = -2.93... E-39
```

On the other hand, it gives very poor results when we combine fractional parts with some nontrivial weight such as $\log(n)$:

```
? Intosc(t -> frac(t) * log(t + 1) / (t + 1)^2, 1)
% = 0.495461... /* Only 4 correct digits */
```

Note that

$$\int_1^\infty \frac{\{t\}\log(t)}{t^2} \, dt = 1 - \gamma - \gamma_1 = 0.495600180582143864254074285792\cdots \, ,$$

where $\gamma_1 = \lim_{N\to\infty} \sum_{1\le n\le N} \log(n)/n - \log^2(N)/2$ is the first Stieltjes constant.

In all the above examples, we have used the `IntoscLagrange` program. For the last example, the `Intoscdelta` program comes to the rescue: first note that all of the above examples (except by construction the ones involving Bessel functions) are treated perfectly with comparable timings by `Intoscdelta`. But so is the last one:

```
? Intoscdelta(t -> log(t + 1) / (t + 1)^2, t -> frac(t), 1)
time = 126 ms.
% = 0.49560018058214386425407428579249888810
? 1 - Euler + polcoef(zeta(1 + x + O(x^4)), 1)
% = 0.49560018058214386425407428579249888810 /* Perfect */
```

The moral is that the $\log(t)$ function confuses `SumLagrange`, but is too mild to confuse `Sumdelta`.

Finally, consider $I = \int_0^\infty J_0(x)^4 \log(x) \, dx$. The $\log(x)$ will confuse our Legendre integration program, so we write the following:

```
? F(x) = besselj(0, x)^4 * log(x);
? I0 = Inta_b(F, 0, PI);
? Intosc(x -> F(x + PI), PI) + I0
time = 502 ms.
% = -0.7675295... /* Only 5 correct decimals */
? Intoscdelta(x -> log(x + PI), \
              x -> besselj(0, x + PI)^4, PI) + I0
time = 319 ms.
% = 7.85 E60 /* Nonsense */
? Intoscpos(x -> F(x + PI), PI) + I0
time = 4,395 ms.
% = -0.76751605277047830731094587022554544210
/* Much slower but perfect. */
```

3.9.3. Integration of oscillatory functions: Extrapolation. There exist other ways to handle this type of integrals, see a series of papers by A. Sidi [**Sid82b**], [**Sid88**], and also by Lucas and Stone [**LS95**]. Their methods are similar to the above, in that they perform an extrapolation of

$$\sum_{0\le j\le n} \int_{j\pi}^{(j+1)\pi} f(x) \, dx, \quad \text{as } n \to \infty$$

using Sidi's mW algorithm. We simply compute the individual integrals by Gauss–Legendre integration and call the LimitSidi program (see §2.5). This leads to the following programs:

——————————————— IntoscSidi.gp ———————————————

```
 1  /* Computes int_0^\infty f(x)dx, where f is a regularly
 2   * oscillating function of half-period H such as a
 3   * trigonometric or Bessel function. H must be given with
 4   * a much larger accuracy than default, or left at the
 5   * default H = 1, with f replaced by f(H*x), in which case
 6   * H need not be given to higher accuracy. */
 7
 8  IntoscSidiinit(safe = 1) =
 9  { my(B = getlocalbitprec());
10
11    if (safe, B *= 3 / 2);
12    localbitprec(B + 32); IntGaussLegendreinit();
13  }
14
15  IntoscSidi(f, H = 1, XW = 0, safe = 1) =
16  {
17    if (!XW, XW = IntoscSidiinit(safe));
18    if (H != 1, f = x -> f(H * x));
19    H * SumSidi(x -> IntGaussLegendre(f, x - 1, x, XW), safe);
20  }
21
22  /* faster but less robust version, setting safe = 0 */
23  IntoscaltSidiinit() = IntoscSidiinit(0);
24  IntoscaltSidi(f, H = 1, XW = 0) = IntoscSidi(f, H, XW, 0);
```

There are two programs: the IntoscSidi program can handle both the case where $\int_{nH}^{(n+1)H}$ is alternating or not alternating. However, the IntoscaltSidi program is faster, but as its name implies, can only be used in the alternating case (otherwise you will have a warning telling you that there are only very few correct bits). Here, the (half) period H must be set analogously to the corresponding alt/pos programs: in the alternating case, H *must* be the half-period, no multiple works, otherwise H can be any multiple of the *period*, not of the half-period. As usual H should be given to a much larger accuracy than the current one, although in the alternating case this is much less important.

As a rule of thumb, if $f(x)$ has no nonoscillating part, you may usually choose IntoscaltSidi. This is the case for instance for $f(x) = g(x)J_0(x)^a \sin(x)^b$ where $g(x)$ is nonoscillating and $a+b$ is *odd*, or if the arguments have different half-periods such as $f(x) = J_0(x)J_0(2x)$, but *not* $f(x) = J_0(x)J_0(3x)$.

Otherwise (including when f is not oscillating at all), you must choose IntoscSidi. Note that in some situations, the program will not give the desired accuracy and a warning will be issued.

EXAMPLES.

```
? A(n) = IntoscaltSidi(x -> sinc(x)^(2*n - 1), PI);
? vector(5, n, bestappr(A(n) / PI, 10^8))
time = 37 ms.
% = [1/2, 3/8, 115/384, 5887/23040, 259723/1146880] /* OK */
? IntoscaltSidi(x -> besselj(0, x) / (x + 1), PI)
time = 58 ms.
% = 0.75461002577097216866261271487098645064 /* perfect */
? IntoscaltSidi(x -> besselj(0, x + 1) / sqrt(x + 1), PI)
time = 63 ms.
% = 0.18869299996431065572003245457785275732 /* perfect */
? IntoscaltSidi(x -> besselj(0, x)^3 * log(x), PI)
time = 114 ms.
% = -0.95850... /* wrong, need to tell about singularity */
? f(x) = besselj(0, x)^3 * log(x);
? Inta_b(f, 0, PI) + IntoscaltSidi(x -> f(x + PI), PI)
time = 120 ms.
% = -0.95971344386790399703194110596480040529 /* perfect */
? IntoscaltSidi(x -> besselj(0, x) * besselj(1, x) * sin(x), PI)
time = 119 ms.
% = 0.34322012515458754117440939375195151453 /* perfect */
? IntoscaltSidi(x -> besselj(0, x) * sin(2 * x), PI) - 1 / sqrt(3)
time = 63 ms.
% = 0.E-38
? IntoscaltSidi(x -> besselj(0, x) * besselj(1, 2 * x), PI) - 1 / 2
time = 116 ms.
% = 0.E-38
? IntoscaltSidi(x -> besselj(0, x) * sin(4 * x), PI) - 1 / sqrt(15)
time = 45 ms.
% = 0.E-38
? IntoscaltSidi(x -> besselj(0, x) * sin(3 * x), PI) - 1 / sqrt(8)
  ***   user warning: reached accuracy of 121 bits.
time = 107 ms.
% = -1.26... E-36 /* 2 decimals lost. */
? IntoscSidi(x -> besselj(0, x) * sin(3 * x), PI) - 1 / sqrt(8)
time = 163 ms.
% = -2.93... E-39 /* Slower but perfect. */
? vector(10, n, bestappr(IntoscSidi(x -> sinc(x)^n, PI) / PI, 10^8))
time = 127 ms.
% = [1/2, 1/2, 3/8, 1/3, 115/384, 11/40, 5887/23040, 151/630,\
    259723/1146880, 15619/72576]
```

As the Intosc program, the IntoscSidi program (but not the IntoscaltSidi program) can also integrate functions involving the fractional part of x, and almost twice faster:

EXAMPLE.

```
? IntoscSidi(t -> frac(t)^2 / (t + 1)^2, 1) - (log(2*Pi)-Euler-1)
time = 6 ms.
```

```
% = -2.93... E-39 /* Perfect */
```

However, worse than the `Intosc` program, it cannot handle at all combinations of fractional parts and logs:

```
? IntoscSidi(t -> frac(t) * log(t + 1) / (t + 1)^2, 1)
  ***   user warning: reached accuracy of 19 bits.
time = 29 ms.
% = 0.495668744...
```

As a last stupid example, note that this program also works (but is much slower) for nonoscillating functions:

```
? IntoscSidi(x -> 1 / (x^2 + 1)) - Pi / 2
time = 1 ms.
% = 0.E-38
? Inta_oo(x -> 1 / (x^2 + 1), 0) - Pi / 2
time = 1 ms.
% = 0.E-38
```

To finish on this method, we once again emphasize the need to have the half-period (here π) given to, say, three times the current accuracy:

```
? \p115
? PI = Pi;
? \p 38
? F(x) = besselj(0, x) * sin(3 * x);
? IntoscSidi(F, Pi) - 1 / sqrt(8)
reached accuracy of 77 bits.
time = 51 ms.
% = -2.09... E-23
? IntoscSidi(F, PI) - 1 / sqrt(8)
time = 160 ms.
% = -2.93... E-39
? Pi * IntoscSidi(x -> F(Pi * x), 1) - 1 / sqrt(8)
time = 163 ms.
% = -2.93... E-39
```

3.9.4. Summary of the possible programs. We have seen four programs for the integration of oscillating functions. First, the DE program, in particular the `IntoscOouraMori` version, leading to the `Intperiodic` program. This program is very fast, but has a severe limitation: it is applicable only if the quasi-periodic part is exactly periodic, with a Fourier expansion having only a reasonably small number of non-negligible coefficients. This excludes functions such as Bessel functions (which are not exactly periodic), and functions involving fractional parts, whose Fourier expansions converge very slowly.

Second, the programs based on simple summation programs: either `Sumalt` or `SumLagrange`. These are an order of magnitude slower than the DE based program, but can treat quasi-periodic functions such as Bessel functions, and in favorable circumstances also functions involving fractional parts.

Third, the program based on more robust but slower summation program, the `Sumoscposdelta` program. This is much slower, and cannot handle quasi-periodic functions such as Bessel functions, but is the only program which treats correctly functions involving fractional parts and logarithms for instance.

Fourth, the `IntoscSidi` and `IntoscaltSidi` programs based on extrapolation. Although much slower than the DE program when it can be applied, it is much more robust and applicable to almost all of our examples (see below). The only ones on which it fails is when there is a combination of fractional parts and logarithms. It is almost certain that a modification of this program could be written to handle this specific case.

One last possibility that we have not yet mentioned is the use of double integrals. For instance, if k is an integer and we want to compute $I = \int_a^\infty f(x) J_k(x)\, dx$, we can use the integral representation of J_k which gives

$$I = \frac{1}{\pi} \int_0^\pi dt \int_a^\infty f(x) \cos(x \sin(t) - kt)\, dx \ ,$$

and the inner integral can be computed using one of the previous methods such as `IntoscOouriMora`. This often gives perfect results, but is orders of magnitude slower than the methods able to deal with Bessel functions since computing a double integral is slow (a few seconds instead of a few hundredth of a second). Nonetheless, this idea should be kept in mind in the case of oscillating functions which are more complicated than Bessel functions and for which the methods seen up to now do not apply.

A final important remark: most methods use some sort of Gaussian integration on compact intervals such as `IntGaussLegendre`. We have seen that in the presence of poles these methods may lose considerable accuracy (see for instance the `IntGaussLegendresplit` program). In this case, it is essential to use some other method such as the DE for the integral from 0 to the period T, and use the methods for oscillating integrals only on $[T, \infty[$ (assuming of course that T is sufficiently large with respect to the possible poles, otherwise we simply choose a larger multiple of T). In other words, instead of `Intoscxxx(f, T)`, we write

```
B = getlocalbitprec(); localbitprec(2 * B); I0 = Inta_b(f, 0, T);
localbitprec(B); return (I0 + Intoscxxx(x -> f(x + T), T))
```

This is what will be done in the timings given below, where for a fair comparison we do not take into account the time for `Inta_b(f, 0, T)`.

3.10. Sample timings for integrals on $[a, b]$

The reader will have noted that we have explained in great detail the DE method, but that we have not given modifications of the other methods to deal for instance with singularities or noncompact intervals. Indeed, the DE seems to be the method of choice in the latter cases, and in fact often also over compact intervals.

Thus, to make useful comparisons, we give the following timings: first, we compare most methods that we have given above for compact intervals. Since $[a, b]$ corresponds to a simple scaling, we always assume that our interval is $[0, 1]$, and that the function does not have any singularity (which, as for DE, can usually be removed by simple changes of variable). Then we give timings for DE in other cases: singularities, noncompact intervals with functions tending to zero rapidly or not.

We consider the following sample integrals:

$$I_1 = \int_0^1 dx/(x+1)\,, \quad I_2 = \int_0^1 dx/(x^2+1)\,, \quad I_3 = \int_0^1 dx/(x^3+x+1)\,,$$

$$I_4 = \int_0^1 x^7 dx/(x^9+x+1)\,, \quad I_5 = \int_0^1 dx/(x+1)^{4/3}\,,$$

$$I_6 = \int_0^1 dx/\sqrt{x^3+x+1}\,, \quad I_7 = \int_0^1 dx/((x+1)\sqrt{x+1})\,,$$

$$I_8 = \int_0^1 dx/(x^4-2x^3+5x^2-4x+1)\,, \quad I_9 = \int_0^1 dx/(x^\pi+x^{1.4}+1)\,,$$

$$I_{10} = \int_0^1 \sin(\pi x)\,dx\,, \quad I_{11} = \int_0^1 (\sin(\pi x)/x)\,dx\,,$$

$$I_{12} = \int_0^1 (\sin(\pi x)/(x+1))\,dx\,, \quad I_{13} = \int_0^1 \tan(x)\,dx\,,$$

$$I_{14} = \int_0^1 \exp(x)\,dx\,, \quad I_{15} = \int_0^1 ((\exp(x)-1)/x)\,dx$$

$$I_{16} = \int_0^1 (x/(\exp(x)-1))\,dx\,, \quad I_{17} = \int_0^1 \log(\Gamma(1+x))\,dx\,,$$

$$I_{18} = \int_0^1 (\log(x+1)/x)\,dx\,, \quad I_{19} = \int_0^1 (\log(x+1)/(x+1))\,dx$$

$$I_{20} = \int_0^1 dx/\log(2+x)\,, \quad I_{21} = \int_0^1 \zeta(x+2)\,dx\,, \quad I_{22} = \int_0^1 \zeta(x+i)\,dx\,.$$

Concerning the above, note that $x^4-2x^3+5x^2-4x+1$ has no real roots and the real part of all its roots is equal to $1/2$.

As we have done in the case of extrapolation methods, we give the timings for $D = 500$ decimal digits, other tables are available from the authors. We recall that the time is given in seconds, where ∞ means more than 5 minutes, $\otimes$ that the program produced an error or nonsense and $*$ if the program cannot handle the integral. If L decimals are lost, this number is given in parentheses.

Note that to avoid stupid errors such as division by 0, we have not written for instance `log(x+1)/x` but `if(!x,1,log(x+1)/x)`. In a proper integration routine, to avoid removable singularities such as this one, one would instead replace $\log(x+1)/x$ by a truncated power series expansion when x is small enough, but we have not done so.

The first row labeled "NFE" gives the Number of Function Evaluations for the method. Usually (but not always) the lower NFE is the faster the method is, but in some cases the method gives wrong answers.

The second row labeled "Init" gives the time for initialization of the integration method (when such an initialization exists).

I	Newton	Chebyshev	Lagrange	Legendre	DE
NFE	1155	1153	43467	417	3163
Init	7.38	1.73	0.00	0.17	0.09
I_1	**0.00**	**0.00**	0.04	**0.00**	0.01
I_2	**0.00**	**0.00**	0.06 (76)	**0.00**	0.01

I_3	0.01	**0.00**	0.08 (80)	**0.00**	0.01
I_4	**0.01**	**0.01**	0.17 (270)	**0.01**	0.02 (97)
I_5	0.02	**0.01**	0.63	**0.01**	0.03
I_6	0.02	0.01	0.20 (79)	**0.00**	0.02
I_7	0.01	0.01	0.15	**0.00**	0.01
I_8	0.03	0.04	$\otimes$	**0.02**	$\otimes$
I_9	$\otimes$	$\otimes$	$\otimes$	$\otimes$	**0.23**
I_{10}	0.13	0.05	2.90	**0.02**	0.07
I_{11}	0.13	0.05	2.93	**0.02**	0.07
I_{12}	0.13	0.05	2.92	**0.02**	0.07
I_{13}	0.26	0.11	3.03 (116)	**0.04**	0.11
I_{14}	0.08	0.06	2.05	**0.02**	0.12
I_{15}	0.08	0.06	2.11	**0.02**	0.13
I_{16}	0.09	0.06	2.15	**0.02**	0.13
I_{17}	2.04	0.73	43.6	**0.26**	1.47
I_{18}	0.09	0.05	2.41	**0.02**	0.06
I_{19}	0.09	0.05	2.40	**0.02**	0.06
I_{20}	0.09	0.05	2.36	**0.02**	0.10
I_{21}	19.7	6.41	∞	**2.32**	17.7
I_{22}	42.6	14.1	∞	**5.34**	41.1

Timings for 500 Decimals.

In some cases (mainly I_4, I_8, and I_9) it is necessary to specify the radius of convergence r, so in those cases the number of function evaluations (NFE) which is given is smaller than the actual value used when $r < 1$.

3.10.1. Conclusion for integration on a compact interval. The sample integrals that we have chosen may not be completely representative (recall that on purpose we choose integrands with no singularity close to the line of integration, and that for now we integrate only on compact intervals), but we can already draw a number of conclusions.

(1) Note that the programs IntGaussLegendre and Inta_b are available in GP under the names intnumgauss and intnum respectively, but we give only the timings of the scripts given in this book. In any case, the speedup would be marginal.

(2) When it can be applied, the fastest method by far is the classical Gaussian integration method IntGaussLegendre. By construction, it cannot handle the artificial integral I_9, it fails completely on $I_8 = \int_0^1 dx/(x^4 - 2x^3 + 5x^2 - 4x + 1)$, unless we give it the correct radius (here $r = 0.133$), we have a slight loss of accuracy for $I_4 = \int_0^1 (x^7/(x^9 + x + 1))\, dx$ (again suppressed by giving $r = 0.39$).

The difference in speed is most notable for the difficult integrals I_{17}, I_{21}, and I_{22} (which involve functions which are slower to compute such as $\log(\Gamma(x))$ and $\zeta(x)$). This is of course due to the much smaller number of function evaluations.

(3) The "primitive" Newton–Cotes methods (or improved using not equally spaced Chebyshev nodes) are still quite usable even at very large accuracies, but are in general not competitive with respect to other methods. For instance, `IntChebyshev` is almost twice slower than `IntGaussLegendre`.

(4) The artificial integral $I_9 = \int_0^1 dx/(x^\pi + x^{1.4} + 1)$ is treated correctly only by the double-exponential program `Inta_b`. Note that it is the mixture of the exponents π and 1.4 which creates the problem, since for instance $\int_0^1 dx/(x+1)^\pi$ is computed perfectly with `IntLagrange`.

(5) As could be expected, extrapolation methods such as `IntLagrange` are not at all competitive.

(6) Nonetheless, the built-in `intnum` program or the homemade `Inta_b` program are very sturdy and handle almost all cases perfectly, with 25% accuracy loss for I_4, and I_8 which is computed wrongly. Note that this is highly dependent on the specific implementation. In fact, both the speed and accuracy of the `Inta_b` program are entirely controlled by the unique instruction `IntgetN(B, 6.18 / mul)`. Decreasing the default value `mul = 1` would increase the speed, but we would lose accuracy on quite simple integrals. On the other hand, increasing it to 2 for instance, would make the program twice slower, but would then enable a perfect computation of I_4, and even enable the correct computation of I_8.

We have also given two methods to perform general Gaussian integration (a priori on a compact interval, but in fact there is no reason to restrict to that case). The first method is based on the computation of the orthogonal polynomials directly by a Gram–Schmidt orthogonalization. The second computes the orthogonal polynomials using continued fractions.

Both methods have three steps: in the first step, if not already given, the moments are computed. In the second step, one computes the orthogonal polynomials. In the final step, one computes the roots of one of these orthogonal polynomials, and the corresponding weights. The methods seem comparable, although the continued fraction method is more elegant and is apparently more stable, so we have preferred using it in our scripts.

To summarize this section:

- If the integral can be correctly handled by Gaussian integration, which can easily be checked, use `IntGaussLegendre` or the built-in GP function `intnumgauss`.
- If you need a variant of Gaussian integration with weights, use the continued fraction method to compute the necessary nodes and weights.
- Otherwise, use either the built-in `intnum` or the `Inta_b` programs.
- When the integrand has poles near the interval of integration, it is usually essential to use a preliminary treatment such as that performed by the `Intgen` program above.

3.11. Sample timings for integrals on $[0, \infty]$

From what we have seen above, to compute $\int_0^\infty f(x)\,dx$ we have four different methods. If f tends to 0 slowly at ∞, we can try the `IntGaussa_oo` program. If f tends to 0 like e^{-x}, i.e., if $f(x) = e^{-x}g(x)$ for some "reasonable" function g, we can apply the Gauss–Laguerre integration program `IntGaussLaguerre` (of course if $f(x) = e^{-ax}g(x)$ we simply replace x by x/a in the integral). Similarly, if $f(x) = e^{-x^2}g(x)$ (or more generally $f(x) = e^{-ax^2}g(x)$), in the same way we apply the Gauss–Hermite method for which we have not given an explicit program, at least if we are dealing with $\int_{-\infty}^{\infty} f(x)\,dx$ or if the function f is even. Finally, we have at our disposal the DE, which is quite generally applicable if the function f is not oscillating, and where we specify the rate of decrease at infinity. Concerning this last method, we have the built-in `intnum` program, as well as the home-made `Inta_oo` program.

So as to be able to compare the different methods, when using functions f tending to 0 at infinity like e^{-x^2}, we will assume that they are even, so that for all integrals we can restrict to integrating on $[0, \infty[$. In any case, we can write $f(x) = f^+(x) + f^-(x)$ with $f^+(x) = (f(x) + f(-x))/2$ and $f^-(x) = (f(x) - f(-x))/2$, and since f^- is odd and f^+ even, $\int_{-\infty}^{\infty} f(x)\,dx = 2\int_0^{\infty} f^+(x)\,dx$, so there is no loss of generality in restricting to even functions.

We will thus choose test functions of two types. First, functions tending to 0 slowly at ∞, typically like $1/x^a$ with $a > 1$. Second, functions tending to 0 like e^{-x}. We could also include a third type of functions tending to 0 as e^{-x^2}, but since these are less frequent and in general much easier to compute, we will omit them (in addition, we can also make the change of variable $y = x^2$). Of course, there are many other types of functions, but we choose those first to be able to make a fair comparison, and second because many integrals can be reduced to those by changes of variable.

3.11.1. Functions tending to 0 slowly. We will first consider the following sample integrals which tend to 0 slowly at infinity. Thus, we have two possible programs: `IntGaussa_oo` and `Inta_oo`. Since there can be a considerable loss of accuracy for `IntGaussa_oo` when the integrand has poles close to the real axis (this having nothing to do with the behavior at infinity), for a few integrands we give both the integral from 0 to ∞ and from 2 to ∞. This will not change any timings, nor any loss of accuracy, except for `IntGaussa_oo` for which the result will now be perfect.

$$J_{s,1} = \int_0^\infty \frac{dx}{(x+1)^3} \ , \quad J_{s,2} = \int_0^\infty \frac{dx}{x^2 + 2x + 2} \ ,$$

$$J_{s,3} = \int_0^\infty \frac{dx}{x^2 + 1} \ , \quad J_{s,4} = \int_0^\infty \frac{x^2}{x^4 + 2x^3 + 3x^2 + 2x + 2}\,dx \ ,$$

$$J'_{s,3} = \int_2^\infty \frac{dx}{x^2 + 1} \ , \quad J'_{s,4} = \int_2^\infty \frac{x^2}{x^4 + 2x^3 + 3x^2 + 2x + 2}\,dx \ ,$$

$$J_{s,5} = \int_0^\infty \frac{dx}{(x+1)^{5/2}} \ , \quad J_{s,6} = \int_0^\infty \frac{dx}{(x+1)^\pi} \ ,$$

$$J_{s,7} = \int_0^\infty \log(1 + 1/(x+1)^2)\,dx \ , \quad J_{s,8} = \int_0^\infty \frac{\log(1 + 1/(x+1))}{x+1}\,dx \ ,$$

$$J_{s,9} = \int_0^\infty \log(1 + 1/(x+1))^2\,dx \ , \quad J_{s,10} = \int_0^\infty \frac{\log(2+x)}{(2+x)^2}\,dx \ ,$$

$$J_{s,11} = \int_0^\infty \frac{dx}{(2+x)^2 \log(2+x)} \ , \quad J_{s,12} = \int_0^\infty \sin(1/(x+1)^2)\,dx \ ,$$

$$J_{s,13} = \int_0^\infty x^3 \log(1 + 1/(x+1)^5)\,dx\,, \quad J_{s,14} = \int_0^\infty \tan(1/(x+1)^2)\,dx\,,$$

$$J'_{s,13} = \int_2^\infty x^3 \log(1 + 1/(x+1)^5)\,dx\,, \quad J'_{s,14} = \int_2^\infty \tan(1/(x+1)^2)\,dx\,,$$

$$J_{s,15} = \int_0^\infty \log(\Gamma(1 + 1/(x+1)^2))\,dx\,.$$

J	IntGaussa_oo	Inta_oo
NFE	416	5597
Init	0.17	0.18
$J_{s,1}$	**0.00**	0.02
$J_{s,2}$	**0.00**	0.02
$J_{s,3}$	0.00 (182)	**0.02**
$J'_{s,3}$	**0.00**	0.02
$J_{s,4}$	0.00 (181)	**0.03**
$J'_{s,4}$	**0.00**	0.04
$J_{s,5}$	**0.01**	0.03
$J_{s,6}$	**0.08**	0.41
$J_{s,7}$	**0.02**	0.12
$J_{s,8}$	**0.02**	0.13
$J_{s,9}$	**0.02**	0.13
$J_{s,10}$	$\otimes$	**0.18**
$J_{s,11}$	$\otimes$	**0.18**
$J_{s,12}$	**0.02**	0.24
$J_{s,13}$	0.02 (113)	**0.12**
$J'_{s,13}$	**0.02**	0.20
$J_{s,14}$	0.02 (150)	**0.25**
$J'_{s,14}$	**0.02**	0.23
$J_{s,15}$	**0.27**	3.24

Timings for 500 Decimals.

Note that all the integrals have been computed with the default parameters $(\alpha, \beta) = (1, 2)$, with the exception of $J_{s,5}$, with $(\alpha, \beta) = (1, 5/2)$ and $J_{s,6}$ with $(\alpha, \beta) = (1, \pi)$ with π computed at double the desired accuracy. This entails an artificially long initialization time, 7.15s and 9.04s respectively. Moreover, in both cases, using directly IntGaussa_oo from 0 to infinity, gives completely wrong results ($\otimes$) because the pole at -1 is too close to the interval of integration. Instead, we used the generalized form IntGaussa_oosplit, which splits the integral into a compact part (handled by IntGaussLegendresplit) and a remainder part far away from the pole (handled by IntGaussa_oo). The timings reported include only the time spent in the latter two functions. The needed call to IntGaussLegendreinit adds 1.81s to the initialization time, i.e, the total initialization time for $J_{s,5}$ is 8.96s. and that for $J_{s,6}$ is 10.85s.

3.11.2. Conclusion for integrals tending to 0 slowly at ∞. The conclusion for this type of integrals is identical to that of the compact case: if the integral can be correctly handled by Gaussian integration, use IntGaussa_oo, possibly by adapting the family of orthogonal polynomials to be used (this is the case for $J_{s,10}$ and $J_{s,11}$), or by a preliminary treatment to avoid the poles (here simply noted in the modified integrals $J'_{s,i}$ for $i = 3$, 4, 13, and 14). Otherwise, use one of the doubly-exponential programs `Inta_oo` or the built-in `intnum`.

3.11.3. Functions tending to 0 like e^{-x}. We now consider the following sample integrals which tend to 0 at infinity like e^{-x}. Thus, we have at least three possible programs: `IntGaussLaguerre` (Gauss–Laguerre), `IntGaussLaguerreinv` (Gaussian integration on $[a, \infty[$ for polynomials in $1/x$) and `Inta_oo` with `fast=1` (doubly-exponential). Gaussian integration programs suffer from a large loss of accuracy and we integrate from 10 to ∞ to reduce this effect. Of course it is trivial to add the integral, e.g., from 1 to 10 computed for instance using Gauss–Legendre. We consider the following integrals which tend to 0 at infinity like e^{-x}.

$$J_{f,1} = \int_{10}^{\infty} x^3 e^{-x}\, dx\ ,\quad J_{f,2} = \int_{10}^{\infty} \frac{e^{-x}}{1+x}\, dx\ ,\quad J_{f,3} = \int_{10}^{\infty} \frac{e^{-x}}{x+10}\, dx\ ,$$

$$J_{f,4} = \int_{10}^{\infty} \frac{e^{-x}}{x+1/10}\, dx\ ,\quad J_{f,5} = \int_{10}^{\infty} e^{-x} \log(x+1)\, dx\ ,$$

$$J_{f,6} = \int_{10}^{\infty} \frac{e^{-x}}{\log(2+x)}\, dx\ ,\quad J_{f,7} = \int_{10}^{\infty} e^{-x} \sin(x)\, dx\ ,\quad J_{f,8} = \int_{10}^{\infty} \frac{\sin(x)}{e^x+1}\, dx\ ,$$

$$J_{f,9} = \int_{10}^{\infty} e^{-x+1/(x+1)}\, dx\ ,\quad J_{f,10} = \int_{10}^{\infty} \log(1+e^{-x})\, dx\ ,$$

$$J_{f,11} = \int_{10}^{\infty} \frac{x}{e^x-1}\, dx\ ,\quad J_{f,12} = \int_{10}^{\infty} \frac{x^2}{e^x-1}\, dx\ ,\quad J_{f,13} = \int_{10}^{\infty} \frac{dx}{e^x+1}\ ,$$

$$J_{f,14} = \int_{10}^{\infty} e^{-x} \tan(1/(x+1))\, dx\ ,\quad J_{f,15} = \int_{10}^{\infty} \sqrt{x+1}\, K_0(x+1)\, dx\ ,$$

$$J_{f,16} = \int_{10}^{\infty} e^{-x/2} \sqrt{x+1}\, K_0(x/2+1)\, dx\ ,\quad J_{f,17} = \int_{10}^{\infty} \frac{\Gamma(x+1)}{(x+1)^{x+1/2}}\, dx\ ,$$

$$J_{f,18} = \int_{10}^{\infty} e^{-x} J_0(x)\, dx\ ,\quad J_{f,19} = \int_{10}^{\infty} \frac{e^{-x}}{\sqrt{x+1}}\, dx\ ,\quad J_{f,20} = \int_{10}^{\infty} \frac{e^{-x}}{\sqrt{x+10}}\, dx\ ,$$

$$J_{f,21} = \int_{10}^{\infty} \frac{e^{-x}}{\sqrt{x+1/10}}\, dx\ ,\quad J_{f,22} = \int_{10}^{\infty} K_0(x+1)\, dx\ ,$$

$$J_{f,23} = \int_{10}^{\infty} e^{-x/2} K_0(x/2+1)\, dx\ ,\quad J_{f,24} = \int_{10}^{\infty} \frac{\Gamma(x+1)}{(x+1)^{x+1}}\, dx\ .$$

For the integrals $J_{f,18}$ to $J_{f,24}$, which tend to infinity like $x^{-1/2} e^{-x}$ times an asymptotic series in $1/x$, we have of course chosen $\alpha = -1/2$ in the program `IntGaussLaguerreinv`.

J	Laguerre	Laguerreinv	Inta_oofast
NFE	832	218	3541
Init	9.84	5.79	0.20
$J_{f,1}$	**0.00**	$\otimes$	0.23
$J_{f,2}$	0.00 (335)	**0.00**	0.23
$J_{f,3}$	0.00 (281)	0.00 (128)	**0.24**
$J_{f,4}$	0.00 (341)	**0.00**	0.24

$J_{f,5}$	0.03 (334)	0.01 (368)	**0.39**
$J_{f,6}$	0.04 (335)	0.01 (366)	**0.39**
$J_{f,7}$	**0.04**	$\otimes$	0.43
$J_{f,8}$	0.09 (341)	$\otimes$	**0.43**
$J_{f,9}$	0.04 (337)	**0.01**	0.24
$J_{f,10}$	0.05 (339)	0.02 (354)	**0.33**
$J_{f,11}$	0.05 (334)	0.01 (372)	**0.23**
$J_{f,12}$	0.05 (337)	$\otimes$	**0.24**
$J_{f,13}$	0.05 (341)	0.01 (354)	**0.23**
$J_{f,14}$	0.03 (339)	**0.01**	0.42
$J_{f,15}$	3.18 (333)	0.24 (241)	**6.29**
$J_{f,16}$	3.87 (332)	0.19 (275)	**6.37**
$J_{f,17}$	0.52 (334)	0.16 (196)	**2.78**
$J_{f,18}$	**0.86**	$\otimes$	2.21
$J_{f,19}$	0.00 (334)	**0.00**	0.24
$J_{f,20}$	0.00 (280)	0.00 (131)	**0.24**
$J_{f,21}$	0.00 (341)	**0.00**	0.24
$J_{f,22}$	3.18 (333)	$\otimes$	**6.28**
$J_{f,23}$	3.87 (326)	$\otimes$	**6.36**
$J_{f,24}$	0.52 (334)	$\otimes$	**2.77**

Timings for 500 Decimals.

As usual, apart from the loss of accuracy, the Gaussian integration programs require much fewer function evaluations, hence are considerably faster.

3.11.4. Conclusion for integrals tending to 0 as e^{-x}. The conclusion here is more delicate. The doubly-exponential programs are sturdy and will compute perfectly all integrals. However, the two Gaussian integration programs IntGauss-Laguerre and IntGaussLaguerreinv are much faster *when they can be applied*, and there lies a slight difficulty.

Ordinary Gauss–Laguerre integration IntGaussLaguerre will be applicable when the integrand is of the form $x^\alpha e^{-x} f(x)$ where $f(x)$ can be well approximated by polynomials of regularly increasing degree on $[0, \infty[$ or power series with infinite radius of convergence. This restricts considerably the type of integrals, and in our chosen examples this corresponds to $J_{f,1} = \int_{10}^\infty x^3 e^{-x}\, dx$, $J_{f,7} = \int_{10}^\infty e^{-x} \sin(x)\, dx$, and $J_{f,18} = \int_{10}^\infty e^{-x} J_0(x)\, dx$.

On the other hand, the IntGaussLaguerreinv program is applicable when the integrand behaves at infinity like $x^\alpha e^{-x} f(x)$, where $f(x)$ has a power series expansion in $1/x$ with a radius of convergence which is not too small, so that $f(1/x)$ can be well approximated by polynomials on $[0, 1]$. In our samples, this corresponds to $J_{f,2} = \int_{10}^\infty (e^{-x}/(1 + x))\, dx$, $J_{f,4} = \int_{10}^\infty \dfrac{e^{-x}}{x + 1/10}\, dx$, $J_{f,9} = \int_{10}^\infty e^{-x+1/(x+1)}\, dx$, $J_{f,14} = \int_{10}^\infty e^{-x} \tan(1/(x + 1))\, dx$, $J_{f,19} = \int_{10}^\infty e^{-x}/\sqrt{1 + x}\, dx$ and $J_{f,21} = \int_{10}^\infty \dfrac{e^{-x}}{\sqrt{x + 1/10}}\, dx$.

As can be seen in the table, it can also give a partially correct answer on some other integrals.

3.12. Sample timings for oscillatory integrals

We consider the following integrals:

$$O_1 = \int_0^\infty dx/(1+x^2) , \quad O_2 = \int_0^\infty \exp(-x)\cos(x)\,dx ,$$

$$O_3 = \int_0^\infty x\sin(x)/(1+x^2)\,dx , \quad O_4 = \int_0^\infty \cos(x)/(1+x^2)\,dx ,$$

$$O_5 = \int_0^\infty \log(4+x^2)\cos(x)/(1+x^2)\,dx , \quad O_6 = \int_0^\infty \sin(x)/x\,dx ,$$

$$O_7 = \int_0^\infty \sin^2(x)/x^2\,dx , \quad O_8 = \int_0^\infty \sin^3(x)/x^3\,dx ,$$

$$O_9 = \int_0^\infty \sin^4(x)/x^4\,dx , \quad O_{10} = \int_0^\infty \cos(x)/\sqrt{x+1}\,dx ,$$

$$O_{11} = \int_0^\infty \sin(x)\log(x+1)\,dx , \quad O_{12} = \int_0^\infty J_0(x)/(x+1)\,dx ,$$

$$O_{13} = \int_0^\infty J_0(x)^4\log(x+1)\,dx , \quad O_{14} = \int_0^\infty J_0(x)^3\log(x+1)\,dx ,$$

$$O_{15} = \int_0^\infty J_0(x)\sin(4x)\,dx , \quad O_{16} = \int_0^\infty J_0(x)\sin(3x)\,dx ,$$

$$O_{17} = \int_0^\infty J_0(x)J_1(x)\sin(x)\,dx , \quad O_{18} = \int_0^\infty J_0(x)J_1(2x)\,dx ,$$

$$O_{19} = \int_1^\infty (\{x\}-1/2)/x\,dx , \quad O_{20} = \int_1^\infty \{x\}^2/x^2\,dx ,$$

$$O_{21} = \int_1^\infty (\{x\}-1/2)\log(x)/x\,dx , \quad O_{22} = \int_1^\infty \{x\}\log(x)/x^2\,dx .$$

REMARKS 3.12.1.

(1) O_1 is a nonoscillating integral that can be naturally evaluated by any method. Similarly, as mentioned above, O_2 would be perfectly evaluated using the DE. Both are given to see the behavior of oscillating methods on such integrals.

(2) Note that O_{11} is a divergent integral, but will nonetheless be correctly evaluated by most of our programs (and equal to $-\gamma$).

(3) Using Proposition 4.2.36 from the next chapter, it is easy to show that

$$O_{19} = \log(2\pi)/2 - 1 , \quad O_{20} = \log(2\pi) - \gamma - 1 , \quad O_{22} = 1 - \gamma_1 - \gamma , \text{ and}$$

$$O_{21} = 1 + \frac{\zeta''(0)}{2} = 1 + \frac{\gamma^2}{4} + \frac{\gamma_1}{2} - \frac{\pi^2}{48} - \frac{\log(2\pi)^2}{4} , \quad \text{where}$$

$$\gamma_1 = \lim_{N\to\infty}\left(\sum_{1\le n\le N} \frac{\log(n)}{n} - \frac{\log^2(N)}{2}\right) = -0.07281584\cdots$$

is the first Stieltjes constant.

O	periodic	alt/Lagr	delta	Sidi	pos
NFE	3347	17056/48960	143416	9880/19304	3470896
Init	0.03	0.01/0.02	0.01	0.02	0.01
O_1	**0.01**	0.09	0.22	0.03	4.2
O_2^*	0.02	0.26	$\otimes$	**0.01**	0.26
O_3^*	**0.01**	0.13	1.04	0.18	40.7
O_4^*	**0.01**	0.13	1.02	0.18	9.51
O_5^*	**0.03**	0.30	2.41	0.69	21.1
O_6^*	**0.01**	0.14	1.01	0.18	40.1
O_7	**0.02**	0.87	1.04	0.20	40.7
O_8^*	**0.02**	0.13	1.07	0.18	10.0
O_9	**0.02**	0.87	1.07	0.19	10.0
O_{10}^*	**0.02**	0.13	1.07 (45)	0.18	*
O_{11}^*	**0.03**	0.29	*	0.71	*
O_{12}^*	*	**1.80**	*	3.27	64.1
O_{13}	*	*	*	*	**151.**
O_{14}^*	*	**1.97**	*	6.24	90.1
O_{15}^*	*	**1.91**	*	2.40	*
O_{16}	*	12.6	*	**5.07**	*
O_{17}^*	*	**3.72**	*	6.72	239.
O_{18}^*	*	**3.20**	*	6.26	208.
O_{19}	*	*	**0.26**	0.44	4.34
O_{20}	*	*	**0.26**	0.44	4.56
O_{21}	*	*	**1.75**	*	45.5
O_{22}	*	*	**1.74**	*	45.6

Timings for 115 Decimals.

The * in the first column indicates that we have an alternating type oscillating integral, so that we can use the faster and more robust functions Intoscalt and IntoscaltSidi.

In the alt/Lagr column, the two NFE correspond to the number of function evaluations for Intoscalt and Intosc respectively (in the delta column they are the same), and similarly in the Sidi column they correspond to the cases $m = 0$ (IntoscaltSidi) and $m = 1$ (IntoscSidi).

Note that, although restricted to exactly periodic functions, and comparatively rather slow, Intoscdelta is the only program among the first four which is able to treat the integrals O_{21} and O_{22} involving both a fractional part and a logarithm (Intoscpos can also treat them, but an order of magnitude slower).

On the other hand, no method among the first four given in the table is able (at least directly without any modification) to evaluate $O_{13} = \int_0^\infty J_0(x)^4 \log(x+1)\,dx$, and only the Intoscpos program will evaluate this integral perfectly, but very slowly since at 115 decimals it already requires 151 seconds. This is the main reason for which we have included this program in the timings, since otherwise it is at least an order of magnitude slower.

The conclusion is quite clear: when the oscillating part is *exactly* periodic (thus excluding Bessel functions), and the Fourier expansion is short (thus excluding functions involving fractional parts), the best method is `Intperiodic`.

When the oscillating part is only approximately periodic, such as Bessel functions, or with long Fourier expansions, `IntoscSidi` is robust and usually faster than `Intoscalt`, and much faster than `Intosc`, and in the rare cases where it is slower, it is not much slower, hence it should be preferred.

3.13. Final conclusion on numerical integration

Whether integrating on a compact interval, or a nonoscillating function on an unbounded interval, the conclusion is the same:

(1) If you want a no nonsense program which will compute reliably and quite fast your integral, use the DE: either the homemade programs or the built-in `intnum`. However, be careful to treat correctly the singularities and the proximity of the poles of the integrand, see the detailed caveats given above.

(2) If speed is crucial (for instance if you must compute millions of integrals, or a double integral), then consider seriously using one form or another of *Gaussian integration*, possibly by modifying the basic programs that we have given to be more adapted to your needs (i.e., by changing the orthogonal polynomials using the continued fraction method).

To emphasize this last point, consider the computation of

$$I = \int_0^1 \int_0^1 e^{xy}\, dx\, dy$$

at 500 decimal digits. Using the DE program `Inta_b` (after the initialization step `vv=Inta_binit()` which requires 0.1 seconds), the two command

`Inta_b(x -> Inta_b(y -> exp(x*y), 0, 1, vv), 0, 1, vv)`

(of course equivalent to `Inta_b(x -> expm1(x) / x, 0, 1)`) requires approximately 5 minutes. But if we use instead `IntGaussLegendre` (after the initialization step `vv = IntGaussLegendreinit()` which requires 0.18 seconds), replacing `Inta_b` by `IntGaussLegendre` now requires only 8.6 seconds, while still giving a perfect result. Similarly, for $\log(\Gamma(1 + xy))$, at 115D the computation requires 19.4 seconds with `Inta_b` but only 0.73 seconds with Gaussian integration.

One last important point needs to be made for integration on unbounded intervals. In all of our tests, we have assumed that the integrand $f(x)$ tends to 0 regularly when $x \to \infty$. We have also considered the case of regularly oscillating functions, and explained which method to choose.

But even when f decreases monotonically and regularly, the methods that we have given do not always apply directly. For instance, the integral $I = \int_0^\infty e^{-\sqrt{x}}\, dx$ will not be computed correctly with any of the methods that we have given. In this precise situation, the best is evidently to make the change of variable $y = \sqrt{x}$, so that $I = 2 \int_0^\infty e^{-y} y\, dy$, which can now be computed trivially (and is equal to 2).

CHAPTER 4

Numerical summation

4.1. Introduction

4.1.1. Aim of this chapter. The problem that we study in this chapter is the following: let f be a well-behaved function on some interval $[a, \infty[$ with $a \in \mathbb{Z}$, either defined only on integers, or for all real numbers in that interval. We want to compute numerically the sum of the infinite series $S = \sum_{n \geq a} f(n)$, and variants such as $S = \sum_{n \geq a} (-1)^n f(n)$, assuming of course that they converge (although for alternating series we will see that it is almost always possible to make sense of sums of divergent series). As usual we want hundreds of decimals.

It will be crucial to make a difference between functions f defined only on integers, and the others: the choice of methods in the former case is evidently more limited. In the case of *alternating* series however, this difference is much less important since we will see that for all practical purposes the best method is the `Sumalt` function, which only uses values of f at integral arguments.

As for numerical integration, there are many methods for numerical summation. We will distinguish three types of methods: those based on extrapolation, those based on the Euler–Maclaurin summation formula, and those based on Gaussian summation. All three types have their advantages/disadvantages (although, as for integration, the Gaussian methods are by far the fastest, when applicable), so we will study the last two types in detail. Indeed, extrapolation has been studied in detail in Chapter 2, and we recall from that chapter that we already have the `SumLagrange` summation program and its variants at our disposal, so we will not study this method anymore, but simply give the corresponding trivial program in Section 4.6 devoted to summation using extrapolation.

4.1.2. Important remarks.

(1) In view of series such as $\zeta(s) = \sum_{n \geq 1} n^{-s}$, we will choose as default lower summation index $n = 1$ (for integrals $\int_a^\infty$ we chose instead $a = 0$ as default). In many cases we do not even let the user specify a if he needs to compute $S = \sum_{n \geq a} u(n)$, but this is of course not necessary since he can use either $S = \sum_{n \geq 1} u(n+a-1)$ or $S = \sum_{n \geq 1} u(n) - \sum_{1 \leq n < a} u(n)$. Note, however, that one must be careful: first, evidently to use the second formula it is of course necessary that $u(n)$ be defined for $1 \leq n < a$. But to use the *first* formula may be dangerous: for instance, if the remainder of the summation of $u(n)$ for $n \leq N$ has a nice expansion of the form $a_0/N + a_1/N^2 + \cdots$, that of $u(n + a - 1)$ will have an expansion of the form $a_0/N + (a_1 - a + 1)/N^2 + \cdots$, which may be bad if a is large (in particular larger than N). The situation is even worse if the asymptotic

expansion of the remainder is of the form $a_0/N^2 + a_1/N^4 + \cdots$, which is not preserved by the change of variable $N \mapsto N + a - 1$.

(2) In many of the summation methods for summands of constant sign, it is necessary to compute an integral of the type $I(N) = \int_N^\infty f(x)\,dx$. At the expense of slightly complicating the scripts, we allow three possibilities for specifying the way in which this integral is computed. The first and simplest is when an antiderivative F of f is known. In that case we choose the one which vanishes at infinity, and then $I(N) = -F(N)$. Otherwise, we need to compute the integral using numerical methods. As we have seen in Chapter 3, there are essentially two types of methods. The most robust is the DE method, which we will choose by default by setting $F = 0$ or F to the parameter `fast` of that method. But we can also choose Gaussian methods, which are much faster but more fragile. For this, we set $F = [\alpha, \beta]$ (by default $[1, 2]$), the parameters of these methods. The command `F = Sumintinit(f, F)` which is systematically used in the scripts does exactly what one needs, by transforming a parameter `F` (a closure, 0, `fast`, or `[al, be]`) into the antiderivative `F` which vanishes at infinity.

(3) To help the programs as much as possible, it is useful (and sometimes essential) to give them the asymptotic behavior at infinity of the summand $f(n)$. We will always assume that it can be written in the form $f(n) = w(n) \sum_{i \geq 1} a_i/n^{\alpha i}$ for some $\alpha > 0$ and where $w(n)$ is some regular function. This can be given to the program as follows: omitted ($w(n) = 1$, $\alpha = 1$), α alone ($w(n) = 1$), w as a closure ($\alpha = 1$), or the pair (α, w). In the special case of Monien summation, if $w(n)$ is given as a closure it will also be necessary to add a parameter a such that $w(n)$ behaves like $n^{a+\varepsilon}$ for all $\varepsilon > 0$.

In addition, the `SumLagrange` program does not need the asymptotic behavior of $f(n)$ itself but of the *remainder* sequence $R(n) = \sum_{m>n} f(m)$: we make the reasonable assumption that $R(n) = nw(n) \sum_{i \geq 1} b_i/n^{\alpha i}$ for some other coefficients b_i.

4.2. Euler–Maclaurin summation methods

We thus begin by a thorough study of Euler–Maclaurin type methods. Note that Euler–Maclaurin is *very* classical: it is as old as Taylor's formula, and there is for instance a full chapter in Bourbaki devoted to it (in fact called "Développements tayloriens généralisés"). Nonetheless, it is so useful that it deserves a fairly long description. We refer to [**Bou51**] and Section 9.2 of [**Coh07b**] for even more details.

4.2.1. Bernoulli polynomials and Bernoulli numbers. We introduce Bernoulli polynomials in the following way:

PROPOSITION 4.2.1. *There exists a unique polynomial $B_n(X)$ such that*

$$\int_x^{x+1} B_n(t)\,dt = x^n.$$

PROOF. Denote by $\mathbb{C}_n[X]$ the $n+1$-dimensional vector space of polynomials of degree less than or equal to n. It is clear that the map ϕ defined by $\phi(P)(x) = \int_x^{x+1} P(t)\,dt$ maps $\mathbb{C}_n[X]$ into itself. If $\phi(P) = 0$, by differentiating we obtain

$P(x + 1) - P(x) = 0$, hence that P is a constant, necessarily equal to 0 since $\phi(P) = 0$. The map is therefore injective, hence is one-to-one, and this implies in particular the proposition. $\square$

The polynomials $B_n(X)$ are called the Bernoulli polynomials.

COROLLARY 4.2.2. *We have* $B_0(X) = 1$, $B_n(X + 1) - B_n(X) = nX^{n-1}$, $\int_0^1 B_n(x)\,dx = 0$ *for* $n \geq 1$, *and* $B_n'(X) = nB_{n-1}(X)$.

PROOF. The first and third formula follow immediately from the proposition, as does the second by differentiating. Since $\int_x^{x+1} B_n'(x)\,dx = B_n(x+1) - B_n(x) = nx^{n-1}$, the last formula follows by uniqueness. $\square$

COROLLARY 4.2.3.

(1) *We have* $B_1(X) = X - 1/2$ *and more generally the recursion*

$$B_n(X) = X^n - \frac{1}{n+1} \sum_{0 \leq m < n} \binom{n+1}{m} B_m(X) .$$

(2) *Equivalently, we have the formal power series expansion*

$$\frac{Te^{TX}}{e^T - 1} = \sum_{n \geq 0} \frac{B_n(X)}{n!} T^n .$$

PROOF. (1) is clear by uniqueness since the integral from x to $x + 1$ of the left-hand side is equal to x^n, while that of the right-hand side is equal to $S/(n+1)$ with

$$S = (x + 1)^{n+1} - x^{n+1} - \sum_{0 \leq m < n} \binom{n+1}{m} x^m = (n+1)x^n$$

by the binomial theorem. For (2) we note that

$$(e^T - 1) \sum_{m \geq 0} (B_m(X)/m!)T^m = \sum_{n \geq 1} T^n/n! \sum_{m \geq 0} (B_m(X)/m!)T^m$$

$$= \sum_{N \geq 1} T^N \sum_{0 \leq m < N} B_m(X)/(m!(N-m)!)$$

$$= \sum_{N \geq 1} (T^N/N!) \sum_{0 \leq m < N} \binom{N}{m} B_m(X) = \sum_{N \geq 1} (T^N/N!)NX^{N-1}$$

by (1), and this last sum is the derivative with respect to X of

$$\sum_{N \geq 1} (T^N/N!)X^N = e^{TX} - 1 ,$$

proving the corollary. $\square$

COROLLARY 4.2.4.

(1) *We have* $B_n(1) = B_n(0)$ *unless* $n = 1$, *in which case* $B_n(0) = -1/2$ *and* $B_n(1) = 1/2$.
(2) *Set* $B_n = B_n(0)$. *We have*

$$B_n(X) = \sum_{0 \leq m \leq n} \binom{n}{m} B_m X^{n-m} .$$

(3) *We have* $B_1 = -1/2$ *and* $B_{2m+1} = 0$ *if* $m \geq 1$.

PROOF. Follows immediately from the above corollaries, and left to the reader. Note that (3) follows from the not completely evident fact that $T/(e^T - 1) + T/2$ is an even function of T. □

The rational numbers $B_n = B_n(0)$ are called the Bernoulli numbers. Note their first few values $B_0 = 1$, $B_1 = -1/2$, $B_2 = 1/6$, $B_4 = -1/30$, $B_6 = 1/42$, $B_8 = -1/30$, $B_{10} = 5/66$, $B_{12} = -691/2730$, $B_{14} = 7/6$.

To be able to give rigorous error terms in formulas involving Bernoulli numbers and polynomials, we need to give bounds for them. These follow from the following important results:

PROPOSITION 4.2.5. *For $k \in \mathbb{Z}_{\geq 1}$ we have*

$$\sum_{n \geq 1} \frac{\cos(2\pi n x)}{n^{2k}} = (-1)^{k-1} \frac{B_{2k}(\{x\})(2\pi)^{2k}}{2 \cdot (2k)!} \ ,$$

where $\{t\} = t - \lfloor t \rfloor$ denotes the fractional part of t. In particular, we have

$$\zeta(2k) = \sum_{n \geq 1} \frac{1}{n^{2k}} = (-1)^{k-1} \frac{B_{2k}(2\pi)^{2k}}{2 \cdot (2k)!} \ ,$$

and for $k \geq 2$ we have $|B_{2k}(\{x\})| \leq |B_{2k}| < 2.17 \cdot (2k)!(2\pi)^{-2k}$.

PROOF. (Sketch.) The function $B_{2k}(\{x\})$ is periodic of period 1 and piecewise C^∞ and continuous since $B_{2k}(1) = B_{2k}(0)$, so it is everywhere equal to the sum of its Fourier series, and a small computation using the properties of the Bernoulli polynomials given above leads to the identities of the proposition. The first inequality of the last formula follows simply from $|\cos(2\pi n x)| \leq 1$ and the second from $2\zeta(2k) \leq 2\zeta(4) = \pi^4/45 < 2.17$ for $k \geq 2$. □

Although not needed in the sequel, note that it is easy to show that for $k \geq 2$ we have $|B_{2k-1}(\{x\})| \leq 2 \cdot (2k-1)!(2\pi)^{-(2k-1)}$.

4.2.2. The basic Euler–Maclaurin formula.

LEMMA 4.2.6. *Let f be a C^∞ function on $[0,1]$. For all $k \geq 1$ we have*

$$\frac{f(0) + f(1)}{2} = \int_0^1 f(t)\, dt + \sum_{1 \leq j \leq \lfloor k/2 \rfloor} \frac{B_{2j}}{(2j)!} \big(f^{(2j-1)}(1) - f^{(2j-1)}(0)\big)$$

$$+ \frac{(-1)^{k-1}}{k!} \int_0^1 f^{(k)}(t) B_k(t)\, dt \ .$$

PROOF. Set

$$u_k = \frac{(-1)^k}{k!} \int_0^1 f^{(k)}(t) B_k(t)\, dt \ .$$

By integration by parts and using $B_k'(X) = k B_{k-1}(X)$ we have

$$(-1)^k k! u_k = \big(f^{(k-1)}(t) B_k(t)\big)\Big|_0^1 - \int_0^1 f^{(k-1)}(t) B_k'(t)\, dt$$

$$= f^{(k-1)}(1) B_k(1) - f^{(k-1)}(0) B_k(0) + (-1)^k k! u_{k-1} \ ,$$

giving the recursion

$$u_k = u_{k-1} + (-1)^k \big(f^{(k-1)}(1) B_k(1) - f^{(k-1)}(0) B_k(0)\big)/k! \ .$$

Using the properties of B_k proved above, we deduce that

$$u_k = u_0 - \big(f(1)B_1(1) - f(0)B_1(0)\big)$$
$$+ \sum_{2 \leq j \leq k} (-1)^j \frac{f^{(j-1)}(1)B_j(1) - f^{(j-1)}(0)B_j(0)}{j!}$$
$$= \int_0^1 f(t)\,dt - \frac{f(1) + f(0)}{2} + \sum_{2 \leq j \leq k} \frac{(-1)^j}{j!} B_j\big(f^{(j-1)}(1) - f^{(j-1)}(0)\big) ,$$

proving the lemma since for $j \geq 2$ we have $B_j = 0$ unless $j = 2j'$ is even. $\qquad\square$

PROPOSITION 4.2.7 (Euler–Maclaurin, first form). *Let a be an integer and f be a C^∞ function on $[a, \infty[$. For all $k \geq 1$ and $N \geq a$ we have*

$$\sum_{a \leq m < N} f(m) = \int_a^N f(t)\,dt + \frac{f(a) - f(N)}{2}$$
$$+ \sum_{1 \leq j \leq \lfloor k/2 \rfloor} \frac{B_{2j}}{(2j)!} \big(f^{(2j-1)}(N) - f^{(2j-1)}(a)\big)$$
$$+ \frac{(-1)^{k-1}}{k!} \int_a^N f^{(k)}(t) B_k(\{t\})\,dt ,$$

where $\{t\} = t - \lfloor t \rfloor$ denotes the fractional part of t.

PROOF. Simply apply the lemma with $f(x)$ replaced by $f(x+m)$, and sum for $a \leq m < N$. The (immediate) details are left to the reader. $\qquad\square$

COROLLARY 4.2.8 (Euler–Maclaurin, second form). *Keep the same assumptions, and assume in addition that $\int_a^\infty |f^{(k)}(t)|\,dt$ converges.*

(1) *There exists a constant $z_k(f; a)$ such that*

$$\sum_{a \leq m < N} f(m) = z_k(f; a) + \int_a^N f(t)\,dt - \frac{f(N)}{2} + \sum_{1 \leq j \leq \lfloor k/2 \rfloor} \frac{B_{2j}}{(2j)!} f^{(2j-1)}(N)$$
$$+ \frac{(-1)^k}{k!} \int_N^\infty f^{(k)}(t) B_k(\{t\})\,dt .$$

(2) *If, in addition, $\int_a^\infty f(t)\,dt$ converges, we have*

$$\sum_{a \leq m < N} f(m) = z'_k(f; a) - \int_N^\infty f(t)\,dt - \frac{f(N)}{2} + \sum_{1 \leq j \leq \lfloor k/2 \rfloor} \frac{B_{2j}}{(2j)!} f^{(2j-1)}(N)$$
$$+ \frac{(-1)^k}{k!} \int_N^\infty f^{(k)}(t) B_k(\{t\})\,dt ,$$

where $z'_k(f; a) = z_k(f; a) + \int_a^\infty f(t)\,dt$.

COROLLARY 4.2.9. *Keep the assumptions of the preceding corollary, and in addition assume that the series $\sum_{m \geq a} f(m)$ converges. We have*

$$\sum_{m > N} f(m) = \int_N^\infty f(t)\,dt - \frac{f(N)}{2} - \sum_{1 \leq j \leq \lfloor k/2 \rfloor} \frac{B_{2j}}{(2j)!} f^{(2j-1)}(N)$$

$$+ \frac{(-1)^{k-1}}{k!} \int_N^\infty f^{(k)}(t) B_k(\{t\})\,dt\ ,$$

or equivalently

$$\sum_{m \geq a} f(m) = \sum_{a \leq m \leq N-1} f(m) + \frac{f(N)}{2} + \int_N^\infty f(t)\,dt$$

$$- \sum_{1 \leq j \leq \lfloor k/2 \rfloor} \frac{B_{2j}}{(2j)!} f^{(2j-1)}(N) + \frac{(-1)^{k-1}}{k!} \int_N^\infty f^{(k)}(t) B_k(\{t\})\,dt\ .$$

PROOF. Immediate from the proposition by writing $\int_a^N = \int_a^\infty - \int_N^\infty$. The absolute convergence of the last integral follows from the assumption since $B_k(\{t\})$ is bounded. $\qquad\square$

The speed of convergence of the Euler–Maclaurin formula is essentially dominated by the size of the B_k, which itself is of the order of $k!/(2\pi)^k$, since the closest singularity to 0 of $1/(e^D - 1)$ is $\pm 2\pi i$. For many functions, such as rational functions, $f^{(k)}(N)/k!$ is of the order of N^{-k}, so the kth term is of the order of $k!/(2\pi N)^k$. If we want this to be of the order of 2^{-B}, by Stirling we must choose $N \geq (k/(2\pi e)) 2^{B/k}$, so if we choose $k \approx \alpha B$ for some α this implies that $N \approx (\alpha/(2\pi e)) 2^{1/\alpha} \cdot B$. In the Euler–Maclaurin formula there is a sum of approximately N values of f, and a sum of $k/2$ derivatives of order up to k, so it seems that we should optimize $N + k/2$ or $N + k$. However this depends very much on how we compute derivatives. Our program below evaluates f at a power series to obtain the Taylor expansion, which makes the computation of the derivatives much more expensive, so the quantity to minimize is more like $N + k^2$ or worse. After some experimentation with different f and different accuracies, we chose $\alpha = 0.1235$ as a good compromise, leading to the following programs:

———————————— SumEulerMaclaurin.gp ————————————

```
/* Antiderivative of f which vanishes at infinity. F is either
 * already such a primitive or indicates expansion at infinity
 * in the format expected by IntGaussa_ooinit [t_VEC] or
 * Inta_ooinit [a scalar]. */
Sumintinit(f, F = 0) =
{ my(tF = type(F), XW);

  if (tF == "t_CLOSURE",, \\ do nothing
      tF == "t_VEC", XW = IntGaussa_ooinit(F);
                     F = N -> -IntGaussa_oo(f, N, F, XW)
                   , XW = Inta_ooinit(F);
                     F = N -> -Inta_oo(f, N, F, XW));
  return (F);
}
```

```
15
16    /* Let f be a closure. Compute sum_{m >= 1} f(m)
17     * under the Euler-MacLaurin assumptions. Derivatives are
18     * computed using power series substitutions. */
19    SumEulerMaclaurin(f, F = 0) =
20    { my(B = getlocalbitprec(), N = 2 * B, k, S, ser);
21
22      F = Sumintinit(f, F);
23      k = ceil(0.1235 * B); if (k % 2, k++);
24
25      bernvec(k \ 2); /* initialize Bernoulli cache */
26      localbitprec(B + 32);
27      ser = f(N + 'x + O('x^(k + 1)));
28      S = sum(m = 1, N - 1, f(m), f(N) / 2.) - F(N)
29          - sum(j = 1, k \ 2, bernfrac(2*j)/(2*j)
30                            * polcoef(ser, 2*j-1), 0.);
31      return (bitprecision(S, B));
32    }
```

REMARKS 4.2.10.

(1) As in integration programs, the argument f is a GP function, given by something like f = (t -> expr(t)).

(2) Consecutive Bernoulli numbers being so useful because of Euler-Maclaurin formulas, GP maintains a cache of all B_{2j} computed so far. The bernvec command computes all Bernoulli numbers up to B_k so that later calls to bernfrac (B_{2j} as a rational number) or bernreal (B_{2j} as a floating point number) become instantaneous.

(3) The optional argument F indicates how we shall compute the integral $\int_N^\infty f(t)\,dt$ and is handled by the Sumintinit program, as mentioned in the introduction. The parameter F is either

- a closure computing directly the antiderivative of f vanishing at infinity,
- or a vector $[a, b]$ for Gaussian methods, indicating that f has an asymptotic expansion of the form $\sum_j a_j/(x^{aj+b})$.
- or a numerical parameter specifying the behavior of f at infinity for a DE method: if the function tends to 0 at infinity like e^{-at} with $a > 0$ set F = a, if it is like t^{-a} with $-2 < a < -1$ set F = a; note that the DE method is the variant given in Chapter 3 under the name Inta_oo because it handles better large values of N. Omitting F uses DE with default parameters.

In the last two cases, the function Sumintinit precomputes a table which will speed up the computation of the integral, but this can also be done once and for all independently of f, at least for a given behavior at infinity: one may compute a single initialization structure XW, then use the closure mechanism to create the antiderivative of each given f (as Sumintint does).

An alternative approach, easier to analyze, replaces Taylor series by numerical differentiation (the DiffM and DiffMinit programs). As in these programs, assume that f can be computed to B bits of accuracy in time $O(B^\omega)$. We can now optimize rigorously the value of α: assuming the weights $w_{m,n}$ in (3.1) are precomputed, we only compute values of f; N values at B bits of accuracy and $k(\omega + 1)$ values at $B(1+1/\omega)+E$ bits for the $k/2$ derivatives $f^{(2j-1)}(N)$ of order up to k, where E is the bit size of the differentiation weights. For the sake of simplicity, let us neglect E: the total cost is thus $B^\omega(N + C_\omega k)$, where $C_\omega = (\omega+1)^{\omega+1}/\omega^\omega$. Replacing k and N by the estimates given above, the optimal α minimizes the expression $\alpha 2^{1/\alpha} + 2\pi e C_\omega \alpha$. Critical points satisfy $2^{1/\alpha}\left(1 - \log(2)/\alpha\right) + 2\pi e C_\omega = 0$. Setting $x = \log(2)/\alpha - 1$ and dividing by e, this becomes $x\exp(x) = 2\pi C_\omega$, or $x = W_0(2\pi C_\omega)$ in terms of the Lambert function. Hence the optimal α is $\log(2)/(1 + W_0(2\pi C_\omega))$. For the default value $\omega = 2$, we obtain $\alpha \approx 0.185$.

────────────────────────── SumEulerMaclaurin2.gp ──────────────────────────

```
1   /* Initialize weights and nodes for SumEulerMaclaurin2,
2    * for functions evaluated to accuracy 2^(-B) in time O(B^om)
3    * If 'a' is set use derivatives up to order k = a * B instead
4    * of computing the optimal value */
5   /* Utility function: choose k, N */
6   SumEulerMaclaurin2kN(om = 2, a = 0) =
7   { my(B = getlocalbitprec(), N, k);
8     localbitprec(32);
9     if (!a,
10      my(C = (om+1)^(om+1) / om^om);
11      a = log(2) / (1 + lambertw(2*Pi*C)));
12    k = ceil(a * B); if (k % 2, k++);
13    N = ceil(k / (2 * Pi * exp(1)) * 2^(B/k));
14    return ([k, N]);
15  }
16  /* Return nodes, weights, N */
17  SumEulerMaclaurin2init(om = 2, a = 0) =
18  { my([k, N] = SumEulerMaclaurin2kN(om, a), k2 = k \ 2);
19    bernvec(k2); /* initialize Bernoulli cache */
20    my([X, W, h] = DiffMinit(k, om));
21    my(H, ih = 1 / h);
22    H = powers(ih^2, k2 - 1, ih / (#X-1)!);
23    X = [ N + h*v | v <- X ];
24    W = sum(j = 1, k2, bernfrac(2*j)/(2*j) * H[j] * W[2*j]);
25    return ([X, W, N]);
26  }
27
28  /* As SumEulerMaclaurin, computing the derivatives using
29   * numerical differentiation */
30  SumEulerMaclaurin2(f, F = 0, XW = 0) =
31  { my(B = getlocalbitprec(), S, X, W, N);
32
33    [X, W, N] = if (XW, XW, SumEulerMaclaurin2init());
```

```
34    F = Sumintinit(f, F);
35    localbitprec(B + 32);
36    S = sum(m = 1, N - 1, f(m), f(N) / 2.) - F(N);
37    localbitprec(bitprecision(X[2]));
38    S -= W * [f(x) | x <- X]~;
39    return (bitprecision(S, B));
40  }
```

REMARKS 4.2.11.

(1) Neglecting E underestimates the cost of computing derivatives. We could estimate it more accurately using asymptotics of Stirling numbers, yielding $E \approx D_\omega k$ as $k \to \infty$, where D_ω is an explicit function of ω only, close to $\omega + 1$. This leads to a more complicated equation for critical points, which we may solve numerically. Since this did not improve the practical efficiency of the function, we do not give the corresponding program.

(2) In the `SumEulerMaclaurin2init` function, we allow specifying directly the parameter α such that $k = \alpha B$, overriding the above choices, to make experiments easier.

There is a major problem with the initialization function, though: it requires a lot of memory. More precisely, `DiffMinit` returns a structure allowing computations of all derivatives of order $\leq k$ at bit accuracy B, and whose size is $O(k^2 B)$; k being linear in B, this uses $O(B^3)$ memory! On the other hand, `SumEulerMaclaurin2init` actually computes nodes and weights such that

$$\sum_k w_k f(x_k) \approx \sum_{1 \leq j \leq \lfloor k/2 \rfloor} \frac{B_{2j}}{(2j)!} f^{(2j-1)}(N)$$

(for fixed k, N depending on B); the size of its output is $O(B^2)$. The solution is to compute directly the needed weights by merging `DiffMinit` and the sum over Bernoulli numbers, which can be done using $O(B^2)$ memory. In fact, we only need derivatives of odd order, which saves half the work, and the symmetric nodes $(N - hj, N + hj)$ occur with weights w_j and $-w_j$, saving another factor 2. This yields the following final program:

─────────────────── | SumEulerMaclaurin3.gp | ───────────────────

```
1   /* merges DiffMinit and SumEulerMaclaurin2init
2    * Set the ALT flag to 1 for the Sumalt variant */
3   DiffEulerMaclaurin(M2, om, ALT = 0) =
4   { my(C, h, W, Q, N2, N, X, x = 'x);
5
6     N2 = ceil((om+1) * M2); N2 += (N2 + 1) % 2; \\ make N2 odd
7     N = N2 \ 2;
8     M = M2 \ 2;
9     my(B1, B = getlocalbitprec());
10    B1 = ceil(B * (1 + 1 / om) + N2 + 32);
11    localbitprec(B1);
12    h = 1.0 >> (ceil(B / (N2 - M2)));
13    my(H, ih = 1 / h);
```

```
14    H = powers(ih^2, M - 1, ih / (N2-1)!);
15    bernvec(M); /* initialize Bernoulli cache */
16    if (ALT, H = vector(M, j, (2^(2*j)-1) * H[j]));
17    H = vector(M, j, bernfrac(2*j)/(2*j) * H[j]);
18    C = binomial(2 * N);uX = vector(N2); W = vector(N2);
19    Q = x * vecprod(vector(N, k, x^2 - k^2));
20    W = vector(N, i,
21      my(L = (Q \ (x - i) + O(x^(M2+1))));
22      my(S = sum(j = 1, M, H[j] * polcoef(L, 2*j-1)));
23      S * C[N - i + 1] * (-1)^(N - i));
24    return ([W, h]);
25  }
26
27  SumEulerMaclaurin3init(om = 2, a = 0) =
28  { my([k, N] = SumEulerMaclaurin2kN(om, a));
29    my([W, h] = DiffEulerMaclaurin(k, om, 0));
30    return ([W, N, h]);
31  }
32  SumEulerMaclaurin3(f, F = 0, XW = 0) =
33  { my(B = getlocalbitprec(), S, W, N, h);
34
35    [W, N, h] = if (XW, XW, SumEulerMaclaurin3init());
36    F = Sumintinit(f, F);
37    localbitprec(B + 32);
38    S = sum(m = 1, N - 1, f(m), f(N) / 2.) - F(N);
39    localbitprec(bitprecision(h));
40    S -= sum(i = 1, #W, W[i] * (f(N + h*i) - f(N - h*i)));
41    return (bitprecision(S, B));
42  }
```

REMARK 4.2.12. This program supersedes SumEulerMaclaurin2. In practice, it should be used only after initialization with SumEulerMaclaurin3init, otherwise the time would be orders of magnitude slower than SumEulerMaclaurin. For instance, with the default parameters the initialization for 154, 308, 616, and 1232 decimals requires respectively 0.010, 0.064, 0.59, and 6 seconds.

The corresponding problem for alternating series is immediate.

PROPOSITION 4.2.13. *Under the Euler–Maclaurin assumptions we have*

$$\sum_{0 \le m < 2N} (-1)^m f(m) = \sum_{1 \le j \le \lfloor k/2 \rfloor} (2^{2j} - 1) \frac{B_{2j}}{(2j)!} \left(f^{(2j-1)}(2N) - f^{(2j-1)}(0) \right)$$

$$+ (-1)^{k-1} \frac{2^{k-1}}{k!} \int_0^{2N} f^{(k)}(t)(B_k(\{t/2\}) - B_k(\{(t+1)/2\})) \, dt \,.$$

PROOF. Trivial from the initial Euler–Maclaurin formula by separating even and odd terms.   □

──────────────── SumaltEulerMaclaurin.gp ────────────────

```
1   /* Compute sum_{m >= 1} (-1)^{m-1} f(m), under the above
2    * assumptions. Assume also that the derivatives can be
3    * computed using power series substitutions. */
4   SumaltEulerMaclaurin(f) =
5   { my(B = getlocalbitprec(), N, k, S, ser);
6
7     k = ceil(0.15 * B); if (k % 2, k++);
8     N = 2 * B; if (N % 2, N++);
9     bernvec(k \ 2); /* initialize Bernoulli cache */
10    localbitprec(B + 32);
11    ser = f(N + 'x + O('x^(k+1)));
12    S = sum(m = 1, N - 1, (-1)^(m-1) * f(m), -f(N)/2.)
13      + sum(j = 1, k \ 2, (2^(2*j)-1) * bernfrac(2*j) / (2*j)
14                         * polcoef(ser, 2*j-1), 0.);
15    return (bitprecision(S, B));
16  }
```

REMARKS 4.2.14.
  (1) Since this is the first program that we give for summing alternating series, we recall from the introduction that **Sumaltxxx(n -> f(n))** computes $\sum_{n \geq 1} (-1)^{n-1} f(n)$.
  (2) Even though the value chosen for N is twice that used in **SumEulerMaclaurin**, this program is much faster than trivially summing $f(2m-1) - f(2m)$, essentially because except in the simplest cases, almost all of the time is spent in computing **ser**, which is done only once.

Here also, we can use numerical differentiation instead of working with Taylor series, similarly to **SumEulerMaclaurin3**. Choosing N and the α parameter is analogous: this time we must have $2^k k!/(2\pi N)^k \leq 2^{-B}$ and this replaces 2π by π in all subsequent computations. The Bernoulli sum in the initialization incorporates the $2^{2j} - 1$ factor.

──────────────── SumaltEulerMaclaurin3.gp ────────────────

```
1   /* As SumaltEulerMaclaurin, computing the derivatives using
2    * numerical differentiation; 'alt' variant of
3    * SumEulerMaclaurin3. */
4
5   SumaltEulerMaclaurin3kN(om = 2, a = 0) =
6   { my(B = getlocalbitprec(), N, k);
7     localbitprec(32);
8     if (!a,
9       my(C = (om+1)^(om+1) / om^om);
10      a = log(2) / (1 + lambertw(Pi*C)));
11    k = ceil(a * B); if (k % 2, k++);
12    N = ceil(k / (Pi * exp(1)) * 2^(B / k)); if (N % 2, N++);
13    return ([k, N]);
14  }
```

```
15   SumaltEulerMaclaurin3init(om = 2, a = 0) =
16   { my([k, N] = SumaltEulerMaclaurin3kN(om, a));
17     my([W, h] = DiffEulerMaclaurin(k, om, 1));
18     return ([W, N, h]);
19   }
20
21   SumaltEulerMaclaurin3(f, XW = 0) =
22   { my(B = getlocalbitprec(), S, X, W, N);
23
24     [W, N, h] = if (XW, XW, SumaltEulerMaclaurin3init());
25     localbitprec(B + 32);
26     S = sum(m = 1, N - 1, (-1)^(m-1) * f(m), -f(N) / 2.);
27     localbitprec(bitprecision(h));
28     S += sum(i = 1, #W, W[i] * (f(N + h*i) - f(N - h*i)));
29     return (bitprecision(S, B));
30   }
```

4.2.3. The constant $z(f;a)$. The constant $z_k(f;a)$ occurring in the Euler–Maclaurin formula is usually independent of k, hence denoted simply $z(f;a)$. For instance, we leave to the reader the proof of the following lemmas:

LEMMA 4.2.15. *Let $k_0 \geq 1$ be an integer. If for all $k \geq k_0$ the sign of $f^{(k)}(t)$ is constant and $f^{(k-1)}(t)$ tends to 0 as $t \to \infty$, the constant $z_k(f;a)$ is independent of $k \geq k_0$.*

LEMMA 4.2.16. *Keep the assumptions of the preceding lemma.*

(1) *We have*

$$z(f; a + 1) = z(f; a) + \int_a^{a+1} f(t)\, dt - f(a) \ .$$

(2) *The quantity*

$$z(f) = z(f; a) + \sum_{1 \leq m < a} f(m) - \int_1^a f(t)\, dt$$

is independent of $a \in \mathbb{Z}_{\geq 1}$. In particular $z(f) = z(f;1)$, and if the series and integral converge we have

$$\sum_{m \geq 1} f(m) = z(f) + \int_1^\infty f(t)\, dt \ .$$

An important result is that one can give explicit expressions for $z(f;a)$. A special, but important, case is as follows:

PROPOSITION 4.2.17. *Let g be a piecewise continuous function on $[0, \infty[$ such that for all $a > 0$ the function $g(t)e^{-at}$ tends to 0 as $t \to \infty$. Assume that the function f is given by the Laplace transform of g, in other words that*

$$f(x) = \int_0^\infty e^{-tx} g(t)\, dt \ .$$

If f and all its derivatives have a constant sign on $[a, \infty[$ for some $a \in \mathbb{Z}_{\geq 1}$, we have

$$z(f; a) = \frac{f(a)}{2} + \int_a^\infty f'(t)\left(\{t\} - \frac{1}{2}\right)dt = \int_0^\infty e^{-at}g(t)\left(\frac{1}{1 - e^{-t}} - \frac{1}{t}\right)dt .$$

In particular, under the assumptions of the previous lemma we have

$$z(f) = \frac{f(1)}{2} + \int_1^\infty f'(t)\left(\{t\} - \frac{1}{2}\right)dt = \int_0^\infty g(t)\left(\frac{1}{e^t - 1} - \frac{e^{-t}}{t}\right)dt .$$

PROOF. First note that the assumption on g implies that f is C^∞ on $]0, \infty[$ and that $f(t)$ and all its derivatives tend to 0 as $t \to \infty$. Thus, since we also assume that they have constant sign, the assumptions of the above lemma are satisfied with $k_0 = 1$. In addition if s is the sign of $f'(t)$ we have

$$\int_a^\infty |f'(t)|\,dt = s\int_a^\infty f'(t)\,dt = s(f(\infty) - f(a)) = -sf(a) ,$$

so we may apply Corollary 4.2.8 with $k = 1$ and $N = a$, which gives

$$f(a) = z(f; a) + f(a)/2 - \int_a^\infty f'(t)(\{t\} - 1/2)\,dt$$

proving the first formula.

To prove the second we could directly use the definition of $z(f; a)$, but it is more instructive to deduce it from the first. Note first the following lemma:

LEMMA 4.2.18. *If $u > 0$ we have*

$$\int_0^\infty e^{-tu}\left(\{t\} - \frac{1}{2}\right)dt = -\frac{1}{u}\left(\frac{1}{e^u - 1} - \frac{1}{u} + \frac{1}{2}\right) .$$

PROOF. We use integration by parts and a very elementary aspect of distributions (in fact here Stieltjes integration): the derivative of $\{t\} - 1/2$ for $t > 0$ is equal to $1 - \sum_{n \geq 1}\delta(t - n)$ where δ is the Dirac distribution. Thus if we denote by I our integral, we have

$$I = -(e^{-tu}/u)(\{t\} - 1/2)\Big|_0^\infty + (1/u)\int_0^\infty e^{-tu}\left(1 - \sum_{n \geq 1}\delta(t - n)\right)dt$$

$$= -1/(2u) + 1/u^2 - (1/u)\sum_{n \geq 1}e^{-nu} = -1/(2u) + 1/u^2 - 1/(u(e^u - 1)) ,$$

as claimed.   □

Note that it is immediate to avoid the use of the Dirac distribution by splitting the integral into intervals $[n, n + 1[$, but the above proof is more elegant.

Resuming our proof, we thus have

$$\int_a^\infty f'(t)(\{t\} - 1/2)\,dt = \int_a^\infty \int_0^\infty (-ue^{-tu}g(u))(\{t\} - 1/2)\,dt\,du$$

$$= -\int_0^\infty ug(u)\left(\int_a^\infty e^{-tu}(\{t\} - 1/2)\,dt\right)du .$$

Since $a \in \mathbb{Z}$ we can change t into $t + a$ in the inner integral, so by the above lemma we obtain

$$\int_a^\infty f'(t)(\{t\} - 1/2)\,dt = \int_0^\infty e^{-au}g(u)(1/(e^u - 1) - 1/u + 1/2)\,du .$$

Using $f(a) = \int_0^\infty e^{-au} g(u)\,du$ and the first formula gives the second, since $1 +$
$1/(e^u - 1) = 1/(1 - e^{-u})$.   □

Note on the numerical computation of $z(f; a)$: if the assumptions of the proposition are satisfied and if the function $g(t)$ is known explicitly, we can use the second
formula and the DE method for rapidly decreasing functions to compute $z(f; a)$, although some care must be taken around $t = 0$. But in general $g(t)$ is not known, and
in addition the conditions of the proposition are not satisfied. Instead, we can use
under much weaker conditions Corollary 4.2.8 with $N = a$, but now not necessarily
with $k = 1$, but with $k = k_0$ sufficiently large so that $z_k(f; a) = z_{k_0}(f; a) = z(f; a)$
for all $k \geq k_0$. Thus

$$z(f; a) = \frac{f(a)}{2} - \sum_{1 \leq j \leq \lfloor k/2 \rfloor} \frac{B_{2j}}{(2j)!} f^{(2j-1)}(a) + \frac{(-1)^{k-1}}{k!} \int_a^\infty f^{(k)}(t) B_k(\{t\})\,dt .$$

It is interesting to note that we can often (although not always) apply one of the
oscillating integration methods seen in Chapter 3 to compute such integrals, and in
particular the fastest and most accurate `IntoscSidi` method. We can for instance
write the following trivial program, which we name after Ramanujan since he was
fascinated by $z(f; a)$:

─────────────────────────── │ SumRamanujanz.gp │ ───────────────────────────

```
1    /* Compute the constant z(f) occurring in Euler-MacLaurin
2     * type formulas (see text), in particular equal to
3     * sum_{n >= 1} f(n) - int_1^oo f(t)dt if series and integral
4     * converge. The integer k >= 1 is such that f^(k)(x) tends
5     * to 0 at infinity. */
6
7    SumRamanujanz(f, k = 1) =
8    { my(k2 = k \ 2, P = bernpol(k), F = derivn(f, k));
9      my(V, S, T);
10
11     V = derivnum(x = 1, f(x), concat(0, vector(k2, j, 2*j-1)));
12     S = V[1] / 2
13         - sum(j = 1, k2, bernfrac(2*j) / (2*j)! * V[j + 1]);
14     T = IntoscSidi(t -> subst(P, 'x, frac(t)) * F(t + 1));
15     return (S + (-1)^(k-1) / k! * T);
16   }
17
```

──

EXAMPLES.
```
? SumRamanujanz(n -> 1 / n^2) + 1 - Pi^2/6
% = -5.8 E-39
? SumRamanujanz(n -> 1 / n) - Euler
% = -2.9... E-39
? SumRamanujanz(n -> 1 / n, 4) - Euler
% = -2.9... E-39 /* Can take k > 1, still perfect, slower. */
? SumRamanujanz(n -> log(n)) + 1 - log(2*Pi) / 2
% = 0.E-38
```

```
? SumRamanujanz(n -> n * log(n)) + 1/4 + zeta'(-1)
reached accuracy of 4 bits.
% = 0.0619... /* Wrong: needs k >= 2. */
? SumRamanujanz(n -> n * log(n), 2) + 1/4 + zeta'(-1)
% = -1.60... E-40
? SumRamanujanz(n -> log(n) / n)
reached accuracy of 20 bits.
% = -0.07483654...
```

In this last example, the `IntoscSidi` program fails, and it cannot be repaired simply. We will mention this phenomenon below when discussing that program. Of course, in this case (as in most of the others), instead of using the oscillating form of the integral for $z(f; a)$, we should use the Laplace transform method.

Exercise: show that `SumRamanujanz(n -> n^k, k+1)` is equal to $\dfrac{1 - B_{k+1}}{k+1}$.

4.2.4. The Abel–Plana formulas. Note that the assumption that f is the Laplace transform of g with g as above implies that $f(z)$ is a holomorphic function of z on $\Re(z) > 0$, and furthermore that $|f(z)|$ is bounded in any vertical strip $0 < \sigma_1 \leq \Re(z) \leq \sigma_2$. If we assume a weaker condition, we can obtain a stronger form of the above proposition known as the *Abel–Plana* formula:

PROPOSITION 4.2.19 (Abel–Plana). *Assume that f is a meromorphic function on $\Re(z) > 0$. Assume that f has no poles on $\mathbb{R}_{\geq a}$, has a finite number of poles on $\Re(z) \geq a$, that $f(z) = o(\exp(2\pi|\Im(z)|))$ as $|\Im(z)| \to \infty$ uniformly in vertical strips of bounded width, and that f and all its derivatives have constant sign and tend to 0 as $x \to \infty$ in $\mathbb{R}$.*

(1) *If $a > 0$ we have*

$$z(f; a) = \frac{f(a)}{2} - \pi R(f) + i \int_0^\infty \frac{f(a + it) - f(a - it)}{e^{2\pi t} - 1}\, dt\,, \quad \text{with}$$

$$R(f) = \sum_{\substack{\alpha \text{ pole of } f \\ \Re(\alpha) \geq a}}{}' \operatorname{Res}_{z=\alpha}((\operatorname{cotan}(\pi z) + i\operatorname{sign}(\Im(\alpha)))f(z))\,,$$

and where $\sum'$ means that the residues of the poles with $\Re(\alpha) = a$ must be counted with coefficient $1/2$.

(2) *If $a > 1/2$ we have*

$$z(f; a) = -\pi R(f) + \int_0^{1/2} f(a - 1/2 + t)\, dt$$

$$- i \int_0^\infty \frac{f(a - 1/2 + it) - f(a - 1/2 - it)}{e^{2\pi t} + 1}\, dt\,.$$

PROOF. The proof is a standard exercise in complex analysis, using the fundamental fact that the function $\phi(z) = \pi \operatorname{cotan}(\pi z)$ has poles at all integers with residue 1. We leave the details as an exercise. □

The point of the slightly more complicated second formula is that the denominator of the integrand does not have poles on $\mathbb{R}_{\geq 0}$, so that techniques of numerical integration can be used more easily.

COROLLARY 4.2.20. *In addition to the above assumptions, assume that the series $\sum_{n \geq a} f(n)$ converges and that $a \in \mathbb{Z}_{\geq 1}$. We have*

$$\sum_{n=a}^{\infty} f(n) = \int_a^{\infty} f(t)\,dt + \frac{f(a)}{2} - \pi R(f) + i \int_0^{\infty} \frac{f(a+it) - f(a-it)}{e^{2\pi t} - 1}\,dt$$

$$= \int_{a-1/2}^{\infty} f(t)\,dt - \pi R(f) - i \int_0^{\infty} \frac{f(a-1/2+it) - f(a-1/2-it)}{e^{2\pi t} + 1}\,dt \ .$$

Remark. By Taylor's formula we can write *formally*

$$f(a+it) - f(a-it) = -2i \sum_{k \geq 1} (-1)^k t^{2k-1} \frac{f^{(2k-1)}(a)}{(2k-1)!} \ ,$$

and since we have the easily proved formula

$$\int_0^{\infty} \frac{t^{2k-1}}{e^{2\pi t} - 1}\,dt = (-1)^{k-1} \frac{B_{2k}}{4k} \ ,$$

we thus have *formally*

$$i \int_0^{\infty} \frac{f(a+it) - f(a-it)}{e^{2\pi t} - 1}\,dt = -\sum_{k \geq 1} \frac{B_{2k}}{(2k)!} f^{(2k-1)}(a) \ ,$$

and so we recover formally the Euler–Maclaurin formula.

Exercise: Write the corresponding formula that one formally obtains using the *second* Abel–Plana formula above.

The following corollary is clear:

COROLLARY 4.2.21. *Keep the above assumptions, let $N \in \mathbb{Z}_{\geq 1}$, and assume that all the poles z of f satisfy $\Re(z) < N$. We then have*

$$\sum_{n \geq 1} f(n) = \sum_{n=1}^{N-1} f(n) + \frac{f(N)}{2} + \int_N^{\infty} f(t)\,dt + i \int_{-\infty}^{\infty} w(t) \frac{f(N+it) - f(N)}{t}\,dt \ ,$$

where $w(t)$ is the weight function defined for $t \geq 0$ by $w(t) = t/(e^{2\pi t} - 1)$ and extended by $w(-t) = w(t)$.

To use this corollary for numerical computation we need several things. First, we must choose N: experiment shows that N should be proportional to the number B of binary digits, and a good choice is $N = \lceil B/3 \rceil$. The main reason for the choice of a large value of N (instead of, for instance, choosing $N = 1$, for which the formula is still valid at least if the poles of f are on the left) is that we can compute the complex integral using much faster methods, see below.

Second, we must compute $I_1 = \int_N^{\infty} f(t)\,dt$. Independently of the method that is used (Gauss or DE), we can write either $I_1 = \int_1^{\infty} f(t + N - 1)\,dt$ or $I_1 = N \int_1^{\infty} f(Nt)\,dt$, and since N is large, in practice this latter expression will be more accurate. To compute the integral, we can of course use the general and robust DE method, but whenever possible since it is much faster, we can try to use the Gaussian integration programs `IntGaussa_oo` on an infinite interval.

Third and most importantly, we must compute

$$I_2 = \int_{-\infty}^{\infty} w(t) \frac{f(N+it) - f(N)}{t}\,dt \ .$$

We may of course write (as it was originally)

$$I_2 = \int_0^\infty ((f(N+it) - f(N-it))/t)t/(e^{2\pi t} - 1)\, dt\ ,$$

and compute this integral using the DE method (taking care of the removable singularity at $t = 0$): however, this will as usual be robust but rather slow. We would thus prefer to use instead Gaussian methods: unfortunately, as we have mentioned in Section 3.7, when the integrand is exponentially decreasing their range of applicability is very limited, and as the example of $\int_0^\infty t/(e^t - 1)\, dt$ (which is rather similar to our needs) shows, they give completely wrong results.

However, if we integrate far away from 0 these errors disappear: consider the following example at 115 decimal digits: the `IntGaussLaguerre` program applied to the integral $\int_0^\infty t/(e^t - 1)\, dt$ gives only 39 correct decimals. On the other hand, for the integral $\int_{35}^\infty t/(e^t - 1)\, dt$ the result is perfect (the integral is around 10^{-14}, so we have at least 100 digits of relative accuracy).

Thus, if N is large enough, we can hope that our complex integral I_2 can be evaluated not only rapidly, but also *correctly*, by Gaussian programs, and this is indeed the case in practice.

It is for this reason that in Chapter 3 we have written a specific Gaussian integration program `IntGaussexp` and its corresponding initialization program `IntGaussexpinit` for this weight $w(x)$. Thus, it is immediate to write the following:

SumAbelPlana.gp

```
/* Compute sum_{n >= 1} f(n) under the Abel-Plana assumptions.
 * We assume for simplicity that f tends
 * to 0 slowly at infinity, and has no pole z with Re(z) >= B/3
 * with B number of desired binary digits. */
alias(SumAbelPlanainit, IntGaussexpinit);

/* Set isreal = 1 if f satisfies f(conj(z) = conj(f(z)). */
SumAbelPlana(f, F = 0, isreal = 0, tab = 0) =
{ my(B = getlocalbitprec(), N, S);

  if (type(tab) != "t_VEC", tab = SumAbelPlanainit(tab));
  N = max(ceil(B / 3), 2);
  F = Sumintinit(f, F);
  S = sum(m = 1, N - 1, f(m), f(N) / 2.) - F(N);
  if (isreal,
    S -= IntGaussexp(t -> imag(f(N+I*t)) / t,
                     1, tab)
  , /* else */
    S += I/2*IntGaussexp(t -> (f(N+I*t) - f(N-I*t)) / t,
                         1, tab));
  return (S);
}
```

Note that it may happen that $(f(N+it) - f(N-it))/t$ loses accuracy around $t = 0$, in which case it should be replaced by its Taylor expansion in a neighborhood of 0 as explained in Section 3.3.2. Since the second Abel–Plana formula does not have this disadvantage, in case of difficulties of this type it may be useful to use it instead, so we leave as an immediate exercise for the reader to write the corresponding program.

There also exists an Abel–Plana formula for alternating series, which has a slightly simpler form:

PROPOSITION 4.2.22. *Under the assumptions of the proposition, but with the stronger growth condition $f(z) = o(\exp(\pi|\Im(z)|))$ as $|\Im(z)| \to \infty$ uniformly in vertical strips, we have*

$$\sum_{n=a}^{\infty}(-1)^{n-a}f(n) = \frac{f(a)}{2} - \pi R^-(f) + \frac{i}{2}\int_0^\infty \frac{f(a+it) - f(a-it)}{\sinh(\pi t)}\,dt$$

$$= -\pi R^-(f) + \frac{1}{2}\int_{-\infty}^\infty \frac{f(a-1/2+it)}{\cosh(\pi t)}\,dt\ ,$$

with

$$R^-(f) = \sum_{\substack{\alpha\ pole\ of\ f\\ \Re(\alpha)\geq a}}' \mathrm{Res}_{z=\alpha}(f(z)/\sin(\pi z))\ ,$$

PROOF. Same proof as the preceding proposition, replacing $\pi\cot(\pi z)$ by $\pi/\sin(\pi z)$. □

For a change, we give the corresponding corollary and program for the second formula:

COROLLARY 4.2.23. *Keep the above assumptions, let $N \in \mathbb{Z}_{\geq 1}$, and assume that all the poles z of f satisfy $\Re(z) < N$. We then have*

$$\sum_{n\geq 1}(-1)^{n-1}f(n) = \sum_{n=1}^{N-1}(-1)^{n-1}f(n) + \frac{(-1)^{N-1}}{2}\int_{-\infty}^\infty \frac{f(N-1/2+it)}{\cosh(\pi t)}\,dt\ .$$

The choice of N will be as before, there is no integral I_1 to compute, and the complex integral I_2 can be computed using the Gaussian weight $1/\cosh(\pi t)$, for which we have given the corresponding programs in Chapter 3. Thus, the corresponding (trivial) program is as follows:

_____ SumaltAbelPlana.gp _____

```
/* Compute sum_{n >= 1} (-1)^{n-1}f(n) under the Abel-Plana
 * assumptions. Assume that f tends to 0 slowly at infinity
 * and has no pole z with large real part. */
alias(SumaltAbelPlanainit, IntGausscoshinit);

SumaltAbelPlana(f, isreal = 0, tab = 0) =
{ my(B = getlocalbitprec(), N2, N, S, I2);

  if (type(tab) != "t_VEC", tab = SumaltAbelPlanainit(tab));
  N = ceil(B / 3); N2 = N - 1/2;
  S = sum(n = 1, N - 1, (-1)^(n-1) * f(n), 0.);
  I2 = if (isreal, IntGausscosh(t -> real(f(N2 + I*t)), 1, tab)
                 , IntGausscosh(t -> f(N2 + I*t), 0, tab));
```

```
14    return (S + (-1)^(N-1) / 2 * I2);
15  }
```

We can deduce from the Abel–Plana formulas an amusing corollary:

COROLLARY 4.2.24. *Under the above assumptions, assume in addition that f is even.*

(1) *If f is holomorphic, we have the "sophomore's dream"*

$$\sum_{n=0}^{\infty}{}' f(n) = \int_0^{\infty} f(t)\,dt\ ,$$

where $\sum'$ means that the term $n = 0$ must have a coefficient $1/2$. (If f is only meromorphic, one must add the correction term $-\pi R(f)$ to the right-hand side).

(2) *We have*

$$\sum_{n=0}^{\infty}{}' (-1)^n f(n) = -\pi R^-(f)$$

(so in particular the sum is zero if f is holomorphic).

For instance for $1 \le k \le 6$ we have

$$\frac{1}{2} + \sum_{m \ge 1} \left(\frac{\sin(m)}{m} \right)^k = \int_0^{\infty} \left(\frac{\sin(t)}{t} \right)^k dt\ ,$$

and for $1 \le k \le 3$ we have

$$\frac{1}{2} + \sum_{m \ge 1} (-1)^m \left(\frac{\sin(m)}{m} \right)^k = 0\ .$$

Note that if you want to check the first formula using numerical integration, you need to use the specific integration methods for oscillatory functions explained in Section 3.9.

This example allows us to check that the growth condition is absolutely essential:

Exercise:

(1) Prove that

$$\frac{1}{2} + \sum_{m \ge 1} \left(\frac{\sin(m)}{m} \right)^7 = \int_0^{\infty} \left(\frac{\sin(t)}{t} \right)^7 dt + \frac{2\pi(7 - 2\pi)^6}{2^7 \cdot 6!}\ .$$

Note that the "error term" is of the order of $9.2 \cdot 10^{-6}$, so small, but not that small.

(2) Prove that

$$\frac{1}{2} + \sum_{m \ge 1} (-1)^m \left(\frac{\sin(m)}{m} \right)^4 = \frac{2\pi(4 - \pi)^3}{2^4 \cdot 3!}\ .$$

A much more spectacular example (due, we believe, to S. Plouffe, as well as many others of the same type) is

$$\frac{1}{2} + \sum_{m \geq 1} e^{-(m/10)^2} = \int_0^\infty e^{-(t/10)^2} \, dt + R = 5\pi^{1/2} + R \,,$$

where R is not zero but is extremely small: $R = 4.14 \cdots 10^{-428}$. The reader may want to use her/his knowledge of theta functions to explain this value.

4.2.5. Variant: χ-Euler–Maclaurin. Before coming to concrete applications, note that there are many variants of the basic Euler–Maclaurin formulas given above. The proofs are essentially identical to those that we have given so are left as exercises for the reader.

PROPOSITION 4.2.25. *Assume $a \leq b$ and that f is C^∞ on $[a, b]$. For all $k \geq 1$ we have*

$$\sum_{\substack{a < m \leq b \\ m \in \mathbb{Z}}} f(m) = \int_a^b f(t) \, dt + \sum_{1 \leq j \leq k} \frac{(-1)^j}{j!} \left(B_j(\{b\}) f^{(j-1)}(b) - B_j(\{a\}) f^{(j-1)}(a) \right)$$

$$+ \frac{(-1)^{k-1}}{k!} \int_a^b f^{(k)}(t) B_k(\{t\}) \, dt \,.$$

This proposition allows us to use Euler–Maclaurin when $a \notin \mathbb{Z}$.
The most important variant is the χ-Euler–Maclaurin formula.

PROPOSITION 4.2.26 (χ-Euler–Maclaurin). *Let χ be a periodic function of period F, say, not necessarily a character. For $j \geq 0$ define the jth χ-Bernoulli functions by*

$$B_j(\chi; t) = F^{j-1} \sum_{1 \leq m \leq F} \chi(m) B_j(\{(t - m)/F\}) \,,$$

and the χ-Bernoulli numbers by $B_j(\chi) = B_j(\chi; 0)$. We have

$$\sum_{1 \leq m \leq NF} \chi(m) f(m) = B_0(\chi) \int_0^{NF} f(t) \, dt$$

$$+ \sum_{1 \leq j \leq k} (-1)^j \frac{B_j(\chi)}{j!} (f^{(j-1)}(NF) - f^{(j-1)}(0))$$

$$+ \frac{(-1)^{k-1}}{k!} \int_0^{NF} f^{(k)}(t) B_k(\chi; t) \, dt \,.$$

PROOF. Write any m as $m = Fm_1 + m_2$ with $1 \leq m_2 \leq F$. By periodicity we have

$$\sum_{1 \leq m \leq NF} \chi(m) f(m) = \sum_{1 \leq m_2 \leq F} \chi(m_2) \sum_{0 \leq m_1 < N} f(Fm_1 + m_2) \,.$$

We now apply Proposition 4.2.25 to the inner sums, replacing the function $f(x)$ by the function $g(x) = f(Fx + m_2)$, and choosing $a = -m_2/F$ and $b = N - m_2/F$. Since $1 \leq m_2 \leq F$ the condition $a < m_1 \leq b$ with $m_1 \in \mathbb{Z}$ is equivalent to

$0 \le m_1 < N$. Thus,

$$\sum_{0 \le m_1 < N} f(Fm_1 + m_2) = \sum_{a < m_1 \le b} g(m_1)$$

$$= \int_a^b g(t)\,dt + \sum_{1 \le j \le k} \frac{(-1)^j}{j!}\left(B_j(\{b\})g^{(j-1)}(b) - B_j(\{a\})g^{(j-1)}(a)\right)$$

$$+ \frac{(-1)^{k-1}}{k!}\int_a^b g^{(k)}(t)B_k(\{t\})\,dt\,,$$

so replacing $g(x)$ by $f(Fx + m_2)$ and making the evident changes of variables in the integrals we obtain

$$\sum_{0 \le m_1 < N} f(Fm_1 + m_2) = \frac{1}{F}\int_0^{NF} f(t)\,dt$$

$$+ \sum_{1 \le j \le k} (-1)^j F^{j-1}\frac{B_j(\{-m_2/F\})}{j!}(f^{(j-1)}(NF) - f^{(j-1)}(0))$$

$$+ \frac{(-1)^{k-1}}{k!}F^{k-1}\int_0^{NF} f^{(k)}(t)B_k(\{(t - m_2)/F\})\,dt\,,$$

so summing on m_2 and using the definitions of the χ-Bernoulli numbers and functions proves the proposition. $\qquad\square$

Note that there are slightly different normalizations for the χ-Bernoulli functions and numbers, which all agree at least if χ is a nontrivial Dirichlet character and $j \ne 1$, so the reader should be careful which exact formula to use depending on the normalizations.

We immediately deduce from the proposition an analogue of the third form of Euler–Maclaurin which we write in the following form:

COROLLARY 4.2.27. *Assume in addition that for some $a \in \mathbb{Z}$ and $N \in \mathbb{Z}_{\ge 1}$, $\sum_{m \ge a} \chi(m)f(m)$ and $\int_{NF}^\infty f(t)\,dt$ converge. We have*

$$\sum_{m \ge a} \chi(m)f(m) = \sum_{a \le m \le NF} \chi(m)f(m) + B_0(\chi)\int_{NF}^\infty f(t)\,dt$$

$$+ \sum_{1 \le j \le k} (-1)^{j-1}\frac{B_j(\chi)}{j!}f^{(j-1)}(NF) + \frac{(-1)^{k-1}}{k!}\int_{NF}^\infty f^{(k)}(t)B_k(\chi;t)\,dt\,.$$

Note that we have clearly

$$B_j(\chi) = F^{j-1}\sum_{1 \le m \le F} \chi(m)B_j(1 - m/F)\,.$$

As we did for Euler–Maclaurin, we can write a corresponding program, where χ is given as a function:

──────────────────────── ChiBernoulli.gp ────────────────────────

```
1  /* Compute B_n(chi), chi given by closure. */
2  ChiBernoulli(chi, F, n) =
3  { my(B = bernpol(n), S = F^(n-1) * 1.);
```

```
4    return (S * sum(m = 1, F, chi(m) * subst(B, 'x, 1 - m/F)));
5  }
```

```
                        ┌─────────────────────────┐
──────────────────────  │ SumchiEulerMaclaurin.gp │  ──────────────────
                        └─────────────────────────┘
1  /* Compute sum_{m >= 1} chi(m)f(m), assuming chi and f are
2   * given by closures. */
3  SumchiEulerMaclaurininit(chi, FCHI) =
4  { my(B = getlocalbitprec(), B0, k, bvec);
5
6    B0 = ChiBernoulli(chi, FCHI, 0);
7    localbitprec(32); k = floor(log(2) * B);
8    localbitprec(B);
9    bvec = vector(k, j, (-1)^j * ChiBernoulli(chi, FCHI, j) / j);
10   return ([B0, bvec]);
11 }
12
13 SumchiEulerMaclaurin(f, F = 0, chi, FCHI, tab = 0) =
14 { my(B = getlocalbitprec(), NF, S1, S2 = 0, S3, ser);
15
16   if (!tab, tab = SumchiEulerMaclaurininit(chi, FCHI));
17   my([B0, bchivec] = tab, k = #bchivec);
18   NF = FCHI * ceil(0.1104*B);
19   S1 = sum(m = 1, NF, chi(m) * f(m), 0.);
20   if (exponent(B0) >= 7 - B,
21     F = Sumintinit(f, F); S2 = -F(NF));
22   ser = f(NF + 'x + O('x^(k + 1)));
23   S3 = sum(j = 1, k, bchivec[j] * polcoef(ser, j - 1), 0.);
24   return (S1 + S2 - S3);
25 }
```

The main difference with the SumEulerMaclaurin program is that $B_n(\chi)$ now grows essentially like $k!F^k/(2\pi)^k$ instead of $k!/(2\pi)^k$.

4.2.6. Variant: Δ-Euler–Maclaurin. An annoying feature of Euler–Maclaurin is that the derivatives $f^{(2j-1)}(N)$ are not always easy to compute directly. Note that although successive derivatives of rational functions are easily computed, we will deal with this special case in a different way in Section 4.4. We have given a generic program using numerical derivation involving only calls to the original f, but it requires increasing the working precision and it will not work in the frequent case of functions known only at integers (for instance, a sequence defined by a recursion).

An alternative is to replace the differentiation operator $D = d/dx$ by some *differencing* operator. Two such operators depending on a positive parameter d are the forward difference operator $\Delta_{d,a}$ and the symmetric difference operator Δ_d defined by

$$\Delta_{d,a}(f)(n) = \frac{f(n+d) - f(n)}{d} \quad and \quad \Delta_d(f)(n) = \frac{f(n+d) - f(n-d)}{2d} .$$

The point of these operators is threefold: first, they can be computed directly from the values of f without any limiting process; second, if $f(n)$ is only known for *integral* values of n, these operators can still be used with d integral, typically $d = 1$; third, they can be trivially iterated:

$$\Delta_{d,a}^k(f)(n) = \frac{1}{d^k} \sum_{0 \leq j \leq k} (-1)^{k-j} \binom{k}{j} f(n + jd) \quad \text{and}$$

$$\Delta_d^k(f)(n) = \frac{1}{(2d)^k} \sum_{0 \leq j \leq k} (-1)^{k-j} \binom{k}{j} f(n - (k - 2j)d) \ .$$

Although we could do a rigorous analysis with error term as we have done for the initial Euler–Maclaurin, we will be content with a more heuristic approach, leaving the rigorous analysis to the reader. Also, since Δ_d is closer to the true derivative than $\Delta_{d,a}$, we will not anymore consider the latter.

The asymptotic series $S = \sum_{j \geq 1} (-1)^j (B_j/j!) f^{(j-1)}(N)$ occurring in Euler–Maclaurin can be written as $\mathcal{D}(f)(N)$, where $\mathcal{D}$ is the differential operator $\mathcal{D} = \sum_{j \geq 1} (-1)^j (B_j/j!) D^{(j-1)}$. By definition of the Bernoulli numbers we have

$$\mathcal{D} = \frac{1}{e^D - 1} - \frac{1}{D} + 1 \ .$$

Thus, if δ is any differential operator with formal power series expansion in D of the form $\delta = D + \sum_{i \geq 2} a_i D^i$, and if $D = \delta + \sum_{i \geq 2} a_i' \delta^i := \phi(\delta)$ is the functionally reverse power series, then we can express $\mathcal{D}$ in terms of δ by using this reverse series, i.e., write

$$\mathcal{D} = \frac{1}{e^{\phi(\delta)} - 1} - \frac{1}{\phi(\delta)} + 1 = \sum_{j \geq 1} (-1)^j \frac{b_j(\delta)}{j!} \delta^{j-1} \ .$$

The Euler–Maclaurin formula, say in its second form, can thus be written

$$\sum_{a \leq m \leq N} f(m) = z_k(f; a) + \int_a^N f(t)\, dt + \sum_{1 \leq j \leq k} (-1)^j \frac{b_j(\delta)}{j!} \delta^{j-1}(f)(N) + R_k(f; N)$$

for some "remainder term" R_k.

Recall that the speed of convergence of the initial Euler–Maclaurin formula is essentially dominated by the size of the B_k, which itself is of the order of $k!/(2\pi)^k$, since the closest singularity to 0 of $1/(e^D - 1)$ is $\pm 2\pi i$, and since $f^{(k)}(N)/k!$ is usually of the order of N^{-k}, the kth term is of the order $k!/(2\pi N)^k$.

For our differencing operator we have $\Delta_d = (e^{dD} - e^{-dD})/(2d) = \sinh(dD)/d$, so that $\phi(z) = \operatorname{asinh}(dz)/d = \log(dz + \sqrt{1 + d^2 z^2})/d$. Using a similar analysis, the order of the $b_k(\Delta_d)$ is $k!/c(d)^k$, where $c(d) = \min(1, \sinh(2d\pi))/d$. In some cases, e.g., when $f(n)$ is apparently only defined for integral n, one can only use Δ_1. However in most cases $f(n)$ is defined also for $n \in \mathbb{R}_{>0}$, so it is preferable to use Δ_d for some $d < 1$. Using the same reasoning as above, we can expect $\Delta_d^{(k)} f(N)/k!$ to be of the order of N^{-k}, so the kth term is of order $k!/(c(d)N)^k$,

We will choose $d \geq \operatorname{asinh}(1)/(2\pi) = \log(1 + \sqrt{2})/(2\pi) = 0.14027 \cdots$, so that $c(d) = 1/d$. We will choose k around $N/(\alpha d)$ for some $\alpha > 1$, so $k \approx B \log(2)/(1 + \log(\alpha))$ and $N \approx \alpha dk$. Once the precomputations made we have exactly $k + N$ evaluations of f, where $k + N \approx (\alpha d + 1)B \log(2)/(1 + \log(\alpha))$. The minimum of this is attained for $\alpha \log(\alpha) = 1/d$, and the solution of this equation is $\alpha = 1/(dW(1/d))$, where W is Lambert's function. Once this is computed, we have $\log(\alpha) = W(1/d)$, so $N \approx B\log(2)/(W(1/d)(1 + W(1/d)))$ and $k \approx NW(1/d)$.

However, this is not the whole story. Computing $\Delta_d^k(f)(n)$ by the binomial formula given above will lead to catastrophic cancellation, which could be overcome only by working at a much larger accuracy. It is immediate to show that, similarly to Bernoulli numbers, $b_1(\Delta_d) = -1/2$ and $b_{2k+1}(\Delta_d) = 0$ for $k \geq 1$. Thus, by simply interchanging summations we have for k even:

$$\sum_{2 \leq j \leq k} (-1)^j \frac{b_j(\Delta_d)}{j!} \Delta_d^{j-1}(f)(N)$$

$$= \sum_{1 \leq j \leq k/2} \frac{b_{2j}(\Delta_d)}{(2d)^{2j-1}(2j)!} \sum_{0 \leq j_1 \leq 2j-1} (-1)^{j_1+1} \binom{2j-1}{j_1} f(N - (2j - 1 - 2j_1)d)$$

$$= \sum_{-k/2+1 \leq m \leq k/2} (-1)^{m+1} f(N - (2m-1)d) \cdot$$

$$\cdot \sum_{\max(m,1-m) \leq j \leq k/2} (-1)^j \frac{b_{2j}(\Delta_d)}{(2d)^{2j-1}(2j)!} \binom{2j-1}{j-m} .$$

Now since we choose d such that $c(d) = 1/d$, $b_{2j}(\Delta_d)$ will be of the order of $(2j)!d^{2j}$, and since $\binom{2j-1}{j-m}$ is roughly of order 2^{2j}, the terms in the inner sum are of the same order of magnitude, and in fact one can show that they all have the same sign.

Using this very heuristic analysis enables us to write for instance the following programs, where we choose $d = 1/4$ (it is immediate to modify them for a different value of d):

──────────────── Sumdelta.gp ────────────────

```
/* Compute S = sum_{n >= 1} f(n) by EM symmetric
 * Delta_{1/4} summation. */
Sumdeltainit(N = 0) =
{ my(B = getlocalbitprec(), C, k, S, X, W, vS);

  localbitprec(32);
  my(E = B * log(2));
  if (!N,
    my(w = lambertw(4));
    N = ceil(E / (w * (1 + w)) + 5);
    k = N * w;
  ,
    if (4 * N < E, error("N too small in Sumdeltainit"));
    k = Solvemulneglog(1 + log(4 * N), E));
  k = ceil(k); if (k%2, k--);
  localbitprec(B + 32);
  C = vector(k \ 2, j, binomial(2 * j - 1));
  S = 4 * log('x / 2 + sqrt(1.0 + 'x^2 / 4 + O('x^(k+3))));
  S = 1 / expm1(S) - 1 / S;
  k \= 2;
  vS = vector(k, j, (-1)^j * polcoef(S, 2 * j - 1));
  W = vector(k, m, (-1)^m*sum(j = m, k, vS[j] * C[j][j-m+1]));
```

```
23      X = vector(k, m, (2 * m - 1) / 4);
24      return ([X, W, N]);
25    }
26
27    /* F:   fast or closure giving integral of f from N to oo;
28     * tab: 0 or vector [X, W, N] output of init. */
29    Sumdelta(f, F = 0, tab = 0) =
30    { my(B = getlocalbitprec(), S);
31
32      if (!tab, tab = Sumdeltainit());
33      my(N = tab[3]);
34      F = Sumintinit(f, F);
35      S = SumXW(x -> f(N - x) - f(N + x), tab);
36      localbitprec(B + 32);
37      S += sum(m = 1, N - 1, f(m), f(N) / 2.) - F(N);
38      return (bitprecision(S, B));
39    }
```

REMARKS 4.2.28.

(1) We have already explained in detail the use of the second parameter F, which is transformed by the Sumintinit program into the antiderivative of f which vanishes at infinity.

(2) To make experiments easier, we allow specifying the parameter N, overriding the above choices. Indeed the latter are optimal only under the assumption that the complexity of evaluating $f(z)$ does not depend on the value of a reasonable z on the real axis (in §4.11, the sum $S_7 = \sum_n \zeta(n)/n^2$ will provide a counterexample).

If N is fixed, we must set k such that $k!/(c(d)N)^k \approx 2^{-B}$, hence $k \log(N/d) - k(\log(k) - 1) \approx E := B \log(2)$ using Stirling and $c(d) = 1/d$. The program Solvemulneglog computes a solution assuming that $1 + \log(N/d) \geq 1 + \log(E)$.

The analogue of SumaltEulerMaclaurin for alternating series using a Δ_d operator, is as follows: the analogue of $\mathcal{D}$ is now $1/(e^D + 1)$ instead of $1/(e^D - 1) - 1/D + 1$. In particular, the closest pole to the origin is now at $i\pi$, so we must choose $d \geq \operatorname{asinh}(1)/\pi = 0.28054 \cdots$, and this leads to the following program, where we choose $d = 1/2$, which entails many simplifications:

―――――――――――――― | Sumaltdelta.gp | ――――――――――――――

```
1    /* Compute sum_{n>=1} (-1)^{n-1} f(n) by EM symmetric
2     * Delta_{1/2} summation. */
3
4    Sumaltdeltainit() =
5    { my(B = getlocalbitprec(), N, k, S, W, w, vS);
6
7      localbitprec(32);
8      my(w = lambertw(2));
9      N = ceil(log(2) * B / (w * (1 + w)) + 5); if (N%2, N++);
```

```
10    k = ceil(N * w); if (k%2, k--);
11    localbitprec(B + 32);
12    C = vector(k, j, binomial(j - 1));
13    k \= 2;
14    W = vector(k, m, (-1)^m * sum(j = m, k,
15              C[2*j-1][j] * C[2*j][j-m+1] / 4^(2*j-1)));
16    X = vector(k, m, m - 1 / 2);
17    return ([X, W, N]);
18  }
19
20  Sumaltdelta(f, tab = 0) =
21  { my(B = getlocalbitprec(), S);
22
23    if (!tab, tab = Sumaltdeltainit());
24    my(N = tab[3]);
25    S = SumXW(x -> f(N - x) - f(N + x), tab);
26    localbitprec(B + 32);
27    S += sum(m = 1, N - 1, (-1)^(m-1) * f(m), -f(N)/2.);
28    return (bitprecision(S, B));
29  }
```

Once again, note that this is faster than summing $f(2m-1) - f(2m)$.
Warning: following our general convention for Sumalt functions, Sumaltdelta(f) computes $\sum_{m\geq 1}(-1)^{m-1}f(m)$.

4.2.7. Euler–Maclaurin and $\zeta(s)$. The original motivation of the Euler–Maclaurin formula is simply the following:

EXAMPLE. For $r \in \mathbb{Z}_{\geq 1}$ we have

$$\sum_{m=1}^{N} m^r = \frac{1}{r+1}\left(N^{r+1} + \frac{r+1}{2}N^r + \sum_{1\leq j\leq \lfloor r/2\rfloor}\binom{r+1}{2j}B_{2j}N^{r+1-2j}\right)$$
$$= N^r + \frac{B_{r+1}(N) - B_{r+1}}{r+1} = \frac{B_{r+1}(N+1) - B_{r+1}}{r+1}.$$

More generally

$$\sum_{0\leq m<N}(m+x)^r = \frac{B_{r+1}(N+x) - B_{r+1}(x)}{r+1}.$$

For instance, we obtain the following well-known formulas:

$$\sum_{m=1}^{N} m = \frac{N(N+1)}{2}, \quad \sum_{m=1}^{N} m^2 = \frac{N(N+1)(2N+1)}{6},$$
$$\sum_{m=1}^{N} m^3 = \frac{N^2(N+1)^2}{4}, \quad \sum_{m=1}^{N} m^4 = \frac{N(N+1)(2N+1)(3N^2+3N-1)}{30}.$$

More interesting are formulas giving the value of the Riemann zeta function $\zeta(s)$. Recall that $\zeta(s)$ is defined for $\Re(s) > 1$ by $\zeta(s) = \sum_{n\geq 1} n^{-s}$, and extended analytically to the whole complex plane into a meromorphic function having a single pole, simple, at $s = 1$, and that around $s = 1$ we have $\zeta(s) = 1/(s-1)+\gamma+O(s-1)$, where $\gamma = 0.577\cdots$ is Euler's constant. A more general function is the *Hurwitz zeta function* $\zeta(s, x)$ defined for $\Re(s) > 1$ and $x \notin \mathbb{Z}_{\leq 0}$ by $\zeta(s, x) = \sum_{n \geq 0}(n+x)^{-s}$, so that for instance $\zeta(s) = \zeta(s, 1)$. With x fixed, as a function of s this can also be analytically continued to the whole of $\mathbb{C}$ into a meromorphic function having a single pole, simple, at $s = 1$, and that around $s = 1$ we have $\zeta(s, x) = 1/(s-1) - \psi(x) + O(s - 1)$, where $\psi(x) = \Gamma'(x)/\Gamma(x)$ is the logarithmic derivative of the gamma function. We will usually assume that $x \in \mathbb{R}_{>0}$.

EXAMPLE. Let $\alpha \in \mathbb{C}$ with $\alpha \neq -1$ and let $x \in \mathbb{R}_{>0}$. For any $k > \Re(\alpha) + 1$, $k \in \mathbb{Z}_{\geq 1}$, we have

$$\sum_{m=0}^{N-1}(m+x)^{\alpha} = \zeta(-\alpha, x) + \frac{(N+x-1)^{\alpha+1}}{\alpha+1} + \frac{(N+x-1)^{\alpha}}{2}$$
$$+ \sum_{1 \leq j \leq \lfloor k/2 \rfloor}\binom{\alpha}{2j-1}\frac{B_{2j}}{2j}(N+x-1)^{\alpha-2j+1} + R_k(\alpha, x, N)\ ,$$

where

$$R_k(\alpha, x, N) = (-1)^k\binom{\alpha}{k}\int_N^\infty (t+x-1)^{\alpha-k}B_k(\{t\})\,dt\ .$$

In addition, if k is even we have

$$|R_k(\alpha, x, N)| \leq \left|2\binom{\alpha}{k+1}\frac{B_{k+2}}{k+2}(N+x-1)^{\alpha-k-1}\right|\ .$$

The upper bound on $R_k(\alpha, x, N)$ follows easily from Euler–Maclaurin and is left to the reader. In fact, the factor of 2 can be removed if α is real (exercise). We obtain a formula involving $\zeta(-\alpha)$ itself by setting $x = 1$, in which case $N+x-1 = N$ and the sum on the left hand side is $\sum_{1 \leq m \leq N} m^{\alpha}$.

Even though this is not the fastest way to compute $\zeta(s, x)$ (see [**Joh15**]), it leads to a reasonably simple algorithm: first choose some relatively large even k, then compute N such that the upper bound for $|R_k(-s, x, N)|$ is less than the desired absolute accuracy, and compute $\zeta(s, x)$ by the above formula. More precisely, if we set $C_k(\alpha) = 2\binom{\alpha}{k+1}B_{k+2}/(k+2)$, the condition $|R_k(\alpha, x, N)| \leq 2^{-B}$ is equivalent to $|N + x - 1| \geq |C_k(\alpha)2^B|^{1/(k+1-\Re(\alpha))}$. Assuming that k is much larger than $\Re(\alpha)$, proposition 4.2.5 shows that $C_k(\alpha)^{1/(k+1-\Re(\alpha))} \approx k/(2\pi e)$ will never be very large nor very small. So we must have essentially $N \geq 2^{B/k}k/(2\pi e)$. The number of summands in the Euler–Maclaurin formula is equal to $N + k/2$, so neglecting the fact that the times to compute each summand is different, we must essentially minimize the quantity $k2^{B/k} + k\pi e$. The `SumEulerMaclaurin2` optimization (formally setting $C_\omega = 1/2$) shows that this minimum is attained when $k = B\log(2)/(1 + W_0(\pi)) \approx 0.334B$. However, this is only realistic if α is a small negative integer; otherwise, the $(m+x)^{\alpha}$ summands involve exponentials and logarithms whereas the $k/2$ Bernoulli summands involve only multiplications and divisions by small integers, assuming the Bernoulli numbers are precomputed and the $(N+x-1)^{\alpha}$ term is factored out of the sum. And the former are asymptotically $O(\log B)$ slower (see for instance [**BZ11**]); a quick measurement under GP shows that a^b is roughly $20\log B$ slower in our implementation (and even $30\log B$ slower for complex inputs). So we should really be optimizing $NC\log(B) + k/2$ for some constant $C \approx 20$, yielding $k = B\log(2)/(1 + W_0(\pi/C\log(B)))$. Our discussion assumes that x is not very large; the comparatively easier case where our computed N satisfies $x \gg N$ is left to the reader: we may then take $N = 1$ and choose the smallest k so that $|R_k(\alpha, x, 1)| \leq 2^{-B}$.

Note that if α is a positive integer we will have $R_k(\alpha, x, N) = 0$ as soon as $k \geq \alpha$, so in that case we can take any value of $N \geq 1$. In fact, using Euler–Maclaurin itself gives the well-known formula $\zeta(-\alpha, x) = -B_{\alpha+1}(x)/(\alpha + 1)$ (and in particular $\zeta(-\alpha) = -B_{\alpha+1}/(\alpha + 1)$ for $\alpha \geq 1$), but we do not need to use this formula and simply allow for the special case.

The case $\alpha = -1$ leads to the computation of the function $\psi(x)$, and in particular of Euler's constant $\gamma = -\psi(1)$:

EXAMPLE. For $k \geq 1$ we have

$$\sum_{m=0}^{N-1} \frac{1}{m+x} = \log(N + x - 1) - \psi(x) + \frac{1}{2(N + x - 1)}$$

$$- \sum_{1 \leq j \leq \lfloor k/2 \rfloor} \frac{B_{2j}}{2j}(N + x - 1)^{-2j} + R_k(-1, x, N) \,,$$

where $R_k(-1, x, N)$ is as above with $\alpha = -1$.

A possible GP program is as follows: the function ZetaHurwitz(s,x) will compute $\zeta(s, x)$ to the current accuracy, except when $s = 1$ (exactly), in which case it computes $-\psi(x)$. In particular for $x = 1$ it computes $\zeta(s)$ and γ respectively.

──────────────────────── ZetaHurwitz.gp ────────────────────────

```
1   /* If s != 1, compute zeta(s, x), otherwise compute -psi(x).
2    * In particular, if x = 1 compute zeta(s), and Euler's
3    * constant if s = 1. */
4   ZetaHurwitz(s, x) =
5   { my(B = getlocalbitprec(), a = -s, ra = real(a));
6     my(rndra = round(ra), k, N, Nx, N2, S1, S2);
7
8     if (x <= 0, error("x <= 0 in ZetaHurwitz"));
9     if (rndra >= 0 && exponent(a - rndra) < 16 - B,
10       /* s ~ negative integer */
11      k = ra + 1;
12      k = 2 * ceil(k/2);
13      N = 1;
14    , /* else */
15      localbitprec(64);
16      my(c = if (type(a) == "t_INT", 1, 20 * log(B)));
17      k = max(ra + 3/2, B * log(2) / (1 + lambertw(Pi / c)));
18      k = 2 * ceil(k/2); /* ensure that k is even */
19      my(C = 2 * binomial(a, k+1) * bernreal(k+2) / (k+2));
20      N = max(1, ceil((abs(C) * 2.^B)^(1 / (k+1-ra)) + 1 - x)));
21    bernvec(k \ 2); /* initialize Bernoulli cache */
22    Nx = N + x - 1; N2 = Nx^(-2);
23    S1 = if (x == 1, dirpowerssum(N, a)
24                   , sum(m = 0, N - 1, (m + x)^a, 0.));
```

```
25    S2 = 0;
26    forstep (j = k, 2, -2,
27      my(aj = a - j);
28      S2 = bernreal(j)/j + aj * (aj+1) / (j*(j+1)) * S2 * N2);
29    S2 = 1/2 + S2 * a / Nx;
30    if (s == 1, S1 -= log(Nx), S2 += Nx / (a + 1));
31    return (S1  - Nx^a * S2);
32  }
```

We have used Horner's rule to compute S_2. For S_1 when $x = 1$, the program uses `dirpowerssum` which only requires p^α for $p \leq N$ prime (see the `Sumpow` program from §9.12.2). We leave as exercises analogous improvements when x is a rational number of small height. Note also that there are faster methods to compute $\psi(x)$ and γ. For instance, the elementary theory of Bessel functions implies that for large x we have

$$\gamma = \frac{\sum_{k \geq 0} x^{2k}(H_k - \log(x))/k!^2}{\sum_{k \geq 0} x^{2k}/k!^2} + O(e^{-4x})$$

with as usual $H_k = \sum_{1 \leq j \leq k} 1/j$, which gives a very fast method for computing γ by a suitable choice of x as a function of the desired accuracy, more precisely by choosing $x = B \log(2)/4$ and stopping the summation at $k \approx 3.6x \approx 0.623B$.

Exercise: Program both this method, as well as the corresponding method with $x^{2k}/k!^2$ replaced by $x^{3k}/k!^3$, and compare their efficiency in high accuracy.

In Chapter 9 we will consider the general problem of the numerical computation of L-functions. However, the above program gives a reasonably efficient way to do this in the special case of *Dirichlet L-functions*. Indeed by definition, if χ is a Dirichlet character of conductor F, for $\Re(s) > 1$ we have

$$L(\chi, s) = \sum_{n \geq 1} \frac{\chi(n)}{n^s} = \sum_{1 \leq r \leq F} \chi(r) \sum_{n \equiv r \pmod{F},\, n \geq 1} \frac{1}{n^s}$$

$$= \sum_{1 \leq r \leq F} \chi(r) \sum_{m \geq 0} \frac{1}{(mF + r)^s} = F^{-s} \sum_{1 \leq r \leq F} \chi(r) \zeta(s, r/F) .$$

By analytic continuation, this is true for all $s \neq 1$, but in fact since the `ZetaHurwitz` program returns $-\psi(x)$ when $s = 1$, this formula will still be valid for $s = 1$ if χ is a nontrivial character.

Thus, the following program computes $L(\chi, s)$ in a simple-minded manner (note that even independently of Chapter 9 there are much more efficient ways to do this):

_____ Lfunchisimple.gp _____

```
1  /* Let chi be a Dirichlet character modulo F, given as a
2   * closure. Compute L(chi,s). */
3  Lfunchisimple(chi, F, s) =
4  { my(S = 0);
5
6    for (r = 1, F,
7      my(c = chi(r));
8      if (c, S += c * ZetaHurwitz(s, r / F)));
```

```
9     return (F^(-s) * S);
10  }
```

EXAMPLE.

```
? \p5009
? Lfunchisimple(x -> kronecker(-4, x), 4, 3) - Pi^3 / 32
time = 91 ms.
% = 7.258... E-5010
? Lfunchisimple(x -> kronecker(-4, x), 4, 1) - Pi / 4
time = 76 ms.
% = 7.258... E-5010
```

In this last example the Hurwitz zeta function is replaced by $-\psi(x)$, but the result is still correct. The above timings assume that Bernoulli numbers are precomputed; if the Bernoulli cache is not yet available, the above examples would require 710 ms each.

Better still, we can obtain Euler–Maclaurin formulas giving a way to compute $\log(\Gamma(x))$. There are at least three equivalent way to so this. The first is to integrate Example 4.2.7 from 1 to x, the second is to use the (almost defining) formula $\log(\Gamma(x)) = \zeta'(0, x) - \zeta'(0, 1)$ (derivative with respect to s), and the third is to apply Euler–Maclaurin directly on the function $f(t) = \log(x + t)$.

EXAMPLE. For $k \geq 1$ we have

$$\sum_{m=0}^{N-1} \log(m + x) = (N + x - 1/2)\log(N + x - 1) - N$$

$$+ \frac{1}{2}\log(2\pi) + 1 - x - \log(\Gamma(x))$$

$$+ \sum_{1 \leq j \leq \lfloor k/2 \rfloor} \frac{B_{2j}}{2j(2j-1)}(N + x - 1)^{1-2j} + R_k(x, N) \,,$$

where

$$R_k(x, N) = -\frac{1}{k}\int_N^\infty (t + x - 1)^{-k} B_k(\{t\})\, dt \,.$$

In addition, if k is even we have

$$|R_k(x, N)| \leq \left|\frac{B_{k+2}}{(k+1)(k+2)}(N + x - 1)^{-k-1}\right| \,.$$

PROOF. Applying Euler–Maclaurin we obtain the above formula with some constant (equal to $z(f; 0) - x\log(x) + 1$ for $f(t) = \log(t + x)$). On the other hand, by definition of the gamma function we have

$$\log(\Gamma(x)) = \lim_{N \to \infty}\left(\log(N!) - \sum_{0 \leq m < N} \log(m + x) + (x - 1)\log(N)\right) \,,$$

and by Stirling's formula (which follows from Euler–Maclaurin apart from the constant $\log(2\pi)/2$), we have $\log(N!) = (N + 1/2)\log(N) - N + \log(2\pi)/2 + o(1)$, so a small computation gives the constant as above. $\qquad\square$

The corresponding GP program is an immediate modification of the one given above, and is in fact simpler. Once again, the program that we give can be improved in several ways, first for efficiency, and second to take into account the range of x (for instance the special cases where $x \in \mathbb{Z}_{\leq 0}$). As usual, we leave this to the reader. Note that the only optimization that we have done is to replace a costly sum of logarithm by the logarithm of the product.

Lngamma.gp

```
1   /* Compute log(Gamma(x)) */
2   Lngamma(x) =
3   { my(B = getlocalbitprec(), k, N, Nx, C, N2, S1, S2, res);
4
5     localbitprec(32);
6     k = max(10, B*log(2) / (1 + lambertw(Pi)));
7     k = 2 * ceil(k / 2); /* ensure that k is even */
8     bernvec(k \ 2 + 1); /* initialize Bernoulli cache */
9     C = bernreal(k+2) / ((k+1) * (k+2));
10    N = max(1, ceil((abs(C) * 2.^B)^(1/(k+1)) + 1 - real(x)));
11    localbitprec(B);
12    Nx = N + x - 1; N2 = Nx^(-2);
13    S1 = log(prod(m = 0, N - 1, m + x, 1.));
14    S2 = 0;
15    forstep (j = k, 2, -2, S2 = bernreal(j) / ((j-1)*j)+S2 * N2);
16    res = (Nx + 1/2) * log(Nx) - N
17          + log(2*Pi)/2 + 1 - x - S1 + S2/Nx;
18    return (res - 2*Pi*I * round(imag(res) / (2*Pi)));
19  } /* principal determination */
20
21  Gamma(x) = exp(Lngamma(x));
```

The above is very well known. Less well-known although easy is that one can obtain formulas for the constant $z(f; a)$ for functions f related to t^{α}, hence to $\zeta(-\alpha)$ or more generally $\zeta(-\alpha, s)$. Indeed, assume that in Euler–Maclaurin (say Corollary 4.2.8) the function $f(t)$ depends on some other parameter α in a differentiable manner, so that we write $f(\alpha, t)$ instead of $f(t)$. Under suitable regularity assumptions all the terms in the formula, except possibly $z(f; a)$, are differentiable in α. It follows that so is $z(f; a)$, and moreover that $z(f'_{\alpha}; a) = z(f; a)'_{\alpha}$ with evident notation.

An example of this is as follows:

PROPOSITION 4.2.29.

(1) *For $\alpha \neq -1$ we have*

$$\sum_{m=1}^{N} m^{\alpha} \log(m) = -\zeta'(-\alpha) + \frac{N^{\alpha+1}}{\alpha+1}\left(\log(N) - \frac{1}{\alpha+1}\right) + \frac{N^{\alpha}\log(N)}{2}$$

$$+ \sum_{1 \leq j \leq \lfloor k/2 \rfloor} \binom{\alpha}{2j-1} \frac{B_{2j}}{2j} N^{\alpha-2j+1}\left(\log(N) + \sum_{0 \leq i < 2j-1} \frac{1}{\alpha-i}\right)$$

$$+ (-1)^{k}\binom{\alpha}{k} \int_{N}^{\infty} t^{\alpha-k}\left(\log(t) + \sum_{0 \leq i < k} \frac{1}{\alpha-i}\right) B_{k}(\{t\})\, dt\,,$$

where it is understood that when α is an integer such that $0 \leq \alpha < n$, we take limits, in other words $\binom{\alpha}{n} = 0$ and

$$\binom{\alpha}{n} \sum_{0 \leq i < n} \frac{1}{\alpha-i} = (-1)^{n-\alpha-1} \frac{\alpha!(n-1-\alpha)!}{n!}\,.$$

(2) *For $\alpha = -1$ the same formula remains valid if we replace*

$$-\zeta'(-\alpha) + \frac{N^{\alpha+1}}{\alpha+1}\left(\log(N) - \frac{1}{\alpha+1}\right)$$

by its limit $\gamma_1 + \log^2(N)/2$ as α tends to -1, where the first Stieltjes constant $\gamma_1 = -0.0728158\cdots$ was computed at the end of Chapter 2.

PROOF. This follows directly from Euler–Maclaurin, using the fact that the derivative of $\binom{\alpha}{n}$ is equal to $\binom{\alpha}{n}\sum_{0 \leq i < n} 1/(\alpha-i)$, except for the constant, essentially $z(f;1)$. To obtain it, we simply choose $x = 1$ in Example 4.2.7, and take the derivative with respect to α. $\qquad\square$

Exercise: Using this proposition, write a program to compute $\zeta'(s)$ for $s \neq 1$, and compare the efficiency of this program with the direct computation of $\zeta'(s)$ using numerical differentiation, as well as a program to compute the constant γ_1.

COROLLARY 4.2.30. *As $N \to \infty$ we have*

$$\prod_{1 \leq m \leq N} m = N! \sim N^{B_1(N+1)} e^{-N} e^{-\zeta'(0)}\,,$$

$$\prod_{1 \leq m \leq N} m^m \sim N^{B_2(N+1)/2} e^{-N^2/4} e^{-\zeta'(-1)+1/12}\,,$$

$$\prod_{1 \leq m \leq N} m^{m^2} \sim N^{B_3(N+1)/3} e^{-N^3/9 - N/12} e^{-\zeta'(-2)}\,,$$

and $\zeta'(0) = -\log(2\pi)/2$ and $\zeta'(-2) = \zeta(3)/(4\pi^2)$.

The constant which enters in the second formula above is known as the Glaisher–Kinkelin constant.

Exercise. Let $r \in \mathbb{Z}_{\geq 1}$, and as usual let $H_k = \sum_{1 \leq j \leq k} 1/j$. Generalizing Corollary 4.2.30, show that

$$\sum_{1 \leq m < N} m^{r-1} \log(m) = \frac{B_r(N)}{r}(\log(N) + H_{r-1})$$

$$- \frac{1}{r} \sum_{0 \leq j \leq r-1} \binom{r}{j} H_{r-j} B_j N^{r-j} - \zeta'(1-r) + o(1) \ .$$

4.2.8. Definite integrals coming from Euler–Maclaurin. Before going any further with these examples, we give some definite integrals for $\zeta(s,x)$ and $\psi(x)$ coming directly from Euler–Maclaurin: recall from Proposition 4.2.17 that there is a definite integral involving $\{t\} - 1/2$ and one involving the (inverse) Laplace transform, and from Proposition 4.2.19 one involving complex integration. In the special case of $\zeta(s,x)$ and $-\psi(x)$, which are the constants $z(f;a)$ occurring in the above examples, this gives the following (recall that one recovers $\zeta(s)$ and γ by setting $x = 1$):

PROPOSITION 4.2.31.

(1) *For* $\Re(s) > 0$ *we have*

$$\zeta(s,x) = \frac{x^{1-s}}{s-1} + x^{-s} - s \int_1^\infty (t+x-1)^{-s-1}\{t\}\,dt \ .$$

(2) *Let* $s \notin \mathbb{Z}_{\leq 1}$. *We have*

$$\zeta(s,x) = \frac{x^{1-s}}{s-1} + \frac{1}{\Gamma(s)} \int_0^\infty t^{s-1} e^{-tx} \left(\frac{1}{1-e^{-t}} - \frac{1}{t} \right) dt$$

$$= \frac{1}{\Gamma(s)} \int_0^\infty \frac{t^{s-1} e^{-tx}}{1-e^{-t}}\,dt \ ,$$

the first integral being valid for $\Re(s) > 0$ *and the second for* $\Re(s) > 1$.

(3) *For all* $s \neq 1$ *we have*

$$\zeta(s,x) = \frac{x^{1-s}}{s-1} + \frac{x^{-s}}{2} + i \int_0^\infty \frac{(x+it)^{-s} - (x-it)^{-s}}{e^{2\pi t} - 1}\,dt \ .$$

(4) *For* $\Re(s) < 0$ *we have*

$$\zeta(s) = 2\sin(\pi s/2) \int_0^\infty \frac{t^{-s}}{e^{2\pi t} - 1}\,dt \ .$$

PROOF. Note that we have

$$\zeta(-\alpha, x) = z(f; 0) - x^{\alpha+1}/(\alpha+1) \ , \quad \text{where } f(t) = (t+x)^\alpha \ .$$

Furthermore, we have $\int_0^\infty t^{s-1} e^{-ta}\,dt = \Gamma(s)a^{-s}$ by definition of the gamma function, so applying Proposition 4.2.17 to $f(u) = (u+x)^\alpha = (u+x)^{-s}$ and $g(t) = t^{-\alpha-1} e^{-tx}/\Gamma(-\alpha)$ gives the first two integrals. Note that

$$\int_0^\infty t^{s-1} e^{-tx}/t, dt = \Gamma(s-1)/x^{s-1} = \Gamma(s)/((s-1)x^{s-1})$$

only for $\Re(s) > 1$. The third integral follows from the Abel–Plana formula (Proposition 4.2.19) applied with $a = 0$. For the fourth, we note that if $\Re(\alpha) > 0$ we may apply Euler–Maclaurin for $f(t) = t^\alpha$ (as opposed to $(t+1)^\alpha$) with $a = 0$, so Abel–Plana gives us an integrand whose numerator is

$$(it)^{-s} - (-it)^{-s} = t^s(e^{-is\pi/2} - e^{is\pi/2}) = -2i\sin(s\pi/2)t^s ,$$

proving the result. The reader may want to apply the second Abel–Plana formula as an exercise. □

Exercise: Show that if $-1 < \Re(s) < 0$ we have

$$\zeta(s) = -s\int_0^\infty t^{-s-1}(\{t\} - 1/2)\, dt .$$

It will in particular be necessary to show the convergence of the integral.

REMARKS 4.2.32.

(1) As we have stated it, the Abel–Plana formula applies only to functions which are real on $\mathbb{R}$, which is not the case if s (or α) is not real. However the integrals that we have given are still valid by analytic continuation.

(2) The last formula immediately implies the well-known result that $\zeta(s)$ vanishes when s is a strictly negative even integer.

(3) As an example of the third formula, we have

$$\frac{\pi^2}{6} = \zeta(2) = \frac{3}{2} + 4\int_0^\infty \frac{t}{(1+t^2)^2(e^{2\pi t} - 1)}\, dt .$$

The analogous results for the function $\psi(x)$ (hence for Euler's constant $\gamma = -\psi(1)$) are as follows:

PROPOSITION 4.2.33. *We have*

$$\psi(x) = \log(x) - \frac{1}{x} + \int_0^\infty \frac{\{t\}}{(t+x)^2}\, dt = \log(x) - \int_0^\infty e^{-xt}\left(\frac{1}{1-e^{-t}} - \frac{1}{t}\right)dt$$

$$= \int_0^\infty \left(\frac{e^{-t}}{t} - \frac{e^{-xt}}{1-e^{-t}}\right)dt = \log(x) - \frac{1}{2x} - 2\int_0^\infty \frac{t}{(x^2+t^2)(e^{2\pi t} - 1)}\, dt .$$

PROOF. Same as above and left to the reader, using the easy formula $\log(x) = \int_0^\infty (e^{-t} - e^{-xt})/t\, dt$. □

Since $\psi(x) = \Gamma'(x)/\Gamma(x)$ is the logarithmic derivative of the gamma function, we can integrate the above to obtain formulas for $\log(\Gamma(x))$:

COROLLARY 4.2.34. *We have*

$$\log(\Gamma(x)) = (x-1)\log(x) - x + 1 + (x-1)\int_0^\infty \frac{\{t\}}{(t+1)(t+x)}\, dt$$

$$= x\log(x) - x + 1 - \int_0^\infty \frac{e^{-t} - e^{-xt}}{t}\left(\frac{1}{1-e^{-t}} - \frac{1}{t}\right)dt$$

$$= \int_0^\infty \left((x-1)\frac{e^{-t}}{t} - \frac{e^{-t} - e^{-xt}}{t(1-e^{-t})}\right)dt$$

$$= \left(x - \frac{1}{2}\right)\log(x) - x + 1 - 2\int_0^\infty \frac{\text{atan}(t) - \text{atan}(t/x)}{e^{2\pi t} - 1}\, dt .$$

PROOF. Since $\log(\Gamma(1)) = 0$, we must integrate from 1 to x the above formulas. For the last three there is no problem. For the first we simply note that the integral in x from of $(t+x)^{-2}$ is equal to $1/(t+1) - 1/(t+x) = (x-1)/((t+1)(t+x))$. □

COROLLARY 4.2.35. *We have*

$$\frac{\log(2\pi)}{2} = 1 + \int_0^\infty \frac{\{t\} - 1/2}{t + 1}\, dt = 1 - \int_0^\infty \frac{e^{-t}}{t} \left(\frac{1}{1 - e^{-t}} - \frac{1}{t} - \frac{1}{2} \right) dt$$

$$= \int_0^\infty \left(\frac{1}{t} - \frac{e^{-t}}{2} - \frac{1}{e^t - 1} \right) \frac{dt}{t} = 1 - 2 \int_0^\infty \frac{\text{atan}(t)}{e^{2\pi t} - 1}\, dt \ .$$

PROOF. Use Stirling's formula, and take care of convergence problems in the first integral. The details are left as an exercise for the reader. $\square$

We leave to the reader the easy task of implementing all of the above integrals using the DE.

4.2.9. Definite integrals involving fractional parts. Although much less efficient, it is amusing to note that instead of using the DE on integrals involving e^{-t}, it is also possible to use the integral representations involving fractional parts, exactly as we did for SumRamanujanz above, as follows:

—————————————— EulersumRamanujan.gp ——————————————

```
1   ZetaHurwitzRam(s, x) =
2   { my(S);
3     S = IntoscSidi(t -> frac(t) / (t + x)^(s + 1));
4     return (x^(1 - s) / (s - 1) + x^(-s) - s * S);
5   }
6   ZetaRam(s) =
7   { my(S = IntoscSidi(t -> frac(t) / (t + 1)^(s + 1)));
8     return (s * (1 / (s - 1) - S));
9   }
10  PsiRam(x) =
11  { my(S = IntoscSidi(t -> frac(t) / (t + x)^2));
12    return (log(x) - 1 / x + S);
13  }
14  LngammaRam(x) =
15  { my(S);
16    S = IntoscSidi(t -> frac(t) / ((t + 1) * (t + x)));
17    return ((x - 1) * (log(x) - 1 + S));
18  }
19  EulerRam() = 1 - IntoscSidi(t -> frac(t) / (t + 1)^2);
20  Log2piRam() =
21  { 2 * (1 + IntoscSidi(t -> (frac(t) - 1/2) / (t + 1))); }
```

Note that there are evident restrictions on the parameters given in the above integral formulas, but also that they need to be given either exactly or to much larger accuracy than the current one because of the restrictions of the IntoscSidi program. In any case, these programs are given as illustrations and are of course not at all competitive.

More generally, note the following proposition which is an immediate consequence of the Euler-Maclaurin formula:

PROPOSITION 4.2.36. *Assume suitable convergence and differentiability conditions on the function f defined on $[1, \infty[$, and set*

$$I_n = \int_1^\infty f^{(n)}(t) B_n(\{t\}) \, dt \ .$$

For $n \geq 2$ we have

$$I_n + n I_{n-1} = -f^{(n-1)}(1) B_n \ ,$$

and for $n = 1$

$$I_1 = \lim_{N \to \infty} \left(\sum_{1 \leq m \leq N} f(m) - \frac{f(1) + f(N)}{2} - \int_1^N f(t) \, dt \right) \ .$$

COROLLARY 4.2.37. *For $n \geq 1$ and $m \geq 1$ set*

$$I_{m,n} = \int_1^\infty \frac{B_n(\{x\})}{x^m} \, dx \ .$$

(1) *For $m \geq 1$ we have*

$$I_{m,1} = \begin{cases} -\dfrac{1}{m-1}\left(\zeta(m-1) - \dfrac{1}{2} - \dfrac{1}{m-2} \right) & \text{for } m \geq 3 \ , \\ 1/2 - \gamma & \text{for } m = 2 \ , \\ \log(2\pi)/2 - 1 & \text{for } m = 1 \ . \end{cases}$$

(2) *For $m \geq n \geq 2$ we have*

$$I_{m,n} = \frac{n}{\binom{m-1}{m-n}} \left(I_{m-n,1} + \sum_{2 \leq j \leq n} \binom{m-n+j-2}{m-n} \frac{B_j}{j(j-1)} \right) \ .$$

PROOF. Simply apply the proposition and Corollary 4.2.8 to $f(x) = 1/x^m$ when $m \geq 1$ and $f(x) = \log(x)$ when $m = 0$.   □

For $m \leq n$, the result is as follows:

PROPOSITION 4.2.38.

(1) *For $n \geq 1$ we have*

$$I_{1,n} = (-1)^n \left(n \left(\zeta'(1-n) - \frac{B_n}{n} H_{n-1} \right) + \sum_{j=0}^n \binom{n}{j} B_j H_{n-j} \right) \ .$$

(2) *For $n \geq m \geq 1$ we have*

$$I_{m,n} = \frac{n!}{(m-1)!} \left(\sum_{0 \leq j \leq m-2} \frac{(m-j-2)!}{(n-j)!} B_{n-j} + \frac{I_{1,n-m+1}}{(n-m+1)!} \right) \ .$$

PROOF. Same, applied to $f(x) = (x^{n-1}/(n-1)!)(\log(x) - H_{n-1})$.   □

COROLLARY 4.2.39. *For $n \geq 0$ and $m \geq 2$ set*

$$J_{m,n} = \int_1^\infty \frac{\{x\}^n}{x^m} \, dx \ .$$

Then $J(m,n)$ is a rational linear combination of 1, $\zeta(j)$ for $2 \leq j \leq m-1$, and $\zeta'(1-j)$ for $0 \leq j \leq n-m+1$ when $n \geq m-1$, where $\zeta'(1)$ is to be interpreted as γ.

PROOF. Follows from the formula $x^n = (1/(n+1)) \sum_{0 \le j \le n} \binom{n+1}{j} B_j(x)$.   □

4.2.10. Other applications of Euler–Maclaurin. The Euler–Maclaurin formula has a great number of applications. Among the most important are the following:

- Find the asymptotic expansion of the Nth partial sum of a divergent series.
- Find the asymptotic expansion of the Nth remainder of a convergent series, and consequently considerably accelerate the convergence of the series.
- Find the asymptotic expansion of the difference between a definite integral and corresponding Riemann sums, which allows us to compute much more accurately and much faster the numerical value of the integral.
- Determine whether a given series converges by comparison with the corresponding integral.

We have already seen a large number of such problems, and examples can of course be varied at will:

Exercise. Using Euler–Maclaurin for $k = 2$, show that if x is a nonzero real number and $\alpha \in \mathbb{R}$ the series $\sum_{m \ge 1} \sin(x \log(m))/m^\alpha$ converges if and only if $\alpha > 1$ (note that this is marked contrast with the series $\sum_{m \ge 1} \sin(xm)/m^\alpha$ which converges for $\alpha > 0$).

It is clear that Euler–Maclaurin gives us a precise estimate of the difference between a sum and an integral. More precisely:

PROPOSITION 4.2.40. *Assume that f is a C^∞ function on some compact interval $[a,b]$. Then for any $N \ge 1$, if we set $h = (b-a)/N$ (the "step size"), for any $k \ge 1$ we have*

$$\int_a^b f(t)\,dt = h \sum_{m=0}^{N-1} f(a+mh) + h\frac{f(b)-f(a)}{2}$$
$$- \sum_{1 \le j \le \lfloor k/2 \rfloor} \frac{B_{2j}}{(2j)!} h^{2j} \left(f^{(2j-1)}(b) - f^{(2j-1)}(a) \right)$$
$$+ \frac{(-1)^k}{k!} h^k \int_a^b f^{(k)}(t) B_k(\{(t-a)/h\})\,dt \ .$$

PROOF. Immediate from Euler–Maclaurin and left to the reader.   □

4.3. Pinelis summation

A different summation method was introduced recently by I. Pinelis [**Pin18**] but in its original form it is applicable only in a special situation. Recall that one of the main difficulties in Euler-Maclaurin summation is that we need to compute a number of derivatives of the summand f, and this is not easy in general. We explained above how to circumvent this by using *finite differences* instead of derivatives, leading to the `Sumdelta` program.

Pinelis's idea is completely different: assume that we have an explicit formula (or fast algorithm for that matter) to compute the *antiderivative* F of f (i.e., such that $F'(x) = f(x)$). He then proves the following:

THEOREM 4.3.1. *For any positive integer m and integer r such that $1 \leq r \leq m$ set*

$$t_{m,r} = (-1)^{r-1} \frac{2}{\binom{2m}{m}} \sum_{j=0}^{\lfloor (m-r)/2 \rfloor} \frac{1}{r+2j} \binom{2m}{m+r+2j} .$$

Let F be the unique antiderivative of f which vanishes at infinity (i.e., $F(x) = \int_\infty^x f(t)\, dt$), and set

$$G_m(n) = t_{m,1} F(n-1/2) + \sum_{1 \leq j < m} t_{m,j+1} (F(n-(j+1)/2) + F(n+(j-1)/2)) .$$

Then for any integer $N \geq 0$ for which the sums are defined we have

$$\sum_{k \geq 0} f(k) = \sum_{0 \leq k < N} f(k) - G_m(N) + R_{m,N} ,$$

where $R_{m,N}$ is an error term which can be explicitly bounded.

Note that the $t_{m,r}$ are the solutions of the linear system coming from the condition that $G_m(n+1) - G_m(n) = f(n)$ for all polynomials f of degree less than or equal to $2m - 1$.

As an important supplement to this theorem, the author shows that for many applications the choice $m = 0.15B$ and $N = 0.5B$ is reasonable, where B denotes the desired bit accuracy. It seems, however, that $N = 0.3B$ is sufficient, but in any case we leave these "magic constants" as optional parameters. This leads to the following simple-minded program:

———————————————— SumPinelis.gp ————————————————

```
1   SumPinelisinit(cm = 0.15) =
2   { my(B = getlocalbitprec(), m = ceil(cm * B));
3     my(V = binomial(2 * m), d = V[m + 1] / 2, D, X, W, S);
4
5     D = vector(m, j, (-1)^(j-1) * V[m + j + 1] / (j * d));
6     W = vector(m);
7     S = 0; forstep (s = m, 1, -2, S += D[s]; W[s] = S);
8     S = 0; forstep (s = m - 1, 1, -2, S += D[s]; W[s] = S);
9     X = vector(m, j, (j - 1) / 2);
10    return ([X, W]);
11  }
12
13  SumPinelis(f, F = 0, cN = 0.3, XW = 0.15) =
14  { my(B = getlocalbitprec(), S, N);
15
16    if (type(XW) != "t_VEC", XW = SumPinelisinit(XW));
17    N = ceil(cN * B) - 1 / 2;
18    F = Sumintinit(f, F);
19    S = SumXW(x -> if (x, F(N-x) + F(N+x), F(N)), XW);
20    localbitprec(B + 32);
21    S = sum(n = 1, N, f(n), 0.) - S;
22    return (bitprecision(S, B));
23  }
```

EXAMPLES.

```
? \p1000
? f(x) = 1 / (1 + x^2);
? F(x) = atan(x) - Pi/2;
? R = SumPinelis(f, F);
time = 318 ms.
? R - Sumdelta(f, F)
time = 1,360 ms.
% = 0.E-1001
? R - sumnum(x = 1, f(x))
time = 2,708 ms.
% = 0.E-1001
? f(x) = log(x + 1) / (x + 1)^2;
? F(x) = -(log(x + 1) + 1) / (x + 1);
? R = SumPinelis(f, F);
time = 227 ms.
? R - Sumdelta(f, F)
time = 1,568 ms.
% = 0.E-1001
? R - sumnum(x = 1, f(x))
time = 4,404 ms.
% = 1.48... E-1002
```

We see that in these cases SumPinelis is an order of magnitude faster than the built-in Euler-Maclaurin sumnum method, and also considerably faster than Sumdelta even when helped by the knowledge of the antiderivative F.

Knowing an explicit antiderivative is quite restrictive. If we really want to use this method when the antiderivative is not known "explicitly", we can simulate it by numerical integration commands, which is done by setting the parameter F to 0 or to a parameter in an integration command, see the Sumintinit program used above in the SumEulerMaclaurin program. Evidently, the resulting program will be much slower, but will at least work in all reasonable cases. For instance:

```
? \p308
? f(x) = 1 / (1 + x^2);
? F(x) = atan(x) - Pi/2;
? R = SumPinelis(f, F);
time = 12 ms.
? R - SumPinelis(f, 0)
time = 1,928 ms.
% = 0.E-307
? R - SumPinelis(f, [1, 2])
time = 353 ms.
% = 0.E-307
? FD = Sumintinit(f, 0); /* Initialize for use with DE */
time = 55 ms.
? FG = Sumintinit(f, [1, 2]); /* Initialize for use with Gauss */
time = 56 ms.
? XW = SumPinelisinit(); /* Initialize coefficients */
time = 1 ms.
```

```
? SumPinelis(f, FD, , XW);
time = 1,863 ms.
? SumPinelis(f, FG, , XW);
time = 296 ms.
```

The time for this last example is of course longer than knowing the explicit antiderivative, but still an order of magnitude faster than using the DE, as is usually the case when one can use Gaussian integration.

One can modify Pinelis's idea by using values of the function f itself instead of its antiderivative for the sum giving the difference between the series and the integral. To do this, we simply solve a linear system coming from the fact that we want the highest possible derivatives to vanish. This is very similar to Pinelis's method, but unfortunately, contrary to the method above, we have not found an explicit formula for the coefficients, so the initialization time is quite long. This method being essentially identical to `Sumdelta` with $d = 1/4$ with no speed advantage, we do not give an explicit program for it.

4.4. Sums and products of rational functions

If the function $f(x)$ to be summed is a *rational function*, there exists an explicit formula given by the following proposition:

PROPOSITION 4.4.1. *Let f be a rational function, and let its decomposition into partial fractions be*

$$f(x) = \sum_{\alpha \ pole} \sum_{1 \leq k \leq -v(\alpha)} \frac{a_{\alpha,k}}{(x - \alpha)^k} \, ,$$

where α runs through the poles of f, $-v(\alpha) \geq 1$ denotes the order of the pole α, and $a_{\alpha,k} \in \mathbb{C}$. Assume that $x^2 f(x)$ is bounded when $x \to \infty$, in other words that $\sum_\alpha a_{\alpha,1} = 0$. Then

$$\sum_{n \geq 0} f(n) = \sum_{\alpha \ pole} \sum_{1 \leq k \leq -v(\alpha)} (-1)^k \frac{a_{\alpha,k}}{(k - 1)!} \psi^{(k-1)}(-\alpha) \, ,$$

where $\psi(x)$ is the usual logarithmic derivative of the gamma function.

Since in Chapter 3 we have given the `Ratdec` program to compute a decomposition into partial fractions, it is thus immediate to write the program corresponding to the above proposition. For simplicity we will assume that the rational function has exact coefficients, otherwise the determination of the multiplicity of the poles is subject to caution:

───────────────────────────── Sumrat.gp ─────────────────────────────

```
1   /* F an exact rational function, compute sum_{n>=1} F(n). */
2   Sumrat(F) =
3   { my(RD, S = 0., X = variable(F));
4
5     if (poldegree(F) >= - 1, error("divergent sum"));
6     RD = Ratdec(F);
7     for (i = 1, #RD,
8       my([z, ct, rr, sz] = RD[i], s = -psi(-X-z+1 + O(X^ct)));
9       S += sum(k = 1, ct,
10        my(t = polcoef(s, k - 1) * polcoef(sz, ct - k));
```

```
11      if (rr == 2, 2 * real(t), t)));
12    if (!imag(F), S = real(S));
13    return (S);
14  }
```

Similarly, we have a (simpler) analogous proposition if we want to compute an infinite *product* of a rational function:

PROPOSITION 4.4.2. *Let f be a rational function, and write*

$$f(x) = C \prod_{\alpha \; zero \; or \; pole} (x - \alpha)^{v(\alpha)} ,$$

where α runs through the zeros and poles of f, $v(\alpha) \in \mathbb{Z}$ is the order of α (positive for a zero, negative for a pole), and $C \in \mathbb{C}^*$. Assume that $x^2(f(x) - 1)$ is bounded as $x \to \infty$, in other words that $C = 1$, $\sum_\alpha v(\alpha) = 0$, and $\sum_\alpha \alpha v(\alpha) = 0$. Then

$$\prod_{n \geq 0} f(n) = \prod_{\alpha \; zero \; or \; pole} \Gamma(-\alpha)^{-v(\alpha)} .$$

The corresponding much simpler program which only uses the `Polrootsspec` program given in Chapter 3 is as follows:

──────────────────────── `Prodrat.gp` ────────────────────────

```
1   /* Z is from Polroots() */
2   Prodaux(Z) =
3   { my(S = 1.);
4
5     for (i = 1, #Z,
6       my([z, ct, rr] = Z[i], g = gamma(1 - z));
7       if (rr == 2, g = norm(g));
8       S *= g^ct);
9     return (S);
10  }
11  /* F an exact rational function, compute prod_{n>=1} F(n). */
12  Prodrat(F) =
13  { my(P = numerator(F), Q = denominator(F));
14
15    if (poldegree(P - Q) >= poldegree(Q) - 1,
16      error("divergent product"));
17    return (Prodaux(Polroots(Q)) / Prodaux(Polroots(P)));
18  }
```

EXAMPLE.
```
? F(k) = Prodrat(((x + 1)^k + 1) / ((x + 1)^k - 1));
? F(2) - sinh(Pi) / Pi
% = 0.E-37
? F(3)
% = 1.5000000000000000000000000000000000000000
```

As an exercise, the reader can prove the above results and generalize at least to all even $k \geq 2$.

4.5. Summation of oscillating series

4.5.1. Summation of alternating series: The CVZ method. We now come to a method which is specific to the summation of alternating series, i.e., which is not a variant of methods for summing positive terms.

It was noticed relatively recently by F. Rodriguez–Villegas, D. Zagier, and the second author, that one can compute the sum of alternating series in a much more efficient and amazingly simple way, see [**CRVZ00**] (a look at the literature shows that the idea had been stated previously, but we give this reference since it is easily accessible). It has the additional advantage that, contrary to all the other methods that we have seen (with the exception of `Sumaltdelta` with $d = 1$), it only requires f to be defined on the integers, and not on the reals.

Assume that we want to compute $S = \sum_{n \geq 0}(-1)^n u(n)$, where $u(n)$ is a sufficiently regular sequence, for instance (but not necessarily) monotonically decreasing and tending to 0 as $n \to \infty$. Under suitable additional assumptions which can be made completely explicit, one can prove that $u(n)$ is the nth moment of a suitable positive weight function $w(x)$ on $[0, 1]$, in other words that

$$u(n) = \int_0^1 x^n w(x)\, dx .$$

It follows immediately that

$$S = \sum_{n \geq 0}(-1)^n u(n) = \int_0^1 \frac{w(x)}{x+1}\, dx .$$

This is the first idea. The second idea is as follows: let $P_n(X)$ be a sequence of polynomials of degree n. Then $(P_n(-1) - P_n(X))/(X+1)$ is now a polynomial of degree $n - 1$, so we can write

$$\frac{P_n(-1) - P_n(X)}{X+1} = \sum_{0 \leq j \leq n-1} c_{n,j} X^j ,$$

say. By definition of w it follows that

$$\int_0^1 \frac{P_n(-1) - P_n(x)}{x+1} w(x)\, dx = \sum_{0 \leq j \leq n-1} c_{n,j} u(j) .$$

On the other hand,

$$\int_0^1 \frac{P_n(-1) - P_n(x)}{x+1} w(x)\, dx = P_n(-1)S - R_n ,$$

with

$$R_n = \int_0^1 \frac{P_n(x)}{x+1} w(x)\, dx .$$

Since clearly $|R_n| \leq \sup_{x \in [0,1]} |P_n(x)| S$, we have shown the following:

PROPOSITION 4.5.1. *With the above notation, and assuming $P_n(-1) \neq 0$ we have*

$$S = \frac{1}{P_n(-1)} \sum_{0 \leq j \leq n-1} c_{n,j} u(j) + R'_n ,$$

with

$$|R'_n| \leq \frac{\sup_{x\in[0,1]}|P_n(x)|}{|P_n(-1)|}S .$$

To take a specific example, let $n = 3$ and

$$P_3(X) = -32x^3 + 48x^2 - 18x + 1 .$$

It is immediate to show that $\sup_{x\in[0,1]}|P_3(x)| = 1$, that $P_3(-1) = 99$, and that $(P_3(-1)-P_3(X))/(X+1) = 32X^2-80X+98$. We thus deduce from the proposition that

$$S \approx (98u(0) - 80u(1) + 32u(2))/99$$

with an error of at most 1%. In other words, knowing only three terms of the sequence we obtain a reasonable approximation to the sum!

The simplest example is $u(n) = 1/(n + 1)$, with $S = \log(2) = 0.693147\cdots$, while $(98u(0) - 80u(1) + 32u(2))/99 = 0.693602\cdots$, even better than expected, quite remarkable.

In fact the choice of polynomial P_3 is a special case of shifted *Chebyshev polynomials*. Recall that one defines the Chebyshev polynomial (of the first kind) $T_n(X)$ by $\cos(nt) = T_n(\cos(t))$. If $x \in [-1, 1]$ we set $t = \text{acos}(x)$, hence $|T_n(x)| = |\cos(n\,\text{acos}(x))| \leq 1$ and with $t = 0$ we see that $T_n(1) = 1$ so $\sup_{x\in[-1,1]}|T_n(x)| = 1$. Since we want the interval $[0, 1]$ instead of $[-1, 1]$, we define $P_n(X) = T_n(1 - 2X)$, so that $\sup_{x\in[0,1]}|P_n(x)| = 1$, and on the other hand $P_n(-1) = T_n(3)$. Since $e^{int} = \left(\cos(t) + i\sqrt{1 - \cos^2(t)}\right)^n$ we deduce that

$$T_n(x) = \left(\left(x + \sqrt{x^2 - 1}\right)^n + \left(x - \sqrt{x^2 - 1}\right)^n\right)/2 .$$

In particular

$$T_n(3) = \left(\left(3 + 2\sqrt{2}\right)^n + \left(3 - 2\sqrt{2}\right)^n\right)/2$$

(note in passing that $3 \pm 2\sqrt{2} = \left(\sqrt{2} \pm 1\right)^2$), proving the following:

COROLLARY 4.5.2. *If in the above proposition we choose $P_n(X) = T_n(1-2X)$, the relative error $|R'_n/S|$ satisfies*

$$|R'_n/S| < \left(3 + 2\sqrt{2}\right)^{-n} = 5.828\cdots^{-n} .$$

REMARKS 4.5.3.

(1) It can be shown that the Chebyshev polynomials are close to optimal for minimizing the quantity

$$\sup_{x\in[-1,1]} |T_n(x)/T_n(3)| ,$$

but of course we do not need optimality to use the above bound.

(2) We have proved the error bound rigorously only for sequences $u(n)$ which are moments of some positive weight function $w(x)$, and one can show that this is equivalent to $u(n)$ being *totally monotonic*, i.e., all forward differences $\Delta^k(u)(n)$ (where $\Delta(a)(n) = a(n + 1) - a(n)$) tend to 0 monotonically with sign $(-1)^k$. However, as we will see in examples below, in practice we can use the method well outside of the range where it is proved (and in fact for nonconvergent series).

(3) When $u(n)$ satisfies some even stronger conditions, by estimating the error term less crudely it is possible to find even better acceleration methods, see the paper [**CRVZ00**] for details. We will not consider these improvements here.

To use the proposition, we also need the coefficients $c_{n,j}$ of the polynomial $\big(P_n(-1) - P_n(X)\big)/(X+1)$. This is given as follows:

LEMMA 4.5.4. *Set* $d_n = \Big(\big(3 + 2\sqrt{2}\big)^n + \big(3 - 2\sqrt{2}\big)^n \Big)/2$. *For* $0 \le j \le n-1$ *we have*

$$
c_{n,j} = (-1)^j \left(d_n - \sum_{m=0}^{j} \frac{n}{n+m} \binom{n+m}{2m} 2^{2m} \right)
$$

$$
= (-1)^j \sum_{m=j+1}^{n} \frac{n}{n+m} \binom{n+m}{2m} 2^{2m} \ .
$$

In addition,

$$
\sum_{j=0}^{n-1} c_{n,j} u(j) = \sum_{m=1}^{n} \frac{n}{n+m} \binom{n+m}{2m} 2^{2m} S_{m-1} \ ,
$$

where $S_{m-1} = \sum_{0 \le j \le m-1} (-1)^j u(j)$ *is the* $(m-1)$st *partial sum of the series.*

PROOF. From $1 - 2x = \cos(t)$ we deduce that $x = (1 - \cos(t))/2 = \sin^2(t/2)$, so the polynomials P_n satisfy $P_n(\sin^2(t)) = \cos(2nt)$. Since

$$
\cos(2(n+1)t) + \cos(2(n-1)t) = 2\cos(2nt)\cos(2t) = 2\cos(2nt)(1 - 2\sin^2(t)) \ ,
$$

we deduce that the P_n satisfy the recursion

$$
P_{n+1}(X) = 2(1 - 2X)P_n(X) - P_{n-1}(X) \ ,
$$

from which it is immediate to show by induction the explicit formula

$$
P_n(X) = \sum_{m=0}^{n} (-1)^m \frac{n}{n+m} \binom{n+m}{2m} 2^{2m} X^m \ .
$$

The formula for the coefficients $c_{n,j}$ follows by performing the Euclidean division by $X + 1$.

For the last formula, note that

$$
\sum_{0 \le j \le n} c_{n,j} u(j) = \sum_{0 \le j \le n} (-1)^j u(j) \sum_{j < m \le n} \frac{n}{n+m} \binom{n+m}{2m} 2^{2m} \ ,
$$

so the result follows by exchanging summations. □

The form of the explicit formula for $c_{n,j}$ implies that we can compute it "on the fly", i.e., as we use the successive terms of the sequence $u(n)$, and the last formula can be used if for instance one knows more explicitly the partial sums S_{m-1} than the sequence $u(n)$ itself.

This leads to the following program:

$$\boxed{\text{Sumalt.gp}}$$

```
1   /* Compute sum_{n >= 1} (-1)^{n-1} u(n). */
2   Sumaltinit() =
3   { my(B = getlocalbitprec(), N, c, d, vc, bic);
4
5     localbitprec(32);
6     N = ceil(B * log(2) / log(3 + sqrt(8)));
7     localbitprec(B + 32);
8     d = (3 + sqrt(8))^N;
9     d = (d + 1 / d) / 2; vc = vector(N); bic = -1.; c = -d;
10    for (j = 0, N - 1,
11      c = bic - c;
12      vc[j+1] = c;
13      bic *= 2 * (j + N) * (j - N) / ((2*j + 1) * (j + 1)));
14    return ([N, d, vc]);
15  }
16
17  Sumalt(u, tab = 0) =
18  { my(B = getlocalbitprec(), N, d, vc, S);
19
20    if (!tab, tab = Sumaltinit());
21    localbitprec(B + 32);
22    [N, d, vc] = tab;
23    S = sum(j = 1, N, vc[j] * u(j));
24    return (bitprecision(S / d, B));
25  }
```

REMARKS 4.5.5.

  (1) As usual u is a GP function, given by something like u=t->expr(t). But now, contrary to previous summation methods, we only need values at successive integer values. It is easy to adapt the interface to the case where those values are given by a vector $v = [u(a), u(a+1), \ldots]$: just set u = n->v[n + a + 1]. This will nicely produce a runtime error if Sumalt tries to access a value beyond the end of the vector.

  (2) An equivalent program is already available in GP under the name sumalt.

It is instructive to give a number of examples showing that this acceleration method can be used well outside of its supposed range of applicability. Of course, Sumalt(n -> 1/n) or Sumalt(n -> 1/n^2) give extremely rapidly and to perfect accuracy the expected results $\log(2)$ and $\pi^2/12$ respectively. More surprising is Sumalt(n -> n), which gives 0.25 almost exactly. And indeed, if we write

```
? \p500
? myzeta(s) = Sumalt(n -> n^(-s)) / (1 - 2^(1-s));
? myzeta(-5)
% = -2.861... E-477
? s = 1/4+10*I; myzeta(s) - zeta(s)
% = 1.014... E-496 + 4.632...E-497*I
```

this gives a rather accurate value of the Riemann zeta function quite far from the domain of convergence of the series: for instance $\zeta(-5) = -1/252$, and working at 500 decimal digits the above function gives the correct result to 479 decimal digits. And it works also for complex values, and even better: for $s = \sigma + it$ with for instance $0 < \sigma < 1$ and $|t| \leq 15$, we lose no more than 4 or 5 decimal digits out of 500.

Similarly, we can compute derivatives of $\zeta(s)$ or other similar functions. For instance `Sumalt(n -> -log(n))` gives the derivative of $(1 - 2^{1-s})\zeta(s)$ at $s = 0$, equal to $\log(\pi/2)/2$, to high accuracy.

Note, however, that for such diverging series the `SumaltAbelPlana` program gives better results. For instance working with the default accuracy of 38 digits, `Sumalt(n -> n^9)` gives the result $31/4$ to only 14 correct digits, while `Sumalt-AbelPlana(n -> n^9)` gives the result with perfect accuracy.

For a different kind of example:

```
? \p 38
? Sumalt(n -> (-1)^n * besselj(0, n*Pi) / n)
% = 0.23129892655550671956636378326870797565
? 1/(2*Pi) * intnum(t = 0, Pi, log(2 - 2*cos(Pi*sin(t))))
% = 0.23129892655550671956636378326870248264
/* Apparently 6 decimals lost, but it is Sumalt which is correct! */
```

An important warning is however necessary. Even though the above is a summation method which uses only the discrete values of a sequence $u(n)$, while in the preceding methods we had $u(n) = f(n)$ for some function f defined over the reals, it is absolutely essential that the sequence $u(n)$ be regular, in some reasonable sense. A typical example is $u(n) = 1/(n + (-1)^n)$, i.e., the computation of

$$S = \sum_{n \geq 1} \frac{(-1)^n}{n + (-1)^{n-1}} \ .$$

This is not a "reasonable" alternating series because of the parasitic term $(-1)^{n-1}$ occurring in the denominator (in fact, note that $\sum_{n \geq 1}(-1)^n/n^\alpha$ converges if and only if $\alpha > 0$, while $\sum_{n \geq 1}(-1)^n/(n^\alpha + (-1)^{n-1})$ converges if and only if $\alpha > 1/2$). Here, one can solve the problem by a simple renumbering of the terms, setting $v(2n) = u(2n)$ and $v(2n - 1) = u(2n + 1)$. But in the case of sums like $S = \sum_{n \geq 0}(-1)^n/(2^n + (-1)^n)$ (which of course converges geometrically, so there is no problem in computing the sum anyway), no rearrangement of terms will solve the problem, so probably the best solution is to add together the even and odd terms, thus obtaining (in general) a series with positive terms, which hopefully may be summed using one of the methods seen above such as Euler–Maclaurin.

Exercise: Although $\sum_{n \geq 1}(-1)^n/n = -\log(2)$, prove that

$$S = \sum_{n \geq 1} \frac{(-1)^n}{n + (-1)^{n-1}} = \log(2) \ .$$

Because of the assumption that $u(n)$ be the nth moment of some (positive) weight function, or not too far from such, the algorithm that we have given above is efficient when $u(n)$ tends to 0 (or not!) polynomially, for instance like $1/n^s$. When $u(n)$ tends to 0 *geometrically*, say like z^{-n}/n^s for some $z \geq 1$, one can do better: indeed, in that case instead of moments on the interval $[0, 1]$ one takes moments on the interval $[0, z^{-1}]$. The exact same analysis can be performed, and shows the following: one simply replaces $3 \pm 2\sqrt{2} = \left(1 + \sqrt{2}\right)^2$ by $2z + 1 + 2\sqrt{z(z+1)} = \left(z + \sqrt{z+1}\right)^2$, and includes a factor z in the recursion formula for b. This is left to the reader.

Of course when $z > 1$, one can evaluate the infinite sum directly since it converges geometrically, but it is less efficient since the above method gives a convergence in $\left(2z + 1 + 2\sqrt{z(z+1)}\right)^{-n}$, much better than z^{-n} (e.g., 9.9^{-n} instead of 2^{-n}, more than 3 times as fast).

One may think of using the above in the evaluation of *continued fractions*

$$S = a(0) + \cfrac{b(0)}{a(1) + \cfrac{b(1)}{a(2) + \cfrac{b(2)}{a(3) + \ddots}}} ,$$

which often correspond to alternating series, for instance when the $a(i)$ and $b(i)$ are positive. Unfortunately, most of the time we are in the pitfall explained in the warning above. To take the very simplest example, that of the continued fraction for the golden ratio $\phi = (1 + \sqrt{5})/2$ which has $a(i) = b(i) = 1$ for all i, the corresponding alternating series is easily shown to be

$$\phi = 1/2 + 5 \sum_{n \geq 1} \frac{(-1)^{n-1}}{\phi^{2n-1} + (-1)^{n-1} + \overline{\phi}^{2n-1}} ,$$

and we are in the same situation as the warning given above because of the parasitic term $(-1)^{n-1}$. Luckily, most of the time (as here), continued fractions converge geometrically, so can be computed directly, but of course this is not always the case. An example, due to Ramanujan, is the formula

$$\frac{\Gamma^2 \left(\dfrac{x+1}{4}\right)}{\Gamma^2 \left(\dfrac{x+3}{4}\right)} = \cfrac{4}{x + \cfrac{1^2}{2x + \cfrac{3^2}{2x + \cfrac{5^2}{2x + \ddots}}}} ,$$

which converges (for $x \in \mathbb{R}$) for $x > 1/2$ like $(-1)^n/n^x$, so not geometrically. Unfortunately, the presence of these parasitic terms prevent the `Sumalt` function or variants of working directly. There are specific methods for accelerating this kind of series which we will not mention here. See Chapter 7 for a study of continued fractions.

4.5.2. Variant for Fourier series. Assume now that we want to sum a series of the form $\sum_{n \geq 0} z^n u(n)$, typically with $z = e^{2\pi i x}$, so corresponding to the summation of (half of) a Fourier series. The analysis that we have done for the initial CVZ method can be reproduced verbatim: we now need the $c_{n,j} = c_{n,j}(z)$ defined by $(P(1/z) - P(X))/(1 - zX) = \sum_{0 \leq j \leq n-1} c_{n,j}(z) X^j$ which can again be computed on the fly if desired, the recursion being essentially the same with an added power of z as coefficient, and we then have for a suitably large N

$$\sum_{n \geq 0} z^n u(n) \approx \sum_{1 \leq j \leq N} w_{N,j}(z) u(j)$$

with $w_{N,j}(z) = c_{N,j}(z)/P_N(1/z)$. The formulas are identical to those for $z = -1$, with some powers of z occurring, and we leave the details to the reader. Note, however, that the coefficients $c_{N,j}(z)$ can become as large as $(1 + \sqrt{2})^{2N}$. When $z = -1$ we have $P_N(1/z) \approx (1 + \sqrt{2})^{2N}$, so the $w_{N,j}(z)$ are small. However, when $z \neq -1$ and especially when z is close to 1, $|P_N(1/z)|$ is much smaller than that, so we have to compensate by increasing the working accuracy, thus giving the following program:

------------------------------------- | Sumosc.gp | -------------------------------------

```
1   /* Compute sum_{n >= 1} z^n u(n), u being a closure.
2    * Since there may be enormous accuracy changes, z should
3    * be given as a closure, but scalar accepted. */
4
5   /* evaluate a quantity given as a scalar or a function: the
6    * evaluation occurs at current dynamic precision */
7   Getz(z) = return (if (type(z) == "t_CLOSURE", z(), z));
8
9   /* Compute weights for Sumoscinit, bit and N come from
10   * SumoscinitBN. Used by Sumchi. */
11  Sumoscinit0(bit, N, z) =
12  { my(c, d, powz, zi, bic, vc = vector(N));
13
14    localbitprec(bit);
15    z = Getz(z); zi = 1 / z;
16    d = (1 - 2 * zi + 2 * sqrt(zi * (zi - 1)))^N;
17    powz = powers(z, N);
18    d = (d + 1 / d) / 2; bic = -1.; c = d;
19    for (j = 0, N - 1,
20      c += bic;
21      vc[j + 1] = c * powz[j + 1];
22      bic *= zi * (j + N) * (j - N) / ((j + 1/2) * (j + 1)));
23    return ([d, vc, bit]);
24  }
25
26  /* Bit accuracy and number of terms needed by Sumoscinit.
27   * Used by Sumchi */
28  SumoscinitBN(z) =
29  { my(B = getlocalbitprec(), N, d, ct, zi, bit);
30
31    localbitprec(32);
32    zi = 1 / Getz(z);
33    d = abs(1 - 2 * zi + 2 * sqrt(zi * (zi - 1)));
34    ct = abs(log(d)) / log(2);
35    if (exponent(ct) < -8, error("non-oscillating function"));
36    N = ceil(B / ct);
```

```
37    bit = ceil(N * (log(3 + sqrt(8)) / log(2))) + 32;
38    return ([bit, N]);
39  }

40

41  /* Initialization function for Sumosc */
42  Sumoscinit(z = -1) =
43  { my([bit, N] = SumoscinitBN(z));
44    return (Sumoscinit0(bit, N, z));
45  }

46

47  /* Compute sum_{n>=1} z^n * u(n). */
48  Sumosc(u, z = -1, tab = 0) =
49  { my(B = getlocalbitprec(), bitnew, d, vc, S);

50

51    if (!tab, tab = Sumoscinit(z));
52    [d, vc, bitnew] = tab;
53    localbitprec(bitnew);
54    S = sum(j = 1, #vc, vc[j] * u(j));
55    return (bitprecision(Getz(z) * S / d, B));
56  }
```

REMARKS 4.5.6.

(1) Note that in general the shifted Chebyshev polynomials are not optimal, but they give a simple program.

(2) The program of course fails when $z = 1$, but even when z is close to 1 we need to considerably increase the working accuracy to obtain a correct result (which of course slows down the program accordingly). As a consequence, it is preferable to give z as a *closure* with no argument, so that the program can compute it to the accuracy that it needs. If you simply give z as a complex number, it may lose some accuracy or even fail completely, unless z is close to -1.

As a direct application, assume that we want to sum Fourier series $C(x) = \sum_{n\geq 0} u(n)\cos(2\pi nx)$, $S(x) = \sum_{n\geq 1} u(n)\sin(2\pi nx)$, or $E(x) = \sum_{n\in\mathbb{Z}} u(n)e^{2\pi inx}$. Thanks to the above program, we can compute

$$F(x) = \sum_{n\geq 0} u(n)e^{2\pi inx} \approx \sum_{0\leq j<N} w_{n,j}(z)u(j)$$

with $z = e^{2\pi ix}$, and since the Chebyshev polynomials have real coefficients we have

$$F(-x) \approx \sum_{0\leq j<N} \overline{w_{n,j}(z)}u(j)\,,$$

so that

$$C(x) \approx \sum_{0 \le j < N} \Re(w_{n,j}(z)) u(j) \, ,$$

$$S(x) \approx \sum_{0 \le j < N} \Im(w_{n,j}(z)) u(j) \, , \quad \text{and}$$

$$E(x) \approx u(0) + \sum_{1 \le j < N} (w_{N,j}(z) u(j) + \overline{w_{N,j}(z)} u(-j)) \, .$$

Note that up to the given approximation we have $w_{n,0}(z) = 1$, if $u(0)$ is not defined in any formula so that we exclude $n = 0$, we simply omit the term $j = 0$ in the approximations. This leads to the following trivial programs:

_____ | SumFourier.gp | _____

```
1   /* Compute sum_{n>=1}u(n)cos(2 pi nx). */
2   Sumcos(u, x = 1/2, tab = 0) =
3   { my(B = getlocalbitprec(), B1, d, vc, S, z);
4
5     z = (() -> exp(2*Pi*I*x));
6     if (!tab, tab = Sumoscinit(z));
7     [d, vc, B1] = tab;
8     localbitprec(B1); vc *= z() / d;
9     S = sum(j = 1, #vc, real(vc[j]) * u(j));
10    return (bitprecision(S, B));
11  }
12
13  /* Compute sum_{n>=1}u(n)sin(2 pi nx). */
14  Sumsin(u, x = 1/2, tab = 0) =
15  { my(B = getlocalbitprec(), B1, d, vc, S, z);
16
17    z = (() -> exp(2*Pi*I*x));
18    if (!tab, tab = Sumoscinit(z));
19    [d, vc, B1] = tab;
20    localbitprec(B1); vc *= z() / d;
21    S = sum(j = 1, #vc, imag(vc[j]) * u(j));
22    return (bitprecision(S, B));
23  }
24
25  /* Compute sum_{n in Z, n not 0}u(n)exp(2 pi nx). */
26  Sumexp(u, x = 1/2, tab = 0) =
27  { my(B = getlocalbitprec(), B1, d, vc, S, z);
28
29    z = (() -> exp(2*Pi*I*x));
30    if (!tab, tab = Sumoscinit(z));
31    [d, vc, B1] = tab;
32    localbitprec(B1); vc *= z() / d;
33    S = sum(j = 1, #vc, my(co = vc[j]);
34                        co * u(j) + conj(co) * u(-j));
```

```
35    return (bitprecision(S, B));
36  }
```

EXAMPLES.
```
? Sumsin(n -> 1/n, 1/5) / Pi
% = 0.30000000000000000000000000000000000000000
? Sumsin(n -> 1/n^3, 1/5) / Pi^3
% = 0.032000000000000000000000000000000000000000
? Sumcos(n -> 1/n^2, 1/5) / Pi^2
% = 0.0066666666666666666666666666666666666666667
? Sumcos(n -> 1/n, 1/5) + log(2 * sin(Pi / 5))
% = 0.E-39
? f(x) = Sumsin(n -> log(n)/n, x);
? g(x) = Pi * (lngamma(x) - 1/2 * log(Pi / sin(Pi * x)) \
               + (x - 1/2) * (log(2*Pi) + Euler));
? f(1/5) - g(1/5)
% = 1.46... E-39
```

4.5.3. Variant for periodic functions. Assume now that we want to compute $\sum_{n \geq 0} \chi(n)u(n)$, where $\chi(n)$ is a periodic function of some period T such that $\sum_{1 \leq n \leq T} \chi(n) = 0$ (not necessarily a Dirichlet character). It is not difficult to generalize the CVZ method to this case, and we leave the details and the implementation to the reader. The following programs were adapted from earlier versions written by Bill Allombert for the special case where $\chi = \left(\frac{D}{\cdot}\right)$. The constant c comes from the analysis of the method, which is similar to the one given above.

———————————————— Sumchiquad.gp ————————————————

```
1   /* Compute sum_{i>=1}(D/i)u(i), D fundamental discriminant. */
2
3   Sumchiquadinit(D) =
4   { my(B = getlocalbitprec(), w, Q, Da = abs(D), R, T, N, c);
5
6     localbitprec(32);
7     w = exp(2*I*Pi / Da);
8     c = (2*log(2)) / log(norm(1 - 2*w + 2*sqrt(w^2 - w)));
9     N = ceil(abs(c) * B);
10    w = Mod('a, polcyclo(Da, 'a));
11    [R,Q] = divrem(polchebyshev(N, 1, 1 - 2*'x), 'x - w);
12    R = lift(Vecrev(R / Q)) * sign(D);
13    T = (lift(sum(i=1,Da-1,kronecker(D,i)*subst(R,'a,w^i))
14            / sum(i=1,Da-1,kronecker(D,i)*w^i)));
15    return ([B + exponent(T), T]);
16  }
17
18  Sumchiquad(u, D, tab = 0) =
19  {
20    if (!tab, tab = Sumchiquadinit(D));
21    my([B1, T] = tab);
```

```
22    localbitprec(B1);
23    return (sum(i = 1, #T, T[i] * u(i), 0.));
24  }
```

Note that the case $D = -4$ can in fact be treated using Sumalt since

$$\sum_{n \geq 0} \left(\frac{-4}{n} \right) u(n) = \sum_{m \geq 0} (-1)^m u(2m+1) .$$

On the other hand, it does not seem possible to use Sumalt directly for $D = -3$.

In the case where χ is a general *primitive* character of conductor F, we can use the formula

$$\chi(n) = \frac{1}{\mathfrak{g}(\overline{\chi})} \sum_{m \bmod {}^* F} \overline{\chi}(m) \zeta_F^{mn}$$

with $\zeta_F = e^{2\pi i / F}$, which combined with the Sumosc program gives

$$\sum_{n \geq 1} \chi(n) u(n) \approx \sum_{1 \leq j \leq N-1} y_{N,j}(\chi) u(j) \quad \text{with}$$

$$y_{N,j} = \frac{1}{\mathfrak{g}(\overline{\chi})} \sum_{m \bmod {}^* F} \overline{\chi}(m) w_{N,j}(\zeta_F^m) ,$$

where $w_{N,j}(z)$ is as in Sumosc above. This gives the following:

─────────────────────────────── Sumchi.gp ───────────────────────────────

```
1   /* sum_{n > 0} \chi(n) u(n), \chi primitive character of
2    * conductor F > 1, given by rational-valued closure chi such
3    * that \chi(n) = exp(2*I*Pi * chi(n)) when (n, F) = 1 */
4
5   Sumchiinit(chi, F) =
6   {
7     my([B, N] = SumoscinitBN(x -> exp(2*I*Pi / F)));
8     my(T = vector(N), G = 0); \\ G = Gauss sum g(conj(chi))
9
10    localbitprec(B);
11    for (m = 1, F - 1,
12      if (gcd(m, F) != 1, next);
13      my([d, vc] = Sumoscinit0(B, N, exp(2*I*Pi * m / F)));
14      my(z = exp(2*I*Pi * (m / F - chi(m))));
15      T += vc * (z / d); G += z);
16    return ([B, G, T]);
17  }
18
19  Sumchi(u, chi, F, tab = 0) =
20  {
21    if (!tab, tab = Sumchiinit(chi, F));
22    my([B, G, T] = tab);
23    localbitprec(B);
24    return (sum(i = 1, #T, T[i] * u(i)) / G);
25  }
```

```
26
27  Sumznchar(u, D) =
28  { my([G, chi] = znchar(D), F = zncharconductor(G, chi));
29    if (G.mod != F, error("non-primitive character"));
30    return (Sumchi(u, n -> chareval(G, chi, n), F));
31  }
```

EXAMPLE. To compute $A = L(\chi_{-3}, 1) = \sum_{n \geq 1} \left(\frac{-3}{n} \right) / n$ (without using the Lfunxxx functions that we will study in Chapter 9), you can use three methods:

```
? A = Sumchiquad(n -> 1 / n, -3)
? A = Sumchi(n -> 1 / n, n -> if (n == 1, 0, n == 2, 1/2), 3)
? A = Sumznchar(n -> 1 / n, -3)
```

the last two, which are equivalent, being much faster.

4.5.4. Non-periodic oscillating series. In Chapter 3 we have seen a large number of methods for integrating oscillating functions, such as periodic functions including functions involving fractional parts, and almost periodic functions such as Bessel functions. The corresponding problem for *summation* is considerably more difficult when the functions are not exactly periodic or involve fractional parts. As a first example, consider

$$S_k(\alpha) = \sum_{n \geq 1} \frac{J_0(n\alpha)}{n^k},$$

with $k \geq 0$ integral (although we could of course also consider k complex). This is a special case of *Schlömilch series*, see for instance [**WW96**, **Wat95**] for details and formulas.[1]

Recall that $J_0(x)$ is almost periodic with half period π, so that $J_0(n\pi)$ should alternate in sign rather regularly, and we can hope that the Sumalt program will treat it correctly. Recalling that our Sumalt program includes an implicit $(-1)^{n-1}$, we write:

```
? S(k) = Sumalt(n -> (-1)^(n-1) * besselj(0, Pi * n) / n^k);
? \p 38
? A = S(0)
% = -0.18169011381620932846223247325497127593
? B = S(1)
% = -0.23129892655550671956636378326870797565
? \p77
? [S(0) - A, S(1) - B]
% = [0.E-39, 0.E-39]
```

Although we do not know if the results are correct, the fact that the results at higher accuracy are identical seems to indicate that they are. In fact, $A = 1/\pi - 1/2$, which checks perfectly.

---

[1]One of the main results is that, under suitable assumptions on f, there exists an expansion $f(x) = a_0/2 + \sum_{n \geq 1} a_n J_0(nx)$ with

$$a_n = 2f(0)\delta_{n,0} + \frac{2}{\pi} \int_0^\pi \int_0^{\pi/2} u \cos(nu) f'(u \sin(\theta)) \, d\theta \, du.$$

Similarly, $J_0(n(2\pi))$ should be regularly decreasing and always positive. We cannot use `Sumpos` which uses exponentially large values of the argument, but we can try for instance `SumLagrange` with parameter $1/2$, since $J_0(x) \sim C/x^{1/2}$ as $x \to \infty$ for a suitable constant C:

```
? S2(k) = SumLagrange(n -> besselj(0, 2 * Pi * n) / n^k, 1/2);
? \p 38
? A = S2(1)
% = 0.58161292163832384991624374289533263762
? B = S2(2)
% = 0.29655096021331926896437390002509498848
```

and we can again check that the results are the same at larger accuracy, giving a good confidence on their correctness. In fact, $B = 2\pi^2/3 - 2\pi$, which checks perfectly.

More generally, we can compute $S_k(\alpha)$ when α is a rational multiple of π with small denominator.

On the other hand, if α is not a rational multiple of π, none of our methods will be able to compute $S_k(\alpha)$. We need to use specific properties of the Bessel function, and the formula which seems the most appropriate is the integral formula

$$J_0(x) = \frac{1}{2\pi} \int_0^{2\pi} e^{ix \sin(t)} \, dt \ .$$

Assume for now that $k \geq 1$. Since by definition $\sum_{n \geq 1} z^n/n^k = \operatorname{Li}_k(z)$ for $|z| \leq 1$, where Li_k is the kth polylogarithm, we have

$$S_k(\alpha) = \frac{1}{2\pi} \int_0^{2\pi} \operatorname{Li}_k\left(e^{i\alpha \sin(t)}\right) dt \ ,$$

which we can hope to compute using one of our integration routines on compact intervals. Note that this formula is a priori valid only for $k \geq 2$, but it is easily shown that it is also true for $k = 1$.

Using the symmetries of the sine function, we thus have

$$S_k(\alpha) = \frac{2}{\pi} \Re\left(\int_0^{\pi/2} \operatorname{Li}_k\left(e^{i\alpha \sin(t)}\right) dt \right) \ .$$

Using our `Inta_b` or the built-in `intnum` program, we thus find instantly for instance that

$$S_1(1) = \sum_{n \geq 1} \frac{J_0(n)}{n} = 0.71411247333998234214104562674402154603 \ ,$$

$$S_2(1) = \sum_{n \geq 1} \frac{J_0(n)}{n^2} = 0.76993406684822643647241516664602518922 \ ,$$

and we can also recover the values for $\alpha = \pi$ and $\alpha = 2\pi$ found above. Note that $S_2(1) = \pi^2/6 - 7/8$, which again checks perfectly.

For $k = 0$ we cannot interchange summations, so the only way that we can see is to use $S_0(-\alpha) = S_0(\alpha)$ and for $\alpha > 0$ the explicit formula

$$S_0(\alpha) = \frac{1}{\alpha} - \frac{1}{2} + 2 \sum_{m=1}^{k} \frac{1}{(\alpha^2 - 4m^2\pi^2)^{1/2}} \ , \quad \text{where } k = \lfloor \alpha/(2\pi) \rfloor \ .$$

This solves our problem, but for very similar sums no such explicit formula exists, and we have not found a satisfactory solution.

As a second example, consider

$$T_k(\alpha) = \sum_{n \geq 1} \frac{\{n\alpha\} - 1/2}{n^k} \,,$$

with $k \geq 1$ integral. If α is a rational number with reasonably small denominator this is periodic, so we can compute it using the methods seen above, and in any case it has an explicit expression as a linear combination of L-function values. We can therefore assume that $\alpha \notin \mathbb{Q}$ or that it has a large denominator. In that case, we have no real answer. If α is a quadratic irrational with small discriminant such as $\sqrt{2}$ or $(1 + \sqrt{5})/2$, it is possible that the methods of Hecke [**Hec22**] will give an algorithm for computing $T_k(\alpha)$. On the contrary, finding a reasonably large number (say 100) of decimals of

$$T_2(\pi) = \sum_{n \geq 1} \frac{\{n\pi\} - 1/2}{n^2} = -0.392598100\cdots$$

seems totally out of reach, but we would be happy to be contradicted.

4.6. Summing by extrapolation

In Chapter 2 we have mentioned that Lagrange extrapolation can be used for computing limits of recursions and in particular sums of infinite series. Immediate programs are as follows:

—————————————— SumLagrange.gp ——————————————

```
1   /* Compute sum_{n >= 1} u(n) using Lagrange extrapolation.
2    * Assume that sum_{j > n} u(j) has a regular asymptotic
3    * expansion f(n)*sum_{i >= 0} a_i/n^(al*i). alf is the scalar
4    * al or a closure f (al = 1) or the pair [al, f].
5    * Increase mul to improve accuracy */
6
7   /* vu(N) = [u(1), ..., u(N)]  */
8   SumLagrangevec(vu, alf = 1, mul = 1, tab = 0) =
9   {
10    Limitvec(N -> my(S = 0., U = vu(N * mul));
11      vector(N, j, S += sum(n = 1 + (j - 1) * mul,
12            j * mul, U[n])), alf, tab);
13  }
14  SumLagrange(u, alf = 1, mul = 1, tab = 0) =
15    SumLagrangevec(N -> vector(N, n, u(n)), alf, mul, tab);
```

Note that, as with recursions, we sum by small blocks of size `mul` instead of one by one, which sometimes increases the accuracy of the result. We allow the same `alf` parameter as in `Limit` to encode the type of asymptotic expansion (α, f) and simply pass it on to `Limitvec`.

Note also that this is evidently an order of magnitude faster than using directly the `Limit` program on the nth partial sum since the latter would use $O(N^2)$ evaluations of u instead of $O(N)$.

EXAMPLE.

```
? \p 2003
   realprecision = 2003 significant digits
? z = zeta(3);
? Limit(n -> sum(j = 1, n, 1/j^3)) - z
time = 2,410 ms.
% = 0.E-2003 /* Perfect and slow. */
? SumLagrange(j -> 1/j^3) - z
time = 98 ms.
% = 0.E-2003 /* Perfect and fast. */
```

For a less trivial example, consider the following problem sent to us by Zhi Wei Sun. As in Section 2.6.2 we consider

$$u(n) = F_n/8^n = \sum_{0 \le j \le n} \binom{n}{j}^3 /8^n ,$$

where the F_n are the Franel numbers. We have seen that $u(n) \sim C/n$ for some explicit constant C, and we want to compute the slowly convergent sum $S = \sum_{n \ge 0} u(n)/(n+1)$.

Apparently none of the methods available in `Pari/GP` nor the other methods given in this chapter will work. But `SumLagrange` works very well, at least if we remember what was done in Section 2.6.2: indeed, we saw that the sequence $u(4n)$ was perfectly behaved with respect to our extrapolation methods, but not $u(n)$ (of course $u(n)$ is just as well behaved as $u(4n)$, but *not* for our methods). Since `SumLagrange` allows such a multiplier, we simply write the following:

```
\p38
? f(n) = sum(j = 0, n, binomial(n, j)^3) / ((n + 1) * 8^n);
? 1 + SumLagrange(f,,4)
% = 1.2922637888778020563745460428952255763
```

REMARKS 4.6.1.

(1) Since all our summation programs begin at $n = 1$, we have to add $f(0) = 1$ to the result of `SumLagrange(f)`.
(2) We can check that the multiplier 5 or the computation to a higher accuracy gives exactly the same 38 decimals, so we can be quite confident in the accuracy of the result. Note that the multiplier 3 would lose 7 decimals, and the multiplier 2 would lose 12, in exact agreement with the accuracy loss shown in Section 2.6.2.
(3) Note that the computation of `SumLagrange(f,,4)` at 38 decimal digits is essentially instantaneous (less than 15 milliseconds).
(4) If we really want to simulate the above computation using the existing `Pari/GP` functions, we can write

```
? limitnum(n -> sum(j = 0, 4 * n, f(j)))
% = 1.2922637888778020563745460428952255763
```

which works perfectly, although much less efficiently (for instance 40 times slower at 115 decimal digits).

(5) In larger accuracies, it is essential to use the single argument `binomial(n)` command of `Pari/GP` which computes all $\binom{n}{j}$ for $0 \le j \le n$ as a vector. Compare:

```
\p115
? g(n) = my(V = binomial(n)); \
     sum(j = 0, n, V[j + 1]^3) / ((n + 1) * 8^n);
? 1 + SumLagrange(f,,5); /* At 115 D need multiplier 5 */
time = 556 ms.
? 1 + SumLagrange(g,,5);
time = 72 ms.
```

When the convergence of the sum is in integral powers of $1/n^2$ we use the following program:

───────────────────────── SumLagrange2.gp ─────────────────────────

```
1   /* Similar, but assume that al = 2 and f(n) = 1/n, so that
2    * u(N)/2 + sum_{n > N} u(n) has a regular asymptotic expansion
3    * in integral powers of 1/n^2. */
4
5   /* vu(N) = [u(1), ..., u(N)]   */
6   SumLagrange2vec(vu, mul = 1) =
7   { my(VU);
8     VU = (N -> my(S = 0., U = vu(N * mul));
9               vector(N, j, my(A = (j-1)*mul + 1, B = A + mul-1);
10                            S += sum(n=A, B, U[n]); S - U[B]/2));
11    return (Limitvec(VU, 2));
12  }
13  SumLagrange2(u, mul = 1) =
14    SumLagrange2vec(M -> vector(M, n, u(n)), mul);
```

───

In this second program, note that we have to subtract $u(N)/2$ to have any chance that a series satisfy the assumptions of `SumLagrange2`: for instance, this will now be the case for any convergent sum of the form $\sum_{n \ge 1} f(n)$ where $f(-n) = -f(n)$ and $f(n)$ is (for instance) a rational function of n.

EXAMPLE. (Continuation of the $\zeta(3)$ example):

```
? SumLagrange2(j -> 1/j^3) - z
time = 69 ms.
% = 0.E-2003 /* Perfect and faster. */
```

Note that the above programs are extremely fast, and as usual, this is counterbalanced by the fact that they are not very robust, and depend on the precise knowledge of the type of asymptotic expansion of the summand.

When the convergence is in integral powers of $1/n^\alpha$ with $\alpha \ne 1$ and $\alpha \ne 2$ (most frequently $\alpha = 1/2$) we of course use the `alf` second argument of `SumLagrange`. Note, however, that depending on the type of asymptotic expansion, if is often better to use the closure version of alf rather than a scalar α. Consider the following:

```
\p115
? f(n) = my(V = binomial(n)); sum(j = 0, n, V[j + 1]^4) / 16^n;
? S = 1 + SumLagrange(f, 1 / 2, 9)
time = 2,708 ms.
% = 1.492446...
? 1 + SumLagrange(f, 1 / 2, 8) - S
time = 1,853 ms.
% = -1.19... E -111 /* A little faster but 4 decimals lost */
? 1 + SumLagrange(f, n -> sqrt(n), 8) - S
time = 274 ms.
% = 0.E-115 /* Almost 10 times faster and perfect. */
```

The reason for the speed difference is that the asymptotic expansion is not only in integral powers of $1/n^{1/2}$, but in *odd* integral powers of $1/n^{1/2}$, which is taken into account in the last command.

A similar method, still based on extrapolation, is to use the LimitSidi program on partial sums of the series, which we described in §2.5.

As for integration, there are two different wrappers: SumaltSidi specific to (approximately) alternating series, and SumSidi for series with fixed sign (with a safe flag set by default), but which can also be used for alternating series, although it will be slower. Both programs perform the exact same operations on $(-1)^{n-1}u(n)$ and $u(n)$ respectively, but the more robust SumSidi uses a higher working accuracy in safe mode.

EXAMPLES.

```
? \p 500
? SumaltSidi(n -> log(n)^2 / n)
time = 850 ms.
% = -0.06537... \\ perfect
? SumaltSidi(n -> 1 / (n^(4/3) + log(n)))
time = 844 ms.
% = 0.798905... \\ perfect
? SumaltSidi(n -> sqrt(n)) \\ divergent!
time = 168 ms.
% = 0.380104... \\
? SumSidi(n -> 1 / n^2, 0) - Pi^2 / 6
time = 141 ms.
   ***   user warning: reached accuracy of 1077 bits.
% = 1.08...E-324 \\ unsafe mode: partially correct
? SumSidi(n -> 1 / n^2, 1) - Pi^2 / 6
time = 385 ms.
% = -2.438...E-501 \\ with safe flag: perfect
? SumSidi(n -> 1/(n^2 + n + 1))
time = 390 ms.
% = 0.7981... \\ perfect
```

As `Sumalt`, `SumaltSidi` assigns a value to the divergent sum $\sum_n (-1)^{n-1}\sqrt{n}$. It actually returns the analytic continuation to $s = -1/2$ of $\sum_n (-1)^{n-1} n^{-s} = (1 - 2^{1-s})\zeta(s)$, where the last equality holds if $\Re(s) > 0$.

```
? s = -1/2; S = (1 - 2^(1-s)) * zeta(s);
? SumaltSidi(n -> sqrt(n)) - S
% = 0.E-501 \\ perfect
```

4.7. Van Wijngaarden's method

A method which can be used in certain cases where none of the others can is due to Van Wijngaarden. Note that if one needs to compute the sum $\sum_{n\geq 1} u(n)$ of a series with *positive* terms $u(n)$, we can use Euler–Maclaurin, Abel–Plana, or Poisson only if $u(n)$ is the restriction to the integers of some natural function defined over the reals. On the other hand, these methods are not applicable (at least not directly) if $u(n)$ is only defined over the integers, for instance by a recursion. The case of continued fractions seen above is an example, but there are many others.

Thus, it is desirable to have a method applicable in that case. One method is to use the extrapolation program `SumLagrange` and variants. A second method is to reduce to the `Sumalt` program, which indeed does not require the sequence to be defined over the reals. This method, due to Van Wijngaarden, comes from the following lemma:

LEMMA 4.7.1. *Let $u(n)$ for $n \geq 1$ be a sequence with positive terms such that $S = \sum_{n\geq 1} u(n)$ converges. If for all $a \in \mathbb{Z}_{\geq 1}$ we set*

$$U(a) = \sum_{j\geq 0} 2^j u(2^j a) \,,$$

then if in addition all the series $U(a)$ converge, we have

$$S = \sum_{a\geq 1} (-1)^{a-1} U(a) \,,$$

so that S is the sum of an alternating series.

PROOF. We have

$$\sum_{a\geq 1} (-1)^{a-1} U(a) = \sum_{a\geq 1,\ j\geq 0} (-1)^{a-1} 2^j u(2^j a)$$
$$= \sum_{N\geq 1} u(N) \sum_{0\leq j\leq v_2(N)} (-1)^{N/2^j - 1} 2^j \,,$$

where as usual $v_2(N)$ is the largest power of 2 dividing N, and where the interchange of summation is justified by the convergence assumptions. Now $N/2^j$ is even for $0 \leq j < v_2(N)$, hence the inner sum is equal to $2^{v_2(N)} - \sum_{0\leq j<v_2(N)} 2^j = 1$, proving the lemma. □

To compute S we can then simply use the program `Sumalt` on the alternating series $(-1)^{a-1} U(a)$. This is preprogrammed in GP under the name `sumpos`, but we give here a possible and slower simple-minded program:

$\boxed{\text{Sumpos.gp}}$

```
1   /* Define U[a] := sum_{n >= 0} 2^n u(a 2^n),
2    * set U[k * 2^i] for all k * 2^i <= N, in place. We assume
3    * u regular to have a reasonable stopping criterion. */
4   SumposU(~U, u, k, E) =
5   { my(N = #U, t = 0, L = exponent(N \ k)); \\ L = [log2(N/k)]
6     for (i = 0, oo, \\ first compute U[k * 2^L]
7       my(z = (u(k << (L + i)) * 1.) << i);
8       if (!z || exponent(z) < E - L, break);
9       t += z);
10    U[k << L] = t; \\ then use U[j] = 2 U[2j] + u(1+j)
11    forstep (i = L-1, 0, -1, U[k<<i] = t = u(k<<i) + t << 1);
12  }
13
14  /* Given a nonnegative sequence u as a closure, compute
15   * S = sum_{n>=1} u(n). */
16  Sumpos(u) =
17  { my(B = getlocalbitprec(), N, U, E = -B - 5);
18    localbitprec(32); N = ceil(B * log(2) / log(3+sqrt(8)));
19    localbitprec(B + 32); U = vector(N);
20    forstep (k = 1, N, 2, SumposU(~U, u, k, E));
21    localbitprec(B); Sumalt(n -> U[n]);
22  }
```

Let us briefly analyze which values of u are needed. For an error bounded by 2^{-B} in Sumalt, we need $n \approx 0.4B$ (since $\log(2)/\log(3 + \sqrt{8}) = 0.393\cdots$). Thus, we need $U(a)$ for $a \leq 0.4B$. Let us assume for simplicity that $u(n) < n^{-\alpha}$ for some $\alpha > 1$. To obtain $U(a)$ to accuracy less than 2^{-B} we need $2^j (2^j a)^{-\alpha} < 2^{-B}$, i.e., $j > B/(a\alpha - 1)$, so we need the values of $u(N)$ for certain values of $N = 2^j a \ll 2^{B/(\alpha-1)} \cdot B$. On the other hand, for a given a we use values of j up to $B/(a\alpha - 1)$. Summing for $a = O(B)$, it follows that the total number of values of $u(N)$ to be computed is $O_\alpha(B \log B)$, which is quite small. So the main inconvenient of this method is that, although not too many values of $u(N)$ need to be computed, some need to be for very large values of N (such as 2^{128}), and for many sequences u this is impossible in practice.

4.8. Monien summation

4.8.1. A naïve approach. In this section we present a method for definite summation due to H. Monien [Mon10]. We first approach the problem in a naïve manner, and then present the algorithm in a more rigorous way.

Recall that Gauss–Legendre integration with N nodes is the unique method which gives exact results for x^k with $0 \leq k \leq 2N - 1$, and is governed by the $(2N)$th derivative for higher powers, which still give very good results for larger k.

We can imitate this for sums, and perhaps surprisingly we can ask that the result is exact or at least essentially perfect for *all* k. Let us see how this is done. Assume for now that we are in the basic situation where the summand is a convergent linear combination of integral powers of $1/n$. We choose some N, and ask that the summation method give exact results for $1/n^k$ with $2 \leq k \leq 2N+1$. In other words, we want nodes x_j and weights w_j such that $\sum_{1 \leq j \leq N} w_j/x_j^k = \zeta(k)$ for $2 \leq k \leq 2N+1$. This is a nonlinear system, but is in fact analogous to a similar nonlinear system that we could write for Gauss–Legendre integration, and it can be solved in a similar manner using orthogonal polynomials, as we will see below. It indeed has a unique solution, and as an added bonus, if N is chosen suitably as a function of the desired bit accuracy B, for $k > 2N+1$ we will have both $\sum_{1 \leq j \leq N} w_j/x_j^k$ and $\zeta(k)$ approximately equal to 1 with an error less than 2^{-B}, so the result is indeed essentially perfect for *all* k.

4.8.2. The basic method. The basic version of Monien summation is based on the use of the psi function, the logarithmic derivative of the gamma function. Since $\psi(z) = \Gamma'(z)/\Gamma(z)$ and $\Gamma(z)$ never vanishes, the poles of $\psi(z)$ are the poles of $\Gamma(z)$, i.e., the negative or 0 integers, and they are simple with residue 1. Thus, if f is some function which is holomorphic in some domain containing the positive reals, and if $\mathcal{C}$ is some contour in this domain going from ∞ in the lower half-plane, around 1, and back to ∞ in the upper half-plane, then from $-\infty$ to $-N$ in the lower half-plane and back from $-N$ to $-\infty$ in the upper half-plane, we have $(1/(2\pi i)) \int_{\mathcal{C}} \psi(1-z) f(z)\,dz = \sum_{n \geq 1} f(n) := S$. Note that we use the fact that $\psi(z)$ does not have poles for $z > 0$. Changing z into $1/z$ we deduce that under reasonable assumptions on f we have

$$S = -\frac{1}{2\pi i} \int_{\mathcal{C}} \frac{1}{z^2} \psi(1 - 1/z) f(1/z)\,dz \;,$$

where now $\mathcal{C}$ is a contour containing the interval $[0,1]$.

We could exploit this formula in two different ways. The first would be to use doubly-exponential integration methods, setting for instance $z = 1/2 + e^{i\pi x}$, so that

$$S = -\frac{1}{2} \int_{-1}^{1} (1/2 + e^{i\pi x})^{-2} e^{i\pi x} \psi(1 - 1/(1/2 + e^{i\pi x})) f(1/(1/2 + e^{i\pi x}))\,dx \;,$$

and then applying the change of variable $x = \phi(t) = \tanh((\pi/2)\sinh(t))$, and even precomputing the weights and the abscissas. However, because of the singularities of the integrand, this will not give a very good accuracy.

H. Monien's idea, which already occurs in Gaussian integration, is to use rational function approximations to $\psi(1 - 1/z)$, and reapply the residue theorem to evaluate S.

Recall that around $z = 0$ we have the power series expansion

$$\psi(1-z) + \gamma = -\sum_{n \geq 1} \zeta(n+1) z^n \;.$$

Using the quotient-difference algorithm (see Section 7.6), we can formally write this as a continued fraction

$$-\sum_{n \geq 1} \zeta(n+1) z^n = \cfrac{c(0)z}{1 + \cfrac{c(1)z}{1 + \cfrac{c(2)z}{1 + \cfrac{c(3)z}{1 + \cdots}}}} \;,$$

where for instance $c(0) = -\zeta(2)$, $c(1) = -\zeta(3)/\zeta(2)$, etc. Experimentally, we have $c(n) = -1/n + O(1/n^2)$. More precisely, using the Asymp program we find that we have two extremely accurate asymptotic expansions:

For n even,

$$c_e(n) := c(n) = -\frac{1}{n} + \frac{3}{n^2} - \frac{22/3}{n^3} + \frac{86/5}{n^4} - \frac{1816/45}{n^5}$$
$$+ \frac{16972/175}{n^6} - \frac{1143104/4725}{n^7} + \frac{710944/1125}{n^8} - \cdots ,$$

and for n odd

$$c_o(n) := c(n) = -\frac{1}{n} + \frac{5/3}{n^3} - \frac{26/5}{n^4} + \frac{587/45}{n^5}$$
$$- \frac{16616/525}{n^6} + \frac{372163/4725}{n^7} - \frac{25906/125}{n^8} + \cdots .$$

Although we do not need it, note that these expansions seem to come from a unique one, since we have:

$$c_e(n+2) = -1/n + 1/n^2 + (2/3)/n^3 + (6/5)/n^4 + (56/45)/n^5 + \cdots \quad \text{and}$$
$$c_o(n+1) = -1/n - 1/n^2 + (2/3)/n^3 - (6/5)/n^4 + (56/45)/n^5 - \cdots .$$

Thus, using Theorem 7.3.7 from Chapter 7, we have $\alpha = 0$, $a_0 = 1$, $a_1 = 0$, $\beta = -1$, $b_0 = -z$, $b_1 = -3/2$ on average, so the convergence is extremely fast, more precisely

$$S - S(n) \sim C \frac{z^n n^{-3/2 + 2z}}{n!} .$$

Incidentally, note that this gives a very fast algorithm for computing $\psi(1-z)$, once the coefficients $c(n)$ computed.

Now if we write $S(n) = p(n)/q(n)$, we have as usual

$$p(-1) = 1 , \ p(0) = 0 , \ p(n+1) = p(n) + c(n)zp(n-1) \text{ for } n \geq 0$$
$$q(-1) = 0 , \ q(1) = 1 , \ q(n+1) = q(n) + c(n)zq(n-1) \text{ for } n \geq 0 .$$

We deduce that $(p(2n), q(2n))$ has degree in z equal to $(n, n-1)$, and $(p(2n+1), q(2n+1))$ has degree equal to (n, n). Thus, if we set $P_n(z) = p(2n+1)$ and $Q_n(z) = q(2n+1)$, the ratio $P_n(z)/Q_n(z)$ will be a very good approximation to $\psi(1-z) + \gamma$, more precisely it will be its (n, n)th Padé approximant.[2] If we let $N_n(z) = z^n P_n(1/z)$ and $D_n(z) = z^n Q_n(1/z)$ be the two corresponding reciprocal polynomials, then $N_n(z)/D_n(z)$ will be a very good approximation to $\psi(1-1/z)+\gamma$.

Now recall that

$$S = -\frac{1}{2\pi i} \int_C \frac{1}{z^2} \psi(1 - 1/z) f(1/z) \, dz .$$

Since f is assumed to be holomorphic around the positive real axis, changing z into $1/z$ as above shows that $\int_C f(1/z)/z^2 \, dz = 0$. It follows that we may replace $\psi(1 - 1/z)$ by $\psi(1 - 1/z) + \gamma$, and hence approximate by $N_n(z)/D_n(z)$. Thus

$$S \approx -\frac{1}{2\pi i} \int_C \frac{N_n(z)}{z^2 D_n(z)} f(1/z) \, dz .$$

---

[2]Note that due to the notation p_n/q_n used for continued fractions, our (P_n, Q_n) corresponds to (Q_n, P_n) in the theory of orthogonal polynomials.

Now we will see below that the polynomials D_n are orthogonal with respect to a certain scalar product, and this automatically implies that all the zeros of $D_n(z)$ are simple and real, so by the residue theorem we have

$$S \approx - \sum_{\substack{\alpha \in]0,1] \\ D_n(\alpha)=0}} \frac{N_n(\alpha)}{\alpha^2 D_n'(\alpha)} f(1/\alpha) := \sum_{1 \le i \le n} w_i f(\beta_i)$$

for weights $w_i = -N_n(\alpha_i)/(\alpha_i^2 D_n'(\alpha_i))$ and abscissas $\beta_i = 1/\alpha_i$, where the α_i are the roots of D_n. Note that in fact we do not need the polynomials D_n and N_n, but only P_n and Q_n: indeed, the β_i are simply the roots of Q_n, and it is immediate to see that $w_i = P_n(\beta_i)/Q_n'(\beta_i)$.

Let us take the example $n = 2$: we find that $\beta_1 = 4.37108\cdots$, $\beta_2 = 1.0228\cdots$, $w_1 = 10.3627\cdots$, and $w_2 = 1.1534\cdots$. When applied to $f(n) = 1/n^k$ for $k = 2, 3,$ 4, and 5 the result is exact, essentially by definition.

Note concerning the error: it is quite difficult to give an estimate for the error in this method. Since it is a Gaussian type method, we could use the general error estimates such as those given by Theorem 3.6.1, but it is easily seen that these are much too pessimistic. Monien himself gives two examples in this paper (where he restricts to functions having asymptotic expansion in integral powers of $1/n^2$ instead of $1/n$), and the error estimates that he gives do not follow a general pattern. Thus, the "magic constants" that we use, in particular to compute the number of nodes, have been obtained purely empirically, but they seem to work reasonably well in actual practice.

There is an essential supplementary ingredient: one can show that the first $\lfloor n/2 \rfloor$ zeros of $Q_n(z)$ are very close to the integers 1, 2, etc. Thus, for these zeros, instead of applying the general `polroots` program (or, in our case, since we know that all the roots are real, the `polrootsreal` program), we simply perform a direct Newton method, finishing by a last iteration at double precision, and obtain half of the roots very fast. We then divide the polynomial Q_n by the roots obtained, and use `polrootsreal` on the quotient. To give an idea of the speed gain, at 1000 decimal digits, this method is approximately 20 times faster than a direct application of `polrootsreal`.

This method will work well for summing functions $f(m)$ which have asymptotic expansions at infinity of the form $f(z) = a_2/z^2 + a_3/z^3 + \cdots$, and more generally if the expansion of f is in integral powers of $1/z^\alpha$ with α integral. On the other hand it is *not* applicable if there are more general terms such as $1/z^\alpha$ with α not integral. We treat this more general case in the next subsection, and we defer to later the explicit GP program. For now, we only give the following specific root finding program, where the parameter `int` should be set to α if α is integral, in which case the roots are αth powers of integers, and otherwise set to 0.

—————————————— SumMonienroots.gp ——————————————

```
1  /* Newton refinement of root z of Q. */
2  SumMonienrefine(Q, z) =
3  { my(Qp = Q', pr, prnew);
4
5    pr = subst(Q, 'x, z);
6    while (1,
```

```
7      z -= pr / subst(Qp, 'x, z); prnew = subst(Q, 'x, z);
8      if (abs(prnew) < abs(pr),
9        pr = prnew
10     , /* else */
11       z = bitprecision(z, 2 * getlocalbitprec());
12       return (z - subst(Q, 'x, z) / subst(Qp, 'x, z))));
13   }
14
15   SumMonienroots(Q, int, roots) =
16   { my(v = []);
17
18     if (int, \\ half the roots are close to integers^int
19       my(m = poldegree(Q) \ 2 - 1);
20       v = vectorv(m, j, SumMonienrefine(Q, j^int));
21       for (j = 1, m,
22         if (exponent(v[j] - j^int) > -5, v = v[1..j]; break));
23       Q \= vecprod(vector(#v, j, 'x - v[j])));
24     return (concat(v, roots(Q)));
25   }
```

REMARK 4.8.1. It is essential to be extremely careful in the implementation of this method as well as similar more general methods: first, we must guarantee that the choice of the working accuracy is sufficient to have the polynomial Q computed using the QD algorithm have all its roots real.

In addition, note that the refinement program SumMonienroots is absolutely specific to the special case of Monien summation with α integral: in the more general cases that we will see below, the first $N/2$ roots may not be in the Newton attraction basin of integers.

As remarked above, this method looks very analogous to Gaussian summation: and indeed, it is immediate to show that the polynomials D_n are orthogonal with respect to the positive definite scalar product

$$\langle f, g \rangle = \sum_{n \geq 1} \frac{1}{n^2} f\left(\frac{1}{n}\right) g\left(\frac{1}{n}\right) .$$

In fact, we can even use the general Gaussian integration program IntGauss to perform Monien summation as follows:

─────────────────── | SumMonienGauss.gp | ───────────────────

```
1   /* Monien summation using Gaussian integration. */
2   SumMonienGaussinit(N = 0) =
3   { my(B = getlocalbitprec(), X, W);
4
5     localbitprec(32);
6     my(c1 = 1 - log(2), c2 = log(2) / 2);
7     N = max(N, ceil(Solvemullog(c1, c2 * B))) + 1;
8     localbitprec(B + 32);
9     [X, W] = IntGaussinit(m -> zeta([2..2*m+2]), , N);
```

```
10    X = vector(#X, j, 1 / X[j]);
11    W = vector(#X, j, W[j] * X[j]^2);
12    return (bitprecision([X, W], B + 32));
13  }
14
15  SumMonienGauss(f, XW = 0) =
16  {
17    if (type(XW) != "t_VEC", XW = SumMonienGaussinit(XW));
18    return (SumXW(f, XW));
19  }
```

REMARK 4.8.2. Although the initialization nodes and weights are in principle identical to those obtained with the ordinary SumMonieninit program that we will see below (in reverse order), the present initialization is much slower and should not be used in actual practice. It is only given to emphasize the analogy with Gaussian integration. For instance, at 1000 decimal digits it requires 18.1 seconds, while the SumMonieninit program requires 2.94 seconds.

4.8.3. Generalized Monien summation. We now first want to generalize the above idea to computing $\sum_{n \geq 1} f(n)$ with f having an expansion at infinity of the form $f(n) = \sum_{m \geq 0} a_m / n^{\alpha m + \beta}$ with $\alpha > 0$ and $\beta > 1$ (the case treated above corresponds to $\alpha = 1$ and $\beta = 2$). This is easily done by generalizing the ψ function:

DEFINITION 4.8.3. Let $\alpha > 0$ and $\beta > 1$. We define

$$\psi_{\alpha,\beta}(z) = \sum_{n \geq 1} \frac{n^{\alpha - \beta} z}{z + n^\alpha} \ .$$

Since $\beta > 1$ this is an absolutely convergent series, so it defines a meromorphic function on $\mathbb{C}$ with poles at all $z = -n^\alpha$ with $n \in \mathbb{Z}_{\geq 1}$, simple with residue $-n^{2\alpha - \beta}$ at $z = -n^\alpha$.

For instance, if $\alpha = 1$ and $\beta = 2$, we have

$$\psi_{1,2}(z) = \sum_{n \geq 1} \frac{z}{n(z + n)} = \sum_{n \geq 1} \left(\frac{1}{n} - \frac{1}{z + n} \right) \ .$$

and since

$$\psi(s) = -\gamma + \sum_{n \geq 1} \left(\frac{1}{n} - \frac{1}{n + s - 1} \right) \ ,$$

it follows that $\psi_{1,2}(z) = \gamma + \psi(z + 1)$.

Thus, generalizing Monien's initial idea, we have on the same contour $\mathcal{C}$ containing $[0, 1]$ as above

$$\sum_{n \geq 1} n^{2\alpha - \beta} F(n^\alpha) = -\frac{1}{2\pi i} \int_{\mathcal{C}} \mathcal{C} \frac{1}{z^2} \psi_{\alpha,\beta}(-1/z) F(1/z) \, dz \ ,$$

hence if as above $f(n) = \sum_{m \geq 0} a_m/n^{\alpha m + \beta}$, and if we set $F(z) = f(z^{1/\alpha})z^{\beta/\alpha - 2}$, we have $n^{2\alpha - \beta} F(n^\alpha) = f(n)$, hence

$$S := \sum_{n \geq 1} f(n) = -\frac{1}{2\pi i} \int_{\mathcal{C}} \frac{1}{z^2} \psi_{\alpha,\beta}(-1/z) F(1/z) \, dz$$

$$= -\frac{1}{2\pi i} \int_{\mathcal{C}} \frac{1}{z^{\beta/\alpha}} \psi_{\alpha,\beta}(-1/z) f(1/z^{1/\alpha}) \, dz \ .$$

Now around $z = 0$ we have

$$\psi_{\alpha,\beta}(-z) = -\sum_{n \geq 0} \zeta(\alpha n + \beta) z^{n+1} \ .$$

Thus, applying the quotient-difference algorithm of Section 7.6 we can formally write

$$-\sum_{n \geq 0} \zeta(\alpha n + \beta) z^{n+1} = c(0)z/(1 + c(1)z/(1 + c(2)z/(1 + \cdots))) \ ,$$

where for instance $c(0) = -\zeta(\beta)$, $c(1) = -\zeta(\alpha + \beta)/\zeta(\beta)$, etc. Experimentally, we again find that $c(n) = -(g(\alpha)/n)^\alpha + O(1/n^{\alpha+1})$ with $g(\alpha) = 2\Gamma(1/\alpha+1)^2/\Gamma(2/\alpha + 1)$, and that we have two extremely accurate asymptotic expansions, and it is not difficult to see that the convergence will now be of the order of $1/n!^\alpha$ instead of $1/n!$. We will give below a general program which includes this as a special case.

Before doing that, note that in the special case $\alpha = 2$ (and $\beta = 2$), which is the only case considered in Monien's paper [**Mon10**], everything is explicit, including the continued fraction. Indeed, it is well-known that

$$\psi_{2,2}(z) = \sum_{n \geq 1} \frac{z}{z + n^2} = \frac{1}{2}(\sqrt{z} \coth(\pi\sqrt{z}) - 1) \ ,$$

and from the continued fraction expansion of coth (which can be immediately obtained from Example 7.4.1), we deduce that

$$-\sum_{n \geq 0} \zeta(2n + 2) z^{n+1} = c(0)z/(1 + c(1)z/(1 + c(2)z/(1 + \cdots))) \ ,$$

with $c(0) = -\pi^2/6$ and $c(n) = -\pi^2/((2n + 1)(2n + 3))$ for $n \geq 1$. This explicit formula avoids the costly use of the `Quodif` algorithm. Before giving the general `SumMonien` program, we give the easier program corresponding to this special case:

──────────────── $\boxed{\text{SumMonien2.gp}}$ ────────────────

```
1   /* Monien summation when the asymptotic expansion of f
2      is in integral powers of 1/n^2. */
3   SumMonien2init(N = 0) =
4   { my(B = getlocalbitprec(), v, vr, PQ, Q, S, X, W);
5
6     localbitprec(32);
7     my(c1 = 1 - log(2) / 2, c2 = log(2) / 4);
8     N = max(N, ceil(Solvemullog(c1, c2 * B))) + 1;
9     localbitprec(1.1 * B + 32);
10    v = vector(2 * N, m, if (m == 1, 1 / 6, 1 / (4 * m^2 - 1)));
11    S = 0; forstep (n = 2 * N, 1, -1, S = v[n] * 'x / (1 + S));
12    PQ = subst(1 + S, 'x, -Pi^2 * 'x);
```

```
13    Q = denominator(PQ);
14    vr = SumMonienroots(Q, 2, polrootsreal);
15    if (poldegree(Q) != #vr,
16      error("incorrect roots in SumMonien2init"));
17    X = vector(N, j, sqrt(vr[j]));
18    W = subst(numerator(PQ) / Q', 'x, vr);
19    W = vector(N, j, W[j] / X[j]^2);
20    return (bitprecision([X, W], B + 32));
21  }
22
23  SumMonien2(f, XW = 0) =
24  {
25    if (type(N) != "t_VEC", XW = SumMonien2init(XW));
26    SumXW(f, XW);
27  }
```

Example of comparison with the ordinary SumMonien given below:
```
? \p1000
? R = SumMonien(n -> lngamma(1 + 1 / n^2));
time = 3,574 ms.
? S = SumMonien2(n -> lngamma(1 + 1 / n^2));
time = 1,058 ms.
? S - R
% = -1.48... E-1002
```
Finally note that, as already mentioned, the SumMonienroots program is fragile and cannot be used when α is not an integer, so we will have to use the general and much slower program polrootsreal instead together with a check.

4.8.4. Monien summation with other weights. Recall that Gaussian integration can be adapted to treat integrals of the form $\int_a^b w(x)f(x)\,dx$ with some fixed "weight function" $w(x)$. Since Monien summation is a discrete Gaussian integration method, this is also the case for such methods.

We assume that we want to compute $\sum_{n\geq 1} f(n)$, where $f(n)$ has an asymptotic expansion of the form $f(n) = w(n)\sum_{m\geq 0} a_m/n^{\alpha m}$, where $w(n)$ is a sufficiently regular weight function (to compute $\sum_{n\geq a} f(n)$, simply replace $f(x)$ by $f(x+a-1)$). The case considered in the previous section corresponds to $w(n) = 1/n^\beta$.

We will need to compute the auxiliary sums $\sum_{n\geq 1} w(n)/n^{\alpha m}$ for $m \geq m_0$ for some initial integral value m_0, and for this we will use the most robust tool available, which is the Sumdelta program. Since the program needs both to know the value of m_0, and also the fast parameter of the Sumdelta program, it is essential that one gives to the program a value fast (which will not be the same as that of Sumdelta) such that $w(n)$ behaves like $n^{\text{fast}+\varepsilon}$ for arbitrarily small ε.

———————————— SumMonien.gp ————————————

```
1  /* c = al, [al, be], [al, n->w(n)] or [al, [n->w(n), fast]]
2   * Return [al, be, n -> w(n)] */
3  Getalbew(c) =
4  {
```

```
5     if (type(c) == "t_CLOSURE", return ([1, 0, c]));
6     if (type(c) != "t_VEC", return ([c, 0, 0]));
7     my([a, b] = c);
8     if (type(a) == "t_CLOSURE", return ([1, -b, a]));
9     if (type(b) == "t_CLOSURE", return ([a, 0, b]));
10    if (type(b) != "t_VEC", return ([a, b, 0]));
11    my([w, fast] = b); return ([a, -fast, w]);
12  }
13
14  /* Initialize weights and abscissas for generalized Monien
15   * summation sum_{n >= 1} f(n) where
16   * f(n) = w(n) * sum_{m >= 0} a_m / n^{al m},  al > 0,
17   * w(n) = O(n^{fast + epsilon}); fast must be set correctly. */
18  SumMonieninit(alff = 1, N = 0) =
19  { my(B = getlocalbitprec(), BN, v, vr, vzeta, PQ, Q);
20    my(X, W, [al, be, w] = Getalbew(alff));
21
22    if (al <= 0, error("divergent SumMonieninit"));
23    if (!w && be == 0, be = al * (1 + floor(1 / al)));
24    if (!w && al == 2 && be == 2, return (SumMonien2init(N)));
25    BN = ceil(B / al);
26    localbitprec(32);
27    my(c1 = 1 - log(2), c2 = log(2) / 2);
28    N = max(N, ceil(Solvemullog(c1, c2 * BN))) + 1;
29    my(NEWB = ceil(1.6 * max(1, al) * BN));
30    my(jin = floor((1 - be) / al) + 1);
31    while (1,
32      localbitprec(NEWB);
33      al = bitprecision(al, NEWB);
34      vzeta = if (!w,
35        -zeta(vector(2 * N, m, al * (m - 1) + be))
36      , /* else */
37        Sumdelta(m -> my(M = m^(-al));
38                      -w(m) * powers(M, 2 * N - 1, M^jin),
39                   -be - jin * al));
40      v = Quodif(vzeta);
41      PQ = CFback(2 * N, n -> v[n+1] * 'x);
42      Q = denominator(PQ);
43      my(int = if (!w && type(al) == "t_INT", al, 0));
44      vr = SumMonienroots(Q, int, polrootsreal);
45      if (poldegree(Q) == #vr, break);
46      /* here accuracy loss, increase it by 50%. */
47      NEWB = (3*NEWB) >> 1);
48    X = vector(N, j, vr[j]^(1/al));
49    W = subst(numerator(PQ) / Q', 'x, vr);
50    if (!w,
51      my(c = be - 2 * al);
52      if (c, W = vector(N, j, W[j] * X[j]^c))
```

```
53      , /* else */
54        W = vector(N, j, W[j] * vr[j]^(jin - 2) / w(X[j])));
55      return (bitprecision([X, W], B + 32));
56    }
57
58    /* Compute sum_{n >= 1} w(n) f(n). If XW a t_INT, recompute XW,
59     * otherwise consider that XW = [X, W]. alff = [al, w, fast]
60     * with everything optional. */
61    SumMonien(f, alff = 1, XW = 0) =
62    {
63      if (type(XW) != "t_VEC", XW = SumMonieninit(alff, XW));
64      return (SumXW(f, XW));
65    }
```

REMARKS 4.8.4.

(1) First and most importantly, note that the Sumdelta command sums a
function with values in $\mathbb{C}^{2N}$. If instead (for al = 1) we had written

```
vzeta = vector(2*N, j, Sumdelta(m -> w(m) / m^(j+jin-1)))
```

this would have been orders of magnitude slower.

(2) We only set the int flag to α in SumMonienroots if $w(n)$ is trivial and
α is integral. In this case, the command approximates half the roots
by consecutive integers to the power α as seen above; otherwise it uses
directly the slow polrootsreal command. For any specific weight, one
can adapt the SumMonienroots program to handle the computation of
the roots, as we shall see for the case of $w(n) = \log(n)$ at the end of this
section.

(3) The constants $c_1 = 1 - \log(2)$ and $c_2 = \log(2)/2$ used to determine N come
from a rough estimate of the error which should be of the order of $1/(2N)!$
for reasonable functions, so we need $(2N)! > 2^B$ and we approximate this
by Stirling's formula.

On the other hand, the "magic constant" 1.6 used to determine the
working accuracy (and 1.1 for SumMonien2init) is more subject to discussion, since it is only there to compensate for the instability of the
quotient-difference algorithm, but it seems to work in practice. In any
case, since we multiply the working accuracy by 1.5 when we do not find
enough real roots, we will eventually use a sufficient accuracy to compensate for the instability.

EXAMPLES. Assume first that we want to compute $S = \sum_{n \geq 1} 1/(n^{3/2} + n^{1/2})$.

```
? \p115
? f(n) = 1 / (n^(3/2) + n^(1/2));
? S = SumMonien(f, [1, 3/2])
time = 69 ms.
% = 1.86... /* Perfect. */
? SumMonien(f, [1, [n -> 1 / n^(3/2), -3/2]]) - S
time = 536 ms.
% = 0.E-115 /* Also perfect, much slower. */
```

```
? SumMonien(f, [1, [n -> 1 / n^(5/2), -5/2]]) - S
time = 544 ms.
% = 0.E-115 /* Same */
? SumMonien(f, [1, [n -> 1 / n^2, -2]]) - S
time = 639 ms.
% = -0.00344... /* Completely wrong since wrong w(n) */
? SumMonien(f, [1, [n -> 1 / n^(3/2), -3]]) - S
time = 721 ms.
% = -1.06... E-43 /* Only 43 correct decimals since
                  * incorrect 'fast' */
? SumMonien(f, [1, [n -> 1 / n^(3/2), -2]]) - S
time = 511 ms.
% = 0.E-115 /* Perfect although incorrect 'fast' */
```

In this case, it is of course much better to use the first syntax since it uses the preexisting `zeta` function instead of the `Sumdelta` summation program.

Assume now that we want to compute $S = \sum_{n\geq1} \log(n^2 + 1)/(n^{3/2} + n^{1/2})$. Here, because of the nontrivial weight we have no real choice:

```
? f(n) = log(n^2 + 1) / (n^(3/2) + n^(1/2));
? SumMonien(f, [n -> log(n^2 + 1) / n^(3/2), -3/2])
time = 616 ms.
% = 7.68... /* Perfect. */
```

In some examples, it is possible to directly compute the sums $\sum_{m\geq1} w(m)/m^j$ instead of using `Sumdelta`. For instance, if $w(m) = \log(m)$, it is simply a matter of replacing $\zeta(j)$ by $-\zeta'(j)$. Also, one can check that the nodes are close to 2, 3, ..., so instead of using `SumMonienroots` directly, which would fail, one could write something like

```
vabs = SumMonienroots(subst(Q, 'x, 'x + 1));
vabs = vector(#vabs, j, vabs[j] + 1);
```

and in addition in the program `SumMonienroots` itself, write `m=floor(n/2)-1` instead of `m=floor(n/2)`. It is clear that this type of initialization will be faster than using the general-purpose program given above.

4.8.5. Monien summation for alternating series. The idea of Monien summation works equally well for the summation of *alternating* series. Here, instead of the function $\psi(1-z)$ used in the basic Monien method, which has poles at $n \in \mathbb{Z}_{\geq1}$ with residues all equal to 1, we use the function $\psi(1 - z) - \psi((1 - z)/2)$ which also has poles at $n \in \mathbb{Z}_{\geq1}$ with residues $(-1)^n$. Using the duplication formula, it is immediate to see that its power series expansion around $z = 0$ is

$$\psi(1 - z) - \psi((1 - z)/2) = 2\log(2) + \sum_{n\geq1} \zeta(n + 1)(1 - 1/2^n)z^n .$$

Thus, once again using the quotient-difference algorithm we write formally

$$\sum_{n\geq1} \zeta(n + 1)(1 - 1/2^n)z^n = \cfrac{d(0)z}{1 + \cfrac{d(1)z}{1 + \cfrac{d(2)z}{1 + \cfrac{d(3)z}{1 + \ddots}}}} ,$$

where for instance $d(0) = \zeta(2)/2$, $d(1) = -(3/2)\zeta(3)/\zeta(2)$, etc.

Experimentally, as above we find that we have two extremely accurate asymptotic expansions:

For n even,

$$d(n) = \frac{d_1}{n} - \frac{2d_1}{n^2} + \frac{d_3}{n^3} - \frac{8d_1 - 4d_3}{n^4} + \frac{d_5}{n^5} - \cdots ,$$

where $d_1 = 0.19967864025773383391636984879\cdots$ and d_3, d_5 are similar, and for n odd

$$d(n) = -\frac{d_1 + 2}{n} + \frac{d_1 + 2}{n^2} - \frac{5d_1 - d_3 + 2}{n^3} + \frac{21d_1 - 5d_3 + 2}{n^4} - \frac{c_5}{n^5} + \cdots .$$

Note that using the Plouffe inverter, one finds that $d_1 + 1$ is the positive root of $x \tanh(x) = 1$.

Once again, using Theorem 7.3.7, we find a similar rapid speed of convergence

$$S - S(n) \sim C \frac{((d_1(d_1 + 2))^{1/2} z)^n n^{-3/2 + 2z}}{n!} .$$

Note that $(d_1(d_1 + 2))^{1/2} = 1/\sinh(d_1 + 1) \approx 0.66274\cdots$.

We now use exactly the same method as above. However, note that since we are dealing with alternating series, we can replace $m \geq 2$ in the expansion of f by $m \geq 1$. This implies that we must include $\log(2)$, and replace $f(z)$ by $zf(z)$, which is equivalent to multiplying by x in the definition of R. Note also that the scalar product is here

$$\langle f, g \rangle = \sum_{n \geq 1} \frac{(-1)^n}{n} f\left(\frac{1}{n}\right) g\left(\frac{1}{n}\right)$$

which is *not* positive definite, so that the roots of DE are not all real (in fact half of them are), so we cannot take the real part, and we cannot use `polrootsreal`. It is of course immediate to generalize to functions $f(n)$ having a more general expansion at infinity.

However, contrary to all the other summation methods for summing positive terms that we have seen, here the best way to use Monien summation for alternating series is *not* to use the above ideas but simply the naïve observation that it suffices to sum $f(2m - 1) - f(2m)$. The main reason is that, although the corresponding programs would require half the number of function evaluations compared with `SumMonien`, we would need to evaluate the function at *complex* values, which is always more than twice as expensive, so we do not give the program explicitly.

4.9. Summing functions defined only on integers

For most of the methods that we have studied, it was necessary to assume that the summand be a function defined on the reals, not only on the integers, and in several places, even defined in a complex neighborhood of $[a, \infty[$. In some cases, one can reduce to this case: for instance if the summand involves the harmonic sum $H_n = \sum_{1 \leq m \leq n} 1/m$ which is a priori defined only for integers n, we can replace H_n by $\psi(n + 1) + \gamma$, where $\psi(x)$ is the logarithmic derivative of the gamma function and $\gamma = 0.57721\cdots$ is Euler's constant. But in most cases it is impossible to find a natural interpolating function, so we must look at what is available.

For alternating series we are in luck: the `Sumalt` program is very efficient (even if it is not always the fastest possible), and involves only the values of the summand at integers. We also recall that there are more complicated variants of that program which are even faster.

For series with positive terms regularly tending to 0 at infinity, the choice is much more limited. In fact, we have seen three such programs, with variants.

(1) The `Sumpos` program (also available as the built-in `sumpos` program). Its main limitation is that it can be used only for summands which can be evaluated at exponentially large values of the argument. For instance, it is out of the question to use it to compute $\sum_{n\geq 1} H_n/n^2$ if H_n is summed naively; in this case, we can replace H_n by $\psi(n+1) + \gamma$ which allows to compute it for exponentially large values of n.
(2) The `SumLagrange` extrapolation program and its variants.
(3) The `SumSidi` program.

We can also think of modifying the other summation programs. Programs such as `SumEulerMaclaurin`, `Sumdelta`, and `SumAbelPlana` involve an integral $\int_N^\infty f(x)\,dx$, so are not applicable to functions f which cannot be extended from the integers to the reals.

On the other hand, the `SumMonien` program does not involve any integrals, but its nodes are not integral. However, we have mentioned that one half of these nodes are very close to integers. This leads to the following possible algorithm:

(1) Apply the `SumMonieninit` algorithm to find the nodes and weights of the ordinary `SumMonien` program, except that first we ask for $10/9$ times more nodes (i.e., multiply the formula giving N by $10/9$), and second discard the weights.
(2) Round the first $9/10$ of the nodes thus obtained to the nearest integer.
(3) By solving a suitable VanderMonde type linear system, find the weights corresponding to these nodes, so that the computation will be exact for $\sum_{n\geq 1} 1/n^k$ for k up to the number of nodes.

We leave to the reader the task of writing the corresponding programs, which give reasonably good results, but in general are not competitive with the other available methods, in particular because of the very long initialization time. Note that, using these modified Monien programs, we also obtain a new *extrapolation* method by summing $u_{n+1} - u_n$, which would not be possible with the methods needing an integral.

4.10. Multiple sums and multizeta values

4.10.1. Double sums. Contrary to the case of multiple integrals, where it was relatively easy to use numerical integration methods to compute them, at the expense of much slower programs, it is not straightforward to do the same for double or multiple *sums*. Let us consider two examples.

Example 1. $S = \sum_{m,n\geq 1} 1/(m^2n^2+1)$ (exercise: why don't we start the sum at $m, n \geq 0$?).

Using the most efficient program `SumMonien`, after initializing weights as in the case of double integrals, we can write

```
tab = SumMonieninit();
f(m,n) = 1 / ((m*n)^2 + 1);
SumMonien(n -> SumMonien(m->f(m,n),, tab),, tab)
```

and even at 500 decimal digits this gives a perfect result in 156 ms.

Example 2. $S = \sum_{m,n \geq 1} 1/(m^2 + n^2)^2$ (exercise: prove that S is equal to $\zeta(2)G - \zeta(4)$, where $G = \sum_{n \geq 0} (-1)^n/(2n+1)^2$ is Catalan's constant, but of course we will not use this). All our summation programs used in the above manner lose considerable accuracy (at 115D `Sumdelta` gives only 45 correct decimals and `SumMonien` only 10). The reason is that when $m = 100$, say, none of the programs is sufficiently clever to understand that $\sum_{n \geq 1} 1/(100^2 + n^2)^2$ must not be computed naively.

The simplest way to deal with this problem is with the usual cheat of increasing accuracy. For instance, let us try to compute S to 115D:

```
? \p 115
? tab = Sumdeltainit();
? f(m,n) = 1 / (m^2+n^2)^2;
? Z = zeta(2) * Catalan - zeta(4);
? S = Sumdelta(m -> Sumdelta(n->f(m,n),, tab),, tab);
time = 13,055 s.
? S - Z
% = -1.16... E-45 \\ only 45 correct decimals
```

This is both slow and inaccurate. The speed is easily improved by precomputing the integration weights once and for all (else the internal `Sumdelta`'s `Sumintinit` calls will recompute them for each new value of m):

```
? XW = Inta_ooinit();
? S = Sumdelta(m -> my(g = n->f(m,n)); \
                my(G = N -> -Inta_oo(g, N,, XW)); \
                Sumdelta(g, G, tab),, tab);
time = 2,371 ms.
? S - Z
% = 1.014... E-38
```

This is indeed faster, but even more inaccurate since in the original command, the external `Sumdelta` increases the accuracy to $B+32$ bits, and the integration weights are computed for this accuracy. We could thus either increase the accuracy of the integrals or, better, the number of nodes to be used thanks to the `mul` parameter of `Inta_ooinit`:

```
? XW = Inta_ooinit(, 4); /* Increase step size for integrals. */
? S = Sumdelta(m -> my(g = n->f(m,n)); \
                my(G = N -> -Inta_oo(g, N,, XW)); \
                Sumdelta(g, G, tab),, tab);
time = 8,877 ms.
? S - Z
% = -2.53... E-116 \\ perfect
```

Of course, in this case, an explicit antiderivative is readily available, and we should have used it in the first place:

```
? F(m,n) = (n*m/(m^2+n^2) - atan(m/n)) / (2*m^3); \\ dF/dn = f
? S = Sumdelta(m -> Sumdelta(n->f(m,n), n->F(m,n), tab),, tab);
time = 358 ms.
? S - Z
% = -2.53... E-116 \\ perfect
```

Note that even if we increase accuracy, `SumMonien` fails completely: at 1000D, we barely get 15 correct digits.

The above method works for many double sums. We can, however, do better (at least for the above sum): it is clear that

$$S = \sum_{m,n \geq 1} \frac{1}{(m^2 + n^2)^2} = 2 \sum_{1 \leq m < n} \frac{1}{(m^2 + n^2)^2} + \frac{1}{4} \sum_{m \geq 1} \frac{1}{m^4} = 2T + \frac{\zeta(4)}{4} ,$$

say, where

$$T = \sum_{n \geq 1} \sum_{1 \leq m \leq n-1} 1/(m^2 + n^2)^2 .$$

Thus, if we use one of the summation programs which use only values at integer arguments such as `SumLagrange`, we can write

```
? U(n) = sum(m = 1, n - 1, 1 / (m^2+n^2)^2, 0.);
? T = SumLagrange(U); S = 2 * T + zeta(4) / 4
```

At 115D, we now obtain 90 correct digits in 4ms (and the requested 115 digits in 9ms after increasing the accuracy to 150D). The above idea can be generalized to arbitrary double (or even multiple) sums: here is a generic driver routine.

──────────────────────── Sumdouble.gp ────────────────────────

```
1   /* Compute sum_{m, n >= 1} f(m, n), where f is given as a
2    * closure with two arguments. Do not assume that f is
3    * defined over the positive reals. */
4   Sumdouble(f, alf = 1, mul = 1) =
5   { my(B = getlocalbitprec(), S1, S2);
6
7     localbitprec(3/2 * B);
8     S1 = SumLagrange(n -> sum(m = 1, n-1, f(m,n), 0.), alf, mul);
9     S2 = SumLagrange(m -> sum(n = 1,   m, f(m,n), 0.), alf, mul);
10    return (bitprecision(S1 + S2, B));
11  }
12
13  /* Summing over diagonals m + n = cst: halve workload */
14  Sumdoublediag(f, alf = 1, mul = 1) =
15  { my(B = getlocalbitprec(), S);
16
17    localbitprec(3/2 * B);
18    S = SumLagrange(n -> sum(k=1, n, f(k,n+1-k), 0.), alf, mul);
19    return (bitprecision(S, B));
20  }
```

──

Thus, the above sum can be computed by the simple command

```
? \p115
? f(m, n) = 1. / (m^2 + n^2)^2;
? S = Sumdouble(f);
time = 21 ms
? S - Z
% = -2.537... E-116
```

The `Sumdoublediag` variant sums over diagonals and is possibly twice as fast. However, it performs badly on the above example, but increasing the multiplier in `SumLagrange` makes it perfect as well (albeit slower):

```
? S = Sumdoublediag(f);
time = 11 ms
? S - Z \\ only 49 correct digits
% = -3.866... E-49
? S = Sumdoublediag(f,, 2); \\ use mul = 2 in SumLagrange
time = 41 ms
? S - Z \\ now perfect
% = -2.537... E-116
```

This generic program must evidently be suitably modified if the inner sums do not have an asymptotic expansion in $1/n$, such as the computation of the multizeta value $\zeta([a, 1])$ that we will encounter below.

4.10.2. Multizeta values. The theory of multizeta values has attracted a lot of attention in the past 20 years, in connection with many different problems in mathematics and theoretical physics. If $\mathbf{a} = (a_1, a_2, \ldots, a_r)$ is a vector of r strictly positive integers, we define

$$\zeta(\mathbf{a}) = \zeta(a_1, a_2, \ldots, a_r) = \sum_{n_1 > n_2 > \cdots > n_r > 0} \frac{1}{n_1^{a_1} n_2^{a_2} \cdots n_r^{a_r}} \ .$$

This converges if and only if $a_1 \geq 2$. For instance,

$$\zeta(a, b) = \sum_{n > m > 0} \frac{1}{n^a m^b} = \sum_{n \geq 1} \frac{H_{n-1}^{(b)}}{n^a} \ ,$$

where $H_n^{(b)}$ is the generalized harmonic sum $H_n^{(b)} = \sum_{1 \leq j \leq n} 1/j^b$.

To compute $\zeta(a, b)$ with our standard summation methods we need $H_n^{(b)}$ to be defined for real n, not only integral n, unless we try to use methods which uses only integral values, see Section 4.9. For instance, to compute $\zeta(a, 2)$ we can write

```
? H2(x) = Pi^2/6 - psi'(x);
? Z2(a) = SumMonien(n -> H2(n) / n^a);
```

Or more generally, for $b \geq 2$:

```
? H(b, n) = sum(m = 1, n, 1 / m^b, 0.);
? Z(a, b) = SumLagrange(n -> H(b, n) / n^a);
```

On the other hand, for $b = 1$ we would need to use the **SumMonienwn** program with a logarithmic weight, or more simply:

```
? H1(x) = psi(x) + Euler();
? Z1(a) = Sumdelta(n -> H1(n) / n^a);
```

Note that by using easy polylogarithmic identities one can also write programs for computing any multizeta value.

In his PhD thesis, P. Akhilesh has found a simpler and faster algorithm to compute multizeta values. The same methods can be applied more generally to alternating multizeta and multipolylogs. The main idea is a recursion which may be used to compute a single multizeta value or simultaneously all multizeta values of weight $k = a_1 + a_2 + \cdots + a_r$ less than or equal to a given bound. Both algorithms are implemented in **Pari/GP** under the evident names **zetamult** and **zetamultall**. We now give and prove his main theorem. First, we extend the definition of multizeta values to $r = 0$ by setting $\zeta(\emptyset) = 1$.

DEFINITION 4.10.1.
(1) To any $\mathbf{a} = (a_1, \ldots, a_r)$ as above (hence with $a_1 \geq 2$) we associate the binary word
$$w(\mathbf{a}) = \{0\}_{a_1-1}1\{0\}_{a_2-1}1\cdots\{0\}_{a_r-1}1 \, ,$$
so that the length of $w(\mathbf{a})$ is equal to $a_1 + \cdots + a_r = k$.
(2) We say that a binary word w is *admissible* if either $w = \emptyset$, or if w begins with a 0 and ends with a 1.

It is thus clear that a word w is admissible if and only if the corresponding multizeta value converges. The following lemma is trivial:

LEMMA 4.10.2. *Let w be nonempty and admissible. There exist unique positive integers a and b and a (possibly empty) admissible word v such that $w = 0\{1\}_{b-1}v\{0\}_{a-1}1$. In addition, the three words $w^{init} = 0\{1\}_{b-1}v$, $w^{mid} = v$, and $w^{fin} = v\{0\}_{a-1}1$ are admissible with the following two exceptions: $v = \emptyset$ and $b = 1$, in which case $w^{init} = 0$, and $v = \emptyset$ and $a = 1$, in which case $w^{fin} = 1$.*

So as to include the exceptions, we will temporarily say that w is quasi-admissible (our notation) if either w is admissible, or if $w = 0$ or $w = 1$. Akhilesh's algorithm is the following result (with slightly modified notation):

THEOREM 4.10.3. *For any quasi-admissible word w there exists a sequence $(\zeta_n(w))_{n\geq0}$ with the following properties:*
(1) *If w is admissible we have $\zeta(w) = \zeta_0(w)$, where $\zeta(w)$ is the multizeta value associated to w.*
(2) $\zeta_n(\emptyset) = 1/\binom{2n}{n}$ *for $n \geq 0$ and $\zeta_n(0) = \zeta_n(1) = 1/(n\binom{2n}{n})$ for $n \geq 1$.*
(3) *With the notation of the lemma, we have the recursion*
$$\zeta_{n-1}(w) = \zeta_n(w) + \frac{\zeta_n(w^{init})}{n^a} + \frac{\zeta_n(w^{fin})}{n^b} + \frac{\zeta_n(w^{mid})}{n^{a+b}} \, .$$
(4) *We have the estimate $\zeta_n(w) \leq (\pi^2/6) \cdot 4^{-n}$.*

PROOF. The proof of this theorem (which can be found in Akhilesh's paper [**Akh17**]) is quite easy and for the convenience of the reader we give a sketch.
The crucial ingredient is the use of *iterated integrals*:

DEFINITION 4.10.4. Let Δ_k denote the simplex $\{(t_i) \in]0,1[^k : 1 > t_1 > t_2 > \cdots > t_k > 0\}$.
(1) If $(\eta_i)_{1\leq i\leq k}$ are differentials of the form $\eta_i(t) = f_i(t)dt$, we define
$$\int_{\Delta_k} \eta_1 \cdots \eta_k = \int_{1>t_1>t_2>\cdots>t_k>0} f_1(t_1)f_2(t_2) \cdots f_k(t_k) \, dt_1 dt_2 \cdots dt_k \, .$$
(2) If $w = \varepsilon_1 \ldots \varepsilon_k$ is a binary word as above, we set
$$\zeta(w) = \int_{\Delta_k} \omega_{\varepsilon_1} \cdots \omega_{\varepsilon_k}$$
where $\omega_0(t) = dt/t$ and $\omega_1(t) = dt/(1-t)$.

The following lemma is easily shown:

LEMMA 4.10.5. *If* $\mathbf{a} = (a_1, \ldots, a_r)$ *is as above and* $w = w(\mathbf{a})$ *is the corresponding binary word, we have* $\zeta(\mathbf{a}) = \zeta(w)$.

We introduce the ordinary n-tail by

$$\zeta_{0,n}(\mathbf{a}) = \sum_{n_1 > n_2 > \cdots > n_r > n} \frac{1}{n_1^{a_1} n_2^{a_2} \cdots n_r^{a_r}} \, ,$$

and the above lemma generalizes to

$$\zeta_{0,n}(\mathbf{a}) = \int_{\Delta_k} \omega_{\varepsilon_1} \cdots \omega_{\varepsilon_{k-1}} (t^n \omega_{\varepsilon_k})$$

This motivates Akhilesh to introduce *double tails*:

DEFINITION 4.10.6. The double tail $\zeta_{m,n}(\mathbf{a})$ is defined by the iterated integral

$$\zeta_{m,n}(\mathbf{a}) = \zeta_{m,n}(w) = \int_{\Delta_k} \left((1-t)^m \omega_{\varepsilon_1} \right) \omega_{\varepsilon_2} \cdots \omega_{\varepsilon_{k-1}} \left(t^n \omega_{\varepsilon_k} \right) \, ,$$

where as usual $w = w(\mathbf{a}) = \varepsilon_1 \ldots \varepsilon_k$.

Although not needed for the proof of the theorem, we note that double tails can also be expressed as multiple sums:

PROPOSITION 4.10.7. *We have*

$$\zeta_{m,n}(\mathbf{a}) = \sum_{n_1 > n_2 > \cdots > n_r > n} \frac{1}{\binom{n_1+m}{m} n_1^{a_1} n_2^{a_2} \cdots n_r^{a_r}} \, .$$

PROOF. This follows from the identity

$$\int_0^1 (1-t)^m t^{n_1 - 1} \, dt = \frac{1}{\binom{n_1+m}{m} n_1} \, .$$

$\square$

Note also the following *duality* relation, whose proof is trivial using the changes of variable $t_k \mapsto 1 - t_k$, but nontrivial from the expression as a sum:

PROPOSITION 4.10.8. *For any binary word* $w = \varepsilon_1 \ldots \varepsilon_k$ *as above, define the dual word* $\overline{w} = \overline{\varepsilon_k} \ldots \overline{\varepsilon_1}$, *where* $\overline{\varepsilon} = 1 - \varepsilon$. *We have the identity* $\zeta(\overline{w}) = \zeta(w)$, *and more generally* $\zeta_{m,n}(\overline{w}) = \zeta_{n,m}(w)$.

The most famous special case of this proposition is the identity $\zeta(2,1) = \zeta(3)$ due to Euler, explicitly

$$\sum_{m > n > 0} \frac{1}{m^2 n} = \sum_{m > 0} \frac{1}{m^3} \, .$$

The main recursion that we need is the following:

PROPOSITION 4.10.9. *For* $\varepsilon = 0$ *or* 1 *set* $s(\varepsilon) = (-1)^{\varepsilon}$. *For* w *admissible, for any* ε *such that the corresponding words are admissible and the indices nonnegative we have*

$$\varepsilon \zeta_{m,n-1}(w\varepsilon) = \zeta_{m,n}(w\varepsilon) - s(\varepsilon)\zeta_{m,n}(w)/n$$
$$(1-\varepsilon)\zeta_{m-1,n}(\varepsilon w) = \zeta_{m,n}(\varepsilon w) + s(\varepsilon)\zeta_{m,n}(w)/m \, .$$

PROOF. Immediate by replacement in the iterated integrals. $\square$

Applying this proposition twice, we deduce the following corollary, again valid whenever everything is defined:

COROLLARY 4.10.10. *We have*

$$(1 - \varepsilon_1)\varepsilon_2\zeta_{n-1,n-1}(\varepsilon_1 w\varepsilon_2) = \zeta(\varepsilon_1 w\varepsilon_2)_{n,n} + s(\varepsilon_1)\frac{\zeta_{n,n}(w\varepsilon_2)}{n}$$

$$- s(\varepsilon_2)\frac{\zeta_{n,n}(\varepsilon_1 w)}{n} - s(\varepsilon_1)s(\varepsilon_2)\frac{\zeta_{n,n}(w)}{n^2} .$$

LEMMA 4.10.11. *If for completeness, we also define for quasi-admissible words*

$$\zeta_{m,n}(\emptyset) = \frac{m!n!}{(m+n)!} = \frac{1}{\binom{m+n}{n}} ,$$

$$\zeta_{m,n}(0) = \int_0^1 (1-t)^m t^n \frac{dt}{t} = \frac{m!(n-1)!}{(m+n)!} = \frac{\zeta_{m,n}(\emptyset)}{n} ,$$

$$\zeta_{m,n}(1) = \int_0^1 (1-t)^m t^n \frac{dt}{1-t} = \frac{(m-1)!n!}{(m+n)!} = \frac{\zeta_{m,n}(\emptyset)}{m} ,$$

the recursions of the corollary remain valid.

The proof of the first three formulas of the theorem easily follow by induction from the corollary after setting (by abuse of notation) $\zeta_n(w) = \zeta_{n,n}(w)$, and the final inequality is an elementary estimate coming from the iterated integral.   □

We can more generally consider *multiple polylogarithms* defined by

$$L(\mathbf{a}, \mathbf{z}) = \sum_{n_1 > n_2 > \cdots > n_r > 0} \frac{z_1^{n_1} \cdots z_r^{n_r}}{n_1^{a_1} \cdots n_r^{a_r}} ,$$

with suitable convergence assumptions. Once again it is easy to express this as an iterated integral over the simplex Δ_k. More precisely, it is convenient to make the change of variable $y_j = (z_1 z_2 \cdots z_j)^{-1}$ ($y_0 = 1$), and the differentials are replaced by $\omega_y = s(y)dt/(t - y)$, where $s(0) = 1$ and $s(y) = -1$ if $y \neq 0$, which generalize both the notation s and the differentials ω_ε. The word $w = w(\mathbf{a}, \mathbf{z})$ corresponding to $(\mathbf{a}, \mathbf{z})$ is now on the alphabet formed by 0 and the y_j. In this situation, such a (nonempty) word $w = \varepsilon_1 \ldots \varepsilon_k$ is admissible if $\varepsilon_1 \neq 1$ and $\varepsilon_k \neq 0$ (instead of $\varepsilon_1 = 0$ and $\varepsilon_k = 1$).

We again have a duality relation generalizing Proposition 4.10.8, two examples being

$$\sum_{n \geq 1} \frac{z^n}{n^2} = - \sum_{n > m > 0} \frac{(z/(z-1))^{n-m}}{nm} , \quad \text{and} \quad \sum_{n > m > 0} \frac{(-1)^{n+m}}{nm^2} = - \sum_{n > m > 0} \frac{2^{-m}}{n^2 m} ,$$

the first being valid for $\Re(z) < 1/2$.

We can again define tails and double tails in exactly the same way. Once we compute the integrals corresponding to words of length 1, the whole recursive theory and algorithm goes through formally. The main difficulty lies in the study of convergence of the algorithm. For instance, it converges (at a geometric rate, as in the original algorithm) when for any admissible sub-word $\varepsilon_{j_1} \ldots \varepsilon_{j_2}$ we have $|1 - \varepsilon_{j_1}||\varepsilon_{j_2}| > 1/4$, but this is only a sufficient condition. We leave this subject to the interested reader, and only mention that the Pari/GP function `polylogmult` implements this.

For simplicity, we now restrict to multizeta values. To compute all of them to a given weight k using Akhilesh's theorem we proceed as follows. First note that thanks to the estimate on $\zeta_n(w)$, we only need to compute all $\zeta_n(w)$ for $n \leq N$, where N is chosen such that $(\pi^2/6) \cdot 4^{-N}$ is less than the desired accuracy. Now assume by induction that such values have been computed for all words w of length less than or equal to k. This is true for $k = 0$ and $k = 1$ since the corresponding values are given by the theorem. If w has length $k + 1 \geq 2$ then w^{init}, w^{mid}, and w^{fin} have length at most k, so their ζ_n values have already been computed. We can thus write the recursion of the theorem in the form $\zeta_{n-1}(w) = \zeta_n(w) + c_n$, where c_n is *known*, so that $\zeta_n(w) = \sum_{n<m\leq N} c_m + \zeta_N(w)$, and since $\zeta_N(w)$ is less than the desired accuracy, we have thus computed the $\zeta_n(w)$, and in particular $\zeta(w) = \zeta_0(w)$.

────────────────── | Zetamult.gp | ──────────────────

```
1   /* Given admissible w = 0e_2....e_{k-1}1, compute a,b,v such
2    * that w=0{1}_{b-1}v{0}_{a-1}1 with v empty or admissible.
3    * Input: binary vector evec=[e_2,...,e_{k-1}]
4    * Output: [a, b, minit, mmid, mfin]. */
5   Fdi(v1, v2) = fromdigits(concat(v1, v2), 2) + 2;
6   Findabm(m) =
7   { my(minit, mmid, mfin);
8     if (!m, return ([1, 1, 2, 1, 2]));
9     my(v = valuation(m, 2), a = 1 + v, b);
10    m >>= v;
11    if (hammingweight(m+1) == 1,
12      my(e = exponent(m+1)); \\ now m = {1}_e
13      mfin = if (a > 1, 2^(a-2) + 2, 2);
14      return ([a, e + 1, 2^e + 1, 1, mfin]));
15    my(evec = binary(m), n = #evec);
16    for (j = 2, n, \\ b = index of first 0
17      if (!evec[j], b = j; break));
18    minit = Fdi(vector(b, j, 1), evec[b..n-1]);
19    mmid  = Fdi([1], evec[b+1..n-1]);
20    mfin  = Fdi([1], evec[b+1..n]) << v;
21    return ([a, b, minit, mmid, mfin]);
22  }
23
24  /* For 1 <= m < 2^{k-1}, 1 <= n <= N + 1, Z[m+2][n] is
25   *   zeta(w)_{n-1}, w corresponding to m (if v=0y1, m=1y).
26   * Z[1] is zeta(emptyset)_{n-1},
27   * Z[2] is zeta({0})_{n-1}=zeta({1})_{n-1} for n >= 2. */
28
29  Zetamultfill(k, N) =
30  { my(K = 2^(k - 1), Z, binvec, vabm, vn);
31
32    binvec = vector(N+1); binvec[1] = 1;
33    for (n = 1, N, binvec[n+1] = (2*(2*n-1) * binvec[n]) / n);
34    Z = vector(K + 1);
```

```
35    Z[1] = vector(N+1, j, 1. / binvec[j]);
36    Z[2] = vector(N+1, j, if (j > 1, Z[1][j] / (j-1), 0));
37    vn = vector(N, n, powers(1. / n, k)); \\ vn[n][a+1] = n^(-a)
38    vabm = vector(2^(k-2), m, Findabm(m-1));
39    for (k1 = 2, k, /* assume length < k1 filled */
40      for (m = 2^(k1-2), 2^(k1-1) - 1,
41        my([a, b, minit, mmid, mfin] = vabm[m - 2^(k1-2) + 1]);
42        my(S = 0);
43        Z[m+2] = vector(N+1);
44        forstep (n = N, 1, -1,
45          S += (Z[minit][n+1] * vn[n][a+1] +
46                Z[mfin][n+1]  * vn[n][b+1] +
47                Z[mmid][n+1]  * vn[n][a+b+1]);
48          Z[m+2][n] = S)));
49    return (Z);
50  }
51
52  Zetamultall(k) =
53  { my(B = getlocalbitprec(), N = B / 2 + 16, Z);
54    localbitprec(B + 32);
55    Z = Zetamultfill(k, N);
56    return (bitprecision(vector(#Z-2, j, Z[j+2][1]), B));
57  }
```

REMARKS 4.10.12.

(1) Since this GP program uses very simple operations, and mainly on vectors of bits, the corresponding C program which is implemented in Pari/GP under the name zetamultall is considerably faster.

(2) This program is incredibly efficient: for instance, at 38D to compute all $524287 = 2^{19} - 1$ multizeta values of weight less than or equal to 20, the zetamultall program requires only 3.7 seconds, and at 115D it requires 15.4 seconds.

(3) The values are output in a specific order; to be useful, the program should be supplemented by a small script extracting the desired multizeta value from the whole output vector, and this is left to the reader. You can also use the existing zetamultconvert program.

(4) As already mentioned, we can compute a *single* multizeta value using the same idea: starting from the word $w = w(\mathbf{a})$, we recursively compute the vectors $Z(\mathbf{w}) = (\zeta_n(\mathbf{w}))_{n \leq N}$ for $\mathbf{w} = w^{\text{init}}, w^{\text{fin}}, w^{\text{mid}}$ and then w. Since the same word may occur many times during the recursion, the vectors $Z(\mathbf{w})$ should be stored in a Map (the GP name for a hash table). In any case, if w has weight k we compute and store $O(k^2)$ values $Z(\mathbf{w})$ instead of 2^k in the Zetamultall program.

(5) As mentioned above, this algorithm can also compute *multiple polylogarithms*, at least in certain ranges.

4.11. Sample timings for summation programs

As for integration, since we have given a large number of methods for numerical summation, some of which being applicable only in certain cases, it is now essential to give a detailed comparison, both in the ranges of applicability, and in their relative speed.

In most cases, we can do a precomputation which does not depend on the specific function to be summed, but only on the accuracy and possibly the rate at which the function tends to 0 at infinity. As in the previous chapter, this precomputation time is given in the tables under the name "Init", and of course in subsequent timings it is assumed that the precomputation has been done once and for all. In the "Init" column for SumMonien, we have included in brackets the initialization time for SumMonieninit, which is much larger, but which is necessary for the sums S_i with $11 \leq i \leq 13$.

In many of the summation methods, we need to compute $\int_N^\infty f(t)\,dt$. This is done using doubly-exponential methods, except for the Sumdelta program with vector second parameter which uses a Gaussian method (as already mentioned, this could also be changed for the other methods). Even though the tables of weights and abscissas will be precomputed, the computation of this integral will often take a large part of the running time. This will *not* be the case for the Sumalt algorithms for which we give timings in the next section.

Although rather arbitrary, we have chosen the following test sums for series with positive and regular terms, for which we use the different sumnum programs at our disposal:

$$S_1 = \sum_{n\geq 1} 1/n^2\;,\quad S_2 = \sum_{n\geq 0} 1/(n^3+n+1)\;,\quad S_3 = \sum_{n\geq 1} n^7/(n^9+n+1)\;,$$

$$S_4 = \sum_{n\geq 1} \sin(\pi/(2n))/n\;,\quad S_5 = \sum_{n\geq 1} (\exp(1/n)-1)/n\;,$$

$$S_6 = \sum_{n\geq 1} \log(\Gamma(1+1/n))/n\;,\quad S_7 = \sum_{n\geq 2} \zeta(n)/n^2\;,$$

$$S_8 = \sum_{n\geq 1} (\mathrm{Li}_2(1/n)-1/n)\;,\quad S_9 = \sum_{n\geq 1} \log(1+1/n)/n\;,$$

$$S_{10} = \sum_{n\geq 2} \log(n)(\log(1-1/n)+1/n) + \log(1-1/n)^2/2\;,$$

$$S_{11} = \sum_{n\geq 1} 1/n^{4/3}\;,\quad S_{12} = \sum_{n\geq 0} 1/\sqrt{n^3+n+1}\;,$$

$$S_{13} = \sum_{n\geq 1} 1/n^{3/2}\;,\quad S_{14} = \sum_{n\geq 1} 1/(n^\pi + n^{1.4}+1)\;,$$

$$S_{15} = \sum_{n\geq 2} \log(n)/n^2\;,\quad S_{16} = \sum_{n\geq 3} \log(\log(n))/n^2\;,$$

$$S_{17} = \sum_{n\geq 2} 1/(n^2\log^2(n))\;,\quad S_{18} = \sum_{n\geq 2} 1/(n^{3/2}\log^2(n))\;,$$

$$S_{19} = \sum_{n\geq 2} 1/(n\log^2(n))\;,\quad S_{20} = \sum_{n\geq 2} \sqrt{\log(n)}/n^2\;,$$

$$S_{21} = -\sum_{n\geq 2} \zeta'(n) \,, \quad S_{22} = \sum_{n\geq 0} \exp(-\sqrt{n}) \,, \quad S_{23} = \sum_{n\geq 0} \exp(-n^{1/4}) \,,$$

where $\mathrm{Li}_2(z) = \sum_{n\geq 1} z^n/n^2$ is the dilogarithm.

Note the mathematical identities $S_8 = S_7$ and $S_{21} = S_9$, and note also that $S_{10} = \gamma_1$, the first Stieltjes constant, negative of the coefficient of $s-1$ in the Taylor expansion of $\zeta(s)$ around $s = 1$ (Exercise: prove these statements).

All the series $\sum_{n\geq a} f(n)$ that we have chosen must satisfy the following conditions:

(1) The function f must be meromorphic in some domain of the complex plane, possibly with cuts, with no poles on $\mathbb{R}_{\geq a}$. In particular, f cannot be an arithmetic function only defined on the integers.

(2) We must have $f(n) \geq 0$ for $n \geq a$, and the derivatives of f on $[a, \infty[$ must be monotonic and tend to 0 for n sufficiently large.

(3) With the exception of S_{21}, which is given for comparison with S_9, and of S_{22} and S_{23}, $f(n)$ must not tend to 0 faster than polynomially as $n \to \infty$.

Note that if f satisfies the first condition but not the second, it is often easy to write f as a difference of two functions that do satisfy it.

In the first column, we have indicated by an asterisk * the series for which an explicit antiderivative of the summand is used, which of course speeds up considerably the Euler–Maclaurin type methods EM, EM3, AP, and delta. The line `Init` indicates the time for the initialization step with default parameters, the line `NFE` indicates the necessary Number of Function Evaluations when no explicit antiderivative is given, and the line `NFE*` when an antiderivative is given.

S	pos	Lag	Sidi	EM	EM3	AP	delta	Monien
Init	0.00	0.00	0.00	0.17	0.48	2.47	0.31	0.35
NFE	542730	553	554	8926	7277	6570	6568	129
NFE*	542730	553	554	3329	1680	973	971	129
S_1^*	1.00	**0.00**	0.39	**0.00**	0.01	**0.00**	**0.00**	**0.00**
S_2	0.51	**0.00**	0.38	0.02	0.01	0.01	0.01	**0.00**
S_3	6.72	**0.00**	0.39	0.17	0.02	0.01	0.01	**0.00**
S_4	3.77	0.05	0.44	0.18	0.20	0.06	0.06	0.01
S_5	2.99	0.03	0.42	0.74	0.23	0.06	0.06	0.01
S_6	12.3	0.90	1.29	19.8	3.68	1.05	0.91	0.08
S_7	1.40	$\otimes$	$\otimes$	∞	0.07	2.40	**0.06**	$\otimes$
S_8^*	4.26	0.38	0.77	0.67	1.20	0.27	0.22	**0.04**
S_9^*	3.63	0.04	0.43	0.30	0.15	0.06	0.04	**0.01**
S_{10}	24.7	$\otimes$	$\otimes$	0.88	0.75	0.57	**0.54**	$\otimes$
S_{11}^*	18.9	0.01	0.41	0.04	0.03	0.06	0.01	**0.00**
S_{12}	6.12	**0.00**	0.39	0.04	0.02	0.01	0.01	**0.00**
S_{13}^*	4.64	**0.00**	0.40	0.01	0.01	**0.00**	**0.00**	**0.00**
S_{14}	27.8	$\otimes$	$\otimes$	1.03	1.07	0.82	**0.77**	$\otimes$
S_{15}^*	21.7	$\otimes$	$\otimes$	0.16	0.14	0.06	0.04	**0.01**
S_{16}	42.0	$\otimes$	$\otimes$	0.79	0.77	0.60	0.57	**0.01**
S_{17}^*	21.6	$\otimes$	$\otimes$	0.16	0.15	0.06	0.04	**0.01**
S_{18}	44.1	$\otimes$	$\otimes$	0.49	0.45	0.36	0.35	**0.01**
S_{19}^*	∞	$\otimes$	$\otimes$	0.16	0.14	0.06	**0.04**	$\otimes$

S_{20}	22.1	$\otimes$	$\otimes$	0.42	0.41	0.32	0.30	**0.01**
S_{21}^*	**12.4**	$\otimes$	42.8	∞	13.7	12.6	15.1	$\otimes$
S_{22}^*	0.26	0.27	$\otimes$	0.21	0.25	**0.05**	**0.05**	$\otimes$
S_{23}^*	0.74	14.4	$\otimes$	0.23	0.27	0.11	**0.06**	$\otimes$

Timings for 500 decimals.

A large number of remarks need to be made concerning this table.

General Remarks

- Note that usually, but not always, the time increases when NFE increases. As already mentioned, some of the entries with a $\otimes$ could be replaced by actual times with additional programming.
- The time needed to initialize Gaussian weights and nodes for computing the antiderivative when it is not given is *not* included in the above timings nor in Init times; see §3.11 for the attached Gaussian integration timings.
- For Lagrange and Monien summation, we have used specific parameters (α or a weight) suited to treat the given examples, otherwise the result would be nonsense. The maximal initialization time for Lagrange and Monien when using these non default parameters is 772s (Lagrange) and 16.4s (Monien).
- We have not included the SumPinelis program. Indeed, it is always considerably slower than some of the other programs and the only sum for which is competitive is S_{21} (about as fast as Sumpos and the built-in naïve summation suminf), where the summand $\zeta'(n)$ is an explicit derivative and which already converges geometrically
- Whenever it is applicable, the Gaussian summation program SumMonien is by far the fastest (and it requires only 129 function evaluations to compute an infinite series to 500 decimal digits!). This is completely analogous to the case of numerical integration, where Gaussian integration is by far the fastest, when applicable.
- SumAbelPlana, Sumdelta and SumEulerMaclaurin3 can handle *all* the sums, so are what we call *robust* programs. Note, however, that Sum-AbelPlana needs *complex* values of the summand, which may limit its use, and that SumEulerMaclaurin3 is on average slower. SumEulerMaclaurin is almost as robust, but needs preprogrammed Taylor expansions, which limits its applicability as well, and is never the fastest program.
- The sum $S_7 = \sum_{n \geq 2} \zeta(n)/n^2$ is very specific and not well suited for our default choices. Indeed, we minimized the number of function evaluations, without taking into account the argument value. On the other hand, the Riemann ζ function being a Dirichlet series, $\zeta(s)$ is much easier to compute as $\Re(s)$ gets large: in fact if $\Re(s) > B$, then $\zeta(s) \approx 1$ with an absolute error bounded by 2^{-B}. Furthermore, Euler's formula at even integers provides half the values in a naïve summation for free, Bernoulli numbers being precomputed. Thus Euler–Maclaurin type methods (EM, EM3, AP, delta) would be slow on this example if using the default optimization choices. In the timings given above, we override the default choices in EM3 and delta to use a much large value of N, above $2B$ to make sure all

evaluations of ζ in the remainder term integral are trivialized. Without this change, the timings for both methods would be above 5 seconds, all spent computing the integral. We did not change AP to allow overriding N (in order not to complicate further its interface) and this method is less severely affected since the default choice is $N \approx B/3$, but it is also improved to about $0.06s$ in this example by choosing $N > 2B$: all three methods then become essentially equivalent since all the time is spent computing $\sum_{n \le N} \zeta(n)/n^2$.

Finally, S_7 is the sum of $\zeta(2) - 1$ with the geometrically convergent series $\sum_{n \ge 2}(\zeta(n) - 1)/n^2$. In this form, it would be very well computed by Sumpos as well.

Since SumMonien is almost always the fastest, let us understand the cases where it fails.

Specific Remarks for SumMonien

- As mentioned above, S_7 is in fact the sum of the two series $\sum_{n \ge 2} 1/n^2$ which by definition is perfect for SumMonien, and the geometrically convergent series $\sum_{n \ge 2}(\zeta(n) - 1)/n^2$, which is not well suited for it: for instance at 500 decimal digits SumMonien(n->1/2^n) has only 75 correct decimals.
- S_{10} has an asymptotic expansion involving both pure integral powers of $1/n$ and these multiplied by $\log(n)$. This explains why SumMonien fails, it would succeed if we separate the summands.
- For the totally artificial sum S_{14} nothing can be done for Monien programs.
- S_{19} fails for SumMonien, but for rather trivial reason. Since we must include the nontrivial weight n->1/log(n)^2, by default we call Sumdelta which computes the moments using Inta_oo, and this fails. The reason that Sumdelta itself succeeds on this example is that we give it explicitly the antiderivative! Equivalently it would also succeed if we gave it explicitly $\int_N^\infty dx/(x \log^2(x)) = 1/\log(N)$, this first moment being the only one which Inta_oo cannot compute correctly.
- S_{21}, S_{22}, S_{23} have convergence behavior which are not at all dealt with by SumMonien, except by introducing artificial weights which would amount to using some other summation programs.

Conclusion of this study:

(1) If $f(n)$ tends exponentially or sub-exponentially fast to 0, use the built-in sumpos program, or the variant Sumpos given above. This is the case for $S_{21} = -\sum_{n \ge 2} \zeta'(n)$, but also for $S_7 = \sum_{n \ge 2} \zeta(n)/n^2$ since we can write $S_7 = \sum_{n \ge 2} 1/n^2 + \sum_{n \ge 2}(\zeta(n) - 1)/n^2$, and for S_{22} and S_{23}. Of course, it is not necessary to accelerate the convergence at all in that case, the series can be computed directly. Note that the built-in program suminf which sums naively until the summand is small enough can also be used if f tends exponentially fast to 0, but *not* in case of sub-exponential decrease such as S_{22} and S_{23}.

(2) Otherwise, if $f(n)$ is a power series in $1/n$, or more generally $1/n^{\alpha m + \beta}$ or $w(n)/n^{\alpha m + \beta}$ as above, use the Monien summation programs SumMonien or SumMonienwn.

(3) In all other cases, use Sumdelta, a variant of Euler–Maclaurin which involves computing symmetric differences instead of derivatives. If no antiderivative is available, make sure to include a second parameter $[\alpha, \beta]$ when the integral involved is correctly computed using Gaussian integration. This is especially useful when the summand takes a lot of time to compute, for instance if it involves the gamma function: the sum S_6 for instance requires 0.91 seconds as indicated when using Gaussian integration; using DE is 4 times slower. Compare:

```
? \p500
? f(n) = lngamma(1+1/n) / n;
? tab = Sumdeltainit();
time = 140 ms.
? F = Sumintinit(f, [1,2]); \\ using Gaussian integration
time = 167 ms.
? Sumdelta(f, F, tab);
time = 909 ms.
? F2 = Sumintinit(f); \\ using DE (default)
time = 183 ms.
? Sumdelta(f, F2, tab);
time = 4,153 ms.
```

As explained above the Sumdeltainit and Sumintinit times are *not* included in the reported timings.

A similar caveat to the one we have given for numerical integration is valid here, and even more so. In all of our examples, we have assumed that the summand tends to 0 monotonically and regularly. However, in the case where the summand tends to 0 *sub-exponentially*, we may be in trouble. Consider the sum $\sum_{n \geq 1} e^{-(n/a)^{\alpha}}$ for $(\alpha, a) = (1/2, 1), (1/2, 10), (1/2, 100), (1/4, 1), (1/4, 10), (1/4, 100)$. The slow function sumpos computes all these sums correctly. The SumEulerMaclaurin and Sumdelta functions need fast=1 for $\alpha = 1/2$, otherwise the exponential overflows, and then they compute the sum correctly for $a = 1$, but lose more and more decimals as a increases (only 5 correct decimals for $a = 100$). For $\alpha = 1/4$, fast=1 gives totally wrong results, but fast=0 gives correct results for $a = 1$ and $a = 10$, but starts to lose a large number of decimals for $a = 100$. In addition, at larger accuracies, there is overflow with fast=0.

In the case of integrals, we could overcome this type of problems by making a change of variable. Here this is not possible, at least directly. To compute this type of sums, a preliminary mathematical treatment in each special case seems to be necessary.

4.12. Sample timings for Sumalt programs

We now give times for the computation of alternating sums. Here, the Sumpos program is replaced by the much faster Sumalt program, SumEulerMaclaurin, SumEulerMaclaurin3 and Sumdelta are replaced by SumaltEulerMaclaurin, SumaltEulerMaclaurin3 and Sumaltdelta which do *not* need any computation of integrals, `SumAbelPlana` is replaced by `SumaltAbelPlana` which only needs one integral instead of two. We can therefore a priori expect the speeds to be much faster for `Sumalt`, `SumaltEulerMaclaurin`, and `Sumaltdelta`, a little faster for `SumaltAbelPlana`. This is indeed what we will see in practice.

We can of course give as test sums exactly the same 31 sums that we have given above, with $(-1)^{n-1}$ included after the summation sign. In addition, we will use the following supplementary sums, which do not converge absolutely:

$$A_1 = \sum_{n \geq 1} (-1)^{n-1}/n , \quad A_2 = \sum_{n \geq 1} (-1)^{n-1}/\sqrt{n} , \quad A_3 = \sum_{n \geq 0} (-1)^n/\sqrt{n^2+1}$$

$$A_4 = \sum_{n \geq 2} (-1)^n \log(n)/n , \quad A_5 = \sum_{n \geq 2} (-1)^n \log(n)/\sqrt{n} ,$$

$$A_6 = \sum_{n \geq 2} (-1)^n / \log(n) , \quad A_7 = \sum_{n \geq 1} (-1)^{n-1}\zeta(n+1)/n ,$$

$$A_8 = \sum_{n \geq 1} (-1)^{n-1} \log(\Gamma(1+1/n)) , \quad A_9 = \sum_{n \geq 2} (-1)^n \operatorname{Li}_2(2/n) .$$

In addition, all the programs give sensible results even for diverging series. We will choose for instance:

$$D_1 = \sum_{n \geq 1} (-1)^{n-1} n , \quad D_2 = \sum_{n \geq 1} (-1)^{n-1} n^2 , \quad D_3 = \sum_{n \geq 1} (-1)^{n-1}\sqrt{n}$$

$$D_4 = \sum_{n \geq 2} (-1)^n \log(n) , \quad D_5 = \sum_{n \geq 2} (-1)^n n \log(n) , \quad D_6 = \sum_{n \geq 2} (-1)^n \log(\Gamma(n)) .$$

S	Sumalt	EM	EM3	AP	delta	Sidi
Init	0.00	0.00	0.43	1.14	0.67	0.00
NFE	655	3329	2135	972	1364	405
S_1	**0.00**	0.01	0.01	**0.00**	0.01	0.16
S_2	**0.00**	0.02	0.01	**0.00**	0.01	0.17
S_3	**0.00**	0.27	0.02	0.01	0.02	0.17
S_4	**0.03**	0.19	0.22	0.04	0.05	0.18
S_5	**0.02**	1.26	0.26	0.04	0.05	0.18
S_6	**0.43**	29.2	4.31	0.77	0.89	**0.43**
S_7	**0.01**	∞	0.02	0.17	0.24	0.60
S_8	**0.15**	0.72	1.40	0.27	0.29	0.27
S_9	**0.03**	0.43	0.18	0.06	0.06	0.19
S_{10}	**0.05**	0.53	0.36	0.11	0.11	0.88
S_{11}	**0.01**	0.04	0.04	0.06	0.02	0.18
S_{12}	**0.00**	0.04	0.02	**0.00**	**0.00**	0.17
S_{13}	**0.00**	0.01	0.02	**0.00**	**0.00**	0.17
S_{14}	**0.07**	0.38	0.53	0.17	0.15	0.85
S_{15}	**0.03**	0.17	0.18	0.06	0.05	0.84
S_{16}	**0.05**	0.31	0.34	0.11	0.11	0.87
S_{17}	**0.03**	0.18	0.18	0.06	0.06	0.84
S_{18}	**0.03**	0.21	0.19	0.06	0.06	0.84

S_{19}	**0.03**	0.18	0.18	0.06	0.06	0.85
S_{20}	**0.03**	0.17	0.18	0.06	0.06	0.84
S_{21}	**10.2**	∞	12.7	11.0	12.2	9.87 (87)
S_{22}	**0.04**	0.22	0.31	0.05	0.07	0.85
S_{23}	**0.04**	0.23	0.34	0.11	0.09	0.88

Timings for 500 decimals (1).

S	Sumalt	EM	EM3	AP	delta	Sidi
Init	0.00	0.00	0.43	1.09	0.67	0.00
A_1	**0.00**	0.01	0.01	**0.00**	0.01	0.16
A_2	**0.00**	0.04	0.01	**0.00**	**0.00**	0.17
A_3	**0.00**	0.04	0.02	**0.00**	**0.00**	0.17
A_4	**0.03**	0.17	0.18	0.05	0.05	0.85
A_5	**0.03**	0.19	0.18	0.06	0.06	0.86
A_6	**0.03**	0.16	0.18	0.06	0.05	0.86
A_7	**0.01**	∞	0.02	0.17	0.24	0.61
A_8	**0.43**	29.2	4.30	0.77	0.89	**0.43**
A_9	**0.18**	0.80	1.57	0.31	0.33	0.29
D_1	**0.00**	**0.00**	**0.00**	**0.00**	**0.00**	0.04
D_2	0.00 (10)	**0.00**	**0.00**	**0.00**	**0.00**	0.04
D_3	**0.00**	0.02	0.01	**0.00**	**0.00**	0.17
D_4	**0.03**	0.13	0.17	0.05	0.05	0.88
D_5	0.02 (6)	0.14	0.18	**0.06**	**0.06**	0.87
D_6	0.03 (6)	0.43 (340)	2.41	0.29	**0.06**	0.88

Timings for 500 decimals (2).

Remarks and Conclusion of this study:

The conclusion of these timings is simpler than for series with positive terms:

- Always use the Sumalt program, or if, for some reason this fails, use Sumaltdelta.
- In the case where the summand takes a long time to compute (e.g., because it involves higher transcendental functions such as $\log(\Gamma(x))$, $\zeta(x)$, or $\zeta'(x)$), the basic Sumalt program can be improved by using other polynomials than the Chebyshev polynomials, see the original article [**CRVZ00**] for details. However the speed gain is at most only of the order of 20%.

CHAPTER 5

Euler products and Euler sums

The preceding chapters dealt with problems of a purely analytic nature. We now come to problems which are much more closely related to number theory. In the present section, we study the numerical computation of sums and products over the set P of prime numbers. Since this set is so complicated, this seems a priori hopeless, until one recalls the fundamental property of the Riemann zeta function for $\Re(s) > 1$:

$$\zeta(s) = \sum_{n \geq 1} \frac{1}{n^s} = \prod_{p \in P} \left(1 - \frac{1}{p^s}\right)^{-1} .$$

Thus if we can express our computation in terms of values of ζ we may be in good shape.

5.1. Euler sums

First, let us show how to compute for $\Re(s) > 1$

$$S(s) = \sum_{p \in P} \frac{1}{p^s} .$$

The idea is the following: thanks to the above product formula for $\zeta(s)$, we have

$$\log(\zeta(s)) = -\sum_{p \in P} \log(1 - 1/p^s) = \sum_{p \in P} \sum_{m \geq 1} 1/(mp^{ms})$$

$$= \sum_{m \geq 1} \sum_{p \in P} 1/p^{ms} = \sum_{m \geq 1} S(ms)/m .$$

Now recall the *second* Möbius inversion formula: if $f(n) = \sum_{m \geq 1} g(mn)$, then $g(n) = \sum_{m \geq 1} \mu(m) f(mn)$. The proof of this is immediate:

$$\sum_{m \geq 1} \mu(m) f(mn) = \sum_{m \geq 1} \mu(m) \sum_{\ell \geq 1} g(\ell mn) = \sum_{N \geq 1} g(Nn) \sum_{m \mid N} \mu(m) = g(n)$$

by the basic property of the Möbius function.

By what we have seen above we have $\log(\zeta(ns))/n = \sum_{m \geq 1} S(mns)/(mn)$, so applying Möbius inversion to $f(n) = \log(\zeta(ns))/n$ and $g(m) = S(ms)/m$ we deduce that $S(ns) = n \sum_{m \geq 1} \mu(m) \log(\zeta(mns))/(mn)$, so that

$$S(s) = \sum_{m \geq 1} \frac{\mu(m)}{m} \log(\zeta(ms)) .$$

Since $\zeta(M) = 1 + O(2^{-M})$ the convergence of this series is at least in $O(2^{-m})$, which is not bad. But we can do much better: it is clear that the proof of the

253

formula obtained above gives more generally

$$S_{>N}(s) = \sum_{m \geq 1} \frac{\mu(m)}{m} \log(\zeta_{>N}(ms)) \,,$$

where $S_{>N}$ means that we restrict the sum to primes $p > N$, and $\zeta_{>N}$ that we take the Euler product for zeta restricted to primes $p > N$ (which is of course *not* the same as restricting the sum defining ζ to $n > N$). We have thus proved the following:

PROPOSITION 5.1.1. *Set* $S(s) = \sum_{p \in P} p^{-s}$. *For any* $N > 0$ *and* s *such that* $\Re(s) > 1$ *we have*

$$S(s) = \sum_{p \in P,\ p \leq N} \frac{1}{p^s} + \sum_{m \geq 1} \frac{\mu(m)}{m} \log(\zeta_{>N}(ms)) \,,$$

where

$$\zeta_{>N}(s) = \zeta(s) \prod_{p \in P,\ p \leq N} (1 - 1/p^s) \,.$$

The whole point of introducing the parameter N is that $\log(\zeta_{>N}(ms)) = O(N^{-ms})$, so the series will converge much faster. Note, however, that N must not be chosen too large because in that case $\zeta_{>N}$ would be extremely close to 1 and there may be cancellation errors. In practice we advise choosing $10 < N < 100$ for instance.

Note in passing that the above proposition proves that $S(s)$ can be analytically continued to the half-plane $\Re(s) > 0$ with the rational points $1/k$ for $k \in \mathbb{Z}_{\geq 1}$ removed, where $S(s)$ has logarithmic singularities, and it is easy to show that it has a natural boundary $\Re(s) = 0$. To give a random example, we compute that

$$S(1/\sqrt{2}) = 0.3354279618920726094118435482977539924\ldots + i\pi \,.$$

For the following, we recall that the valuation of a power series is its order at $x = 0$:

COROLLARY 5.1.2. *Let* $A(x) = \sum_{m \geq 1} a(m)x^m$ *be a power series with no constant term, nonzero radius of convergence* r, *and valuation* v, *and set* $S(A; s) = \sum_{p \in P} A(1/p^s)$. *Define*

$$c(n) = \sum_{d|n} \frac{\mu(d)}{d} a\left(\frac{n}{d}\right) \,.$$

Then for all $N \geq 1$ *and* $\Re(s) > \max(-\log(r)/\log(N), 1/v)$ *we have*

$$S(A; s) = \sum_{p \in P,\ p \leq N} A(1/p^s) + \sum_{n \geq 1} c(n) \log(\zeta_{>N}(ns)) \,.$$

In particular, if $a(1) = 0$ *and* $r > 1/2$ *we have*

$$S(A; 1) = \sum_{p \in P,\ p \leq N} A(1/p) + \sum_{n \geq 2} c(n) \log(\zeta_{>N}(n)) \,.$$

PROOF. Simply write $S_{>N}(A; s) = \sum_{m \geq 1} a(m) S_{>N}(ms)$ and use the proposition. The details are left to the reader. $\qquad\square$

Since it is not immediate to estimate the speed of convergence in the general case, we restrict to the case of a rational function F. If we write $F = N/D$ with N and D coprime polynomials, the expansion at infinity of F is the expansion at 0 of

$F(1/x) = N(1/x)/D(1/x)$, so the radius of convergence r is equal to the modulus of the zero of $D(1/x)$ closest to the origin, hence $1/r$ is equal to the modulus of the zero of $D(x)$ furthest from the origin, and this is approximately given by the command `polrootsbound` which gives an upper bound for this largest modulus, which we will use since it is considerably faster than computing the roots using `polroots`.

In addition, note that when computing $\log(\zeta_N(s))$ by the evident formula given above, there is considerable cancellation, so it is necessary to increase the working accuracy. All this is done in the following program:

────────────────────────────── | SumEulerrat.gp | ──────────────────────────────

```
1   LogzetaN(s, N) =
2   { my(B = getlocalbitprec());
3
4     B += ceil(abs(real(s) - 1) * exponent(2 * N));
5     localbitprec(B);
6     if (bitprecision(s) < B, s = bitprecision(s, B));
7     log(zeta(s) * prodeuler(p = 2, N, 1 - p^(-s)));
8   }
9
10  Sdmob(ser, n) =
11    sumdiv(n, d, moebius(d) * polcoef(ser, n / d) / d);
12
13  /* Compute sum_{p >= a, prime} F(p^s), F rational function. */
14  SumEulerrat(F, s = 1, a = 2) =
15  { my(B = getlocalbitprec(), vx = variable(F), rs, N, r, lim);
16    my(FI, vF, sal, S);
17
18    FI = subst(F, vx, 1 / vx); vF = valuation(FI, vx);
19    rs = real(s);
20    if (rs <= 1 / vF, error("real(s)  <=  1/v"));
21    localbitprec(32);
22    r = 1 / max(polrootsbound(denominator(F)), 1);
23    N = ceil(max(30, a + 1));
24    if (rs <= -log(r) / log(N), error("real(s) too small"));
25    lim = ceil(B * log(2) / log(N^rs * r)) + 1;
26    localbitprec(B);
27    sal = bitprecision(1., B + 32) * FI + O(vx^(lim+1));
28    bernvec(floor((lim * s + 1) / 2)); /* cache the B_{2n} */
29    S = sum(n = vF, lim,
30            my(m = Sdmob(sal, n)); if (m, m * LogzetaN(n*s, N)));
31    forprime (p = a, N, S += subst(F, vx, p^s));
32    return (bitprecision(S, B));
33  }
```

EXAMPLES. (All essentially instantaneous):

```
? SumEulerrat(1 / x^2)
% = 0.45224742004106549850654336483224793417
```

```
? SumEulerrat(1 / (x^2 + 1))
% = 0.38905955531696837171041438969249635814
? /* SumEulerrat(1 / (x^(3/2) + 1)) would be illegal, so: */
? SumEulerrat(1 / (x^3 + 1), 1/2)
% = 0.71426973554924155313504391134630993732
```

5.2. Euler products

A completely similar method can be used to compute Euler *products* numerically. The result is as follows:

COROLLARY 5.2.1. *Let* $B(x) = 1 + \sum_{m \geq 1} b(m)x^m$ *be a power series with constant term* 1 *with nonzero radius of convergence* r, *and assume that* $B(z) \neq 0$ *for* $|z| < r$. *Let* v *be the valuation of* $B(x) - 1$ *and set* $P(B; s) = \prod_{p \in P} B(1/p^s)$. *Write* $\log(B(x)) = \sum_{m \geq 1} a(m)x^m$ *and define as above*

$$c(n) = \sum_{d|n} \frac{\mu(d)}{d} a\left(\frac{n}{d}\right) .$$

Then for all $N \geq 1$ *and* $\Re(s) > \max(-\log(r)/\log(N), 1/v)$ *we have*

$$P(B; s) = \prod_{p \in P,\ p \leq N} B(1/p^s) \prod_{n \geq 1} \zeta_{>N}(ns)^{c(n)} .$$

In particular, if $b(1) = 0$ *and* $r > 1/2$ *we have*

$$P(B; 1) = \prod_{p \in P,\ p \leq N} B(1/p) \prod_{n \geq 2} \zeta_{>N}(n)^{c(n)} .$$

Furthermore $c(n)$ *satisfies the recurrence*

$$c(n) = b(n) - \frac{1}{n} \sum_{1 \leq k \leq n-1} kc(k) \sum_{1 \leq q \leq n/k} b(n - qk) .$$

PROOF. The first part of the corollary is immediate by applying the preceding corollary to $A(x) = \log(B(x))$, which has the same valuation as $B(x) - 1$, and radius of convergence at least r since we assume that B has no zeros in $|z| < r$. For the recurrence, we note that

$$B'(x) = \sum_{m \geq 1} mb(m)x^{m-1} = B(x)\log(B(x))' = B(x) \sum_{m \geq 1} ma(m)x^{m-1} ,$$

and the result follows by identification of coefficients and the formula $\sum_{d|n} dc(d) = na(n)$ which defines $c(n)$. $\quad\square$

REMARKS 5.2.2.

(1) The coefficients $c(n)$ are the unique exponents such that we have the formal expansion

$$1 + \sum_{m \geq 1} b(n)x^m = \prod_{n \geq 1} (1 - x^n)^{-c(n)} .$$

(2) It is usually preferable to use the formula giving $c(n)$ in terms of $a(n)$. However it may happen that $a(n)$ is not easy to compute directly, and one can then use the recurrence for $c(n)$.

A possible program is as follows, essentially copied from the preceding one:

—————————————— | ProdEulerrat.gp | ——————————————

```
1   /* Compute prod_{p >= a, prime} F(p^s), F rational function. */
2   ProdEulerrat(F, s = 1, a = 2) =
3   { my(B = getlocalbitprec(), vx = variable(F));
4     my(FIm1, rs, N, r, lim, vF, sal, S);
5
6     FIm1 = subst(F, vx, 1/vx) - 1;
7     vF = valuation(FIm1, vx);
8     rs = real(s);
9     if (rs <= 1/vF, error("real(s) <= 1/v"));
10    localbitprec(32);
11    r = 1 / max(polrootsbound(numerator(F)),
12                polrootsbound(denominator(F)));
13    N = ceil(max(30, a + 1));
14    if (rs <= -log(r) / log(N), error("real(s) too small"));
15    lim = ceil(B * log(2) / log(N^rs * r)) + 1;
16    localbitprec(B);
17    sal = log(1 + bitprecision(1., B + 32)*FIm1 + O(vx^(lim+1)));
18    bernvec(floor((lim * s + 1) / 2)); /* cache the B_{2n} */
19    S = sum(n = vF, lim,
20            my(m = Sdmob(sal, n)); if (m, m * LogzetaN(n*s, N)));
21    S = exp(S);
22    forprime (p = a, N, S *= subst(F, vx, p^s));
23    return (bitprecision(S, B));
24  }
```

EXAMPLES. (Again essentially instantaneous):

```
? ProdEulerrat(1 + 1 / x^3) - zeta(3) / zeta(6)
% = 0.E-38
? ProdEulerrat(1 - 1 / (x - 1)^2, 1, 3) /* Twin prime constant */
% = 0.66016181584686957392781211001455577843
```

Note that both of these programs are already implemented in GP under the respective names sumeulerrat and prodeulerrat.

5.3. Variants involving $\log(p)$ or $\log(\log(p))$

A large number of similar sums or products over primes can be computed in an analogous manner. A simple example is $\sum_{p \in P} \log(p)/p^s$ for $\Re(s)$, for which we can modify Proposition 5.1.1 essentially by replacing $\log(\zeta(s)) = -\sum_{p \in P} \log(1 - p^{-s})$ by its derivative $\zeta'(s)/\zeta(s) = -\sum_{p \in P} \log(p)p^{-s}/(1 - p^{-s})$. In this way we immediately compute for instance that

$$\sum_{p \in P} \frac{\log(p)}{p^2} = 0.4930911093687644621978262\cdots$$

We can also compute *limits* such as

$$\lim_{s \to 1^+} \left(\sum_{p \in P} \frac{1}{p^s} - \log(\zeta(s)) \right) = -0.31571845205389007685 \cdots ,$$

simply by suppressing the term $m = 1$ in the formula expressing $S(s)$ in terms of $\log(\zeta(ms))$, and

$$\lim_{x \to \infty} \left(\sum_{\substack{p \in P \\ p \le x}} \frac{1}{p} - \log(\log(x)) \right) = 0.26149721284764278375 \cdots ,$$

using the easily proved formula

$$\lim_{x \to \infty} \left(\sum_{\substack{p \in P \\ p \le x}} \frac{1}{p} - \log(\log(x)) \right) = \gamma + \lim_{s \to 1^+} \left(\sum_{p \in P} \frac{1}{p^s} - \log(\zeta(s)) \right) ,$$

where as usual γ is Euler's constant.

A little more challenging is the computation of sums over primes p when $\log(p)$ is in the *denominator* of the summand, for instance sums like $T(s) = \sum_{p \in P} 1/(p^s \log(p))$ for $\Re(s) \ge 1$. When $\log(p)$ is on the numerator, we have to use the derivative of $\log(\zeta(s))$. Analogously, here we have to use the *integral* of $\log(\zeta(s))$, in other words

$$\int_s^\infty \log(\zeta(t)) \, dt .$$

Even though there are faster methods for computing this, for simplicity we use the doubly-exponential method, and to compute say, the 30 values of this integral for integer $s \in [1, 30]$ to 38 decimals requires only 0.81 seconds. The following simple-minded program computes $T(s)$, using the formula given in the proposition below:

--------------------------------- SumEulerlog.gp ---------------------------------

```
1   /* Compute sum_{p prime} 1/(p^s log(p)). */
2   SumEulerlog(s, N = max(2, 30 / abs(s))) =
3   { my(B = getlocalbitprec(), S = 0, LN, T, lim);
4
5     localbitprec(32); LN = log(N);
6     lim = ceil(B*log(2) / LN);
7     localbitprec(B + 32);
8     forprime (p = 2, N, S += 1 / (p^s * log(p)));
9     LN = bitprecision(LN, B + 32);
10    T = intnuminit(0, [oo, 1]);
11    forsquarefree (K = 1, lim,
12      my([k] = K, m = moebius(K) / k, a = 1 / (k * LN));
13      /* Tk = intnuminit(0, [oo, 1/a]) */
14      my(Tk = vector(#T, i, if (i == 1, T[1], T[i] * a)));
15      S += m * intnum(t = 0, oo, LogzetaN(k * (t + s), N), Tk));
16    return (S);
17  }
```

In a few seconds we obtain for instance

$$\sum_{p\in P} \frac{1}{p\log(p)} = 1.63661632335126086856956\ 58\cdots$$

$$\sum_{p\in P} \frac{1}{p^2\log(p)} = 0.507782187859199318774375\cdots$$

Note that the series $\sum 1/(p\log(p))$ barely converges (and as usual is over the irregular set of primes), so it is remarkable that there is no problem in computing its sum.

We leave as an exercise for the reader the task of writing a program to compute $\sum_{p\in P} 1/(p^a\log^b(p))$ with $a > 1$ or $a = 1$ and $b \geq 1$, arbitrary a, and arbitrary integer b.

An even harder problem is when $\log(\log(p))$ (or worse) occurs. This can also be dealt with, but is more technical. The idea is as follows: recall that we have a linear formula expressing $S(s) = \sum_{p\in P} 1/p^s$ in terms of $f(s) = \log(\zeta(s))$. By derivation, it follows that we also have a formula for $S^{(k)}(s) = (-1)^k \sum_{p\in P} (\log(p))^k/p^s$ in terms of $f^{(k)}(s)$. For instance this is how we computed $\sum_{p\in P} \log(p)/p^2$. We now would like to differentiate *with respect to k*, since the derivative with respect to k of $\log(p)^k$ is $\log(\log(p))\log(p)^k$, which is exactly what is needed after setting $k = 0$. A priori this does not seem realistic, but in fact is quite easy:

PROPOSITION 5.3.1.

(1) *For $\Re(s) > 0$ and $\Re(x) \geq 1$ we have*

$$\sum_{p\in P} \frac{1}{p^x\log(p)^s} = \frac{1}{\Gamma(s)} \sum_{k\geq 1} \frac{\mu(k)}{k} \int_0^\infty t^{s-1}\log(\zeta(k(t+x)))\,dt\ .$$

(2) *For $\Re(s) > -1$ and $\Re(x) \geq 1$ we have*

$$\sum_{p\in P} \frac{1}{p^x\log(p)^s} = -\frac{1}{\Gamma(s+1)} \sum_{k\geq 1} \mu(k) \int_0^\infty t^s \frac{\zeta'}{\zeta}(k(t+x))\,dt\ .$$

PROOF. (Sketch.) For (1) it is easy to check the absolute convergence of the series and of the integral. We then proceeds as above for the evaluation of $S(s)$: we express $\log(\zeta)$ as a sum over primes thanks to the Euler product and use Möbius inversion. For (2), change s into $s + 1$ and take the derivative with respect to x. The details of these proofs are left to the reader. □

Using this proposition, one can then compute the Taylor series expansion of the left hand side around $s = 0$ for instance, and obtain hundreds of decimals of expressions such as

$$\sum_{p\in P} \frac{\log(\log(p))}{p\log(p)}$$

and

$$\lim_{x\to\infty}\left(\sum_{\substack{p\in P \\ p\leq x}} \frac{\log(\log(p))^n}{p} - \frac{\log(\log(x))^{n+1}}{n+1}\right)\ .$$

All these computations are left as exercises for the reader. For instance, we can write the following program, which is a simple modification of the previous one:

─────────────────────────   SumEulerloglog.gp   ─────────────────────────

```
1   /* Compute sum_{p prime} log(log(p))/(p^s log(p)). */
2   SumEulerloglog(s, N = max(2, 30 / abs(s))) =
3   { my(B = getlocalbitprec(), S = 0, LN, T, lim, E);
4
5     localbitprec(32); LN = log(N);
6     lim = ceil(B*log(2) / LN);
7     localbitprec(B + 32);
8     forprime (p = 2, N, S += log(log(p)) / (p^s * log(p)));
9     LN = bitprecision(LN, B + 32);
10    T = intnuminit(0,[oo, 1]);
11    E = Euler();
12    forsquarefree (K = 1, lim,
13      my([k] = K, m = moebius(K) / k, a = 1 / (k * LN));
14      /* Tk = intnuminit(0, [oo, 1/a]) */
15      my(Tk = vector(#T, i, if (i == 1, T[1], T[i] * a)));
16      S -= m * intnum(t = 0, oo, (log(t) + E)
17                                 * LogzetaN(k * (t+s) , N), Tk));
18    return (S);
19  }
```

In a few seconds we compute that

$$\sum_{p\in P} \frac{\log(\log(p))}{p\log(p)} = 0.6410802156599846604833518891513999518913451\cdots$$

Once again we leave as an exercise for the reader the task of writing a program to compute $\sum_{p\in P}\log^k(\log(p))/(p^a\log^b(p))$ with $a > 1$ or $a = 1$ and $b \geq 1$, arbitrary a, arbitrary integer b, and arbitrary nonnegative integer k.

5.4. Variants involving quadratic characters

We now would like to compute prime sums and Euler products which, in addition to involving regular functions over primes, also involve a quadratic character $\left(\frac{D}{n}\right)$ where D is a not necessarily fundamental discriminant.

A natural idea is to introduce, in addition to the Riemann ζ function, the Dirichlet L-functions $L(\chi, s)$ for Dirichlet characters χ, in particular for $\chi(n) = \left(\frac{D}{n}\right)$.

We can easily generalize the results of the previous sections to this case. If for any character χ we set

$$S(\chi, s) = \sum_p \frac{\chi(p)}{p^s}$$

then, exactly as before and with evident notation Möbius inversion gives

$$S(\chi, s) = \sum_{p\in P,\ p\leq N} \frac{\chi(p)}{p^s} + \sum_{m\geq 1} \frac{\mu(m)}{m}\log(L_{>N}(\chi^m, ms))\,.$$

We will briefly study below the computation of $L_{>N}(\chi^m, ms)$.

Let us specialize to the case where χ is a quadratic character $\left(\frac{D}{n}\right)$ for some (not necessarily fundamental) discriminant D. Then $\chi^k = \chi$ if k is odd and $\chi^k = \chi_0$, the trivial character modulo D ($\chi_0(n) = 1$ if $(n, D) = 1$ and $\chi_0(n) = 0$ otherwise) if $k \geq 2$ is even. Thus,

$$S(\chi, s) = \sum_{\chi(p)=1} \frac{1}{p^s} - \sum_{\chi(p)=-1} \frac{1}{p^s} \,,$$

hence

$$\sum_{\chi(p)=1} \frac{1}{p^s} = \frac{1}{2}(S(\chi_0, s) + S(\chi, s)) \quad \text{and}$$

$$\sum_{\chi(p)=-1} \frac{1}{p^s} = \frac{1}{2}(S(\chi_0, s) - S(\chi, s)) \,,$$

where

$$S(\chi_0, s) = S(s) - \sum_{p|D} \frac{1}{p^s}$$

with the notation of the previous section.

For $j = -1$, 0, 1 corresponding to the three possible values of χ, let $A_j(x) = \sum_{m \geq 1} a_j(m)x^m$ be power series satisfying the same assumptions as in Corollary 5.1.2. We set

$$S(A; s) = \sum_{j \in \{-1,0,1\}} \sum_{p \in P, \, \chi(p)=j} A_j(1/p^s) \,.$$

A generalization of Corollary 5.1.2, which is an immediate consequence of the above formulas is as follows:

PROPOSITION 5.4.1. *Define*

$$c_j(n) = \sum_{d|n, \, 2\nmid d} \frac{\mu(d)}{d} a_j\left(\frac{n}{d}\right) \,.$$

Then for all $N \geq 1$ and $\Re(s) > \max(-\log(r)/\log(N), 1/v)$ we have

$$S(A; s) = \sum_{j \in \{-1,0,1\}} \sum_{p \in P, \, p \leq N \, \chi(p)=j} A_j(1/p^s)$$

$$+ \frac{1}{2} \sum_{n \geq 1}(c_1(n) - c_{-1}(n)) \log(L_{>N}(\chi, ns))$$

$$+ \frac{1}{2} \sum_{n \geq 1}(c_1(n) + c_{-1}(n) - c_1(n/2)) \log(\zeta_{p\nmid D, \, >N}(ns))$$

$$+ \sum_{p|D, \, p>N} A_0(1/p^s) \,,$$

with the usual convention that $c_1(x) = 0$ if $x \notin \mathbb{Z}$ and where we set

$$\zeta_{p\nmid D, \, >N}(s) = L_{>N}(\chi_0, s) = \zeta_{>N}(s) \prod_{p|D, \, p>N} \left(1 - \frac{1}{p^s}\right) \,.$$

It is clear that such formulas can be generalized to Dirichlet characters χ of higher order r, and the result involves in a simple manner the functions $L(\chi^b, N)$ for $0 \leq b < r$. The quadratic case is especially simple since it involves $L(\chi, N)$ and $L(\chi^0, N) = L(\chi_0, N)$ which is essentially the Riemann ζ function.

We must also explain how to compute the quantities $L_{>N}(\chi, ns)$, or equivalently $L(\chi, ns)$, at least when $\chi = \left(\frac{D}{\cdot}\right)$. When $|D|$ is very small (for instance $D = -3$ or $D = -4$), we can use the Sumchiquad algorithm explained to us by B. Allombert and described in Section 4.5.3.

When $\Re(ns)$ is large and we do not need too much accuracy, we can also directly use the definition (either as a sum or as an Euler product). Otherwise, we can trivially reduce to the case of a fundamental discriminant D, and we can use one of several methods for computing $L(\chi, s)$ such as the one described in Chapter 9, which in GP is simply lfun(D,n*s) (note that the lfun command can be used with general Dirichlet characters, not only quadratic, and vastly more general types of L-functions; however, it does require the existence of a functional equation of standard type, which explains the necessity in our situation of reducing to fundamental discriminants).

5.5. Variants involving congruences

Let a and k be coprime positive integers. Instead of considering Euler sums or products over all primes, we can consider the same but restricted to primes congruent to a modulo k. This is easily done as follows. For all a coprime to k and s with $\Re(s) > 1$ we define

$$S_a(s) = \sum_{p \in P, \; p \equiv a \pmod{k}} \frac{1}{p^s} \quad \text{and} \quad T(\chi, s) = \sum_{a \bmod k} \chi(a) S_a(s) \, .$$

For any character χ modulo k we have

$$\log(L(\chi, s)) = \sum_{p \in P} -\log\left(1 - \frac{\chi(p)}{p^s}\right) = \sum_{m \geq 1} \frac{1}{m} \sum_{p \in P} \frac{\chi^m(p)}{p^{ms}}$$

$$= \sum_{m \geq 1} \frac{1}{m} \sum_{a \bmod k} \chi^m(a) S_a(ms) = \sum_{m \geq 1} \frac{T(\chi^m, ms)}{m} \, .$$

It follows that for any integer $n \geq 1$

$$\frac{\log(L(\chi^n, ns))}{n} = \sum_{m \geq 1} \frac{T(\chi^{mn}, mns)}{mn} \, ,$$

so by the second Möbius inversion formula we have in particular

$$T(\chi, s) = \sum_{a \bmod k} \chi(a) S_a(s) = \sum_{m \geq 1} \frac{\mu(m)}{m} \log(L(\chi^m, ms)) \, .$$

Now by orthogonality of characters, if we denote by X_k the group of characters of $(\mathbb{Z}/k\mathbb{Z})^*$ we have for b coprime to k

$$\sum_{\chi \in X_k} \overline{\chi}(b) T(\chi, s) = \sum_{a \bmod k} S_a(s) \sum_{\chi \in X_k} \chi(ab^{-1}) = \phi(k) S_b(s) \, ,$$

hence

$$S_a(s) = \frac{1}{\phi(k)} \sum_{\chi \in X_k} \overline{\chi}(a) \sum_{m \geq 1} \frac{\mu(m)}{m} \log(L(\chi^m, ms)) \, .$$

As above, it is then immediate to compute any convergent Euler sum of the form $\sum_{p \equiv a \pmod{k}} f(p)$ or any convergent Euler product of the form $\prod_{p \equiv a \pmod{k}} f(p)$.

As mentioned above, to compute $L(\chi^n, ns)$ (hence $L_{>N}(\chi^n, ns)$) the simplest is to use the preprogrammed GP lfun program.

As an application, if we set $P_a = \prod_{p \equiv a \pmod 5} 1/(1 - 1/p^2)$, we immediately compute that

$$P_1 = 1.0109151606010195226049565842895149209845386275817385237\cdots$$
$$P_2 = 1.3685720538766490858607638904831099901702078288858952050\cdots$$
$$P_3 = 1.1357648786689216268686643009472082289511936413005468744\cdots$$
$$P_4 = 1.0049603239222975589937496248102521847955102941880228801\cdots.$$

Of course $P_1 P_2 P_3 P_4 = (1 - 1/5^2)\zeta(2) = 4\pi^2/25$.

The product $P_2 P_3$ was computed in [**ERS19**] using a different and more complicated method which apparently does not allow the computation of P_2 and P_3 individually, however see [**Ram19**].

5.6. Hardy–Littlewood constants: Quadratic polynomials

Let $A(X) \in \mathbb{Z}[X]$ be a polynomial with integer coefficients and assume that $A(X)$ is irreducible in $\mathbb{Q}[X]$ and has content 1. For any prime p, we let $\omega(p) = \omega(A, p)$ be the number of solutions in $\mathbb{F}_p$ of $A(x) = 0$. Then conjecturally, the number of integers n with $1 \leq n \leq N$ such that $|A(n)|$ is prime should be asymptotic to

$$\frac{H(A)}{\deg(A)} \cdot \frac{N}{\log N}, \quad \text{where} \quad H(A) = \prod_p \frac{p - \omega(p)}{p - 1}$$

is the so-called *Hardy–Littlewood constant* of the polynomial A. It is therefore interesting to compute this Euler product. In particular, polynomials with a large $H(A)$ should have asymptotically more prime values.

The case of linear polynomials is trivial: $H(aX + b) = a/\phi(a)$ since we assume that a and b are coprime. In the present section we first treat the next simplest case of quadratic polynomials, and consider polynomials of larger degree later. Thus, let $A(X) = aX^2 + bX + c$ be an irreducible quadratic polynomial with $\gcd(a, b, c) = 1$, and let $D = b^2 - 4ac$ be its discriminant. Writing $A(X) = 2aX(X - 1)/2 + (a + b)X + c$, we see that to have infinitely many primes we must assume the stronger condition $\gcd(2a, a + b, c) = 1$.

It is easy to see that $\omega(p)$ is given by the following formulas.
- If $p \nmid a$, then $\omega(p) = 1 + \left(\frac{D}{p}\right)$.
- If $p \mid a$ and $p \nmid b$, then $\omega(p) = 1$.
- If $p \mid a$ and $p \mid b$ (hence $p \nmid c$), then $\omega(p) = 0$.

A little computation left to the reader shows the following:

PROPOSITION 5.6.1. *Let $A(X) = aX^2 + bX + c$ be an irreducible polynomial of degree 2 with $\gcd(2a, a + b, c) = 1$. The Hardy–Littlewood constant of A is given by the formula*

$$H(A) = c_2 \prod_{p>2} \left(1 - \frac{\left(\frac{D}{p}\right)}{p - 1}\right) \prod_{p|a,\ p\nmid 2b} \frac{p - 1}{p - 2} \prod_{p|\gcd(a,b),\ p>2} \frac{p}{p - 1},$$

where $c_2 = 1$ if $a + b$ is odd, or $c_2 = 2$ if $a + b$ is even (hence c odd).

Write $D = D_0 f^2$ with D_0 a fundamental discriminant. Since

$$\prod_{p>2}\left(1 - \frac{\left(\frac{D}{p}\right)}{p-1}\right) = \prod_{p>2}\left(1 - \frac{\left(\frac{D_0}{p}\right)}{p-1}\right) \prod_{p|D,\ p>2}\left(1 - \frac{\left(\frac{D_0}{p}\right)}{p-1}\right)^{-1},$$

we must therefore compute the Euler product

$$C(D_0) = \prod_{p>2}\left(1 - \frac{\left(\frac{D_0}{p}\right)}{p-1}\right)$$

for a fundamental discriminant D_0. With a slight abuse of notation, set $C(D_0, s) = \prod_{p>2}(1 - \left(\frac{D_0}{p}\right))/(p^s - 1))$, so that $C(D_0) = C(D_0, 1)$, and set $\chi(n) = \left(\frac{D_0}{n}\right)$. Then for $\Re(s) > 1$

$$-\log(C(D_0, s)) = -\sum_{p,\ \chi(p)=1} \log\frac{p^s - 2}{p^s - 1} - \sum_{p,\ \chi(p)=-1} \log\frac{p^s}{p^s - 1},$$

hence with the notation of Section 5.4, $C(D_0, s) = e^{-S(A;s)}$ with $A_0(1/p^s) = 0$,

$$A_1(1/p^s) = -\log\left(\frac{p^s - 2}{p^s - 1}\right) = \sum_{m\geq 1}\frac{2^m - 1}{mp^{ms}},$$

$$A_{-1}(1/p^s) = -\log\left(\frac{p^s}{p^s - 1}\right) = -\sum_{m\geq 1}\frac{1}{mp^{ms}},$$

so that $a_1(m) = (2^m - 1)/m$ and $a_{-1}(m) = -1/m$. Thus,

$$c_1(n) = \frac{1}{n}\sum_{d|n,\ 2\nmid d}\mu(d)(2^{n/d} - 1)$$

and

$$c_{-1}(n) = -\frac{1}{n}\sum_{d|n,\ 2\nmid n}\mu(d).$$

Set $a(n) = (c_1(n) - c_{-1}(n))/2$ and $b(n) = (c_1(n) + c_{-1}(n) - c_1(n/2))/2$, so that

$$S(A; s) = \sum_{n\geq 1}a(n)\log(L(\chi, ns)) + \sum_{n\geq 1}b(n)\log(\zeta_{p\nmid D_0}(ns)).$$

Then by the above we have

$$a(n) = \frac{1}{2n}\sum_{d|n,\ 2\nmid d}\mu(d)2^{n/d}$$

and one easily checks that $b(n) = a(n) - a(n/2)$ if $n > 1$, while $b(1) = 0$ (recall that we set $a(x) = 0$ if x is not integral), so we can make $s \to 1$ by limiting the second sum to $n \geq 2$. We can thus compute $C(D_0)$, using as above $L_{p>N}(\chi, n)$ and $\zeta_{p\nmid D_0,\ p>N}(n)$. The following program implements this:

```
──────────────────────  HardyLittlewood2.gp  ──────────────────────
1   /* Auxiliary functions. */
2   ZetaDN(P, s) = zeta(s) * prod(j = 1, #P, 1 - P[j]^(-s));
3
4   LchiN(L, Ebad, s) =
```

```
5    { my([P, E] = Ebad);
6      lfun(L, s) * prod(j = 1, #P, subst(E[j], 'x, P[j]^(-s)));
7    }
8
9    LchiNinit(D, P) =
10   { my(Ebad = [], Pbad = []);
11     foreach (P, p,
12       my(s = kronecker(D, p));
13       if (s, Ebad = concat(Ebad, 1 - s*'x);
14             Pbad = concat(Pbad, p)));
15     return ([Pbad, Ebad]);
16   }
17
18   Oddpart(n) = n >> valuation(n, 2);
19
20   /* The real work; D is fundamental. */
21   HLW2(D, N) =
22   { my(B = getlocalbitprec(), lim, S1, S2, L, P, v, Ebad);
23
24     localbitprec(32); lim = ceil(B*log(2)/log(N/2));
25     localbitprec(B + lim + exponent(lim));
26     L = lfuninit(D, [1/2, lim, 0]);
27     v = vector(lim);
28     forfactored (X = 1, lim,
29       my([n, fan] = X, S = 0, P = fan[,1]);
30       if (n % 2 == 0, P = P[^1]);
31       X = matconcat([P, vectorv(#P,i,1)]);
32       fordivfactored (X, Y,
33         my([d] = Y); \\ odd squarefree divisor of X
34         S += moebius(Y) << (n/d));
35       v[n] = S / (2*n));
36     P = setunion(factor(abs(D))[,1]~, primes([2, N]));
37     Ebad = LchiNinit(D, P);
38     S1 = sum(n = 1, lim, v[n] * log(LchiN(L, Ebad, n)));
39     S2 = sum(n = 2, lim, (v[n] - if (n%2 == 0, v[n/2]))
40                            * log(ZetaDN(P, n)));
41     return (S1 + S2);
42   }
43
44   /* Compute the Hardy-Littlewood constant of aX^2+bX+c. */
45   HardyLittlewood2(A, N = 50) =
46   { my(D = poldisc(A), S, P);
47
48     if (poldegree(A) != 2, error("polynomial of degree != 2"));
49     my([a, b, c] = Vec(A));
50     if (issquare(D) || gcd([2 * a, a + b, c]) > 1, return (0));
51     N = max(N, 3);
52     /* Take care of the prime p = 2. */
```

```
53    S = if ((a + b) % 2, 1., 2.);
54    /* Take care of odd primes dividing a. */
55    P = factor(Oddpart(a))[,1];
56    foreach (P, p,
57      S *= if (b % p, (p - 1) / (p - 2), p / (p - 1)));
58    /* Take care of odd primes dividing the index f. */
59    my([D0, f] = coredisc(D, 1));
60    P = factor(Oddpart(f))[,1];
61    S /= vecprod([1 - kronecker(D0, p) / (p - 1) | p <- P]);
62    /* Take care of the primes p <= N. */
63    S *= prodeuler(p = 3, N, 1 - kronecker(D0, p) / (p - 1));
64    /* Do the real work */
65    return (S * exp(-HLW2(D0, N)));
66    }
```

For example, using this program we compute that

$$H(X^2 + X + 41) = 6.6395463549428433306471137152997759329371091\cdots$$

$$H(X^2 + X + 75) = 0.6219533598519743400087125748592568290582438\cdots$$

$$H(2X^2 - 199) = 7.3291180993696071658232761749275362031861488\cdots.$$

Thus, we can reasonably expect that the first and third polynomials will produce many primes (and in some sense are record-breaking), while the second will produce much fewer primes (but it is easy to find polynomials of the same shape which produce even less).

REMARKS 5.6.2.

(1) The use of the preliminary lfuninit command in the HLW2 program avoids recomputing some data for each value of n, and speeds up the program considerably.

(2) Note the use of the forfactored command, which uses a sieve to factor consecutive integers, and fordivfactored which avoids factoring integers whose prime decomposition is known by construction (refer to the Pari/GP manual for the use of these functions).

(3) The bit accuracy B needs to be increased by two increments: first a quantity lim which measures the difference between relative and absolute accuracy (GP works with relative, but we need absolute), and second a quantity exponent(lim) which measures approximately the number of bits of accuracy lost when summing lim terms. In other scripts this additional quantity was often arbitrarily set to 5, 10, or more often 32, but here it is essential to optimize since the speed of the lfuninit command is extremely sensitive to the bit accuracy.

As an amusing exercise, we can search for quadratic polynomials whose Hardy–Littlewood constant is large, and in view of the above examples, larger than 6, say. We may of course assume that $a > 0$, and since changing X into $X + 1$ changes b into $b + 2a$ and changing X into $-X$ changes b into $-b$ we may assume that $0 \le b \le a$. This of course only excludes the most basic polynomial transformations. In addition, since we are looking for *large* $H(A)$, we may assume that c is odd (otherwise all even x are excluded), hence that $a + b$ is even (otherwise all odd x are excluded), and this exactly corresponds to the case $c_2 = 2$ in the formula for $H(A)$. We can thus write the following:

———————————— | HardyLittlewoodsearch.gp | ————————————

```
 1  HardyLittlewoodsearch(lima, limc) =
 2  { my(V = List());
 3
 4    localbitprec(64);
 5    if (limc % 2 == 0, limc++);
 6    for (a = 1, lima,
 7      forstep (b = a%2, a, 2,
 8        my(g = gcd(a, b));
 9        forstep (c = -limc, limc, 2,
10          if (gcd(c, g) > 1, next);
11          my(r = HardyLittlewood2(Pol([a, b, c]), 50));
12          if (r > 6.5, listput(V, [[a, b, c], r]);
13                       print(V[#V])))));
14    \\ sort by decreasing 'r', remove duplicates
15    return (vecsort(Vec(V), 2, 4 + 8));
16  }
```

Note that the flag $4 + 8$ in the `vecsort` command means first that we want to sort by descending instead of ascending order, and second that we want to remove duplicate entries.

Running this program with `lima=12` and `limc=1000`, we see that in that range the largest $H(A)$ is indeed obtained for $A(X) = 2X^2 - 199$ (and a few equivalent polynomials) with $H(A) = 7.3291 \cdots$ as given above, followed by $A(X) = 6X^2 + 6X + 31$ (and equivalent) with $H(A) = 6.9208 \cdots$.

On the other hand, by making a less systematic search using a truncated form of the formula for $H(A)$, it is easy to find quadratic polynomials with $H(A) > 8$, for instance $H(37X^2 + 23X - 8863) = 8.097818 \cdots$. In any case, it is easy to show that $H(A)$ is unbounded.

5.7. Hardy–Littlewood constants: General polynomials

5.7.1. Cubic polynomials. Let $A(X) = aX^3 + bX^2 + cX + d$ be a cubic polynomial. In order for A to represent infinitely many primes it is evidently necessary that A be irreducible and that $\gcd(a, b, c, d) = 1$. But writing

$$A(X) = 6a(X(X-1)(X-2)/6) + (6a + 2b)(X(X-1)/2) + (a + b + c)X + d$$

we see that a stronger condition is $\gcd(6a, 2b, a + b + c, d) = 1$ (similar to the condition $\gcd(2a, a+b, c) = 1$ used in the previous script). Let D be the discriminant of the polynomial A. The primes p dividing a or D are finite in number, so for such primes $\omega(p)$ can be computed directly (we will see below the GP commands). Thus, let p be a prime not dividing a or D, and let K be the cubic number field defined by the polynomial A. Since $p \nmid aD$ the decomposition of the polynomial A modulo p reflects the decomposition of the prime p in the extension $K/\mathbb{Q}$, hence $\omega(p)$ is equal to the number of prime ideals of K above p which are of residual degree 1.

In other words, with evident notation, if p splits as 3, 21, or 111 then $\omega(p) = 0, 1$, or 3 respectively. Thus again with evident notation we have

$$H(A) = \prod_{p|aD} \frac{p - \omega(p)}{p - 1} (P_3 P_{111})(1) \quad \text{with}$$

$$P_3(s) = \prod_{\substack{p \nmid aD \\ p\mathbb{Z}_K = \mathfrak{p}_3}} \frac{p^s}{p^s - 1} \quad \text{and} \quad P_{111}(s) = \prod_{\substack{p \nmid aD \\ p\mathbb{Z}_K = \mathfrak{p}_1 \mathfrak{p}_1' \mathfrak{p}_1''}} \frac{p^s - 3}{p^s - 1} .$$

Thus $\log(P_3(s)) = \sum_{k \geq 1} S_3(ks)/k$ and $\log(P_{111}(s)) = -\sum_{k \geq 1}(3^k - 1)S_{111}(ks)/k$ with $S_3(s) = \sum_{p|aD, \ p\mathbb{Z}_K = \mathfrak{p}_3} 1/p^s$ and similarly for $S_{111}(s)$.

We must now distinguish the Galois type of the number field K. Assume first that K is a cyclic cubic field, i.e., that D is a square. This case is essentially identical with the quadratic case, with 2 replaced by 3 (this is true more generally for C_ℓ-fields with ℓ prime). More precisely, we have

$$S_3(s) = \frac{1}{3} \sum_{3 \nmid n} \frac{\mu(n)}{n} \log((\zeta^3/\zeta_K)_{p \nmid aD}(ns))$$

$$S_{111}(s) = \sum_{n \geq 1} \frac{\mu(n)}{n} \log(\zeta_{p \nmid aD}(ns)) - S_3(s) .$$

Thus, if we set

$$c_1(n) = \frac{1}{n} \sum_{d|n} \mu(d)(3^{n/d} - 1) \quad \text{and} \quad c_2(n) = \frac{1}{3n} \sum_{d|n, \ 3 \nmid d} \mu(d)3^{n/d}$$

we have

$$\log((P_3 P_{111})(s)) = -\sum_{n \geq 1} (c_1(n) \log(\zeta_{p \nmid aD}(ns)) + c_2(n) \log((\zeta^3/\zeta_K)_{p \nmid aD}(ns)))$$

$$= \sum_{n \geq 1} a(n) \log((\zeta_K/\zeta)_{p \nmid aD}(ns)) + \sum_{n \geq 1} c(n) \log(\zeta_{p \nmid aD}(ns))$$

with $a(n) = -c_2(n)$ and $c(n) = 2c_2(n) - c_1(n)$. Since $c_1(1) = 2$ and $c_2(1) = 1$, we have $c(1) = 0$ (otherwise our Euler product would not converge), and for $n \geq 2$ we have $c_1(n) = (1/n) \sum_{d|n} \mu(d)3^{n/d}$. An easy computation shows that $c_1(n) = 3c_2(n) - c_2(n/3)$, so that $c(n) = -c_2(n) + c_2(n/3) = a(n) - a(n/3)$, exactly analogous to the quadratic case.

Assume now that K is a noncyclic cubic field, and let k its quadratic resolvent field, in other words $k = \mathbb{Q}(\sqrt{D})$. We first note that it follows from the quadratic case that (with evident notation)

$$S_{21}(s) = \frac{1}{2} \sum_{2 \nmid n} \frac{\mu(n)}{n} \log((\zeta^2/\zeta_k)_{p \nmid aD}(ns))$$

(although a priori we do not need S_{21}, we will need it for S_{111}). A completely similar computation gives

$$S_3(s) = \frac{1}{3} \sum_{3 \nmid n} \frac{\mu(n)}{n} \log((\zeta \zeta_k/\zeta_K)_{p \nmid aD}(ns)) ,$$

hence

$$S_{111}(s) = \sum_{n \geq 1} \frac{\mu(n)}{n} \log(\zeta_{p\nmid aD}(ns)) - S_{21}(s) - S_3(s) \ .$$

Thus, if in addition to c_1 and c_2 defined above we set

$$c_3(n) = \frac{1}{2n} \sum_{d \mid n, \ 2 \nmid d} \mu(d)(3^{n/d} - 1) \ ,$$

we have

$$\begin{aligned}
\log((P_3 P_{111})(s)) = &-\sum_{n \geq 1} (c_1(n) \log(\zeta_{p\nmid aD}(ns)) \\
&+ c_2(n) \log((\zeta\zeta_k/\zeta_K)_{p\nmid aD}(ns)) + c_3(n) \log((\zeta^2/\zeta_k)_{p\nmid aD}(ns))) \\
= &\sum_{n \geq 1} a(n) \log((\zeta_K/\zeta)_{p\nmid aD}(ns)) + \\
&\sum_{n \geq 1} b(n) \log((\zeta_k/\zeta)_{p\nmid aD}(ns)) + \sum_{n \geq 1} c(n) \log(\zeta_{p\nmid aD}(ns)) \ ,
\end{aligned}$$

with $a(n) = -c_2(n)$, $b(n) = c_2(n) - c_3(n)$, and $c(n) = -c_1(n) + c_2(n) + c_3(n)$, and As usual $c_1(1) = 2$, $c_2(1) = 1$, and $c_3(1) = 1$, so $c(1) = 0$ as it should for convergence. As above, since $c_1(n) = 3c_2(n) - c_2(n/3)$ we check that $c(n) = a(n) - b(n) - a(n/3)$.

This shows that the formulas in the cyclic and noncyclic case can be unified by setting $b(n) = 0$ (and $\zeta_k/\zeta = 1$) in the cyclic case.

This leads to the following program:

───────────────────── | HardyLittlewood3.gp | ─────────────────────

```
/* In the cyclic cubic case, we need z^3/z_K=z^2/(z_K/z) */
/* In the noncyclic cubic case, we need z^2/z_k=z/L_D
   and zz_k/z_K=z_k/(z_K/z)=zL_D/(z_K/z). */

/* Auxiliary functions. */
/* ZetaDN and LchiN are defined in Hardylittlewood2.gp. */

LchiKNinit(nf, P) =
{
  my(Ebad = [], Pbad = [], D = nf.disc);
  foreach (P, p,
    my(E, v = idealprimedec(nf, p));
    E = if (#v == 1, if (D % p == 0, next);
                     1 + 'x + 'x^2,
            #v == 2, if (D % p == 0, 1 - 'x, 1 - 'x^2),
            #v == 3, (1 - 'x)^2);
    Ebad = concat(Ebad, E);
    Pbad = concat(Pbad, p));
  return ([Pbad, Ebad]);
}

HLW3(A, N, P, D) =
{ my(flnoncyc = (D != 1), nf, lim);
```

```
24      my(LK, va, vb, vc, S1, S2, S3);
25      my(B = getlocalbitprec(), EKbad);
26
27      localbitprec(32);
28      lim = ceil(B*log(2)/log(N/3)); nf = nfinit(A);
29      localbitprec(B + ceil(1.585*lim) + exponent(lim));
30      LK = lfundiv(lfuncreate(A), lfuncreate(1));
31      LK = lfuninit(LK, [1/2, lim, 0]);
32      va = vector(lim); vb = vector(lim); vc = vector(lim);
33      forfactored (X = 1, lim,
34        my([n] = X);
35        S2 = S3 = 0;
36        fordivfactored (X, Y,
37          my([d] = Y, n3 = 3^(n/d), mob = moebius(Y));
38          if (mob,
39            if (d % 2 && flnoncyc, S2 += mob * (n3 - 1));
40            if (d % 3, S3 += mob * n3)));
41        va[n] = -S3 / (3 * n);
42        if (flnoncyc, vb[n] = -(va[n] + S2 / (2 * n)));
43        vc[n] = va[n] - vb[n] - if (n%3 == 0, va[n / 3], 0));
44      EKbad = LchiKNinit(nf, P);
45      S1 = sum(n = 1, lim, va[n] * log(LchiN(LK, EKbad, n)));
46      S2 = 0;
47      if (flnoncyc,
48        my(LD, Echibad = LchiNinit(D, P));
49        LD = lfuninit(D, [1/2, lim, 0]);
50        S2 = sum(n = 1, lim, vb[n] * log(LchiN(LD, Echibad, n))));
51      S3 = sum(n = 2, lim, vc[n] * log(ZetaDN(P, n)));
52      return (S1 + S2 + S3);
53    }
54
55    /* Compute the Hardy-Littlewood constant of aX^3+bX^2+cX+d. */
56    HardyLittlewood3(A, N = 50) =
57    { my(DA = poldisc(A), S = 1., Da6, P, v = variable(A));
58
59      if (poldegree(A) != 3, error("polynomial of degree != 3"));
60      my([a, b, c, d] = Vec(A));
61      if (!polisirreducible(A) ||
62          gcd([6 * a, 2 * b, a + b + c, d]) > 1, return (0));
63      N = max(N, 5);
64      /* Take care of bad primes and p <= N. */
65      Da6 = abs(6 * a * DA);
66      P = setunion(factor(Da6)[,1]~, primes([5, N]));
67      foreach (P, p, S *= (p - #polrootsmod(A, p)) / (p - 1));
68      /* Do the real work. */
69      if (a != 1, A = a^2 * subst(A, v, v / a));
70      return (S * exp(HLW3(A, N, P, coredisc(DA))));
71    }
```

```
72
73  HardyLittlewood(A, N = 50) =
74  { my(d = poldegree(A));
75
76      if (d <= 0, return (0),
77          d == 1,
78              my(a = abs(polcoef(A, 1))); return (a / eulerphi(a)),
79          d == 2, return (HardyLittlewood2(A, N)),
80          d == 3, return (HardyLittlewood3(A, N)),
81          error("HardyLittlewood not implemented for d >= 4"));
82  }
```

This program illustrates a number of important GP commands:

(1) As in the quadratic case, we need the `lfun` command, but now not only for $\zeta_k/\zeta = L(\chi_D)$ (obtained in the function LchiN), but also for the nontrivial ζ_K/ζ obtained as follows: `lfuncreate(A)` creates ζ_K, `lfuncreate(1)` creates ζ in the necessary format for `lfun` functions, and finally `lfundiv` performs the division ζ_K/ζ.

(2) Note the use of `#polrootsmod(A, p)` which computes $\omega(p)$ directly, but is used only for a small number of primes (those dividing $6aD$ and those less than or equal to N).

(3) Note the use of `#idealprimedec(nf, p)` which allows the computation of the Euler factor of ζ_K/ζ at these same primes, using the function LchiKNEulerbad to determine this factor for the five possible splitting types of primes in K. In fact, by a basic result of algebraic number theory, for the primes p not dividing Da6, we could use `#idealprimedec(nf, p)` instead of `#polrootsmod(A, p)` (of course, `idealprimedec` is necessary for the Euler factors).

5.7.2. C_ℓ polynomials. What we did in the cyclic cubic case (not in the S_3 case) can immediately be generalized to the cyclic C_ℓ case for ℓ prime, generalizing both the quadratic and the cyclic cubic case.

Let $A(X) = \sum_{0 \le n \le \ell} a_n X^n$ be an irreducible polynomial of degree exactly equal to ℓ, and assume that the number field K defined by A is abelian with Galois group C_ℓ. We can immediately copy what we did above. First, setting $D(A) = \operatorname{disc}(A)$ and $E(A) = \ell! a_\ell D(A)$, with evident notation we have

$$H(A) = \prod_{p \mid E(A)} \frac{p - \omega(p)}{p - 1} (P_\ell P_{1^\ell})(1) \quad \text{with}$$

$$P_\ell(s) = \prod_{\substack{p \nmid E(A) \\ p\mathbb{Z}_K = \mathfrak{p}_\ell}} \frac{p^s}{p^s - 1} \quad \text{and} \quad P_{1^\ell}(s) = \prod_{\substack{p \nmid E(A) \\ p\mathbb{Z}_K = \mathfrak{p}_1 \mathfrak{p}_1' \cdots}} \frac{p^s - \ell}{p^s - 1},$$

and performing exactly the same computation we find that if we set

$$a(n) = -\frac{1}{\ell n} \sum_{d \mid n,\ \ell \nmid d} \mu(d) \ell^{n/d}$$

and $c(n) = a(n) - a(n/\ell)$, we have

$$\log((P_\ell P_{1^\ell})(s)) = \sum_{n \geq 1} a(n) \log((\zeta_K/\zeta)_{p \nmid E(A)}(ns)) + \sum_{n \geq 2} c(n) \log(\zeta_{p \nmid E(A)})(ns) \ .$$

We leave to the reader the immediate task of writing the corresponding GP script which will be essentially identical to the cyclic cubic one.

5.7.3. General polynomials. The case of general polynomials of degree $d \geq 4$ is more complicated, and it is not completely clear how to treat them, at least in a straightforward manner. As in the quadratic and cubic case, we can evidently reduce to the computation of what one can call the Hardy–Littlewood constants of *number fields* (as opposed to polynomials). When the number field is *abelian* over $\mathbb{Q}$, there is no problem since we have d Artin L-functions which are simply Hecke L-series, and the splitting of the primes is completely understood in terms of these series (we have mentioned the case of C_ℓ-number fields above, when ℓ is prime).

If it is not abelian, or even Galois, it is not clear to the authors how to proceed in general, although it may not be too difficult. What we explained above for S_3-cubic fields can also easily be done for A_4-quartic fields, but it is not completely clear how to do it for D_4- or S_4-quartic fields. We leave to the reader to explore this avenue of computation.

CHAPTER 6

Gauss and Jacobi sums

We now come to a completely different type of computation, since it consists in computing numerically a finite sum, say having N terms, in nontrivial time, in other words in time $O(N^\alpha)$ for some $\alpha < 1$.

Note that we are asking a purely numerical problem: we are *not* asking for exact formulas nor for (upper) bounds (although we will see such results), but for ways to compute the sum, exactly as an algebraic number, or numerically in general, either as a complex or a p-adic approximation, and in this later case we may be interested only in low accuracy, for instance a computation modulo p^2.

We may also want to compute *many* sums simultaneously, a problem which usually has solutions which are faster than computing each sum independently of the others.

We will begin by considering what are probably the most important types of exponential sums: Gauss and Jacobi sums, and we will consider other types later.

6.1. Gauss and Jacobi sums over $\mathbb{F}_q$

6.1.1. Basic definitions: Finite fields and Gauss sums. We can study Gauss and Jacobi sums in two different contexts: first, and most importantly, over finite fields $\mathbb{F}_q$, with $q = p^f$ a prime power, and second, over the ring $\mathbb{Z}/N\mathbb{Z}$ (or more generally over finite rings such as $\mathbb{Z}_K/I$ for some ideal I in the ring of integers $\mathbb{Z}_K$ of a number field). The two notions coincide when $N = q = p$ is prime, but the methods and applications are quite different. Since almost everything is completely standard, we usually do not give proofs and instead refer to numerous textbooks dealing with the subject such as [**Coh07a**] or [**BEW98**].

To give the definitions over $\mathbb{F}_q$ we need to recall some fundamental (and easy) results concerning finite fields.

PROPOSITION 6.1.1. *Let p be a prime, $f \geq 1$, and $\mathbb{F}_q$ be the finite field with $q = p^f$ elements, which exists and is unique up to isomorphism.*

(1) *The multiplicative group $\mathbb{F}_q^*$ is cyclic.*
(2) *The map ϕ such that $\phi(x) = x^p$ is a field isomorphism from $\mathbb{F}_q$ to itself leaving $\mathbb{F}_p$ fixed. It is called the* Frobenius *map.*
(3) *The extension $\mathbb{F}_q/\mathbb{F}_p$ is a Galois (i.e., normal and separable) field extension, with Galois group which is cyclic of order f generated by ϕ.*

In particular, we can define the *trace* $\mathrm{Tr}_{\mathbb{F}_q/\mathbb{F}_p}$ and the *norm* $\mathcal{N}_{\mathbb{F}_q/\mathbb{F}_p}$, and we have the formulas (where from now on we omit $\mathbb{F}_q/\mathbb{F}_p$ for simplicity):

$$\mathrm{Tr}(x) = \sum_{0 \leq j \leq f-1} x^{p^j} \quad \text{and} \quad \mathcal{N}(x) = \prod_{0 \leq j \leq f-1} x^{p^j} = x^{(p^f-1)/(p-1)} = x^{(q-1)/(p-1)} .$$

DEFINITION 6.1.2. Let χ be a character from $\mathbb{F}_q^*$ to an algebraically closed field C of characteristic 0. For $a \in \mathbb{F}_q$ we define the *Gauss sum* $\mathfrak{g}(\chi, a)$ by

$$\mathfrak{g}(\chi, a) = \sum_{x \in \mathbb{F}_q^*} \chi(x) \zeta_p^{\mathrm{Tr}(ax)} \,,$$

where ζ_p is a fixed primitive pth root of unity in C. We also set $\mathfrak{g}(\chi) = \mathfrak{g}(\chi, 1)$.

Recall that a character on $\mathbb{F}_q^*$ is simply a group homomorphism from $\mathbb{F}_q^*$ to C^*, and that strictly speaking this definition depends on the choice of ζ_p. However, if ζ_p' is some other primitive pth root of unity we have $\zeta_p' = \zeta_p^k$ for some $k \in \mathbb{F}_p^*$, so

$$\sum_{x \in \mathbb{F}_q^*} \chi(x) \zeta_p'^{\mathrm{Tr}(ax)} = \mathfrak{g}(\chi, ka) \,.$$

In fact it is trivial to see (this follows from the next proposition) that $\mathfrak{g}(\chi, ka) = \chi^{-1}(k) \mathfrak{g}(\chi, a)$.

DEFINITION 6.1.3. We define ε to be the trivial character, i.e., such that $\varepsilon(x) = 1$ for all $x \in \mathbb{F}_q^*$. We extend characters χ to the whole of $\mathbb{F}_q$ by setting $\chi(0) = 0$ if $\chi \neq \varepsilon$ and $\varepsilon(0) = 1$.

Note that this apparently innocuous definition of $\varepsilon(0)$ is *crucial* because it simplifies many formulas. Note also that the definition of $\mathfrak{g}(\chi, a)$ is a sum over $x \in \mathbb{F}_q^*$ and not $x \in \mathbb{F}_q$, while for Jacobi sums we will use all of $\mathbb{F}_q$.

Exercise:

(1) Show that $\mathfrak{g}(\varepsilon, a) = -1$ if $a \in \mathbb{F}_q^*$ and $\mathfrak{g}(\varepsilon, 0) = q - 1$.
(2) If $\chi \neq \varepsilon$, show that $\mathfrak{g}(\chi, 0) = 0$, in other words that

$$\sum_{x \in \mathbb{F}_q} \chi(x) = 0$$

(here it does not matter if we sum over $\mathbb{F}_q$ or $\mathbb{F}_q^*$).
(3) Deduce that if $\chi_1 \neq \chi_2$ then

$$\sum_{x \in \mathbb{F}_q^*} \chi_1(x) \chi_2^{-1}(x) = 0 \,.$$

This relation is called for evident reasons *orthogonality of characters*.

Because of this exercise, if necessary we may assume that $\chi \neq \varepsilon$ and/or that $a \neq 0$.

Exercise: Let χ be a character of $\mathbb{F}_q^*$ of exact order n.

(1) Show that $n \mid (q-1)$ and that $\chi(-1) = (-1)^{(q-1)/n}$. In particular, if n is odd and $p > 2$ we have $\chi(-1) = 1$.
(2) Show that $\mathfrak{g}(\chi, a) \in \mathbb{Z}[\zeta_n, \zeta_p]$, where as usual ζ_m denotes a primitive mth root of unity.

PROPOSITION 6.1.4.

(1) *If $a \neq 0$ we have*

$$\mathfrak{g}(\chi, a) = \chi^{-1}(a) \mathfrak{g}(\chi) \quad \text{and} \quad \mathfrak{g}(\chi^p, a) = \chi^{1-p}(a) \mathfrak{g}(\chi, a) \,.$$

(2) *We have $\mathfrak{g}(\chi^{-1}) = \chi(-1) \overline{\mathfrak{g}(\chi)}$, and if $\chi \neq \varepsilon$ we have $|\mathfrak{g}(\chi)| = q^{1/2}$.*

6.1.2. Finite fields and Jacobi sums. Recall that we have extended characters of $\mathbb{F}_q^*$ by setting $\chi(0) = 0$ if $\chi \neq \varepsilon$ and $\varepsilon(0) = 1$.

DEFINITION 6.1.5. *For $1 \leq j \leq k$ let χ_j be characters of $\mathbb{F}_q^*$. We define the Jacobi sum*

$$J_k(\chi_1, \ldots, \chi_k; a) = \sum_{x_1 + \cdots + x_k = a} \chi_1(x_1) \cdots \chi_k(x_k)$$

and $J_k(\chi_1, \ldots, \chi_k) = J_k(\chi_1, \ldots, \chi_k; 1)$.

Note that, as mentioned above, we do not exclude the cases where some $x_i = 0$, using the convention of Definition 6.1.3 for $\chi(0)$.

The following lemma shows that it is only necessary to study $J_k(\chi_1, \ldots, \chi_k)$:

LEMMA 6.1.6. *Set $\chi = \chi_1 \cdots \chi_k$.*

(1) *If $a \neq 0$ we have*

$$J_k(\chi_1, \ldots, \chi_k; a) = \overline{\chi}(a) J_k(\chi_1, \ldots, \chi_k) \,.$$

(2) *If $a = 0$, abbreviating $J_k(\chi_1, \ldots, \chi_k; 0)$ to $J_k(0)$ we have*

$$J_k(0) = \begin{cases} q^{k-1} & \text{if } \chi_j = \varepsilon \text{ for all } j \,, \\ 0 & \text{if } \chi \neq \varepsilon \,, \\ \chi_k(-1)(q-1)J_{k-1}(\chi_1, \ldots, \chi_{k-1}) & \text{if } \chi = \varepsilon \text{ and } \chi_k \neq \varepsilon \,. \end{cases}$$

As we have seen, a Gauss sum $\mathfrak{g}(\chi)$ belongs to the rather large ring $\mathbb{Z}[\zeta_{q-1}, \zeta_p]$ (and in general not to a smaller ring). The advantage of Jacobi sums is that they belong to the smaller ring $\mathbb{Z}[\zeta_{q-1}]$, and as we are going to see, that they are closely related to Gauss sums. Thus, when working *algebraically*, it is almost always better to use Jacobi sums instead of Gauss sums. On the other hand, when working *analytically* (for instance in $\mathbb{C}$ or $\mathbb{C}_p$), it may be better to work with Gauss sums: we will see below the use of root numbers (suggested by Louboutin), and of the Gross–Koblitz formula.

Note that $J_1(\chi_1) = 1$. Outside of this trivial case, the close link between Gauss and Jacobi sums is given by the following easy proposition, whose apparently technical statement is only due to the trivial character ε: if none of the χ_j nor their product is trivial, we have the simple formula given by (3).

PROPOSITION 6.1.7. *Denote by t the number of χ_j equal to the trivial character ε, and as above set $\chi = \chi_1 \cdots \chi_k$.*

(1) *If $t = k$ then $J_k(\chi_1, \ldots, \chi_k) = q^{k-1}$.*

(2) *If $1 \leq t \leq k-1$ then $J_k(\chi_1, \ldots, \chi_k) = 0$.*

(3) *If $t = 0$ and $\chi \neq \varepsilon$ then*

$$J_k(\chi_1, \ldots, \chi_k) = \frac{\mathfrak{g}(\chi_1) \cdots \mathfrak{g}(\chi_k)}{\mathfrak{g}(\chi_1 \cdots \chi_k)} = \frac{\mathfrak{g}(\chi_1) \cdots \mathfrak{g}(\chi_k)}{\mathfrak{g}(\chi)} \,.$$

(4) *If $t = 0$ and $\chi = \varepsilon$ then*

$$J_k(\chi_1, \ldots, \chi_k) = -\frac{\mathfrak{g}(\chi_1) \cdots \mathfrak{g}(\chi_k)}{q}$$

$$= -\chi_k(-1) \frac{\mathfrak{g}(\chi_1) \cdots \mathfrak{g}(\chi_{k-1})}{\mathfrak{g}(\chi_1 \cdots \chi_{k-1})} = -\chi_k(-1) J_{k-1}(\chi_1, \ldots, \chi_{k-1}) \,.$$

In particular, in this case we have

$$\mathfrak{g}(\chi_1) \cdots \mathfrak{g}(\chi_k) = \chi_k(-1) q J_{k-1}(\chi_1, \ldots, \chi_{k-1}) \,.$$

COROLLARY 6.1.8. *With the same notation, assume that $k \geq 2$ and all the χ_j are nontrivial. Setting $\psi = \chi_1 \cdots \chi_{k-1}$, we have the following recursive formula:*

$$J_k(\chi_1, \ldots, \chi_k) = \begin{cases} J_{k-1}(\chi_1, \ldots, \chi_{k-1}) J_2(\psi, \chi_k) & \text{if } \psi \neq \varepsilon \,, \\ \chi_{k-1}(-1) q J_{k-2}(\chi_1, \ldots, \chi_{k-2}) & \text{if } \psi = \varepsilon \,. \end{cases}$$

The point of this recursion is that the definition of a k-fold Jacobi sum J_k involves a sum over q^{k-1} values for $x_1, \ldots, x_{k-1}$, the last variable x_k being determined by $x_k = 1 - x_1 - \cdots - x_{k-1}$, so neglecting the time to compute the $\chi_j(x_j)$ and their product (which is a reasonable assumption), using the definition takes time $O(q^{k-1})$. On the other hand, using the above recursion boils down at worst to computing $k - 1$ Jacobi sums J_2, for a total time of $O((k-1)q)$. Nonetheless, we will see that in some cases it is still better to use directly Gauss sums and formula (3) of the proposition.

Since Jacobi sums J_2 are the simplest and the above recursion in fact shows that one can reduce to J_2, we will drop the subscript 2 and simply write $J(\chi_1, \chi_2)$. Note that

$$J(\chi_1, \chi_2) = \sum_{x \in \mathbb{F}_q} \chi_1(x) \chi_2(1 - x) \,,$$

where the sum is over the whole of $\mathbb{F}_q$ and *not* $\mathbb{F}_q \setminus \{0, 1\}$ (which makes a difference only if one of the χ_i is trivial). More precisely it is clear that $J(\varepsilon, \varepsilon) = q^2$, and that if $\chi \neq \varepsilon$ we have $J(\chi, \varepsilon) = \sum_{x \in \mathbb{F}_q} \chi(x) = 0$, which are special cases of Proposition 6.1.7.

Exercise: Let $n \mid (q - 1)$ be the order of χ. Prove that $\mathfrak{g}(\chi)^n \in \mathbb{Z}[\zeta_n]$.

Exercise: Assume that none of the χ_j is equal to ε, but that their product χ is equal to ε. Prove that (using the same notation as in Lemma 6.1.6):

$$J_k(0) = \left(1 - \frac{1}{q}\right) \mathfrak{g}(\chi_1) \cdots \mathfrak{g}(\chi_k) \,.$$

Exercise: Prove the following reciprocity formula for Jacobi sums: if the χ_j are all nontrivial and $\chi = \chi_1 \cdots \chi_k$, we have

$$J_k(\chi_1^{-1}, \ldots, \chi_k^{-1}) = \frac{q^{k-1-\delta}}{J_k(\chi_1, \ldots, \chi_k)} \,,$$

where $\delta = 1$ if $\chi = \varepsilon$, and otherwise $\delta = 0$.

6.1.3. Applications of $J(\chi, \chi)$. In this short subsection we give without proof a couple of applications of the special Jacobi sums $J(\chi, \chi)$. Once again the proofs are not difficult. We begin by the following result, which is a special case of the Hasse–Davenport relations that we will give below.

LEMMA 6.1.9. *Assume that q is odd, and let ρ be the unique character of order 2 on $\mathbb{F}_q^*$. For any nontrivial character χ we have*

$$\chi(4) J(\chi, \chi) = J(\chi, \rho) \,.$$

Equivalently, if $\chi \neq \rho$ we have

$$\mathfrak{g}(\chi) \mathfrak{g}(\chi \rho) = \chi^{-1}(4) \mathfrak{g}(\rho) \mathfrak{g}(\chi^2) \,.$$

Exercise:

(1) Prove this lemma.

(2) Show that $\mathfrak{g}(\rho)^2 = (-1)^{(q-1)/2}q$.

PROPOSITION 6.1.10.

(1) *Assume that $q \equiv 1 \pmod 4$, let χ be one of the two characters of order 4 on $\mathbb{F}_q^*$, and write $J(\chi,\chi) = a + bi$. Then $q = a^2 + b^2$, $2 \mid b$, and $a \equiv -1 \pmod 4$.*

(2) *Assume that $q \equiv 1 \pmod 3$, let χ be one of the two characters of order 3 on $\mathbb{F}_q^*$, and write $J(\chi,\chi) = a + b\rho$, where $\rho = \zeta_3$ is a primitive cube root of unity. Then $q = a^2 - ab + b^2$, $3 \mid b$, $a \equiv -1 \pmod 3$, and $a + b \equiv q - 2 \pmod 9$.*

(3) *Let $p \equiv 2 \pmod 3$, $q = p^{2m} \equiv 1 \pmod 3$, and let χ be one of the two characters of order 3 on $\mathbb{F}_q^*$. We have*

$$J(\chi,\chi) = (-1)^{m-1}p^m = (-1)^{m-1}q^{1/2} \ .$$

COROLLARY 6.1.11.

(1) *(Fermat.) Any prime $p \equiv 1 \pmod 4$ is a sum of two squares.*

(2) *Any prime $p \equiv 1 \pmod 3$ is of the form $a^2 - ab + b^2$ with $3 \mid b$, or equivalently $4p = (2a - b)^2 + 27(b/3)^2$ is of the form $c^2 + 27d^2$.*

(3) *(Gauss.) $p \equiv 1 \pmod 3$ is itself of the form $p = u^2 + 27v^2$ if and only if 2 is a cube in $\mathbb{F}_p^*$.*

Exercise: Assuming the proposition, prove the corollary.

6.1.4. The Hasse–Davenport relations. All the results that we have given up to now on Gauss and Jacobi sums have rather simple proofs, which is one of the reasons we have not given them. However, there exist other important relations which are considerably more difficult to prove. Before giving them, it is instructive to explain how one can "guess" their existence, if one knows the classical theory of the gamma function $\Gamma(s)$.

Recall that $\Gamma(s)$ is defined (at least for $\Re(s) > 0$) by

$$\Gamma(s) = \int_0^\infty e^{-t}t^s dt/t \ ,$$

and the beta function $B(a,b)$ by $B(a,b) = \int_0^1 t^{a-1}(1-t)^{b-1}\,dt$. The function e^{-t} transforms sums into products, so is an *additive* character, analogous to ζ_p^t. The function t^s transforms products into products, so is a multiplicative character, analogous to $\chi(t)$ (dt/t is simply the Haar invariant measure on $\mathbb{R}_{>0}$). Thus $\Gamma(s)$ is a continuous analogue of the Gauss sum $\mathfrak{g}(\chi)$.

Similarly, the Jacobi sum $J(\chi_1, \chi_2) = \sum_t \chi_1(t)\chi_2(1-t)$, is an analogue of the beta function. Thus, it does not come too much as a surprise that analogous formulas are valid on both sides. To begin with, it is not difficult to show that $B(a,b) = \Gamma(a)\Gamma(b)/\Gamma(a+b)$, exactly analogous to $J(\chi_1,\chi_2) = \mathfrak{g}(\chi_1)\mathfrak{g}(\chi_2)/\mathfrak{g}(\chi_1\chi_2)$. The analogue of $\Gamma(s)\Gamma(-s) = -\pi/(s\sin(s\pi))$ is

$$\mathfrak{g}(\chi)\mathfrak{g}(\chi^{-1}) = \chi(-1)q \ .$$

But the gamma function has a duplication formula

$$\Gamma(s)\Gamma(s + 1/2) = 2^{1-2s}\Gamma(1/2)\Gamma(2s) \ ,$$

and more generally a multiplication (or distribution) formula for $m \geq 1$:

$$\prod_{0 \leq a < m} \Gamma\left(s + \frac{a}{m}\right) = m^{1/2-ms}(2\pi)^{(m-1)/2}\Gamma(ms) \ .$$

The duplication formula is clearly the analogue of the formula

$$\mathfrak{g}(\chi)\mathfrak{g}(\chi\rho) = \chi^{-1}(4)\mathfrak{g}(\rho)\mathfrak{g}(\chi^2)$$

given above. The *Hasse–Davenport product relation* is the analogue of the distribution formula for the gamma function.

THEOREM 6.1.12. *Let ρ be a character of exact order m dividing $q - 1$. For any character χ of $\mathbb{F}_q^*$ we have*

$$\prod_{0 \leq a < m} \mathfrak{g}(\chi\rho^a) = \chi^{-m}(m)k(p, f, m)q^{(m-1)/2}\mathfrak{g}(\chi^m) \ ,$$

where $k(p, f, m)$ is the fourth root of unity given by

$$k(p, f, m) = \begin{cases} \left(\dfrac{p}{m}\right)^f & \text{if } m \text{ is odd,} \\ (-1)^{f+1}\left(\dfrac{(-1)^{m/2+1}m/2}{p}\right)^f \left(\dfrac{-1}{p}\right)^{f/2} & \text{if } m \text{ is even,} \end{cases}$$

where $(-1)^{f/2}$ is to be understood as i^f when f is odd.

Remark: For some reason, in the literature this formula is always stated in the weaker form where the constant $k(p, f, m)$ is not given explicitly.

Contrary to the proof of the distribution formula for the gamma function, the proof of this theorem is quite long. There are essentially two completely different proofs: one using classical algebraic number theory, and one using p-adic analysis. The latter is simpler and gives directly the value of $k(p, f, m)$. See [**Coh07a**] and [**Coh07b**] for both detailed proofs.

Gauss sums satisfy another type of nontrivial relation, also due to Hasse–Davenport, the so-called *lifting relation*, as follows:

THEOREM 6.1.13. *Let $\mathbb{F}_{q^n}/\mathbb{F}_q$ be an extension of finite fields, let χ be a character of $\mathbb{F}_q^*$, and define the* lift *of χ to $\mathbb{F}_{q^n}$ by the formula $\chi^{(n)} = \chi \circ \mathcal{N}_{\mathbb{F}_{q^n}/\mathbb{F}_q}$. We have*

$$\mathfrak{g}(\chi^{(n)}) = (-1)^{n-1}\mathfrak{g}(\chi)^n \ .$$

This relation is essential in the initial proof of the Weil conjectures for diagonal hypersurfaces done by Weil himself.

6.2. Practical computations of Gauss and Jacobi sums

6.2.1. Introduction and motivation. Since $\mathbb{F}_q^*$ is a cyclic group, so is its group of characters (which is noncanonically isomorphic), and in the sequel we will let ω denote a generator of this group, hence of exact order $q - 1$. Note that $\omega^{q-1} = \varepsilon$, so beware that the relation $\chi_1\chi_2(x) = \chi_1(x)\chi_2(x)$ may not be true if $x = 0$, for instance if $\chi_1 = \omega$ and $\chi_2 = \omega^{q-2}$.

For notational simplicity, we will write $J(r_1, \ldots, r_k)$ instead of $J(\omega^{r_1}, \ldots, \omega^{r_k})$.

One of the motivations for the computation of such Jacobi sums is the following theorem, which is quite easy to prove, see [**BCM15**] for much more general results:

THEOREM 6.2.1. *For $m \geq 2$ consider the hypersurface $H = H(a_1, \ldots, a_m, b)$ defined in $\mathbb{P}^{m-1}(\mathbb{F}_q)$ by the equation*

$$\sum_{1 \leq i \leq m} a_i x_i^m - b \prod_{1 \leq i \leq m} x_i = 0 \,,$$

where $b \in \mathbb{F}_q^*$ and $a_i \in \mathbb{F}_q^*$ for all i. Denote by $N_q(H)$ the number of solutions in $\mathbb{F}_q^m$, so that the number of projective points is given by $|H(\mathbb{F}_q)| = (N_q(H) - 1)/(q - 1)$. If $\gcd(m, q - 1) = 1$, we have

$$N_q(H) = (-1)^{m-1} + \sum_{0 \leq n \leq q-2} \omega^{-n}(z) J_m(n, n, \ldots, n) \,,$$

where $z = \prod_{1 \leq i \leq m}(a_i/b)$.

REMARKS 6.2.2.
   (1) There exists a more complicated formula valid also when $\gcd(m, q-1) > 1$.
   (2) The case $m = 2$ is trivial (solve a quadratic equation in a finite field), the case $m = 3$ corresponds to counting points on an elliptic curve, the case $m = 4$ on the symmetric square of an elliptic curve, and the case $m = 5$ on a Calabi–Yau threefold, the Dwork quintic.

From now on, we will consider the specific example $m = 5$, so that our goal is to compute the quantity

$$S(q; z) = \sum_{0 \leq n \leq q-2} \omega^{-n}(z) J_5(n, n, n, n, n) \,,$$

where we do not necessarily assume that $q \not\equiv 1 \pmod 5$. We will see methods which compute the Jacobi sums individually, and others more globally.

6.2.2. Elementary methods. By the recursion of Corollary 6.1.8, we have *generically* (i.e., except for special values of n which will be considered separately):

$$J_5(n, n, n, n, n) = J(n, n) J(2n, n) J(3n, n) J(4n, n) \,.$$

Since $J(n, an) = \sum_x \omega^n(x) \omega^{an}(1 - x)$, the cost of computing J_5 as written is $\widetilde{O}(q)$, where here and after we write $\widetilde{O}(q^\alpha)$ to mean $O(q^{\alpha+\varepsilon})$ for all $\varepsilon > 0$ (soft-O notation). Thus computing $S(q; z)$ by this direct method requires time $\widetilde{O}(q^2)$.

We can however do much better. Since the values of the characters are all in $\mathbb{Z}[\zeta_{q-1}]$, we work in this ring. In fact, even better, we work in the ring with zero divisors $R = \mathbb{Z}[X]/(X^{q-1} - 1)$, together with the natural surjective map sending the class of X in R to ζ_{q-1}. Indeed, let g be the generator of $\mathbb{F}_q^*$ such that $\omega(g) = \zeta_{q-1}$. We have, again *generically*:

$$J(n, an) = \sum_{1 \leq u \leq q-2} \omega^n(g^u) \omega^{an}(1 - g^u) = \sum_{1 \leq u \leq q-2} \zeta_{q-1}^{nu + an \log_g(1 - g^u)} \,,$$

where $\log_g$ is the *discrete logarithm* to base g defined modulo $q - 1$, i.e., such that $g^{\log_g(x)} = x$. If $(q - 1) \nmid n$ but $(q - 1) \mid an$ we have $\omega^{an} = \varepsilon$ so we must add the contribution of $u = 0$, which is 1, and if $(q - 1) \mid n$ we must add the contribution of $u = 0$ *and* of $x = 0$, which is 2 (recall the *essential* convention that $\chi(0) = 0$ if $\chi \neq \varepsilon$ and $\varepsilon(0) = 1$, see Definition 6.1.3).

In other words, if we set

$$P_a(X) = \sum_{1 \le u \le q-2} X^{(u+a \log_g(1-g^u)) \bmod (q-1)} \in R \,,$$

we have

$$J(n, an) = P_a(\zeta_{q-1}^n) + \begin{cases} 0 & \text{if } (q-1) \nmid an \,, \\ 1 & \text{if } (q-1) \mid an \text{ but } (q-1) \nmid n \,, \text{ and} \\ 2 & \text{if } (q-1) \mid n \,. \end{cases}$$

Thus, if we set finally

$$P(X) = P_1(X)P_2(X)P_3(X)P_4(X) \bmod X^{q-1} \in R \,,$$

we have (still generically) $J_5(n, n, n, n, n) = P(\zeta_{q-1}^n)$. Assume for the moment that this is true for all n (we will correct this below), let $\ell = \log_g(z)$, so that $\omega(z) = \omega(g^\ell) = \zeta_{q-1}^\ell$, and write

$$P(X) = \sum_{0 \le j \le q-2} a_j X^j \,.$$

We thus have

$$\omega^{-n}(z) J_5(n, n, n, n, n) = \zeta_{q-1}^{-n\ell} \sum_{0 \le j \le q-2} a_j \zeta_{q-1}^{nj} = \sum_{0 \le j \le q-2} a_j \zeta_{q-1}^{n(j-\ell)} \,,$$

hence

$$S(q; z) = \sum_{0 \le n \le q-2} \omega^{-n}(z) J_5(n, n, n, n, n) = \sum_{0 \le j \le q-2} a_j \sum_{0 \le n \le q-2} \zeta_{q-1}^{n(j-\ell)}$$

$$= (q-1) \sum_{0 \le j \le q-2,\ j \equiv \ell \pmod{q-1}} a_j = (q-1)a_\ell \,.$$

The result is thus immediate as soon as we know the coefficients of the polynomial P. Since there exist fast methods for computing discrete logarithms, this leads to a $\widetilde{O}(q)$ method for computing $S(q; z)$.

To obtain the correct formula, we need to adjust for the special n for which $J_5(n, n, n, n, n)$ is not equal to $J(n, n)J(n, 2n)J(n, 3n)J(n, 4n)$, which are the same for which $(q-1) \mid an$ for some a such that $2 \le a \le 4$, together with $a = 5$. This is easy but boring, and should be skipped on first reading.

(1) For $n = 0$ we have $J_5(n, n, n, n, n) = q^4$, and on the other hand $P(1) = (J(0,0) - 2)^4 = (q-2)^4$, so the correction term is $q^4 - (q-2)^4 = 8(q-1)(q^2 - 2q + 2)$.

(2) For $n = (q-1)/2$ (if q is odd) we have

$$J_5(n, n, n, n, n) = \mathfrak{g}(\omega^n)^5 / \mathfrak{g}(\omega^{5n}) = \mathfrak{g}(\omega^n)^4 = \mathfrak{g}(\rho)^4$$

since $5n \equiv n \pmod{q-1}$, where ρ is the character of order 2, and we have $\mathfrak{g}(\rho)^2 = (-1)^{(q-1)/2} q$, so $J_5(n, n, n, n, n) = q^2$. On the other hand

$$P(\zeta_{q-1}^n) = J(\rho, \rho)(J(\rho, 2\rho) - 1)J(\rho, \rho)(J(\rho, 2\rho) - 1)$$

$$= J(\rho, \rho)^2 = \mathfrak{g}(\rho)^4 / q^2 = 1 \,,$$

so the correction term is $\rho(z)(q^2 - 1)$.

(3) For $n = \pm(q-1)/3$ (if $q \equiv 1 \pmod 3$), writing $\chi_3 = \omega^{(q-1)/3}$, which is one of the two cubic characters, we have

$$J_5(n,n,n,n,n) = \mathfrak{g}(\omega^n)^5/\mathfrak{g}(\omega^{5n}) = \mathfrak{g}(\omega^n)^5/\mathfrak{g}(\omega^{-n})$$
$$= \mathfrak{g}(\omega^n)^6/(\mathfrak{g}(\omega^{-n})\mathfrak{g}(\omega^n)) = \mathfrak{g}(\omega^n)^6/q$$
$$= qJ(n,n)^2$$

(check all this). On the other hand

$$P(\zeta_{q-1}^n) = J(n,n)J(n,2n)(J(n,3n)-1)J(n,4n)$$
$$= \frac{\mathfrak{g}(\omega^n)^2}{\mathfrak{g}(\omega^{2n})}\frac{\mathfrak{g}(\omega^n)\mathfrak{g}(\omega^{2n})}{q}\frac{\mathfrak{g}(\omega^n)^2}{\mathfrak{g}(\omega^{2n})}$$
$$= \frac{\mathfrak{g}(\omega^n)^5}{q\mathfrak{g}(\omega^{-n})} = \frac{\mathfrak{g}(\omega^n)^6}{q^2} = J(n,n)^2 \ ,$$

so the correction term is $2(q-1)\Re(\chi_3^{-1}(z)J(\chi_3,\chi_3)^2)$.

(4) For $n = \pm(q-1)/4$ (if $q \equiv 1 \pmod 4$), writing $\chi_4 = \omega^{(q-1)/4}$, which is one of the two quartic characters, we have

$$J_5(n,n,n,n,n) = \mathfrak{g}(\omega^n)^5/\mathfrak{g}(\omega^{5n}) = \mathfrak{g}(\omega^n)^4 = \omega^n(-1)qJ_3(n,n,n) \ .$$

In addition, we have

$$J_3(n,n,n) = J(n,n)J(n,2n) = \omega^n(4)J(n,n)^2 = \rho(2)J(n,n)^2 \ ,$$

so

$$J_5(n,n,n,n,n) = \mathfrak{g}(\omega^n)^4 = \omega^n(-1)q\rho(2)J(n,n)^2 \ .$$

Note that

$$\chi_4(-1) = \chi_4^{-1}(-1) = \rho(2) = (-1)^{(q-1)/4} \ ,$$

(Exercise: prove it!), so that $\omega^n(-1)\rho(2) = 1$ and the above simplifies to $J_5(n,n,n,n,n) = qJ(n,n)^2$.

On the other hand,

$$P(\zeta_{q-1}^n) = J(n,n)J(n,2n)J(n,3n)(J(n,4n)-1)$$
$$= \frac{\mathfrak{g}(\omega^n)^2}{\mathfrak{g}(\omega^{2n})}\frac{\mathfrak{g}(\omega^n)\mathfrak{g}(\omega^{2n})}{\mathfrak{g}(\omega^{3n})}\frac{\mathfrak{g}(\omega^n)\mathfrak{g}(\omega^{3n})}{q}$$
$$= \frac{\mathfrak{g}(\omega^n)^4}{q} = \omega^n(-1)\rho(2)J(n,n)^2 = J(n,n)^2$$

as above, so the correction term is $2(q-1)\Re(\chi_4^{-1}(z)J(\chi_4,\chi_4)^2)$.

(5) For $n = a(q-1)/5$ with $1 \le a \le 4$ (if $q \equiv 1 \pmod 5$), writing $\chi_5 = \omega^{(q-1)/5}$ we have $J_5(n,n,n,n,n) = -\mathfrak{g}(\chi_5^a)^5/q$, while abbreviating $\mathfrak{g}(\chi_5^{am})$ to $g(m)$ we have

$$P(\zeta_{q-1}^n) = J(n,n)J(n,2n)J(n,3n)J(n,4n)$$
$$= -\frac{g(n)^2}{g(2n)}\frac{g(n)g(2n)}{g(3n)}\frac{g(n)g(3n)}{g(4n)}\frac{g(n)g(4n)}{q}$$
$$= -\frac{g(n)^5}{q} \ ,$$

so there is no correction term.

Summarizing, we have shown the following:

PROPOSITION 6.2.3. *Let* $S(q; z) = \sum_{0 \le n \le q-2} \omega^{-n}(z) J_5(n, n, n, n, n)$. *Let* $\ell = \log_g(z)$ *and let* $P(X) = \sum_{0 \le j \le q-2} a_j X^j$ *be the polynomial defined above. We have*

$$S(q; z) = (q-1)(T_1 + T_2 + T_3 + T_4 + a_\ell),$$

where $T_m = 0$ *if* $m \nmid (q-1)$ *and otherwise*

$$T_1 = 8(q^2 - 2q + 2), \quad T_2 = \rho(z)(q+1),$$

$$T_3 = 2\Re(\chi_3^{-1}(z) J(\chi_3, \chi_3)^2), \quad and \quad T_4 = 2\Re(\chi_4^{-1}(z) J(\chi_4, \chi_4)^2),$$

with the above notation.

Thanks to Proposition 6.1.10, these supplementary Jacobi sums $J(\chi_3, \chi_3)$ and $J(\chi_4, \chi_4)$ can be computed in logarithmic time using Cornacchia's algorithm (this is not quite true, one needs an additional slight computation, do you see why?).

Note also for future reference that the above proposition *proves* that $(q-1) \mid S(q, z)$, which is not clear from the definition, although it also follows from Theorem 6.2.1 when $q \not\equiv 1 \pmod 5$, since $m = 5$.

6.2.3. Implementations. For simplicity, assume that $q = p$ is prime. We have written simple programs for computing $S(q; z)$. In the first one, we use the naïve formula expressing J_5 in terms of $J(n, an)$ and sum on n, except that we use the reciprocity formula which gives $J_5(-n, -n, -n, -n, -n)$ in terms of $J_5(n, n, n, n, n)$ to sum only over $(p-1)/2$ terms instead of $p-1$. Of course to avoid recomputation, we precompute a discrete logarithm table.

The timings for $p \approx 10^k$ for $k = 2$, 3, and 4 are 0.01, 1.30, and 157. seconds respectively, compatible with $\widetilde{O}(q^2)$ time.

On the other hand, implementing in a straightforward manner the algorithm given by the above proposition gives timings for $p \approx 10^k$ for $k = 2$, 3, 4, 5, 6, and 7 of 0.00, 0.01, 0.07, 1.00, 14.4, and 103 seconds respectively, of course much faster and compatible with $\widetilde{O}(q)$ time.

The main drawback of this method is that it requires $O(q)$ storage: it is thus applicable only for $q \le 10^8$, say, which is more than sufficient for many applications, but of course not for all. For instance, the case $p \approx 10^7$ mentioned above already required a few gigabytes of storage.

6.2.4. Using theta functions. A completely different way of computing Gauss and Jacobi sums has been suggested by S. Louboutin [**Lou02**, §4]. It is related to the theory of L-functions of Dirichlet characters that we will study later, and in our context is valid only for $q = p$ prime, not for prime powers, but in the context of Dirichlet characters it is valid in general (simply replace p by N and $\mathbb{F}_p$ by $\mathbb{Z}/N\mathbb{Z}$ in the following formulas when χ is a *primitive character* of conductor N, see below for definitions):

DEFINITION 6.2.4. *Let* χ *be a character on* $\mathbb{F}_p$, *and let* $e = 0$ *or 1 be such that* $\chi(-1) = (-1)^e$. *The theta function associated to* χ *is the function defined on the upper half-plane* $\Im(\tau) > 0$ *by*

$$\Theta(\chi, \tau) = \sum_{m \in \mathbb{Z}} m^e \chi(m) e^{i\pi m^2 \tau / p} = \chi(0) + 2 \sum_{m \ge 1} m^e \chi(m) e^{i\pi m^2 \tau / p}.$$

The main property of this function, which is a direct consequence of the *Poisson summation formula*, and is equivalent to the functional equation of Dirichlet L-functions, is as follows:

PROPOSITION 6.2.5. *We have the functional equation*

$$\Theta(\chi, -1/\tau) = W(\chi)(\tau/i)^{(2e+1)/2}\Theta(\chi^{-1}, \tau) ,$$

with the principal determination of the square root, and where $W(\chi) = \mathfrak{g}(\chi)/(i^e p^{1/2})$ *is the so-called* root number.

COROLLARY 6.2.6. *If* $\chi(-1) = 1$ *we have*

$$\mathfrak{g}(\chi) = p^{1/2} \frac{\sum_{m\geq 1} \chi(m)\exp(-\pi m^2/pt)}{t^{1/2}\sum_{m\geq 1} \overline{\chi(m)}\exp(-\pi m^2 t/p)}$$

and if $\chi(-1) = -1$ *we have*

$$\mathfrak{g}(\chi) = p^{1/2}i \frac{\sum_{m\geq 1} \chi(m)m\exp(-\pi m^2/pt)}{t^{3/2}\sum_{m\geq 1} \overline{\chi(m)}m\exp(-\pi n^2 t/p)}$$

for any t *such that the denominator does not vanish.*

Note that the optimal choice of t is $t = 1$, and (at least for p prime) it seems that the denominator never vanishes (there are counterexamples when p is not prime, but only four are known, see [**CZ13**]).

It follows from this corollary that $\mathfrak{g}(\chi)$ can be computed numerically as a complex number in $\widetilde{O}(p^{1/2})$ operations. Thus, if χ_1 and χ_2 are nontrivial characters such that $\chi_1\chi_2 \neq \varepsilon$ (otherwise $J(\chi_1, \chi_2)$ is trivial to compute), the formula $J(\chi_1, \chi_2) = \mathfrak{g}(\chi_1)\mathfrak{g}(\chi_2)/\mathfrak{g}(\chi_1\chi_2)$ allows the computation of J_2 *numerically* as a complex number in $\widetilde{O}(p^{1/2})$ operations.

To recover J itself as an algebraic number we could either compute all its conjugates, but this would require more time than the direct computation of J, or possibly use the LLL algorithm, which although fast, would also require some time. In point-counting applications, we only need J to sufficient accuracy: we perform all the elementary operations in $\mathbb{C}$, and since we know that at the end the result will be an integer for which we know an upper bound, we obtain a proven exact result.

More generally, we have generically $J_5(n, n, n, n, n) = \mathfrak{g}(\omega^n)^5/\mathfrak{g}(\omega^{5n})$, which can thus be computed in $\widetilde{O}(p^{1/2})$ operations. It follows that $S(p; z)$ can be computed in $\widetilde{O}(p^{3/2})$ operations, which is slower than the elementary method seen above. The main advantage is that we do not need much storage: more precisely, we want to compute $S(p; z)$ to sufficiently small accuracy that we can recognize it as an integer, so a priori up to an absolute error of 0.5. However, we have seen that $(p-1) \mid S(p; z)$: it is thus sufficient to have an absolute error less than $(p-1)/2$ thus at worse each of the $p-1$ terms in the sum to an absolute error less than $1/2$. Since generically $|J_5(n, n, n, n, n)| = p^2$, we need a relative error less than $1/(2p^2)$, so less than $1/(10p^2)$ on each Gauss sum. In practice this is overly pessimistic, but it does not matter: for $p \leq 10^9$, this bound means that 19 decimal digits suffice.

The main term in the theta function computation (with $t = 1$) is $\exp(-\pi m^2/p)$, so we need $\exp(-\pi m^2/p) \leq 1/(100p^2)$, say, in other words $\pi m^2/p \geq 4.7 + 2\log(p)$, so $m^2 \geq p(1.5 + 0.7\log(p))$.

This means that we will need the values of $\omega(m)$ only up to this limit, of the order of $O((p\log(p))^{1/2})$, considerably smaller than p. Thus, instead of computing a full discrete logarithm table, which takes some time but more importantly a lot of space, we compute only discrete logarithms up to that limit, using specific algorithms for doing so which exist in the literature, some of which being quite easy.

A straightforward implementation of this method gives timings for $k = 2, 3$, and 4 of 0.02, 0.93, and 39.5 seconds respectively, compatible with $\widetilde{O}(p^{3/2})$ time. This is faster than the completely naïve method, but slower than the method explained above. Its advantage is that it requires much less space. For p around 10^8, however, it becomes too slow to be practical. We will see that its usefulness is mainly in the context where it was invented, i.e., for L-functions of Dirichlet characters.

6.3. Using the Gross–Koblitz formula

6.3.1. Introduction. This section is more demanding than the preceding ones, but is very important since it gives by far the best method for computing Gauss (and Jacobi) sums over finite fields. We refer to [**Coh07b**] for complete details.

In the preceding sections, we have considered Gauss sums as belonging to a number of different rings: the ring $\mathbb{Z}[\zeta_{q-1}, \zeta_p]$ or the field $\mathbb{C}$ of complex numbers, and for Jacobi sums the ring $\mathbb{Z}[\zeta_{q-1}]$, but also the ring $\mathbb{Z}[X]/(X^{q-1} - 1)$, and again the field $\mathbb{C}$.

In number theory there exist other algebraically closed fields which are useful in many contexts, the fields $\mathbb{C}_\ell$ of ℓ-adic numbers, one for each prime number ℓ. These fields come with a topology and analysis which are rather special: one of the main things to remember is that a sequence of elements tends to 0 if and only the ℓ-adic valuation of the elements (the largest exponent of ℓ dividing them) tends to infinity. For instance 2^m tends to 0 in $\mathbb{C}_2$, but in no other $\mathbb{C}_\ell$, 15^m tends to 0 in $\mathbb{C}_3$ and in $\mathbb{C}_5$, and $m!$ tends to 0 in every $\mathbb{C}_\ell$ (in particular, the exponential series has only a *finite* radius of convergence in any $\mathbb{C}_\ell$).

The most important subrings of $\mathbb{C}_\ell$ are the ring $\mathbb{Z}_\ell$ of ℓ-adic integers, the elements of which can be written as $x = a_0 + a_1\ell + \cdots + a_k\ell^k + \cdots$ with $a_j \in [0, \ell-1]$, and its field of fractions $\mathbb{Q}_\ell$, which contains $\mathbb{Q}$.

In dealing with Gauss and Jacobi sums over $\mathbb{F}_q$ with $q = p^f$, the only $\mathbb{C}_\ell$ which is of use for us is the one with $\ell = p$ (in highbrow language, we are going to use implicitly *crystalline* p-adic methods, while for $\ell \neq p$ it would be *étale* ℓ-adic methods).

6.3.2. Preliminaries to the Gross–Koblitz formula. Apart from this relatively strange topology, many definitions and results valid on $\mathbb{C}$ have analogues in $\mathbb{C}_p$. The main object that we will need in our context is the analogue of the gamma function, naturally called the p-adic gamma function, in the present case due to Morita (there is another one, see [**Coh07b**]), and denoted Γ_p. Its definition is in fact quite simple:

DEFINITION 6.3.1. For $s \in \mathbb{Z}_p$ we define

$$\Gamma_p(s) = \lim_{m \to s} (-1)^m \prod_{\substack{0 \le k < m \\ p \nmid k}} k \, ,$$

where the limit is taken over any sequence of positive integers m tending to s for the p-adic topology.

It is of course necessary to show that this definition makes sense, but this is not difficult, and most of the important properties of $\Gamma_p(s)$, analogous to those of $\Gamma(s)$, can be deduced from it.

However we need a much deeper property of $\Gamma_p(s)$ known as the Gross–Koblitz formula: it is in fact an analogue of a formula for $\Gamma(s)$ known as the Chowla–Selberg formula, and it is also closely related to the Davenport–Hasse relations that we have seen above.

The proof of the Gross–Koblitz formula was initially given using tools in crystalline cohomology, but an elementary proof due to A. Robert now exists, see for instance [**Coh07b**] once again.

The Gross–Koblitz formula tells us that certain products of p-adic gamma functions at *rational* arguments are in fact *algebraic numbers*, more precisely *Gauss sums* (explaining their importance for us). This is quite surprising since usually transcendental functions such as Γ_p take transcendental values.

To give a specific example, we have $\Gamma_5(1/4)^2 = -2 + \sqrt{-1}$, where $\sqrt{-1}$ is the square root of -1 in $\mathbb{Z}_5$ congruent to 3 modulo 5. This can easily be checked *numerically* thanks to the p-adic capabilities of `Pari/GP` which include the computation of the p-adic gamma function:

```
? G = gamma(1/4 + O(5^100))
% = 1 + 4*5 + 3*5^4 + 5^6 + 5^7 + ... + O(5^100)
? P = algdep(G, 4)
% = x^4 + 4*x^2 + 5 /* It does seem algebraic. */
? subst(P, x, gamma(1/4 + O(5^1000)))
% = O(5^1000) /* Still true at much larger p-adic accuracy. */
? G^2 + 2
% = 3 + 3*5 + 2*5^2 + 3*5^3 + 5^4 + 2*5^6 + 5^7 + ...
? % == -sqrt(-1 + O(5^100))
% = 1 /* Same. */
```

Before stating the formula we need to collect a number of facts, both on classical algebraic number theory and on p-adic analysis. None are difficult to prove, see [**Coh07a**] and [**Coh07b**]. Recall that $q = p^f$.

• We let $K = \mathbb{Q}(\zeta_p)$ and $L = K(\zeta_{q-1}) = \mathbb{Q}(\zeta_{q-1}, \zeta_p) = \mathbb{Q}(\zeta_{p(q-1)})$, so that L/K is an extension of degree $\phi(q-1)$. There exists a unique prime ideal $\mathfrak{p}$ of K above p, and we have $\mathfrak{p} = (1 - \zeta_p)\mathbb{Z}_K$, $\mathfrak{p}^{p-1} = p\mathbb{Z}_K$, and $\mathbb{Z}_K/\mathfrak{p} \simeq \mathbb{F}_p$. The prime ideal $\mathfrak{p}$ splits into a product of $g = \phi(q-1)/f$ prime ideals $\mathfrak{P}_j$ of degree f in the extension L/K, i.e., $\mathfrak{p}\mathbb{Z}_L = \mathfrak{P}_1 \cdots \mathfrak{P}_g$, and for any prime ideal $\mathfrak{P} = \mathfrak{P}_j$ we have $\mathbb{Z}_L/\mathfrak{P} \simeq \mathbb{F}_q$.

Exercise: Prove directly that for any f we have $f \mid \phi(p^f - 1)$.

• Fix one of the prime ideals $\mathfrak{P}$ as above. There exists a unique group isomorphism $\omega = \omega_{\mathfrak{P}}$ from $(\mathbb{Z}_L/\mathfrak{P})^*$ to the group of $(q-1)$st roots of unity in L, such that for all $x \in (\mathbb{Z}_L/\mathfrak{P})^*$ we have $\omega(x) \equiv x \pmod{\mathfrak{P}}$. It is called the *Teichmüller character*, and it can be considered as a character of order $q-1$ on $\mathbb{F}_q^* \simeq (\mathbb{Z}_L/\mathfrak{P})^*$. We can thus *instantiate* the definition of a Gauss sum over $\mathbb{F}_q$ by defining it as $\mathfrak{g}(\omega_{\mathfrak{P}}^{-r}) \in L$.

• Let ζ_p be a primitive pth root of unity in $\mathbb{C}_p$, fixed once and for all. There exists a unique $\pi \in \mathbb{Z}_p[\zeta_p]$ satisfying $\pi^{p-1} = -p$, $\pi \equiv 1 - \zeta_p \pmod{\pi^2}$, and we set $K_{\mathfrak{p}} = \mathbb{Q}_p(\pi) = \mathbb{Q}_p(\zeta_p)$, and $L_{\mathfrak{P}}$ the *completion* of L at $\mathfrak{P}$. The field extension

$L_{\mathfrak{P}}/K_{\mathfrak{p}}$ is Galois, with Galois group isomorphic to $\mathbb{Z}/f\mathbb{Z}$ (which is the same as the Galois group of $\mathbb{F}_q/\mathbb{F}_p$, where $\mathbb{F}_p$ (resp., $\mathbb{F}_q$) is the so-called *residue field* of K (resp., L)).

• We set the following:

DEFINITION 6.3.2. *We define the p-adic Gauss sum by*

$$\mathfrak{g}_q(r) = \sum_{x \in L_{\mathfrak{P}}, \; x^{q-1}=1} x^{-r} \zeta_p^{\mathrm{Tr}_{L_{\mathfrak{P}}/K_{\mathfrak{p}}}(x)} \in L_{\mathfrak{P}} \;.$$

Note that this depends on the choice of ζ_p, or equivalently of π. Since $\mathfrak{g}_q(r)$ and $\mathfrak{g}(\omega_{\mathfrak{P}}^{-r})$ are algebraic numbers, it is clear that they are equal, although viewed in fields having different topologies. Thus, results about $\mathfrak{g}_q(r)$ translate immediately into results about $\mathfrak{g}(\omega_{\mathfrak{P}}^{-r})$, hence about general Gauss sums over finite fields.

6.3.3. The Gross–Koblitz formula: Statement and applications. The Gross–Koblitz formula is as follows:

THEOREM 6.3.3 (Gross–Koblitz). *Denote by $s(r)$ the sum of digits to base p of the integer r mod $(q-1)$, i.e., of the unique integer r' such that $r' \equiv r \pmod{q-1}$ and $0 \le r' < q - 1$. We have*

$$\mathfrak{g}_q(r) = -\pi^{s(r)} \prod_{0 \le i < f} \Gamma_p\left(\left\{ \frac{p^{f-i}r}{q-1} \right\} \right) \;,$$

where $\{x\}$ denotes the fractional part of x.

Let us show how this can be used to compute Gauss or Jacobi sums, and in particular our sum $S(q; z)$. First, let us *prove* the identity given above for $\Gamma_5(1/4)$, which we only checked numerically. We choose of course $f = 1$, $p = q = 5$, and $r = 1$. The theorem tells us that $\Gamma_5(1/4) = -\pi^{-1}\mathfrak{g}_5(1)$. Let i be the unique square root[1] of -1 in $\mathbb{Z}_5$ congruent to 3 modulo 5 . Since $f = 1$ we have $L_{\mathfrak{P}} = K_{\mathfrak{p}}$, so $\mathfrak{g}_5(1) = \zeta_5 - i\zeta_5^3 - \zeta_5^4 + i\zeta_5^2$ (the exponents, which are powers of i, are taken modulo 5). We compute that $\mathfrak{g}_5(1)^2 = (2-i)u$ with $u^2 = -5$, hence $u = \pm\pi^2$, so it follows that $\Gamma_5(1/4)^2 = \pm(2-i)$, and computing both sides modulo 5 shows that the correct sign is $-$, hence $\Gamma_5(1/4)^2 = -2 + i$, proving our identity.

Let us now compute our sum $S(q; z)$, and assume for simplicity that $f = 1$, in other words that $q = p$: the right hand side is thus equal to $-\pi^{s(r)}\Gamma_p(\{pr/(p-1)\})$. Since we can always choose r such that $0 \le r < p - 1$, we have $s(r) = r$ and $\{pr/(p-1)\} = \{r + r/(p-1)\} = r/(p-1)$, so the right-hand side is equal to $-\pi^r\Gamma_p(r/(p-1))$. Now an easy property of Γ_p is that it is differentiable: recall that p is "small" in the p-adic topology, so $r/(p-1)$ is close to $-r$, more precisely $r/(p-1) = -r + pr/(p-1)$ (this is how we obtained it in the first place!). Thus in particular, if $p > 2$ we have the Taylor expansion

$$\Gamma_p(r/(p-1)) = \Gamma_p(-r) + (pr/(p-1))\Gamma_p'(-r) + O(p^2)$$
$$= \Gamma_p(-r) - pr\Gamma_p'(-r) + O(p^2) \;.$$

---

[1]This is the opposite of the one given by GP, which is specified as the one whose first p-adic digit lies in $[0, p/2[$.

Since $0 \le r < p - 1$, it is easy to show from the definition that

$$\Gamma_p(-r) = 1/r! \quad \text{and} \quad \Gamma'_p(-r) = (-\gamma_p + H_r)/r! \,,$$

where $H_r = \sum_{1 \le n \le r} 1/n$ is the harmonic sum, and $\gamma_p = -\Gamma'_p(0)$ is the p-adic analogue of Euler's constant.

Exercise: Prove these formulas, as well as the congruence for γ_p given below.

There exist infinite (p-adic) series enabling accurate computation of γ_p, but since we only need it modulo p, we use the easily proved congruence

$$\gamma_p \equiv \big((p-1)! + 1\big)/p = W_p \pmod{p} \,,$$

the so-called *Wilson quotient*.

Thus the Gross–Koblitz formula tells us that for $0 \le r < p - 1$ we have

$$\mathfrak{g}_q(r) = -\frac{\pi^r}{r!}\big(1 - pr(H_r - W_p) + O(p^2)\big) \,.$$

For notational simplicity, set $\mathcal{J}(-r) = J(-r, -r, -r, -r, -r)$. Thus, for $(p-1) \nmid 5r$ we have

$$\mathcal{J}(-r) = \frac{\mathfrak{g}(\omega_\mathfrak{P})^5}{\mathfrak{g}(\omega_\mathfrak{P}^5)} = \frac{\mathfrak{g}_q(r)^5}{\mathfrak{g}_q(5r)} = \pi^{f(r)}(a + bp + O(p^2)) \,,$$

where a and b will be computed below and

$$f(r) = 5r - (5r \bmod p - 1) = 5r - (5r - (p-1)\lfloor 5r/(p-1)\rfloor)$$
$$= (p-1)\lfloor 5r/(p-1)\rfloor \,,$$

so that $\pi^{f(r)} = (-p)^{\lfloor 5r/(p-1)\rfloor}$ since $\pi^{p-1} = -p$. Since we want the result modulo p^2, we consider three intervals together with special cases:

(1) If $r > 2(p-1)/5$ but $(p-1) \nmid 5r$, we have

$$\mathcal{J}(-r) \equiv 0 \pmod{p^2} \,.$$

(2) If $(p-1)/5 < r < 2(p-1)/5$ we have

$$\mathcal{J}(-r) \equiv (-p)\frac{(5r - (p-1))!}{r!^5} \pmod{p^2} \,.$$

(3) If $0 < r < (p-1)/5$ we have $f(r) = 0$ and $0 < 5r < (p-1)$ hence

$$\mathcal{J}(-r) = \frac{(5r)!}{r!^5}\big(1 - 5pr(H_r - W_p)\big)\big(1 + 5pr(H_{5r} - W_p)\big) \pmod{p^2}$$
$$\equiv \frac{(5r)!}{r!^5}\big(1 + 5pr(H_{5r} - H_r)\big) \pmod{p^2} \,.$$

(4) Finally, if $r = a(p-1)/5$ we have $\mathcal{J}(-r) = p^4 \equiv 0 \pmod{p^2}$ if $a = 0$, and otherwise $\mathcal{J}(-r) = -\mathfrak{g}_q(r)^5/p$. Recall that the p-adic valuation of $\mathfrak{g}_q(r)$ is equal to $r/(p-1) = a/5$, hence that of $\mathcal{J}(-r)$ is equal to $a - 1$, which is greater or equal to 2 as soon as $a \ge 3$. For $a = 2$, i.e., $r = 2(p-1)/5$, we thus have

$$\mathcal{J}(-r) \equiv p\frac{1}{r!^5} \equiv (-p)\frac{(5r - (p-1))!}{r!^5} \pmod{p^2} \,,$$

which is the same formula as for $(p-1)/5 < r \le 2(p-1)/5$. For $a = 1$, i.e., $r = (p-1)/5$, we thus have

$$\mathcal{J}(-r) \equiv -\frac{1}{r!^5}\big(1 - 5pr(H_r - W_p)\big) \pmod{p^2} \,,$$

while on the other hand

$$(5r)! = (p-1)! = -1 + pW_p \equiv -1 - p(p-1)W_p \equiv -1 - 5prW_p ,$$

and $H_{5r} = H_{p-1} \equiv 0 \pmod{p}$ (Wolstenholme's congruence), so

$$\frac{(5r)!}{r!^5}\left(1 + 5pr(H_{5r} - H_r)\right) \equiv -\frac{1}{r!^5}(1 - 5prH_r)(1 + 5prW_p)$$

$$\equiv -\frac{1}{r!^5}\left(1 - 5pr(H_r - W_p)\right) \pmod{p^2} ,$$

which is the same formula as for $0 < r < (p-1)/5$.

An important point to note is that we are working p-adically, but the final result $S(p; z)$ being an integer, it does not matter at the end. There is one small additional detail to take care of: we have

$$S(p; z) = \sum_{0 \le r \le p-2} \omega^{-r}(z)J(r,r,r,r,r) = \sum_{0 \le r \le p-2} \omega^r(z)\mathcal{J}(-r) ,$$

so we must express $\omega^r(z)$ in the p-adic setting. Since $\omega = \omega_{\mathfrak{P}}$ is the *Teichmüller character*, it is easy to show that $\omega(z)$ is the p-adic limit of z^{p^k} as $k \to \infty$. in particular $\omega(z) \equiv z \pmod{p}$, but more precisely $\omega(z) \equiv z^p \pmod{p^2}$.

Exercise: Let $p \ge 3$. Assume that $z \in \mathbb{Z}_p \setminus p\mathbb{Z}_p$ (for instance that $z \in \mathbb{Z} \setminus p\mathbb{Z}$). Prove that z^{p^k} has a p-adic limit $\omega(z)$ when $k \to \infty$, that $\omega^{p-1}(z) = 1$, that $\omega(z) \equiv z \pmod{p}$, and $\omega(z) \equiv z^p \pmod{p^2}$.

We have thus proved the following

PROPOSITION 6.3.4. *We have*

$$S(p; z) \equiv \sum_{0 < r \le (p-1)/5} \frac{(5r)!}{r!^5}\left(1 + 5pr(H_{5r} - H_r)\right)z^{pr}$$

$$- p \sum_{(p-1)/5 < r \le 2(p-1)/5} \frac{\left(5r - (p-1)\right)!}{r!^5} z^r \pmod{p^2} .$$

In particular

$$S(p; z) \equiv \sum_{0 < r \le (p-1)/5} \frac{(5r)!}{r!^5} z^r \pmod{p} .$$

Remarks.
(1) Note that, as must be the case, all mention of p-adic numbers has disappeared from this formula. We used the p-adic setting only in the proof. It can be proved "directly", but with some difficulty.
(2) We used the Taylor expansion only to order 2. It is of course possible to use it to any order, thus giving a generalization of the above proposition modulo any power of p.

The point of giving all these details is as follows: it is easy to show that $(p-1) \mid S(p; z)$ (in fact we have seen this in the elementary method above). We can thus easily compute $S(p; z)$ modulo $p^2(p-1)$. On the other hand, it is possible to prove (but not easy, it is part of the Weil conjectures proved by Deligne), that $|S(p; z) - p^4| < 4p^{5/2}$. It follows that as soon as $8p^{5/2} < p^2(p-1)$, in other words $p \ge 67$, the computation that we perform modulo p^2 is sufficient to determine

$S(p; z)$ exactly. It is clear that the time to perform this computation is $\widetilde{O}(p)$, and in fact much faster than any that we have seen.

The corresponding GP program is quite simple:

$\boxed{\texttt{GrossKoblitz.gp}}$

```
1   /* Compute S(p;z) using Gross-Koblitz, p >= 67 */
2   Srp(p, z) =
3   { my(p2 = p^2, p5 = (p-1) \ 5, zp, Z, H, S1, S2, r0);
4     my(x = 'x, N = 1250*x^3 - 1750*x^2 + 750*x - 96);
5
6     zp = Mod(z, p2)^p; Z = 1; S1 = 0; H = 0;
7     for (r = 1, p5,
8       my(R = 5*r, D = (R * (R-1) * (R-2) * (R-3) * (R-4)) % p2);
9       Z *= zp / r^5 * D;
10      H = (H + subst(N, x, r) / D) % p;
11      S1 += Z * (1 + p * R * H));
12    zp = Mod(z, p); Z = Mod(Z, p); S2 = 0; r0 = p5 + 1;
13    my(b = 5 * r0 - (p-1)); /* 2 <= b <= 5 */
14    Z *= b! * (5 - b)! * (-1)^b;
15    for (r = r0, (2*(p-1)) \ 5,
16      Z *= zp / r^5;
17      if (r != r0, my(R = 5 * r);
18                   Z *= (R+1) * R * (R-1) * (R-2) * (R-3));
19      S2 += Z);
20    my(a = lift(S1 - p * lift(S2))); /* S = a (mod p^2) */
21    return (p2^2 + (a - (1 + a) * p2) % ((p - 1) * p2));
22  }
```

REMARKS 6.3.5.

(1) The program assumes $p \geq 67$ and directly implements the proposition. It uses a Horner scheme to compute the hypergeometric part of the two sums (the quotient of factorials and power of z). We obtain the difference $h_r = H_{5r} - H_r$ of harmonic numbers from the recurrence

$$h_r - h_{r-1} = \frac{1}{5r} + \cdots + \frac{1}{5r-4} - \frac{1}{r}$$

$$= (1250r^3 - 1750r^2 + 750r - 96) \Big/ \prod_{i=0}^{4} (5r - i).$$

(2) At the end of the first loop computing S_1, we have $Z = (5r)! z^{pr}/r!^5$ for $r = \lfloor (p-1)/5 \rfloor = r_0 - 1$. The variable $b = 5r_0 - (p-1)$ in the program is a tiny non-negative integer. Before the second loop, we multiply Z by a suitable factor to reach $(5r_0 - (p-1))! z^{r_0}/r_0!^5$, which is slightly more efficient than recomputing z^{r_0}. This uses $z^p \equiv z \pmod{p}$, Wilson's

theorem $(p-1)! \equiv -1 \pmod{p}$ and

$$(p-1)!/(5r!) = (p-1)\ldots(5r+1) = (p-1)\ldots(5r_0-4)$$
$$\equiv (-1)\ldots(5r_0-5-(p-1)) \pmod{p}$$
$$\equiv (-1)^{5-b}(5-b)! \pmod{p}$$

This program gives timings for $p \approx 10^k$ for $k = 2$, 3, 4, 5, 6, 7, and 8 of 0, 0, 0.007, 0.07, 0.76, 8.06, and 81.9 seconds respectively, of course much faster and compatible with $\widetilde{O}(p)$ time. The great additional advantage is that we use very little memory, so this is the best method known to us (note, however, that if we need to compute $S(p; z)$ for *millions* of primes p and the same z, there are much faster methods, see for instance [**CKR20**]).

Numerical example: Choose $p = 10^8 + 7$ and $z = 2$. In 80 seconds we find that $S(p; z) \equiv a \pmod{p^2}$ with $a = 2740416964447728$. Using the Chinese remainder formula

$$S(p; z) = p^4 + ((a - (1+a)p^2) \bmod ((p-1)p^2)),$$

we immediately deduce that

$$S(p; z) = 100000028999775932708758063337094.$$

Here is a summary of the timings that we have mentioned, where as usual ∞ means more than 5 minutes:

k	2	3	4	5	6	7	8
Naïve	0.01	1.30	157.	∞	∞	∞	∞
Theta	0.02	0.93	39.5	∞	∞	∞	∞
Mod $X^{q-1}-1$	0.00	0.01	0.07	1.00	14.4	202.	∞
Gross–Koblitz	0.00	0.00	0.007	0.07	0.76	8.06	81.9

Time for computing $S(p; z)$ for $p \approx 10^k$, in seconds.

6.4. Gauss and Jacobi sums over $\mathbb{Z}/N\mathbb{Z}$

6.4.1. Definitions. Another context in which one encounters Gauss sums is over finite rings such as $\mathbb{Z}/N\mathbb{Z}$. The theory coincides with that over $\mathbb{F}_q$ when $q = p = N$ is prime, but is rather different otherwise. These other Gauss sums enter in the important theory of *Dirichlet characters*:

DEFINITION 6.4.1. Let χ be a (multiplicative) character from the multiplicative group $(\mathbb{Z}/N\mathbb{Z})^*$ of invertible elements of $\mathbb{Z}/N\mathbb{Z}$ to the complex numbers $\mathbb{C}$. We denote by abuse of notation again by χ the map from $\mathbb{Z}$ to $\mathbb{C}$ defined by $\chi(x) = \chi(x \bmod N)$ when x is coprime to N, and $\chi(x) = 0$ if x is not coprime to N, and call it the Dirichlet character modulo N associated to χ.

It is clear that a Dirichlet character satisfies $\chi(xy) = \chi(x)\chi(y)$ for all x and y, that $\chi(x+N) = \chi(x)$, and that $\chi(x) = 0$ if and only if x is not coprime with N. Conversely, it immediate that these properties characterize Dirichlet characters.

A crucial notion (which has no equivalent in the context of characters of $\mathbb{F}_q^*$) is that of *primitivity*:

Assume that $M \mid N$. If χ is a Dirichlet character modulo M, we can transform it into a character χ_N modulo N by setting $\chi_N(x) = \chi(x)$ if x is coprime to N, and $\chi_N(x) = 0$ otherwise. We say that the characters χ and χ_N are *equivalent*. Conversely, if ψ is a character modulo N, it is not always true that one can find χ modulo M such that $\psi = \chi_N$. If it is possible, we say that *ψ can be defined modulo M*.

DEFINITION 6.4.2. Let χ be a character modulo N. We say that χ is a *primitive character* if χ cannot be defined modulo M for any proper divisor M of N, i.e., for any $M \mid N$ such that $M \neq N$.

Exercise: Assume that $N \equiv 2 \pmod 4$. Show that there do not exist any primitive characters modulo N.

Exercise: Let $p > 2$ be a prime number and assume that $p^a \mid N$. Show that if χ is a primitive character modulo N, the *order* of χ (the smallest k such that χ^k is a trivial character) is *divisible* by p^{a-1}.

As we will see, questions about general Dirichlet characters can always be reduced to questions about primitive characters, and the latter have much nicer properties.

PROPOSITION 6.4.3. *Let χ be a character modulo N. There exists a divisor f of N called the* conductor *of χ (this f has nothing to do with the f used above such that $q = p^f$), having the following properties:*

(1) *The character χ can be defined modulo f, in other words there exists a character ψ modulo f such that $\chi = \psi_N$ using the notation above.*
(2) *f is the smallest divisor of N having this property.*
(3) *The character ψ is a primitive character modulo f.*

There is also the notion of *trivial character modulo N*: however we must be careful here, and we set the following:

DEFINITION 6.4.4. The trivial character modulo N is the Dirichlet character associated with the trivial character of $(\mathbb{Z}/N\mathbb{Z})^*$. It is usually denoted by χ_0 (but be careful, the index N is implicit, so χ_0 may represent different characters), and its values are as follows: $\chi_0(x) = 1$ if x is coprime to N, and $\chi_0(x) = 0$ if x is not coprime to N.

In particular, $\chi_0(0) = 0$ if $N \neq 1$. The character χ_0 can also be characterized as the only character modulo N of conductor 1.

DEFINITION 6.4.5. Let χ be a character modulo N. The *Gauss sum* associated to χ and $a \in \mathbb{Z}$ is

$$\mathfrak{g}(\chi, a) = \sum_{x \bmod N} \chi(x)\zeta_N^{ax} \,,$$

and we write simply $\mathfrak{g}(\chi)$ instead of $\mathfrak{g}(\chi, 1)$.

The most important results concerning these Gauss sums is the following:

PROPOSITION 6.4.6. *Let χ be a character modulo N.*

(1) *If a is coprime to N we have*
$$\mathfrak{g}(\chi, a) = \chi^{-1}(a)\mathfrak{g}(\chi) = \overline{\chi(a)}\mathfrak{g}(\chi) \,,$$
and more generally $\mathfrak{g}(\chi, ab) = \chi^{-1}(a)\mathfrak{g}(\chi, b) = \overline{\chi(a)}\mathfrak{g}(\chi, b)$.

(2) *If χ is a primitive character, we have*
$$\mathfrak{g}(\chi, a) = \overline{\chi(a)}\mathfrak{g}(\chi)$$
for all a, in other words, in addition to (1), we have $\mathfrak{g}(\chi, a) = 0$ if a is not coprime to N.

(3) *If χ is a primitive character, we have $|\mathfrak{g}(\chi)|^2 = N$.*

Note that (1) is trivial, and that since $\chi(a)$ has modulus 1 when a is coprime to N, we can write indifferently $\chi^{-1}(a)$ or $\overline{\chi(a)}$. On the other hand, (2) is not completely trivial.

We leave to the reader the easy task of defining Jacobi sums and of proving the easy relations between Gauss and Jacobi sums.

6.4.2. Reduction to prime Gauss sums: Odoni's theorem. A fundamental and little-known fact is that in the context of Gauss sums over $\mathbb{Z}/N\mathbb{Z}$ (as opposed to $\mathbb{F}_q$), one can in fact always reduce to prime N. First note (with proof) the following easy result:

PROPOSITION 6.4.7. *Let $N = N_1 N_2$ with N_1 and N_2 coprime, and let χ be a character modulo N.*

(1) *There exist unique characters χ_i modulo N_i such that $\chi = \chi_1 \chi_2$ in an evident sense, and if χ is primitive, the χ_i will also be primitive.*

(2) *We have the identity (valid even if χ is not primitive):*
$$\mathfrak{g}(\chi) = \chi_1(N_2)\chi_2(N_1)\mathfrak{g}(\chi_1)\mathfrak{g}(\chi_2) \,.$$

PROOF. (1). Since N_1 and N_2 are coprime there exist u_1 and u_2 such that $u_1 N_1 + u_2 N_2 = 1$. We define $\chi_1(x) = \chi(xu_2 N_2 + u_1 N_1)$ and $\chi_2(x) = \chi(xu_1 N_1 + u_2 N_2)$. We leave to the reader to check (1) using these definitions.

(2). When x_i ranges modulo N_i, $x = x_1 u_2 N_2 + x_2 u_1 N_1$ ranges modulo N (check it, in particular that the values are distinct!), and $\chi(x) = \chi_1(x)\chi_2(x) = \chi_1(x_1)\chi_2(x_2)$. Furthermore,
$$\zeta_N = \exp(2\pi i/N) = \exp(2\pi i(u_1/N_2 + u_2/N_1)) = \zeta_{N_1}^{u_2}\zeta_{N_2}^{u_1} \,,$$
hence
$$\begin{aligned}
\mathfrak{g}(\chi) &= \sum_{x \bmod N} \chi(x)\zeta_N^x \\
&= \sum_{\substack{x_1 \bmod N_1 \\ x_2 \bmod N_2}} \chi_1(x_1)\chi_2(x_2)\zeta_{N_1}^{u_2 x_1}\zeta_{N_2}^{u_1 x_2} \\
&= \mathfrak{g}(\chi_1; u_2)\mathfrak{g}(\chi_2; u_1) = \chi_1^{-1}(u_2)\chi_2^{-1}(u_1)\mathfrak{g}(\chi_1)\mathfrak{g}(\chi_2) \,,
\end{aligned}$$
so the result follows since $N_2 u_2 \equiv 1 \pmod{N_1}$ and $N_1 u_1 \equiv 1 \pmod{N_2}$. □

Thanks to the above result, the computation of Gauss sums modulo N is reduced to the computation of Gauss sums modulo prime powers.

Here a remarkable simplification occurs, due to Odoni: Gauss sums modulo p^a for $a \geq 2$ can be "explicitly computed", in the sense that there is a direct formula not involving a sum over p^a terms for computing them. Although the proof is not difficult, we do not give it, and refer instead to [**Coh**]. We use the classical notation $\boldsymbol{e}(x)$ to mean $e^{2\pi i x}$.

THEOREM 6.4.8 (Odoni et al.). *Let χ be a* primitive *character modulo p^n.*

(1) *Assume that $p \geq 3$ is prime and $n \geq 2$. Write $\chi(1+p) = \boldsymbol{e}(-b/p^{n-1})$ with $p \nmid b$. Define*

$$A(p) = \frac{p}{\log_p(1+p)} \quad and \quad B(p) = A(p)\big(1 - \log_p(A(p))\big) \,,$$

except when $p^n = 3^3$, in which case we define $B(p) = 10$. Then

$$\mathfrak{g}(\chi) = p^{n/2}\boldsymbol{e}(bB(p)/p^n)\chi(b) \cdot \begin{cases} 1 & \text{if } n \geq 2 \text{ is even,} \\ \left(\dfrac{b}{p}\right) i^{p(p-1)/2} & \text{if } n \geq 3 \text{ is odd.} \end{cases}$$

(2) *Let $p = 2$ and assume that $n \geq 4$. Write $\chi(1+p^2) = \boldsymbol{e}(b/p^{n-2})$ with $p \nmid b$. Define*

$$A(p) = -\frac{p^2}{\log_p(1+p^2)} \quad and \quad B(p) = A(p)\big(1 - \log_p(A(p))\big) \,,$$

except when $p^n = 2^4$, in which case we define $B(p) = 13$. Then

$$\mathfrak{g}(\chi) = p^{n/2}\boldsymbol{e}(bB(p)/p^n)\chi(b) \cdot \begin{cases} \boldsymbol{e}(b/8) & \text{if } n \geq 4 \text{ is even,} \\ \boldsymbol{e}(((b^2-1)/2 + b)/8) & \text{if } n \geq 5 \text{ is odd.} \end{cases}$$

(3) *If $p^n = 2^2$, or $p^n = 2^3$ and $\chi(-1) = 1$, we have $\mathfrak{g}(\chi) = p^{n/2}$, and if $p^n = 2^3$ and $\chi(-1) = -1$ we have $\mathfrak{g}(\chi) = p^{n/2}i$.*

Thanks to this theorem, we see that the computation of Gauss sums in the context of Dirichlet characters can be reduced to the computation of Gauss sums modulo p for prime p. This is of course the same as the computation of a Gauss sum for a character of $\mathbb{F}_p^*$.

We recall the available methods for computing a single Gauss sum of this type:

(1) The naïve method, time $\widetilde{O}(p)$ (applicable in general, time $\widetilde{O}(N)$).
(2) Using the Gross–Koblitz formula, also time $\widetilde{O}(p)$, but the implicit constant is much smaller, and also computations can be done modulo p or p^2 for instance, if desired (applicable only to $N = p$, or in the context of finite fields).
(3) Using theta functions, time $\widetilde{O}(p^{1/2})$ (applicable in general, time $\widetilde{O}(N^{1/2})$).

6.4.3. Other complete exponential sums. A complete exponential sum is an expression of the type

$$S(f; N) = \sum_{x \bmod N} \chi(x)e^{2\pi i f(x)/N} \,,$$

where typically χ is a Dirichlet character modulo a divisor of N, and f is an integer-valued function such that $f(x + N) \equiv f(x) \pmod{N}$, typically a polynomial with integer coefficients.

The reductions that we have made for Gauss sums are also valid for this type of more general exponential sums: by using the Chinese remainder theorem we can reduce to N being a prime power, and a generalization of Odoni's theorem is also valid, in other words $S(f; p^n)$ can be computed "explicitly" when $n \geq 2$, using what we call the *p-adic stationary phase theorem*. Since it is not so well known, we state here without proof a special case (see [**Coh**]).

Let p be a prime, and let F be a function from $\mathbb{Z}_p$ to $\mathbb{Q}_p$ having radius of convergence at least 1 (typically but not necessarily a polynomial). We want to compute

$$S = S(F; p^n) = \sum_{z \bmod p^n} e(F(z)/p^n)$$

for $n \geq 2$ (whenever this makes sense, i.e., when $F(z + p^n) \equiv F(z) \pmod{p^n}$ for all z). We set $G_k(z) = F^{(k)}(z)/(k-1)!$, and recall that we denote by v_p the *p*-adic valuation. We set $q_p = p$ if $p \geq 3$ and $q_2 = 4$.

THEOREM 6.4.9. *Assume that for all $z \in \mathbb{Z}_p$, for $k \geq 1$ we have $v_p(G_k(z)) \geq 0$ (which trivially implies that S makes sense), that for $k \geq 3$ we have $v_p(G_k(z)) \geq 1$, and that for all α such that $F'(\alpha) \equiv 0 \pmod{q_p}$, the so-called* stationary points, *we have $v_p(F''(\alpha)) = 0$. Then either there do not exist such stationary points, in which case $S = 0$, or such α exist, in which case:*

(1) *There exists a unique $u \in \mathbb{Z}_p$ such that $F'(u) = 0$ and $v_p(F''(u)) = 0$.*
(2) *For $n \geq 2$, $p^n \neq 2^4$, and $p^n \neq 2^2$, we have*

$$S = p^{n/2} e(F(u)/p^n) g(u, p, n) ,$$

where

$$g(u, p, n) = \begin{cases} 1 & \text{if } p > 2 \text{ and } n \text{ is even} , \\ \left(\dfrac{F''(u)}{p}\right) i^{p(p-1)/2} & \text{if } p > 2 \text{ and } n \text{ is odd} , \\ e(F''(u)/8) & \text{if } p = 2 \text{ and } n \text{ is even} , \\ \left(\dfrac{F''(u)}{2}\right) e(F''(u)/8) & \text{if } p = 2 \text{ and } n \text{ is odd} . \end{cases}$$

It is easy to see that Odoni's theorem is a consequence of the above. Another example is that of *Kloosterman sums*

$$K(a, b; N) = \sum_{\substack{x \bmod N \\ \gcd(x, N) = 1}} e^{(ax + bx^{-1})/N} ,$$

where x^{-1} denotes an inverse of x modulo N. In this case, the reduction to N prime is done thanks to the following results left as easy exercises for the reader:
• If $\gcd(N_1, N_2) = 1$ and $u_1 N_1 + u_2 N_2 = 1$ then

$$K(a, b; N_1 N_2) = K(u_2^2 a, b; N_1) K(u_1^2 a, b; N_2) .$$

This reduces to the computation of $K(a, b; p^n)$.
• If $\gcd(a, b, p^n) = p^m$, we have

$$K(a, b; p^n) = p^b K(ab/p^{2m}, 1; p^{n-m}) .$$

This reduces to the computation of $K(a, 1; p^n)$.

LEMMA 6.4.10.

(1) *If $p \geq 3$ and $n \geq 2$ (resp., $p = 2$ and $n \geq 6$) we have $K(a, 1; p^n) = 0$ if $\left(\frac{a}{p}\right) \neq 1$ (resp., if $a \not\equiv 1$ (mod 8)), and otherwise there exists b such that $b^2 \equiv a$ (mod p^n), and $K(a, 1; p^n) = K(b, b; p^n)$.*

(2) *If $p = 2$ and $n \leq 5$ then $K(a, 1; p^n) = 0$ unless either $n = 4$ and $a \equiv 1$ (mod 4), in which case $K(a, 1; p^4) = (-1)^{(a-1)(a+3)/32} 2^{5/2}$, or $n = 3$ and $a \equiv 3$ (mod 4), in which case $K(a, 1; p^3) = (-1)^{(a^2-1)/8} 4$, or $n = 2$ and a odd, in which case $K(a, 1; p^2) = (-1)^{(a+1)/4} 2$, or $n = 1$ in which case $K(a, 1; p) = (-1)^{a+1}$.*

This reduces to the computation of $K(b, b; p^n)$ for $p \nmid b$ if $p \geq 3$ and $n \geq 2$ or $p = 2$ and $n \geq 6$, together with the computation of $K(a, 1; p)$ for $p \geq 3$ and $p \nmid a$. The value of $K(b, b; p^n)$ for $p = 3$ and $n \geq 2$ or $p = 2$ and $n \geq 6$ is given by the p-adic stationary phase theorem given above, which in this case is due to Salié.

THEOREM 6.4.11 (Salié). *Assume that $p \nmid b$.*

(1) *If p is odd and $n \geq 2$, we have*

$$K(b, b; p^n) = \begin{cases} 2p^{n/2} \cos(4\pi b/p^n) & \text{if } n \text{ is even,} \\ 2p^{n/2} \left(\dfrac{b}{p}\right) \cos(4\pi b/p^n) & \text{if } n \text{ is odd and } p \equiv 1 \pmod 4, \\ -2p^{n/2} \left(\dfrac{b}{p}\right) \sin(4\pi b/p^n) & \text{if } n \text{ is odd and } p \equiv 3 \pmod 4. \end{cases}$$

(2) *If $p = 2$ and $n \geq 6$ we have*

$$K(b, b; 2^n) = \begin{cases} 2^{n/2+3/2} \left(\dfrac{b}{2}\right) \cos(b(\pi/4 + \pi/2^{n-2})) & \text{if } n \text{ is even,} \\ 2^{n/2+3/2} \cos(b(\pi/4 + \pi/2^{n-2})) & \text{if } n \geq 9 \text{ is odd,} \\ -2^{n/2+3/2} \cos(b(\pi/4 + \pi/2^{n-2})) & \text{if } n = 7. \end{cases}$$

As for Gauss sums, and all complete exponential sums in general, we are now reduced to the computation of

$$K(a, 1; p) = \sum_{1 \leq x \leq p-1} e^{2\pi i (ax + x^{-1})/p},$$

for $p \geq 3$ prime and $p \nmid a$. Note that $ax + x^{-1} = y$ is equivalent to $ax^2 - xy + 1 = 0$, and since $p \nmid a$, for given y there are $1 + \left(\frac{y^2-4a}{p}\right)$ values of x, and we deduce that

$$K(a, 1; p) = \sum_{x \bmod p} \left(\frac{x^2 - 4a}{p}\right) e^{2\pi i x/p}.$$

This looks superficially similar to a Gauss sum, but is much deeper. Recall the following: if χ is a nontrivial character modulo p, it is easy to show that $\mathfrak{g}(\chi)$ has modulus $p^{1/2}$, and using for instance theta functions we can compute it in time $O(p^{1/2+\varepsilon})$ for all $\varepsilon > 0$.

For Kloosterman sums, it is easy to show that $K(a, 1; p) = O(p^{3/4})$ (see for instance Section 9.3 of [CS17]), and it is quite a difficult result of Weil that in fact $|K(a, 1; p)| < 2p^{1/2}$. However, to the authors' knowledge there are no algorithms for computing numerically an individual value of $K(a, 1; p)$ in time $O(p^{1-\delta})$ for some $\delta > 0$.

Exercise: Show that

$$K(1, 1; p) = \frac{1}{p-1} \sum_{\chi} \mathfrak{g}(\chi)^2 \,,$$

where the sum is over all characters modulo p. Unfortunately, this result does not seem to help in computing $K(1, 1, p)$ efficiently.

CHAPTER 7

Numerical computation of continued fractions

7.1. Generalities

Recall that a continued fraction is an expression of the form

$$S = a(0) + \cfrac{b(0)}{a(1) + \cfrac{b(1)}{a(2) + \cfrac{b(2)}{a(3) + \ddots}}} \; .$$

In the present section, we are going to study their numerical evaluation, mostly from a nonrigorous point of view, but which in practice works very well. We assume the reader familiar with the elementary theory of continued fractions. In particular, we use the following notation:

$$S(n) = a(0) + \cfrac{b(0)}{a(1) + \cfrac{b(1)}{a(2) + \ddots + \cfrac{b(n-1)}{a(n)}}} = \frac{p(n)}{q(n)} \; ,$$

so that

$$p(-1) = 1 \; , \;\; p(0) = a(0) \; , \;\; p(n+1) = a(n+1)p(n) + b(n)p(n-1) \text{ for } n \geq 0$$
$$q(-1) = 0 \; , \;\; q(0) = 1 \; , \;\; q(n+1) = a(n+1)q(n) + b(n)q(n-1) \text{ for } n \geq 0 \; .$$

From the matrix identity $\left(\begin{smallmatrix} p(n+1) & p(n) \\ q(n+1) & q(n) \end{smallmatrix}\right) = \left(\begin{smallmatrix} p(n) & p(n-1) \\ q(n) & q(n-1) \end{smallmatrix}\right) \left(\begin{smallmatrix} a(n+1) & 1 \\ b(n) & 0 \end{smallmatrix}\right)$, it is immediate to deduce that for $n \geq 0$

$$p(n+1)q(n) - p(n)q(n+1) = (-1)^n b(0) \cdots b(n) \; ,$$

or equivalently,

$$S(n+1) - S(n) = (-1)^n \frac{b(0) \cdots b(n)}{q(n)q(n+1)} \; ,$$

so that

$$S(N) = a(0) + \sum_{n=1}^{N} (-1)^{n-1} \frac{b(0) \cdots b(n-1)}{q(n-1)q(n)} \; ,$$

or, if $S(n)$ has a limit S when $n \to \infty$,

$$S - S(N) = \sum_{n \geq N} (-1)^n \frac{b(0) \cdots b(n)}{q(n)q(n+1)} \; .$$

It is also immediate to show in the same way that for $n \geq 0$

$$S(n+2) - S(n) = (-1)^n \frac{b(0) \cdots b(n)a(n+2)}{q(n)q(n+2)} \; ,$$

so that for instance

$$S(2N) = a(0) + \sum_{n=1}^{N} \frac{b(0) \cdots b(2n-2)a(2n)}{q(2n-2)q(2n)}$$

and

$$S - S(2N) = \sum_{n \geq N} \frac{b(0) \cdots b(2n)a(2n+2)}{q(2n)q(2n+2)} \ .$$

In addition, if we let

$$\rho(n) = \cfrac{b(n)}{a(n+1) + \cfrac{b(n+1)}{a(n+2) + \cdots}}$$

be the nth "tail" of the continued fraction, then it is clear that

$$S = \frac{(a(n+1) + \rho(n+1))p(n) + b(n)p(n-1)}{(a(n+1) + \rho(n+1))q(n) + b(n)q(n-1)} = \frac{p(n+1) + \rho(n+1)p(n)}{q(n+1) + \rho(n+1)q(n)} \ ,$$

and furthermore we evidently have $\rho(n) = b(n)/(a(n+1)+\rho(n+1))$, in other words

$$\rho(n)(a(n+1) + \rho(n+1)) - b(n) = 0 \ .$$

7.2. Naïve numerical computation

The formulas given above provide several methods for computing the finite continued fraction $S(N)$. One method is the direct (or backwards) method using the definition of $S(N)$, for instance with the following simple-minded program:

────────────────────── CFback.gp ──────────────────────

```
1   /* Given two closures a and b, compute S(N) directly.
2    * Allow omitting a for the important special case a(n) = 1 */
3   CFback(N, b, a = n -> 1)=
4   { my(S = 0.);
5     forstep (n = N, 1, -1, S = b(n-1) / (a(n) + S));
6     return (a(0) + S);
7   }
```

A second type of methods are *forward* methods, using the recursions for $p(n)$ and $q(n)$ to compute $S(N)$. There are many variants: we can compute $p(n)$ and $q(n)$ separately using the recursions, and either compute $S(N)$ or store all the values of $S(n)$ for $n \leq N$. We can also use the formula for $S(n+1) - S(n)$ (which avoids computing the $p(n)$), or that for $S(2n+2) - S(2n)$. We give these variants as three different programs, where all is set to 0 if only $S(N)$ is desired, and if it is different from 0 the vector of values of $S(n)$ (or of $S(2n)$ in the third program) is output.

────────────────────── CFforward.gp ──────────────────────

```
1   /* Compute S(n) using recursions for p(n) and q(n). Output
2    * [S(1)...S(N)] if 'all' is set, otherwise S(N). */
3   CFforward0(a, b, N, all = 0) =
4   { my(pm1 = 1., p0 = a(0) * 1., qm1 = 0., q0 = 1., v);
```

```
5
6      if (all, v = vector(N));
7      for (n = 1, N,
8        my(A = a(n), B = b(n - 1));
9        my(p1 = A * p0 + B * pm1); pm1 = p0; p0 = p1;
10       my(q1 = A * q0 + B * qm1); qm1 = q0; q0 = q1;
11       if (all, v[n] = p0 / q0));
12     return (if (all, v, p0 / q0));
13   }
14
15   /* Compute S(n) using the formula for S(n+1) - S(n). Output
16    * [S(1)...S(N)] if 'all' is set, otherwise S(N). */
17   CFforward1(a, b, N, all = 0) =
18   { my(qm1 = 0., S = a(0) * 1., q0 = 1., prodb = 1., v);
19
20     if (all, v = vector(N));
21     for (n = 0, N-1,
22       my(B = b(n));
23       my(q1 = a(n+1) * q0 + B * qm1); qm1 = q0; q0 = q1;
24       prodb *= B;
25       S += (-1)^n * prodb / (q0 * qm1);
26       if (all, v[n+1] = S));
27     return (if (all, v, S));
28   }
29
30   /* Compute S(2n) using the formula for S(2n+2) - S(2n). Output
31    * [S(2)...S(2N)] if 'all' is set, otherwise S(2N). */
32   CFforward2(a, b, N, all = 0) =
33   { my(qm1 = 0., q0 = 1., S = a(0) * 1., prodb = b(0) * 1.);
34     my(n, v);
35
36     if (all, v = vector(N));
37     n = 0;
38     while (n < 2*N-1,
39       my(q1 = a(n+1) * q0 + b(n) * qm1); qm1 = q0; q0 = q1; n++;
40       my(A = a(n+1), B = b(n));
41       q1 = A * q0 + B * qm1;
42       S += prodb * A / (q1 * qm1);
43       if (all, v[(n+1)/2] = S);
44       qm1 = q0; q0 = q1; prodb *= b(n+1) * B; n++);
45     return (if (all, v, S));
46   }
```

The backward method is usually numerically more stable, but since we will need to compute many successive $S(n)$ (for instance for extrapolation purposes), we will still mainly use the forward methods. In practice, if we set all to 1 in the forward methods to return a vector of values, the CFback program is the fastest, followed by CFforward2 (computing half as many values to stop at the same $S(N)$), then CFforward0, which are approximately twice slower, then CFforward1, approximately 2.5 times slower. In fact, as we will see below, the big advantage of the CFforward2 method is that it can usually be used in conjunction with Lagrange's extrapolation method to give the limit of $S(n)$ to excellent accuracy.

Note that it is possible to improve on the stability of the forward methods. One program, taken directly from [**TB86**], is as follows:

—————————————— CFforwardstable.gp ——————————————

```
 1   /* Compute S(n) in a stable manner. Output [S(1), ..., S(N)]
 2    * if 'all' is set, otherwise only S(N). Early abort if S(n)
 3    * stabilizes for n < N. */
 4   CFforwardstable(a, b, N, all = 0) =
 5   { my(E = - getlocalbitprec(), e = 2.^(7+E), v, h, C, D);
 6
 7     if (all, v = vector(N));
 8     h = a(0) * 1.; if (exponent(h) < E, h = e);
 9     D = 0.; C = h;
10     for (n = 1, N,
11       my(A = a(n), B = b(n-1));
12       D = A + B * D; if (exponent(D) < E, D = e);
13       C = A + B / C; if (exponent(C) < E, C = e);
14       D = 1 / D; my(z = D * C);
15       h *= z; if (all, v[n] = h);
16       if (exponent(z - 1) < E, return (if (all, v[1..n], h))));
17     return (if (all, v, h));
18   }
```

This final program with all $= 1$ is about 3 times slower than CFback when early abort does not occur. We will not expand further upon this difficult stability problem in the evaluation of continued fractions. On the other hand, it is useful to make a *heuristic* but very practical study of the type of convergence of continued fractions occurring in practice.

7.3. Speed of convergence of infinite continued fractions

In the preceding section, we have only considered the evaluation of *finite* continued fractions. However, almost all interesting continued fractions are infinite. The problem of *convergence* of $S(n)$ to a limit S is quite difficult, so we will restrict to cases where this convergence can not only be proved, but its speed quite accurately estimated, at least heuristically. We emphasize that all the statements are heuristic, but valid in practical cases.

Since this section is rather tedious although completely elementary, we advise the reader to skip to the statement of the main result (Theorem 7.3.7) after understanding the assumptions and notation.

We will make the following fundamental assumptions, of course not always satisfied: there exist *integers* α and β and *real* constants a_i, b_i for $i \geq 0$ with

$a_0 > 0$ and $b_0 \neq 0$, such that as $n \to \infty$ we have the asymptotic expansions

$$a(n) = n^\alpha a_0 \left(1 + \frac{a_1}{n} + \frac{a_2}{n^2} + \cdots\right) , \quad b(n) = n^\beta b_0 \left(1 + \frac{b_1}{n} + \frac{b_2}{n^2} + \cdots\right) .$$

REMARKS 7.3.1. A number of remarks concerning these assumptions.

(1) The restriction that α and β are integers is almost always satisfied in practice, but the (heuristic) analysis that we will give can easily be generalized. In addition, the restriction $a_0 > 0$ is not serious, since if S is given as above, then $-S$ is given by the continued fraction where $a(n)$ is changed into $-a(n)$ and $b(n)$ is preserved. On the other hand, if one studies continued fraction of complex functions then the a_i and b_i may be nonreal, but for simplicity we exclude that case.

(2) Note that the convergents of the continued fraction remain the same if we replace $a(n)$ by $a(n)u(n)$ and $b(n)$ by $b(n)u(n)u(n+1)$ for any sequence $u(n)$ of nonzero numbers with $u(0) = 1$. In particular, if the $a(n)$ are all nonzero the continued fraction may be written as

$$a(0) + \cfrac{b(0)/a(1)}{1 + \cfrac{b(1)/(a(1)a(2))}{1 + \cfrac{b(2)/(a(2)a(3))}{1 + \ddots}}} ,$$

and for $n \geq 1$ we have the asymptotic expansion

$$\frac{b(n)}{a(n)a(n+1)} = n^{\beta - 2\alpha} \frac{b_0}{a_0^2} \left(1 + \frac{B_1}{n} + \frac{B_2}{n^2} + \cdots\right) ,$$

with $B_1 = b_1 - 2a_1 - \alpha$ and

$$B_2 = b_2 - (\alpha + 2a_1)b_1 - 2a_2 + 3a_1^2 + (2\alpha + 1)a_1 + \alpha(\alpha + 1)/2 .$$

We will use this reduction to analyze the convergence.

(3) It often happens in practice that the $a(n)$ and $b(n)$ are given by different formulas according to whether n is even or odd. For completeness, we give the corresponding computation as a lemma:

LEMMA 7.3.2. Let $S(n) = p(n)/q(n)$ be a (formal) continued fraction as above. Then $S(2n) = P(n)/Q(n)$ is given by the continued fraction

$$S(2n) = A(0) + B(0)/(A(1) + B(1)/(A(2) + \cdots + B(n-1)/A(n))) ,$$

with

$$A(0) = a(0) \quad A(1) = a(2)a(1) + b(1) \quad B(0) = a(2)b(0) ,$$

$$A(n+1) = a(2n+2)a(2n+1) + b(2n+1) + \frac{a(2n+2)b(2n)}{a(2n)} \quad and$$

$$B(n) = -\frac{a(2n+2)b(2n)b(2n-1)}{a(2n)} \quad for \ n \geq 1 .$$

PROOF. Immediate computation left to the reader.   □

Thus, when $a(2n)$ and $a(2n-1)$ (resp., $b(2n)$ and $b(2n-1)$) are given by different asymptotic expansions, we can contract the continued fraction by using only its even terms, and compute the asymptotic expansions of $A(n)$ and $B(n)$ using the formulas of the lemma. Note in particular the following:

COROLLARY 7.3.3. *We have the identity*

$$1 + \cfrac{b(0)z}{1 + \cfrac{b(1)z}{1 + \cdots}} = 1 + \cfrac{b(0)z}{1 + b(1)z - \cfrac{b(1)b(2)z^2}{1 + (b(2) + b(3))z - \cfrac{b(3)b(4)z^2}{1 + (b(4) + b(5))z - \cdots}}} .$$

Our first goal is to study *generic* sequences $u(n)$ like $p(n)$ and $q(n)$ satisfying the recursion

$$u(n+1) = a(n+1)u(n) + b(n)u(n-1) .$$

A word concerning genericity: consider for instance the case $a(n) = b(n) = 1$. The general solution of the recursion is then $u(n) = A\phi^n + B\overline{\phi}^n$, with A and B arbitrary constants, and $\phi = (1 + \sqrt{5})/2$, $\overline{\phi} = (1 - \sqrt{5})/2$. When $n \to \infty$ the dominant term is ϕ^n, so that $u(n) \sim A\phi^n$, *except* when $A = 0$, in which case $u(n) \sim B\overline{\phi}^n$ tends exponentially fast to 0. Thus $A \neq 0$ is the generic case.

To study the speed of convergence, we will mainly use the formula $S(n+1) - S(n) = (-1)^n b(0) \cdots b(n)/(q(n)q(n+1))$. We first introduce some useful notation:

Notation.

(1) In the sequel, $\mathcal{C}(1/n^\alpha)$ will denote an asymptotic expansion of the form $C(1 + a_1/n^\alpha + a_2/n^{2\alpha} + \cdots)$, where $\alpha > 0$, C is a nonzero constant, which may differ from one formula to the next (as will the a_i).

(2) All of our asymptotics will be of the form

$$u(n)^{\pm 1} = n!^F E^n e^{(Dn)^{1/2}} n^P \mathcal{C}(1/n^\alpha)$$

with $D \geq 0$, $F \geq 0$, and $\alpha = 1$ or $1/2$. To avoid cumbersome notation, we will simply write $u(n)^{\pm 1} = [F, E, D, P]$, and omit the α which will almost always be equal to $1/2$; more precisely, we will have $q(n) = [F, E, D, P]$ and $S - S(n)$ (or similar quantities such as $S(n+1) - S(n)$) equal to $1/[F, E, D, P]$. This will also be the main part of the output of the CFtype program that we will give below.

As a first illustration of this notation we have the following:

LEMMA 7.3.4. *We have* $\prod_{0 \leq j \leq n} b(j) = [\beta, b_0, 0, b_1]$.

PROOF. Indeed, upon taking logarithms we see that

$$\prod_{1 \leq j \leq n} (1 + b_1/j + \cdots) = n^{b_1} \cdot \mathcal{C}(1/n) ,$$

so the result follows.   □

We now need to study sequences $u(n)$ satisfying the same recursions as $q(n)$. To do this, it is much simpler to set $v(n) = u(n)/u(n-1)$, which satisfies the recursion $v(n+1) = a(n+1) + b(n)/v(n)$.

We will make the (easily justified heuristically) *Ansatz* that

$$v(n) = n^\gamma c_0(1 + c_1/n^{1/2} + c_2/n + \cdots) ,$$

and our goal is to find γ and the c_i. Once this is done, we can easily obtain $u(n)$. Indeed:

LEMMA 7.3.5. *Assume that $v(n) = u(n)/u(n-1)$ has the above asymptotic expansion. Then $u(n) = [\gamma, c_0, 4c_1^2, c_2 - c_1^2/2]$. In particular, generically we have*

$$S(n+1) - S(n) = 1/[2\gamma - \beta, -c_0^2/b_0, 16c_1^2, \gamma + 2c_2 - b_1 - c_1^2] , \text{ and}$$
$$S(n+2) - S(n) = 1/[2\gamma - \beta, c_0^2/b_0, 16c_1^2, 2\gamma + 2c_2 - b_1 - c_1^2] .$$

PROOF. Note that $u(n) = u(0) \prod_{1 \le j \le n} v(j)$. Now

$$\log(1 + c_1/j^{1/2} + c_2/j + \cdots) = c_1/j^{1/2} + (c_2 - c_1^2/2)/j + \cdots ,$$

hence

$$\sum_{1 \le j \le n} \log(1 + c_1/j^{1/2} + c_2/j + \cdots) = 2c_1 n^{1/2} + (c_2 - c_1^2/2) \log(n) + \mathcal{C}(1/n^{1/2}) ,$$

so $u(n) = [\gamma, c_0, 4c_1^2, c_2 - c_1^2/2]$ as desired. In addition, generically, $q(n)$ will behave like $u(n)$, so $q(n)q(n+1) = [2\gamma, c_0^2, 16c_1^2, 2c_2 - c_1^2 + \gamma]$, and using Lemma 7.3.4 we deduce that

$$S(n+1) - S(n) = 1/[2\gamma - \beta, -c_0^2/b_0, 16c_1^2, \gamma + 2c_2 - b_1 - c_1^2] ,$$

proving the first result. The second is proved similarly. □

LEMMA 7.3.6. *Assume that $S(n+1) - S(n) = 1/[F, E, D, P]$. Then:*
  (1) *If $F > 0$, or $F = 0$ and $|E| > 1$, or $F = 0$, $E = -1$, and $D > 0$, we have $S - S(n) = 1/[F, E, D, P]$.*
  (2) *If $F = 0$, $E = 1$, and $D > 0$ then $S - S(n) = 1/[0, 1, D, P - 1/2]$.*
  (3) *If $F = 0$, $E = -1$, $D = 0$, and $P > 0$, then if $S(2n+2) - S(2n)$ has a similar but nonalternating asymptotic expansion (as we will see below), $S - S(n) = 1/[0, -1, 0, P]$.*
  (4) *If $F = 0$, $E = 1$, $D = 0$, and $P > 1$, then $S - S(n) = 1/[0, 1, 0, P - 1]$.*

PROOF. Immediate and left to the reader. □

To find γ and the c_i, first note that

$$\frac{b(n)}{v(n)} = n^{\beta-\gamma} \frac{b_0}{c_0}(1 + \sum_{i \ge 1} d_i/n^{i/2})$$

for explicit d_i which are best computed by a CAS ($d_1 = -c_1$, $d_2 = b_1 + c_1^2 - c_2$, etc.). To solve the recursion, we assume that $a(n) = 1$, since we have seen that we can reduce to that case by replacing $b(n)$ by $b(n)/(a(n)a(n+1))$. We will thus make the study in this case, then summarize the results for the general case using the expansion of $b(n)/(a(n)a(n+1))$ given above.

We have $v(n+1) = n^\gamma c_0(1 + \sum_{i \ge 1} e_i/n^{i/2})$, where again the e_i are computed by a CAS ($e_1 = c_1$, $e_2 = c_2 + \gamma$, etc.). The recursion is thus

$$n^\gamma c_0(1 + e_1/n^{1/2} + e_2/n + \cdots) = 1 + n^{\beta-\gamma}(b_0/c_0)(1 + d_1/n^{1/2} + d_2/n + \cdots) .$$

We now reason by successive approximations. To a first approximation, we have $v(n+1) \approx v(n)$ (i.e., we neglect the terms $d_i/n^{i/2}$ and $e_i/n^{i/2}$), so $v(n)$ is a root of the second degree equation $X^2 - X - n^\beta b_0 = 0$, so generically $n^\gamma c_0 = (1 + \sqrt{1 + 4n^\beta b_0})/2$. We are thus led to consider three cases (as mentioned, the reader is advised to skip these tedious and easy computations and to go directly to the statement of Theorem 7.3.7):

Case $\beta < 0$. In that case we obtain $n^{\gamma} c_0 = 1$, so $\gamma = 0$ and $c_0 = 1$, hence by Lemma 7.3.5 $S(n+1) - S(n)$ is generically of the order of $n!^{\beta}$ with $\beta < 0$ (the other terms have much smaller order of magnitude). It follows that the continued fraction converges very fast (like a power of $1/n!$). More precisely, recalling that β is an integer:

(1) If $\beta = -1$, then $e_1 = 0$ and $e_2 = b_0$, hence $c_1 = 0$ and $c_2 = b_0$, so by Lemmas 7.3.5 and 7.3.6 we have $q(n) = [0, 1, 0, b_0]$ and $S - S(n) = 1/[1, -1/b_0, 0, 2b_0 - b_1]$.

(2) If $\beta \leq -2$ then $e_1 = e_2 = 0$, so $q(n) = [0, 1, 0, 0]$ and $S - S(n) = 1/[-\beta, -1/b_0, 0, -b_1]$.

Case $\beta > 0$. In that case we have $n^{\gamma} c_0 = \sqrt{n^{\beta} b_0}$, so $\gamma = \beta/2$ and $c_0 = b_0^{1/2}$. This is however possible only if $b_0 > 0$, which we therefore assume (if $b_0 < 0$ then either the continued fraction does not converge, or it has an oscillating behavior which is difficult to analyze).

Writing the above recursion and simplifying, we obtain

$$e_1/n^{1/2} + e_2/n + \cdots = b_0^{-1/2}/n^{\beta/2} + d_1/n^{1/2} + d_2/n + \cdots .$$

Now recall that β is assumed to be an integer, so we consider the first values.

(1) Case $\beta = 1$, hence $\gamma = 1/2$. We thus have $e_1 = b_0^{-1/2} + d_1$ and $e_i = d_i$ for $i \geq 2$. Since $d_1 = -c_1$ and $e_1 = c_1$, this gives $c_1 = b_0^{-1/2}/2$. Since $d_2 = b_1 + c_1^2 - c_2 = b_1 - c_2 + 1/(4b_0)$ and $e_2 = c_2 + \gamma = c_2 + 1/2$, this gives $c_2 = b_1/2 + 1/(8b_0) - 1/4$, so that by Lemmas 7.3.5 and 7.3.6 we have $q(n) = [1/2, b_0^{1/2}, 1/b_0, b_1/2 - 1/4]$ and both $S(n+1) - S(n)$ and $S - S(n)$ are of the form $1/[0, -1, 4/b_0, 0]$ (the powers of n cancel). It follows that the convergence of the continued fraction is sub-exponential.

(2) Case $\beta = 2$, hence $\gamma = 1$. Here we obtain $e_1 = d_1$, hence $c_1 = 0$, and $e_2 = b_0^{-1/2} + d_2$, hence $c_2 + 1 = b_0^{-1/2} + b_1 - c_2$, so $c_2 = b_1/2 + b_0^{-1/2}/2 - 1/2$, so that by Lemma 7.3.5 $q(n) = [1, b_0^{1/2}, 0, b_1/2 + b_0^{-1/2}/2 - 1/2]$ and $S(n+1) - S(n) = 1/[0, -1, 0, b_0^{-1/2}]$, In this case the continued fraction converges slowly, if at all. Note that this is far from proving the convergence of the continued fraction, even heuristically (think of the series $\sum_{n \geq 1} (-1)^{n-1}/(n^{1/2} + (-1)^n)$, which diverges). To look at the convergence and its speed, we compute $S(n+2) - S(n)$: by the second formula of Lemma 7.3.5 we have $S(n+2) - S(n) = \mathcal{C}(1/n^{1/2})/n^{b_0^{-1/2}+1}$, so that for instance for n even $S - S(n) = \mathcal{C}(1/n^{1/2})/n^{b_0^{-1/2}}$, and since both $S - S(n)$ and $S(n+1) - S(n)$ tend to 0 it follows that the continued fraction converges, although slowly.

(3) Case $\beta \geq 3$. Here we obtain $e_1 = d_1$ and $e_2 = d_2$, so that $c_1 = 0$ and $b_1 - c_2 = c_2 + \beta/2$, so $c_2 = b_1/2 - \beta/4$, and Lemma 7.3.5 tells us that $S(n+1) - S(n) = (-1)^n \mathcal{C}(1/n^{1/2})$, so generically the continued fraction diverges. However, the second formula of that lemma tells us that $S(n+2) - S(n) = \mathcal{C}(1/n^{1/2})/n^{\beta/2}$, so separately the even and odd terms converge to a limit, and $q(n) = [\beta/2, b_0^{1/2}, 0, b_1/2 - \beta/4]$ and $S_e - S(2n) = \mathcal{C}(1/n^{1/2})/n^{\beta/2-1}$, where S_e is the limit of the even terms, and similarly for S_o, the limit of the odd terms.

Case $\beta = 0$. The last remaining case is $\beta = 0$. From the first approximation seen above, we have $\gamma = 0$ and $c_0 = (1 + \sqrt{4b_0 + 1})/2$, so we will assume in this case that $b_0 \geq -1/4$, otherwise the continued fraction will have an oscillating behavior. We separate the cases $b_0 > -1/4$ and $b_0 = -1/4$.

(1) $\beta = 0$ and $b_0 > -1/4$. The recursion gives

$$c_0(1 + e_1/n^{1/2} + e_2/n + \cdots) = 1 + \frac{b_0}{c_0}(1 + d_1/n^{1/2} + d_2/n + \cdots) \,.$$

Identification of the constant coefficients gives $c_0 = (1 + \sqrt{4b_0 + 1})/2$ as seen above. Since $e_1 = c_1$ and $d_1 = -c_1$, the coefficients of $1/n^{1/2}$ give $c_1(c_0 + b_0/c_0) = 0$. Now $c_0 + b_0/c_0 = 2c_0 - 1 = \sqrt{4b_0 + 1} \neq 0$ in the present case, so $c_1 = 0$. Finally, identification of the coefficients of $1/n$ give $c_0 e_2 = (b_0/c_0)d_2 = (c_0 - 1)d_2$, and since $e_2 = c_2$ and $d_2 = b_1 - c_2$, we obtain $(2c_0 - 1)c_2 = (c_0 - 1)b_1$, so $c_2 = b_1(1 - 1/\sqrt{4b_0 + 1})/2$. We deduce from Lemma 7.3.5 that $q(n) = [0, (1 + \sqrt{4b_0 + 1})/2, 0, b_1(1 - 1/\sqrt{4b_0 + 1})/2]$ and $S(n + 1) - S(n) = 1/[0, -c_0^2/b_0, 0, -b_1/\sqrt{4b_0 + 1}]$. We claim that $|c_0^2/b_0| > 1$. Indeed, since $c_0^2/b_0 = 1 + c_0/b_0$ and $c_0 > 0$, the result is clear when $b_0 > 0$. Thus, assume $b_0 < 0$. Then $|c_0^2/b_0| > 1$ iff $c_0/b_0 < -2$ iff $c_0 > -2b_0$, iff $1 + \sqrt{4b_0 + 1} > -4b_0$, iff $\sqrt{4b_0 + 1} > -(4b_0 + 1)$, which is true since the left hand side is positive and the right hand side negative by assumption.

Thus we always have exponential convergence, that is $S - S(n) = 1/[0, -c_0^2/b_0, 0, -b_1/\sqrt{4b_0 + 1}]$.

(2) $\beta = 0$, $b_0 = -1/4$, and $b_1 < 0$ (we will soon see why we distinguish this case). Here again c_0 is a root of the same second degree equation, hence $c_0 = 1/2$. The recursion gives

$$\frac{1}{2}(1 + e_1/n^{1/2} + e_2/n + \cdots) = 1 - \frac{1}{2}(1 + d_1/n^{1/2} + d_2/n + \cdots) \,.$$

The constant coefficients of course agree, and for $i \geq 1$ we have $e_i = -d_i$. For $i = 1$ this is a tautology, for $i = 2$ it gives $c_2 = -(b_1 + c_1^2 - c_2)$, so $c_1^2 = -b_1$. This is why we assumed $b_1 < 0$ (the other case will be $b_1 = 0$). For $i = 3$, our CAS tells us that $e_3 + d_3 = c_1(2c_2 + \gamma - 1/2 - c_1^2 - b_1) = c_1(2c_2 - 1/2)$, and since $b_1 \neq 0$, hence $c_1 \neq 0$, we deduce that $c_2 = 1/4$. Thus, by Lemma 7.3.5 we have $q(n) = [0, 1/2, 4(-b_1), 1/4 + b_1/2]$ and $S(n+1) - S(n) = 1/[0, 1, 16(-b_1), 1/2]$, hence by Lemma 7.3.6 $S - S(n) = 1/[0, 1, 16(-b_1), 0]$. (Note that here the partial quotients do not alternate in sign since the $(-1)^n$ has disappeared). It follows that we have subexponential convergence.

(3) $\beta = 0$, $b_0 = -1/4$, and $b_1 = 0$. Note that since we still have $c_1^2 = -b_1$ we cannot have $b_1 > 0$ otherwise we would get an oscillating behavior difficult to analyze. Thus $c_1 = 0$, $e_3 + d_3 = 0$ automatically, and now for $i = 4$ we must have $e_4 + d_4 = 0$ and our CAS gives us a one line long second degree equation for c_2 which dramatically simplifies since $\beta = \gamma = 0$ and $b_1 = c_1 = 0$, and reduces simply to $c_2^2 - c_2 + b_2 = 0$. Thus if $b_2 > 1/4$ this equation has no real solution, so again we reject this case as being difficult, while if $b_2 \leq 1/4$ we have $c_2 = (1 + \sqrt{1 - 4b_2})/2$. By Lemma 7.3.5, we deduce that $q(n) = [0, 1/2, 0, (1 + \sqrt{1 - 4b_2})/2]$ and $S(n + 1) - S(n) = 1/[0, 1, 0, \sqrt{1 - 4b_2} + 1]$. Note that this is again nonalternating, and also

shows that if $b_2 = 1/4$ we will have a logarithmic divergence, so we must assume that $b_2 < 1/4$, and in that case $S - S(n) = 1/[0, 1, 0, \sqrt{1 - 4b_2}]$.

We have now finished our heuristic study of the speed of convergence of continued fractions of the type given above. We summarize what we have found, coming back to the general case where $a(n)$ is not necessarily the constant 1. In that case, we must replace

$$\beta \quad \text{by} \quad \beta - 2\alpha \,,$$
$$b_0 \quad \text{by} \quad b_0/a_0^2 \,,$$
$$b_1 \quad \text{by} \quad B_1 = b_1 - 2a_1 - \alpha \,,$$
$$b_2 \quad \text{by} \quad B_2 \,,$$

where B_2 is given above before Lemma 7.3.2, and if $q(n) = [F, E, D, P]$ we must replace this by $q(n) = [F + \alpha, a_0 E, D, a_1 + P]$. This gives the following:

THEOREM 7.3.7. *With the above notation, in the following cases the continued fraction converges generically with the stated speed (recall that we always assume that $a_0 > 0$, $b_0 \neq 0$, and α and β integral).*

(1) $\beta \leq 2\alpha - 2$. *In this case* $q(n) = [\alpha, a_0, 0, a_1]$ *and*

$$S - S(n) = (-1)^n \frac{(b_0/a_0^2)^n n^{b_1 - 2a_1 - \alpha}}{n!^{2\alpha - \beta}} \mathcal{C}(1/n^{1/2}) \,,$$

so extremely fast convergence.

(2) $\beta = 2\alpha - 1$. *In this case* $q(n) = [\alpha, a_0, 0, a_1 + b_0/a_0^2]$ *and*

$$S - S(n) = (-1)^n \frac{(b_0/a_0^2)^n n^{b_1 - 2a_1 - \alpha - 2b_0/a_0^2}}{n!} \mathcal{C}(1/n^{1/2}) \,,$$

so extremely fast convergence.

(3) $\beta = 2\alpha + 1$ *and* $b_0 > 0$. *In this case* $q(n) = [\alpha + 1/2, b_0^{1/2}, a_0^2/b_0, b_1/2 - \alpha/2 - 1/4]$ *and*

$$S - S(n) = (-1)^n \frac{\mathcal{C}(1/n^{1/2})}{e^{2a_0(n/b_0)^{1/2}}} \,,$$

hence alternating sub-exponential convergence, quite fast.

(4) $\beta = 2\alpha + 2$ *and* $b_0 > 0$. *In this case* $q(n) = [\alpha + 1, b_0^{1/2}, 0, b_1/2 - \alpha/2 + a_0 b_0^{-1/2}/2 - 1/2]$ *and*

$$S - S(n) = (-1)^n \frac{\mathcal{C}(1/n^{1/2})}{n^{a_0/b_0^{1/2}}} \,,$$

so the continued fraction converges, but slowly: we will say that it has alternating polynomial convergence.

(5) $\beta \geq 2\alpha + 3$. *In this case usually $S(n)$ does not tend to a limit, but the odd and even terms separately do, and* $q(n) = [\beta/2 - \alpha, (b_0/a_0^2)^{1/2}, 0, b_1/2 - a_1 - \beta/4]$ *and*

$$S_e - S(2n) = \frac{\mathcal{C}(1/n^{1/2})}{n^{\beta/2 - \alpha - 1}} \,,$$

and similarly for $S_o - S(2n+1)$.

(6) $\beta = 2\alpha$ and $a_0^2 + 4b_0 > 0$. In this case we have

$$q(n) = \left[\alpha, c_0, 0, a_1 + (b_1/2 - a_1 - \alpha/2)(1 - a_0/\sqrt{a_0^2 + 4b_0})\right] \quad and$$

$$S - S(n) = \frac{n^{(b_1 - 2a_1 - \alpha)a_0/\sqrt{a_0^2 + 4b_0}}}{(-c_0^2/b_0)^n}\mathcal{C}(1/n^{1/2})$$

with $c_0 = (a_0 + \sqrt{a_0^2 + 4b_0})/2$, and since $|c_0^2/b_0| > 1$ we have exponential convergence, alternating if $b_0 > 0$ and nonalternating if $b_0 < 0$.

(7) $\beta = 2\alpha$, $a_0^2 + 4b_0 = 0$, and $2a_1 > b_1 - \alpha$. In this case, we have

$$q(n) = [\alpha, a_0/2, 4(2a_1 - (b_1 - \alpha)), 1/4 + (b_1 - \alpha)/2] \quad and$$

$$S - S(n) = \frac{\mathcal{C}(1/n^{1/2})}{e^{4((2a_1 - (b_1 - \alpha))n)^{1/2}}},$$

so nonalternating sub-exponential convergence.

(8) $\beta = 2\alpha$, $a_0^2 + 4b_0 = 0$, $2a_1 = b_1 - \alpha$, and $B_3 > 0$, with

$$B_3 = (b_1 + \alpha - 1)^2 - 2\alpha(\alpha - 1) - 4(b_2 - 2a_2).$$

In this case $q(n) = [\alpha, a_0/2, 0, a_1 + (1 + \sqrt{B_3})/2]$ and

$$S - S(n) = \frac{\mathcal{C}(1/n^{1/2})}{n^{\sqrt{B_3}}}.$$

We have nonalternating polynomial convergence, hence slow.

In all other cases, the continued fraction either does not converge or is difficult to study.

Since this theorem involves quite a number of cases, we will write a program which tells the expected precise rate of convergence. The input to this program are the two functions $a(n)$ and $b(n)$, assumed to be rational functions in the specific variable n, without other variables (like z, c, etc.), or closures whose values are rational functions (if they are not rational functions, simply replace them by the first terms of their asymptotic expansions in integral powers of $1/n$). The result is a 3-component vector [FEDP(Q), FEDP(S), num], where the FEDP are as usual 4-component vectors [F,E,D,P] for $Q(n)$ and $S - S(n)$ with the above notation, so that $Q(n) \sim \text{FEDP}(Q) \cdot \mathcal{C}(1/n^{1/2})$ and $S - S(n) \sim \mathcal{C}(1/n^{1/2})/\text{FEDP}(S)$, and num is a number from 1 to 8 giving the type of convergence in the numbering of the above theorem. The reader can try this program as well as the next ones with all the examples given in the next subsection.

—————————————— CFtype.gp ——————————————

```
/* Return exact result if possible, else floating point */
SQRT(x) = my(a); if (issquare(x, &a), a, sqrt(x));

/* Give the convergence type of a continued fraction as above.
 * fa and fb are rational functions of the variable 'n.
 * Except in certain cases, the coefficients of fa and fb must
 * be numbers, not other variables. */
CFerr() = error("cannot conclude, probably diverges");
CFtype(fa, fb) =
{ my([al, a0, a1, a2] = CFtype_i(fa));
```

```
11    my([be, b0, b1, b2] = CFtype_i(fb), c0, r1, D, D2, D3, t0);
12
13    r1 = 2*al - be;
14    if (b0 <= 0 && r1 <= -1, CFerr());
15    if (a0 < 0, a0 = -a0);
16    t0 = a0^2 / b0;
17    if (r1 > 1,   return ([[al, a0, 0, a1],
18                           [r1, -t0, 0, al+2*a1-b1], 1]),
19      r1 == 1, return ([[al, a0, 0, a1 + 1/t0],
20                        [1,-t0, 0, 2/t0 + al + 2*a1 - b1], 2]),
21      r1 ==-1, return ([[be/2, SQRT(b0), t0, b1/2-be/4],
22                        [0, -1, 4*t0, 0], 3]),
23      r1 ==-2, my(s0 = SQRT(t0));
24               return ([[be/2, a0/s0, 0, (b1+s0)/2-be/4],
25                        [0, -1, 0, s0], 4]),
26      r1 < -2, return ([[-r1/2, SQRT(1/t0), 0, b1/2-a1-be/4],
27                        [0, 1, 0, r1/2+1], 5]));
28    /* Here r1 = 0: be = 2al */
29    D = a0^2 + 4*b0;
30    if (D < 0, CFerr());
31    if (D > 0,
32      my(s = SQRT(D));
33      c0 = (a0+s)/2;
34      return ([[al, c0, 0, a1 + (b1/2-a1-al/2)*(1-a0/s)],
35              [0, -c0^2/b0, 0, (2*a1-b1+al)*a0/s], 6]));
36    /* Here be = 2al and a0^2+4b0 = 0 */
37    D2 = 2*a1 - b1 + al;
38    if (D2 < 0, CFerr());
39    if (D2 > 0,
40      return ([[al, a0/2, 4*D2, 1/4 + (b1 - al)/2],
41              [0, 1, 16*D2, 0], 7]));
42    /* Here be = 2al, a0^2 + 4b0 = 0, and 2a1 - b1 + al = 0 */
43    D3 = (b1 + al - 1)^2 - be*(al - 1) - 4*(b2 - 2*a2);
44    if (D3 <= 0, CFerr());
45    my(s3 = SQRT(D3));
46    return ([[al, a0/2, 0, a1+(1+s3)/2], [0, 1, 0, s3], 8]);
47 }
48
49 /* auxiliary function for CFtype */
50 CFtype_i(f) =
51 { my(n = 'n, al, a0, a1, a2);
52
53   if (type(f) == "t_CLOSURE", f = f(n));
54   al = poldegree(f, n);
55   [a0, a1, a2] = Vec(subst(f / n^al, n, 1 / n) + O(n^3));
56   return ([al, a0, a1 / a0, a2 / a0]);
57 }
```

Using the extrapolation programs given in Chapter 2, it is easy to compute numerically the constants C implicit in all the asymptotic estimates given above, both for $q(n)$ (program CFqnasymp) and for $S - S(n)$ (program CFasymp), and also the limit S of the continued fraction itself (program CFlim):

─────────────────────────── | CFasymp.gp | ───────────────────────────

```
1   /* Let [F,E,D,P] code an asymptotic expansion of type
2    * n!^F * E^n * e^{(D*n)^{1/2}} * n^P.
3    * Compute value at n. */
4
5   CFtypeeval(FEDP, n) =
6   { my(S, [f, e, d, p] = FEDP());
7
8     S = if (f, factorial(n)^f, 1);
9     if (e != 1, S *= e^n);
10    if (d, S *= exp(sqrt(d * n)));
11    if (p, S *= n^p);
12    return (S);
13  }
14
15  /* Let a(0)+b(0)/(a(1)+b(1)/(a(2)+...)) be a continued
16   * fraction; compute [[p1,q1], ..., [pn,qn]]. */
17  CFpnqn(a, b, N) =
18  { my(V = vector(N));
19
20    V[1] = [a(0) * a(1) + b(0) * 1., a(1) * 1.];
21    V[2] = a(2) * V[1] + b(1) * [a(0), 1];
22    for (n = 2, N-1, V[n+1] = a(n+1) * V[n] + b(n) * V[n-1]);
23    return (V);
24  }
25
26  /* Compute limit. */
27  CFlim(a, b, al = 1/2, mul = 10) =
28  { my(VU);
29    VU = (N -> my(V = CFpnqn(a, b, mul * N));
30              vector(N, n, my([pn, qn] = V[mul * n]); pn / qn));
31    Limitvec(VU, al);
32  }
33
34  /* f(n) known function such that q(n) is asymptotic to
35   * C.[F,E,D,P](n) (syntax as above) for some constant C
36   * with asymptotic behavior in integral powers of 1/n^al.
37   * Compute C. All parameters MUST be given as closures. */
38  CFqnasymp(a, b, FEDP, al = 1/2, mul = 10) =
39  { my(VU);
40
41    VU = (N -> my(V = CFpnqn(a, b, mul * N));
42      vector(N, n, my(qn = V[mul * n][2]);
```

```
43                    qn / CFtypeeval(FEDP, mul * n)));
44    Limitvec(VU, al);
45  }
46
47  /* Assume known limit S, and f(n) a known function such that
48   * S-p(n)/q(n) is asymptotic to C/f(n) = C/[F,E,D,P](n)
49   * (syntax as above) for some constant C with asymptotic
50   * behavior in integral powers of 1/n^al. Compute C.
51   * All parameters MUST be given as closures. */
52  CFasymp(a, b, FEDP, S, al = 1/2, mul = 10) =
53  { my(VU);
54
55    VU = (N -> my(B = getlocalbitprec());
56      localbitprec(32);
57      localbitprec(B + exponent(CFtypeeval(FEDP, mul * N)));
58      my(s = S(), V = CFpnqn(a, b, mul * N));
59      vector(N, n, my([pn,qn] = V[mul * n]);
60                  CFtypeeval(FEDP, mul * n) * (s - pn / qn)));
61    Limitvec(VU, al);
62  }
```

An important comment concerning the above programs: since the extrapolation programs such as `Limitvec` need to increase the accuracy to perform the computation, we cannot give the necessary constants such as FEDP and S as simple real or complex numbers, unless they are exact (integers or fractions), otherwise they would be used at the given accuracy and not at the accuracy necessary for extrapolation. Thus for instance, instead of using the constant `Pi`, which is inexact, we must use the *closure* `x->Pi`, which will automagically compute the given constant (here π) to the necessary accuracy. This can of course only be done for constants for which there is a computable formula for any accuracy (for instance `x->exp(1)`), but something like `x->yourconstant` would of course not be of any use since the program cannot know how to compute `yourconstant` to a higher accuracy than what you give, unless it is an exact (as opposed to approximate) value, or if you give it yourself to a much larger accuracy than the desired one.

7.4. Examples of each convergence case

To test numerically the above derivations, as well as the acceleration methods that we give below in the case of slow convergence, it is useful to have a small set of typical examples for each case, if possible with an explicit limiting value. We begin with hypergeometric functions, which already provide examples for quite a number of cases. Recall that

$$
{}_pF_q(a_1,\ldots,a_p;b_1,\ldots,b_q;z) = \sum_{n\geq 0} \frac{\prod_{1\leq i\leq p}(a_i)_n}{\prod_{1\leq j\leq q}(b_j)_n} \frac{z^n}{n!} \ ,
$$

where $(a)_n = a(a+1)\cdots(a+n-1)$ is the Pochhammer symbol, and that this series has radius of convergence ∞, 1, or 0, according to whether $p \leq q$, $p = q+1$, or $p \geq q+2$ respectively.

EXAMPLE 7.4.1. $_0F_0(z) = e^z$. One of the continued fractions for e^z is

$$e^z = 1 + \cfrac{2z}{2 - z + \cfrac{z^2}{6 + \cfrac{z^2}{10 + \cfrac{z^2}{14 + \cdots}}}},$$

so that $a(0) = 1$, $a(1) = 2 - z$, $a(n) = 4n - 2$ for $n \geq 2$, $b(0) = 2z$, $b(n) = z^2$ for $n \geq 1$, so that $\alpha = 1$ and $\beta = 0$. This corresponds to Case (1) of the proposition. We have $a_0 = 4$, $a_1 = -1/2$, $b_0 = z^2$, $b_1 = 0$, so by the proposition

$$S - S(n) \sim (-1)^n C \frac{(z/4)^{2n}}{n!^2},$$

hence the convergence is extremely fast. One can show that $C = (\pi/2)ze^z$. In fact, let us check all this numerically using the programs given above, say for $z = 3$:

```
? a = (n -> if (n == 0, 1, if (n == 1, -1, 4 * n - 2)));
? b = (n -> if (n == 0, 6, 9));
? CFtype(a, b)
% = [[1, 4, 0, -1/2], [2, -16/9, 0, 0], 1]
/* q(n) ~ C*n!*4^n*n^{-1/2}, S-S(n) ~ C'/(n!^2*(-16/9)^n) */
? CFlim(a, b) - exp(3)
% = 0.E-37
? CFasymp(a, b, z -> [2, -16/9, 0, 0], z -> exp(3))
% = 94.650862861939374274941043134968781919
? (Pi / 2) * 3 * exp(3)
% = 94.650862861939374274941043134968781919 /* Perfect. */
? CFqnasymp(a, b, z -> [1, 4, 0, -1/2])
% = 0.12588771213108679131966309531069167590
? exp(-3/2) / sqrt(Pi)
% = 0.12588771213108679131966309531069167590
```

Similarly,

$$_0F_1(a; z) = 1 + \frac{z}{1!a} + \frac{z^2}{2!a(a+1)} + \cdots$$

The standard continued fraction is

$$a\frac{_0F_1(a; z)}{_0F_1(a + 1; z)} = a + \cfrac{z}{a + 1 + \cfrac{z}{a + 2 + \cfrac{z}{a + 3 + \cdots}}},$$

so that $a(n) = n + a$ and $b(n) = z$, so that again $\alpha = 1$ and $\beta = 0$. We have $a_0 = 1$, $a_1 = a$, $b_0 = z$, $b_1 = 0$, so that

$$S - S(n) \sim (-1)^n C \frac{z^n n^{-2a-1}}{n!^2},$$

again extremely fast convergence.

A special case of the above continued fraction for $_0F_1$ is

$$z\frac{I_{b/a-1}(2z/a)}{I_{b/a}(2z/a)} = b + \cfrac{z^2}{a+b+\cfrac{z^2}{2a+b+\ddots}} \,,$$

where I_ν is the I-Bessel function. Here $a(n) = an + b$ and $b(n) = z^2$, and the result is similar:

$$S - S(n) \sim (-1)^n C\frac{(z/a)^{2n}n^{-2b/a-1}}{n!^2} \,.$$

EXAMPLE 7.4.2.

$$_1F_1(a;b;z) = 1 + \frac{a}{1!b}z + \frac{a(a+1)}{2!b(b+1)}z^2 + \cdots$$

Here there are three standard continued fractions. First,

$$\frac{_1F_1(a+1;b+1;z)}{_1F_1(a;b;z)} = \cfrac{b}{b-z+\cfrac{(a+1)z}{b+1-z+\cfrac{(a+2)z}{b+2-z+\cfrac{(a+3)z}{b+3-z+\ddots}}}} \,,$$

so $a(0) = 0$, $a(n) = n+b-z-1$ for $n \geq 1$, $b(0) = b$, $b(n) = (a+n)z$ for $n \geq 1$. Here $\alpha = 1$, $\beta = 1$, so this is Case (2) of the proposition. We have $a_0 = 1$, $a_1 = b-z-1$, $b_0 = z$, $b_1 = a$, so

$$S - S(n) \sim (-1)^n C\frac{z^n n^{a-2b+1}}{n!} \,,$$

so now the convergence is in $1/n!$, still extremely fast.

The second continued fraction is

$$\frac{_1F_1(a+1;b+1;z)}{_1F_1(a;b;z)} = \cfrac{b}{b-\cfrac{(b-a)z}{b+1+\cfrac{(a+1)z}{b+2-\cfrac{(b-a+1)z}{b+3+\cfrac{(a+2)z}{b+4-\ddots}}}}} \,,$$

so that $a(0) = 0$, $a(n) = n + b - 1$ for $n \geq 1$, $b(0) = b$, $b(2n) = (n + a)z$, and $b(2n + 1) = -(b - a + n)z$ for $n \geq 1$. Using Lemma 7.3.2, we deduce that

$$A(n) = (2n + b - 1)(2n + b - 2) - (b - a + n - 1)z$$
$$+ (2n + b - 1)(n - 1 + a)z/(2n + b - 3) \,, \quad \text{and}$$
$$B(n) = (2n + b + 1)(n + a)(b - a + n - 1)z^2/(2n + b - 1) \,,$$

so that $\alpha = 2$ and $\beta = 2$, $a_0 = 4$, $a_1 = b - 3/2$, $b_0 = z^2$, $b_1 = b$. We deduce from Case (1) that

$$S - S(2n) = (-1)^n \frac{(z/4)^{2n}n^{1-b}}{n!^2}\mathcal{C}(1/n^{1/2}) \,,$$

which by Stirling's formula is equivalent to

$$S - S(n) = (-1)^{n/2}\frac{(z/2)^n n^{1/2-b}}{n!}\mathcal{C}(1/n^{1/2})$$

for n even, so that this continued fraction converges slightly faster than the preceding one (z replaced by $z/2$).

The third continued fraction is for the special case where one of the variables is equal to 1:

$$_1F_1(1; c+1; z) = \frac{e^z\Gamma(c+1)}{z^c} - \cfrac{c}{z+1-c-\cfrac{1(1-c)}{z+3-c-\cfrac{2(2-c)}{z+5-c-\cdots}}} \,,$$

so $a(0) = e^z\Gamma(c+1)/z^c$, $a(n) = z-c+2n-1$ for $n \geq 1$, $b(0) = -c$, $b(n) = -n(n-c)$ for $n \geq 1$, hence $\alpha = 1$, $\beta = 2$, $a_0 = 2$, $b_0 = -1$, $a_1 = (z-c-1)/2$, $b_1 = -c$, so $\beta = 2\alpha$, $a_0^2 + 4b_0 = 0$, $2a_1 - (b_1 - \alpha) = z$. We are thus in Case (7) of the proposition, and

$$S - S(n) \sim Ce^{-4(nz)^{1/2}} \,,$$

nonalternating sub-exponential convergence. Note that this is much slower than the preceding types of convergence.

EXAMPLE 7.4.3.

$$_2F_0(a, b; z) = 1 + \frac{ab}{1!}z + \frac{a(a+1)b(b+1)}{2!}z^2 + \cdots$$

Note that the radius of convergence is 0, but this is nonetheless the asymptotic expansion of many important functions.

The standard continued fraction is

$$\frac{_2F_0(a, b; z)}{_2F_0(a, b+1; z)} = 1 + \cfrac{az}{(b+1)z-1-\cfrac{(a+1)(b+1)z^2}{(a+b+3)z-1-\cfrac{(a+2)(b+2)z^2}{(a+b+5)z-1-\cdots}}} \,,$$

so that $a(0) = 1$, $a(1) = (b+1)z-1$, $a(n) = (a+b+2n-1)z-1$ for $n \geq 2$, $b(0) = az$, and $b(n) = -(a+n)(b+n)z^2$ for $n \geq 1$. Thus $\alpha = 1$, $\beta = 2$, $a_0 = 2z$, $a_1 = (a+b-1)/2 - 1/(2z)$, $a_2 = 0$, $b_0 = -z^2$, $b_1 = a+b$, $b_2 = ab$, so $\beta = 2\alpha$, $a_0^2 + 4b_0 = 0$, $2a_1 - (b_1 - \alpha) = -1/z$, so we need $z < 0$ to be able to apply the proposition, and we are in Case (7) with

$$S - S(n) \sim Ce^{-4(-n/z)^{1/2}} \,.$$

It follows that we have relatively fast sub-exponential convergence, although the initial series did not converge for any z. (Note that to apply the proposition we need $a_0 > 0$, but to achieve this it suffices to change S into $-S$, $a(n)$ into $-a(n)$, and leave $b(n)$ unchanged, which changes a_0 into $-a_0$ and does not change a_1 and b_1.)

An important special case of the above is the *incomplete gamma function*, defined for $x \in \mathbb{R}_{>0}$ by

$$\Gamma(s, x) = \int_x^\infty t^{s-1}e^{-t}\, dt \,.$$

It is immediate to show by integration by parts that we have the asymptotic expansion

$$\Gamma(s, x) = x^{s-1}e^{-x}{_2F_0}(1-s, 1; -1/x) \,.$$

Since trivially $_2F_0(a, 0; z) = 1$, it is easy to deduce from above the well-known continued fraction expansion

$$\Gamma(s, x) = \cfrac{x^s e^{-x}}{x + 1 - s - \cfrac{1(1 - s)}{x + 3 - s - \cfrac{2(2 - s)}{x + 5 - s + \ddots}}},$$

which satisfies

$$S - S(n) \sim C e^{-4(nx)^{1/2}}.$$

This is of course the same continued fraction as that occurring in $_1F_1(1; s + 1; x)$ given above. Note also that one can show that $C = 2\pi/\Gamma(1 - s)$ (see Exercise 24 of Chapter 8 of [**Coh07b**]).

EXAMPLE 7.4.4. A special case of the preceding example is the function

$$\mathrm{erfc}(x) = \frac{2}{\sqrt{\pi}} \int_x^\infty e^{-t^2}\, dt = \frac{\Gamma(1/2, x^2)}{\sqrt{\pi}}.$$

The above continued fraction gives

$$\frac{\sqrt{\pi}}{2} \mathrm{erfc}(x) = \cfrac{x e^{-x^2}}{2x^2 + 1 - \cfrac{1 \cdot 2}{2x^2 + 5 - \cfrac{3 \cdot 4}{2x^2 + 9 - \cfrac{5 \cdot 6}{2x^2 + 13 - \ddots}}}}.$$

However, we have another continued fraction

$$\frac{\sqrt{\pi}}{2} \mathrm{erfc}(x) = \cfrac{e^{-x^2}}{2x + \cfrac{2}{2x + \cfrac{4}{2x + \cfrac{6}{2x + \ddots}}}},$$

and here $a(0) = 0$, $a(n) = 2x$ for $n \geq 1$, $b(0) = e^{-x^2}$, $b(n) = 2n$ for $n \geq 1$, so $\alpha = 0$, $\beta = 1$, $a_0 = 2x$, $b_0 = 2$. It follows that we have alternating sub-exponential convergence

$$S - S(n) \sim (-1)^n C e^{-2x(2n)^{1/2}}.$$

Since $2^{3/2} < 4$, it follows that this continued fraction converges slightly less fast than the preceding one. This corresponds to Case (3) of the proposition.

EXAMPLE 7.4.5.

$$_2F_1(a, b; c; z) = 1 + \frac{ab}{c1!}z + \frac{a(a + 1)b(b + 1)}{c(c + 1)2!}z^2 + \cdots$$

This is the most famous hypergeometric function, first studied in detail by Gauss. Here there exist many continued fraction expansions, most coming from the so-called *contiguity relations* of Gauss. We select three. The first is

$$c\frac{{}_2F_1(a,b;c;z)}{{}_2F_1(a,b+1;c+1;z)} = c - \cfrac{a(c-b)z}{c+1 - \cfrac{(b+1)(c-a+1)z}{c+2 - \cfrac{(a+1)(c-b+1)z}{c+3 - \ddots}}} \, ,$$

so that $a(n) = c+n$, $b(2n) = -(a+n)(c-b+n)z$, $b(2n-1) = -(b+n)(c-a+n)z$. Thus, by Lemma 7.3.2, the contracted continued fraction has

$$A(n) = (c+2n)(c+2n-1) - (b+n)(c-a+n)z$$
$$\qquad - (c+2n)(a+n-1)(c-b+n-1)z/(c+2n-2) \, , \quad \text{and}$$
$$B(n) = -(c+2n+2)(a+n)(c-b+n)(b+n)(c-a+n)z^2/(c+2n) \, .$$

Thus $\alpha = 2$, $\beta = 4$, $a_0 = -2z+4$, $a_1 = c-1/2$, $b_0 = -z^2$, $b_1 = 2c+1$, so $\beta = 2\alpha$ and $a_0^2 + 4b_0 = 16(1-z)$. Thus, if $z < 1$ we are in Case (6) of the proposition, i.e., we have nonalternating exponential convergence for n even

$$S - S(n) \sim C\frac{1}{((1+\sqrt{1-z})^2/|z|)^n} \, .$$

On the other hand, for $z = 1$ we have $a_0^2 + 4b_0 = 0$, and $2a_1 = b_1 - \alpha$, and we compute that $a_2 = ab - ac/2 - bc/2 + c^2/2 - c/4$ and $b_2 = -a^2 + ac + bc - b^2 + c^2 + 3c/2$, so we compute that $B_3 = 4(a+b-c)^2$, hence if $c \neq a+b$ we are in Case (8) of the proposition. It follows that the continued fraction will converge as soon as $c \neq a+b$, with nonalternating polynomial convergence. However, our reasoning was heuristic, and it is easily shown that if $c < a+b$ the continued fraction "tends to infinity", so in fact we must assume that $c > a+b$, and in that case

$$S - S(n) \sim C\frac{1}{n^{2(c-a-b)}} \, .$$

Note that since ${}_2F_1(a,b;c;1) = \Gamma(c)\Gamma(c-a-b)/(\Gamma(c-a)\Gamma(c-b))$, when $z = 1$ we have $S = c-a$.

Note also that in the special case where $a = 1$ (or $b = 1$) we have the continued fraction expansion

$$_2F_1(1,b;c;z) = \cfrac{1}{1 + \cfrac{-bz}{c + \cfrac{(b-c)z}{c+1 + \cfrac{-c(b+1)z}{c+2 + \cfrac{2(b-c-1)z}{c+3 + \cfrac{-(c+1)(b+2)z}{c+4 + \ddots}}}}}} \, ,$$

which has the same convergence properties.

A second continued fraction is of "Euler" type:

$$c\frac{{}_2F_1(a,b;c;z)}{{}_2F_1(a,b+1;c+1;z)} = c + (b-a+1)z$$

$$- \cfrac{(c-a+1)(b+1)z}{c+1+(b-a+2)z - \cfrac{(c-a+2)(b+2)z}{c+2+(b-a+3)z - \cdots}} ,$$

so that $a(n) = c+n+(b-a+1+n)z$, $b(n) = -(b+1+n)(c-a+1+n)z$, hence $\alpha = 1$, $\beta = 2$, $a_0 = z+1$, $a_1 = ((b-a+1)z+c)/(z+1)$, $a_2 = 0$, $b_0 = -z$, $b_1 = c-a+b+2$, and $b_2 = (b+1)(c-a+1)$. Thus $\beta = 2\alpha$, $a_0^2 + 4b_0 = (z-1)^2$. Note that to apply the proposition we need $a_0 > 0$, hence $z > -1$. Thus, if $z > -1$ and $z \neq 1$ we are in Case (6) of the proposition, with exponential convergence (alternating if $z < 0$, nonalternating if $z > 0$), and

$$S - S(n) \sim C\frac{n^{(a-b+c-1)s}}{z^{sn}} ,$$

where $s = +1$ if $|z| > 1$ and $s = -1$ if $|z| < 1$ (strictly speaking we have proved this only for $z > 0$).

If $z = 1$ we have $2a_1 = b_1 - \alpha$, and we compute that $B_3 = (a+b-c)^2$, so as above, if $c > a+b$ we are in Case (8) of the proposition, i.e., we have nonalternating polynomial convergence

$$S - S(n) \sim \frac{C}{n^{c-a-b}} ,$$

again with $S = c - a$.

A third continued fraction is of "Nördlund" type:

$$c\frac{{}_2F_1(a,b;c;z)}{{}_2F_1(a+1,b+1;c+1;z)} = c - (a+b+1)z$$

$$+ \cfrac{(a+1)(b+1)(z-z^2)}{c+1-(a+b+3)z + \cfrac{(a+2)(b+2)(z-z^2)}{c+2-(a+b+5)z + \cdots}} .$$

Here $a(n) = c+n-(a+b+2n+1)z$, $b(n) = (a+n+1)(b+n+1)(z-z^2)$, so $\alpha = 1$, $\beta = 2$, $a_0 = 1 - 2z$, $a_1 = (c-(a+b+1)z)/(1-2z)$, $b_0 = z - z^2$, $b_1 = a+b+2$. Thus $\beta = 2\alpha$ and $a_0^2 + 4b_0 = 1$. We are thus always in Case (6) of the proposition, but since we need $a_0 > 0$ to apply it, we must assume $z < 1/2$. In this case

$$S - S(n) \sim C\frac{n^{a+b-2c+1}}{((z-1)/z)^n} ,$$

so exponential convergence, alternating if $0 < z < 1/2$, nonalternating if $z < 0$.

EXAMPLE 7.4.6. Another ${}_2F_1$ example is as follows:

$$\mathrm{atan}(z) = \cfrac{z}{1 + \cfrac{(1z)^2}{3 + \cfrac{(2z)^2}{5 + \cdots}}} ,$$

so $a(0) = 0$, $a(n) = 2n - 1$ for $n \geq 1$, $b(0) = z$, $b(n) = n^2z^2$ for $n \geq 1$, so that $\alpha = 1$, $\beta = 2$, $a_0 = 2$, $a_1 = -1/2$, $a_2 = 0$, $b_0 = z^2$, $b_1 = b_2 = 0$. Here $\beta = 2\alpha$

and $a_0^2 + 4b_0 = 4(z^2 + 1)$. It follows that the convergence is always alternating exponential, with

$$S - S(n) \sim (-1)^n C \frac{1}{((1 + \sqrt{z^2 + 1})/|z|)^{2n}} \ .$$

On the other hand, if we change z to iz, i.e., if we consider

$$\operatorname{atanh}(z) = \cfrac{z}{1 - \cfrac{(1z)^2}{3 - \cfrac{(2z)^2}{5 - \ddots}}}$$

with z real, we have $b(n) = -n^2 z^2$ so $b_0 = -z^2$, hence $a_0^2 + 4b_0 = 4(1 - z^2)$. Thus we first need $|z| \le 1$, otherwise the continued fraction will be oscillating and need not converge (and indeed it does not). If $|z| < 1$, we have $a_0^2 + 4b_0 > 0$, so we now have nonalternating exponential convergence with

$$S - S(n) \sim C \frac{1}{((1 + \sqrt{1 - z^2})/|z|)^{2n}} \ .$$

On the other hand, if $|z| = 1$ we have $a_0^2 + 4b_0 = 0$, $2a_1 = b_1 - \alpha$, and $B_3 = 0$, so the continued fraction will not converge. Once again these examples correspond to Cases (6) and (8) of the proposition.

EXAMPLE 7.4.7. S. Ramanujan was a master in continued fractions. One of them (in fact discovered earlier by Barnes in 1872) is as follows:

$$\frac{\Gamma^2\left(\dfrac{x+1}{4}\right)}{\Gamma^2\left(\dfrac{x+3}{4}\right)} = \cfrac{4}{x + \cfrac{1^2}{2x + \cfrac{3^2}{2x + \cfrac{5^2}{2x + \ddots}}}} \ .$$

Here $a(0) = 0$, $a(1) = x$, $a(n) = 2x$ for $n \ge 2$, $b(0) = 4$, $b(n) = (2n - 1)^2$ for $n \ge 1$, so $\alpha = 0$, $\beta = 2$, $a_0 = 2x$, and $b_0 = 4$. We have $\beta = 2\alpha + 2$ and $b_0 > 0$, so for $x > 0$ the continued fraction has alternating polynomial convergence:

$$S - S(n) \sim (-1)^n C \frac{1}{n^x} \ .$$

This corresponds to Case (4) of the proposition. We will see below that it is possible to *accelerate* this continued fraction, and obtain one which converges exponentially fast.

EXAMPLE 7.4.8. Define the function $J(x)$ for $x > 0$ by the formula

$$\log(\Gamma(1 + x)) = (x + 1/2)\log(x) - x + \log(2\pi)/2 + J(x) \ .$$

By Euler–Maclaurin, $J(x)$ has the nonconvergent asymptotic expansion

$$J(x) = \sum_{n \ge 1} \frac{B_{2n}}{(2n - 1)(2n)} \frac{1}{x^{2n-1}} = \frac{1}{12x} - \frac{1}{360x^3} + \cdots$$

As usual with Euler–Maclaurin, we can hope to obtain an accuracy of 10^{-D} only if $x > D/3$ approximately, so at the default 38 decimals, $x \geq 13$. Nonetheless, we can expand this asymptotic expansion into a continued fraction, more precisely, we can formally write

$$S(X) = \sum_{n \geq 1} \frac{B_{2n}}{(2n-1)(2n)} X^n = \cfrac{X/12}{1 + \cfrac{X/30}{1 + \cfrac{53X/210}{1 + \cdots}}}$$

(this can be done automatically using the Quodif algorithm below), so that $J(x) = xS(1/x^2)$. An experimental study (which should be easy to prove) shows that the numerators $b(n)$ are very close to $n^2 X/16$, hence those of $J(x)$ close to $n^2/(16x^2)$. We are thus again in Case (4) of the proposition, so that

$$J(x) - S(n) \sim (-1)^n \frac{C}{n^{4x}} \ .$$

We have thus transformed an everywhere divergent series into a (slowly) convergent continued fraction for all $x > 0$.

Consider for instance $x = 10$. Using Euler–Maclaurin the optimal value of $2n$ that we can choose is $n = 32$, which gives 28 decimals. On the other hand, taking only 10 terms of the continued fraction gives 40 decimals, and we would obtain 60 decimals by taking 32 terms.

EXAMPLE 7.4.9. Recall from Theorem 3.5.5 and Corollary 3.5.6 that any family of orthogonal polynomials gives rise to *finite* continued fractions. Even though the main point is to use these finite expressions, we can ask whether the corresponding *infinite* continued fraction converges, when the variable X is given some fixed value.[1] The reader can first check that in all the special cases that we have given, the continued fractions involving the c_n given by Theorem 3.5.7 do *not* converge. On the other hand, those involving the a_n and b_n do converge for suitable values of the variable X. As an example, you can check that the continued fraction used in Example 1 of Section 3.6.7 (special Jacobi polynomials) converges like $1/n^{4(\alpha^2/4 - X)^{1/2}}$ when $X < \alpha^2/4$, and that of Example 2 ($1/\cosh(\pi x)$) converges like $1/n^{2(-X)^{1/2}}$ when $X < 0$, both corresponding to Case (8).

Exercise: in this last case, guess an expression for the limiting value as a Laplace transform, and check its numerical validity. Note that, even though the convergence is slow, the CFlim program instantly gives the correct limiting value. For instance, check that the limiting value for $X = -1$ is equal to

$$-2 \int_0^\infty \frac{e^{-2t}}{\cosh(t)} \, dt \ ,$$

which gives a value to the alternating but diverging series $\sum_{k \geq 0} E_{2k}/2^{2k}$.

---

[1]Warning: because of the numbering used for orthogonal polynomials, the sequence (a_n, b_n) using the notation of continued fractions is in fact the sequence $(X - a_{n-1}, -b_n)$ in the notation of orthogonal polynomials.

7.5. Convergence acceleration of continued fractions

In the preceding subsections, we have seen that in many cases we have super-exponential, exponential, or sub-exponential convergence, so we can compute numerically the value of the continued fraction using one of the naïve methods explained above. However, in two cases the convergence is slow (i.e., polynomial): first when $\beta = 2\alpha+2$ and $b_0 > 0$ (Case (4) of the proposition), where we have alternating polynomial convergence, and second when $\beta = 2\alpha$, $a_0^2 + 4b_0 = 0$, $2a_1 = b_1 - \alpha$, and $B_3 > 0$ with B_3 as above (Case (8) of the proposition), where we have nonalternating polynomial convergence. In these cases we must find a method to compute the continued fraction to high accuracy. In addition, we have Case (5) where usually the continued fraction does not converge, but the odd and even terms do separately, and we may want also to accelerate that case.

Let us consider a typical example of the first case (Case (4) of the proposition): the continued fraction for $\Gamma((x+1)/4)^2/\Gamma((x+3)/4)^2$ given above. We have $a(n) = 2x$ and $b(n) = (2n+1)^2$. Thus $\alpha = 0$, $\beta = 2$, $a_0 = 2x$, $b_0 = 4$, so we are indeed in that case, and we have seen that $S(n+1) - S(n) \sim (-1)^n C/n^x$, $S(n+2) - S(n) \sim C/n^{x+1}$, and $S - S(n) \sim (-1)^n C/n^x$. A natural idea is to use the `Sumalt` program, since S is the sum of an alternating series. However the reader can check that in this way we only obtain two or three decimals of accuracy. The main reason is due to the fact that the alternating series is not "reasonable", as explained above.

The next idea is to use variants of the Lagrange extrapolation method, which is the program `CFlim` given above. And in the present case, we obtain perfect results when x has a small denominator q, using `al=1/q`, and still good results otherwise. Warning: all parameters such as x must be computed to large accuracy beforehand. For instance, if $x = 1/\pi$ and $a(x)$ and $b(x)$ are the coefficients of the above continued fraction, `CFlim(a(1/Pi), b(1/Pi), 1/Pi)` gives nonsense, while

```
default(realprecision, 1000);
P = 1/Pi;
default(realprecision, 38);
CFlim(a(P), b(P), P)
```

gives 25 correct decimals out of 38.

A different method for convergence acceleration is called the *Bauer–Muir* transform. Recall that if $\rho(n)$ is the nth tail of the continued fraction with limit S, we have $S = (p(n+1) + \rho(n+1)p(n))/(q(n+1) + \rho(n+1)q(n))$ and the recursion $\rho(n)(a(n+1) + \rho(n+1)) - b(n) = 0$. Let $r(n)$ be an approximation to $\rho(n)$, and set $p'(n) = p(n+1) + r(n+1)p(n)$ and $q'(n) = q(n+1) + r(n+1)q(n)$. If $r(n)$ is a sufficiently good approximation, then $p'(n)/q'(n)$ will be a much better approximation to S than $p(n)/q(n)$. It is also clear that $r(n)$ is a good approximation if and only if $d(n) := r(n)(a(n+1) + r(n+1)) - b(n)$ is close to 0. Let us find precise formulas and quantify all this.

LEMMA 7.5.1. *As above, set*

$$p'(n) = p(n+1) + r(n+1)p(n) \,,$$
$$q'(n) = q(n+1) + r(n+1)q(n) \,,$$
$$d(n) = r(n)(a(n+1) + r(n+1)) - b(n) \,,$$

and assume for simplicity that $d(n) \neq 0$ for all n. We have the formal identity

$$\frac{p'(n)}{q'(n)} = a'(-1) + \cfrac{b'(-1)}{a'(0) + \cfrac{b'(0)}{a'(1) + \ddots + \cfrac{b'(n-1)}{a'(n)}}} \ ,$$

where

$$a'(-1) = a(0) + r(0) \ , \quad a'(0) = a(1) + r(1) \ ,$$
$$a'(n) = a(n+1) + r(n+1) - r(n-1)d(n)/d(n-1) \quad for \ n \geq 1 \ ,$$

and

$$b'(-1) = -d(0) \ , \quad b'(n) = b(n)d(n+1)/d(n) \quad for \ n \geq 0 \ .$$

PROOF. Immediate formal computation left to the reader. □

Let us now look at the different slow convergence types and see what we can do.

(1) In Case (4), $\beta = 2\alpha + 2$ and $b_0 > 0$. Here we choose naturally $r(n) = b_0^{1/2}n^{\beta/2}(1 + r_1/n)$, so that

$$d(n) = b_0^{1/2}n^{\beta/2}(1 + r_1/n) \times$$
$$\left(a_0 n^{\beta/2-1} + b_0^{1/2}n^{\beta/2}(1 + (r_1 + \beta/2)/n)\right) - b_0 n^\beta (1 + b_1/n) \ ,$$

so identifying the coefficients of $1/n$ gives $r_1 = (b_1 - \beta/2 - a_0/b_0^{1/2})/2$, and then $d(n) = b_0 n^\beta C/n^2$. Now by our previous analysis, we know that $q(n+1)/q(n) \sim b_0^{1/2}n^{\beta/2}$, so that $q'(n) \sim 2b_0^{1/2}n^{\beta/2}q(n)$, and since $b'(0) \cdots b'(n) = b(0) \cdots b(n)d(n+1)/d(0)$, it follows that

$$\frac{b'(0) \cdots b'(n)}{q'(n)q'(n+1)} = \frac{b(0) \cdots b(n)}{q(n)q(n+1)}C/n^2 \ .$$

We deduce from Case (4) that

$$S - S'(n) = (-1)^n \frac{\mathcal{C}(1/n^{1/2})}{n^{a_0/b_0^{1/2}+2}} \ .$$

We have thus notably improved the convergence. Of course, if we also identify the coefficients of $1/n^2$, $1/n^3$, etc. in the above formula for $d(n)$, we can accelerate the convergence as much as we want.

(2) Case (8) is similar but more complicated. After a little computation, we find that to gain a factor $1/n^2$ as above we must choose $r(n) = -(a_0/2)n^\alpha(1 + r_1/n)$ with $r_1 = (b_1 + 1 - \alpha + \sqrt{B_3})/2$, using the notation of Case (8).

(3) Recall that in Case (5) only the odd and even terms of the continued fraction converge, but in general not the continued fraction itself. This is because we find that $p(n+1)/q(n+1) - p(n)/q(n)$ tends to a constant (which a priori is nonzero) for a given parity of n. However, if we perform the Bauer–Muir transform with $r(n) = b_0^{1/2}n^{\beta/2}(1 + (b_1 - \beta/2)/(2n))$, we again find an acceleration of a factor of $1/n^2$, so that $p'(n+1)/q'(n+1) - p'(n)/q'(n)$ will now be asymptotic to $(-1)^n C/n^2$, so the new continued fraction will converge. Note also that it is not always easy to compute β,

b_0, etc., so we note that in the present case we can take instead $r(n) = (b(n-1)b(n))^{1/4}$.

(4) Note that in two cases of the proposition (Cases (3) and (7)) we have sub-exponential convergence in $e^{-Cn^{1/2}}$ for some $C > 0$. Even though this is relatively fast convergence, we could hope to accelerate it since the number of terms to be computed for a given accuracy will be proportional to the *square* of that accuracy. Note that this is particularly important in the context of the computation of inverse Mellin transforms, essential for computing L-functions, by using the method of Dokchitser which (at least conjecturally) always gives continued fractions with sub-exponential convergence.

Unfortunately, the reader can check that Bauer–Muir transforms are essentially useless here since at best they improve convergence by a small power of n, which gives a negligible gain in view of the sub-exponential convergence.

We can in principle iterate the Bauer–Muir transform and obtain more and more accelerations. However the formulas become in general more complicated. In some cases however, we can compute the successive accelerations explicitly and even use diagonal processes to speed up the continued fraction even more: this is the basis of Apéry's proof of the irrationality of $\zeta(3)$. Let us explain how this works:

By abuse of notation, set $a(n,0) = a(n)$, $b(n,0) = b(n)$, $p(n,0) = p(n)$, and $q(n,0) = q(n)$. Assume that we give ourselves a sequence in two variables $r(n,\ell)$, and apply by induction the Bauer–Muir transformation using this sequence for ℓ fixed, so that for $\ell \geq 0$:

$$p(n, \ell + 1) = p(n + 1, \ell) + r(n + 1, \ell)p(n, \ell)$$

and similarly for $q(n, \ell + 1)$. We thus obtain two variable sequences $a(n,\ell)$, $b(n,\ell)$, and $d(n,\ell)$, and with a suitable choice of $r(n,\ell)$, the quotient $p(n,\ell)/q(n,\ell)$ will progressively accelerate the continued fraction. But better, we can consider the diagonal $p(n,n)/q(n,n)$ (or possibly $p(n, n + \varepsilon)/q(n, n + \varepsilon)$ for $\varepsilon = \pm 1$), and hope for even faster convergence.

Let us give what is probably the simplest example: consider Ramanujan's continued fraction given above, in the form

$$\frac{4\Gamma^2((x+3)/4)}{\Gamma^2((x+1)/4)} = x + \cfrac{1^2}{2x + \cfrac{3^2}{2x + \cfrac{5^2}{2x + \ddots}}} \, .$$

We thus have $a(0,0) = a(0) = x$, $a(n,0) = a(n) = 2x$ for $n \geq 1$, and $b(n,0) = b(n) = (2n+1)^2$. For $n \geq 0$ we want $d(n) = r(n)(a(n+1) + r(n+1)) - b(n)$ to be small, so we look for $r(n) = 2n + r_1$, and replacing gives $d(n) = 4(r_1 + x)n + r_1^2 + (2x + 2)r_1 - 1$, so we choose $r_1 = -x$, i.e., $r(n) = 2n - x$, hence $d(n) = -(x + 1)^2$. What is remarkable is that $d(n)$ does not depend on n, so the formulas simplify considerably. Inserting the index, we have $r(n,0) = 2n - x$, $d(n,0) = -(x+1)^2$, hence by Lemma 7.5.1, if we set $a(n,1) = a'(n)$ and $b(n,1) = b'(n)$, we have $a(0,1) = x + 2$, $a(n,1) = 2x + 4$ for $n \geq 1$, and $b(n,1) = (2n+1)^2$. Thus we have simply changed x into $x + 2$ (this of course follows from the functional equation of the gamma function). We have seen in Example 7.4.7 above that the convergence of the initial continued fraction was in $1/n^x$, so it is now in $1/n^{x+2}$, consistent with the fact that we gain a factor $1/n^2$ as mentioned above.

But this is not the point. Continuing in this way, we evidently have $a(n, \ell) = 2(x + 2\ell)$, $b(n, \ell) = (2n + 1)^2$, $r(n, \ell) = 2n - (x + 2\ell)$, and $d(n, \ell) = -(x + 2\ell + 1)^2$. Note that we did *not* set $a(n, 1) = a'(n - 1)$ and $b(n, 1) = b'(n - 1)$, as would be more natural in view of Lemma 7.5.1, since otherwise we would not be able to proceed by induction as we are doing now. The following lemma is a restatement of Lemma 7.5.1, and the corollaries follow by an immediate computation:

LEMMA 7.5.2. *Let $u(n, \ell)$ denote $p(n, \ell)$ or $q(n, \ell)$. We have the following formulas:*

$$R(n, \ell) := a(n, \ell) + r(n, \ell) ,$$
$$d(n, \ell) := r(n, \ell)R(n + 1, \ell) - b(n, \ell) ,$$
$$u(n + 1, \ell) = a(n + 1, \ell)u(n, \ell) + b(n, \ell)u(n - 1, \ell) ,$$
$$u(n, \ell + 1) = u(n + 1, \ell) + r(n + 1, \ell)u(n, \ell) ,$$
$$a(0, \ell + 1) = R(1, \ell) ,$$
$$a(n, \ell + 1) = R(n + 1, \ell) - r(n - 1, \ell)d(n, \ell)/d(n - 1, \ell) ,$$
$$b(n, \ell + 1) = b(n, \ell)d(n + 1, \ell)/d(n, \ell) .$$

COROLLARY 7.5.3. *We have*

$$u(n, \ell + 1) = R(n + 1, \ell)u(n - 1, \ell + 1) - d(n, \ell)u(n - 1, \ell) \quad and$$
$$u(n, \ell + 1) = R(n + 1, \ell)u(n, \ell) + b(n, \ell)u(n - 1, \ell) .$$

COROLLARY 7.5.4. *We have*

$$\frac{p(n, n)}{q(n, n)} = a(0) + \cfrac{b(0)}{R(1, 0) + \cfrac{-d(1, 0)}{R(2, 0) + \cfrac{b(1, 1)}{R(2, 1) + \cfrac{-d(2, 1)}{R(3, 1) + \cdots + \cfrac{-d(n, n - 1)}{R(n + 1, n - 1)}}}}} .$$

In the Ramanujan example, we obtain:

PROPOSITION 7.5.5. *We have*

$$\frac{4\Gamma^2((x + 3)/4)}{\Gamma^2((x + 1)/4)} = x + \cfrac{1^2}{x + 2 + \cfrac{(x + 1)^2}{x + 4 + \cfrac{3^2}{x + 6 + \cfrac{(x + 3)^2}{x + 8 + \cfrac{5^2}{x + 10 + \cdots}}}}} .$$

In particular, we have

$$\frac{\Gamma^2(1/4)}{\Gamma^2(3/4)} = \cfrac{8}{1 - \cfrac{1^2}{12 - \cfrac{3^2}{24 - \cfrac{5^2}{36 - \ddots}}}} \ .$$

PROOF. The first formula follows from the corollary, and the second from Lemma 7.3.2 after a small computation left to the reader.   □

Using Theorem 7.3.7, we see that for the special case of the second continued fraction (but this is true in general) we are in Case (6), and $c_0 = 2(1 + \sqrt{2})^2$, so we have nonalternating exponential convergence with

$$S - S(n) \sim \frac{C}{\left(1 + \sqrt{2}\right)^{4n}} \ .$$

We have thus transformed a continued fraction which converged polynomially into one converging exponentially. One can prove that $C = 4(\Gamma^2(1/4)/\Gamma^2(3/4))$.

A slightly more complicated example, which leads to Apéry's work, is as follows. If $f(n)$ is some arithmetic function, we will denote by $f!(n)$ for $n \geq 0$ the function defined by $f!(0) = 1$ and $f!(n) = f(n)f!(n-1)$. We then have the following lemma due to Euler:

LEMMA 7.5.6. *Let f and g be two arithmetic functions and assume that $f(n) \neq 0$ for all n. We have*

$$\sum_{1 \leq j \leq n} \frac{g!(j-1)}{f!(j)} = a(0) + \cfrac{b(0)}{a(1) + \cfrac{b(1)}{a(2) + \cdots + \cfrac{b(n-1)}{a(n)}}} = \frac{p(n)}{q(n)} \ ,$$

with $a(0) = 0$, $b(0) = 1$, $a(1) = f(1)$, $a(n) = f(n) + g(n-1)$ for $n \geq 2$ $b(n) = -f(n)g(n)$ for $n \geq 1$, $q(n) = f!(n)$, and $p(n) = f!(n)\sum_{1 \leq j \leq n} g!(j-1)/f!(j)$.

PROOF. Immediate formal computation left to the reader.   □

This allows the formal transformation of a series into a continued fraction, so that instead of using methods to accelerate series, we can use methods to accelerate continued fractions, which may be rather different, and in particular the Bauer–Muir transform. For instance:

COROLLARY 7.5.7. *We have*

$$\zeta(k) = \cfrac{1}{1^k + 0^k - \cfrac{1^{2k}}{2^k + 1^k - \cfrac{2^{2k}}{3^k + 2^k - \ddots}}} \ .$$

This follows from the lemma by taking $f(n) = g(n) = n^k$. Of course we have not changed the convergence: we are in Case (8), $B_3 = (k-1)^2$, and $S - S(n) \sim C/n^{k-1}$ as in the original series.

For $k = 2$, 3, and 4, we may, however, apply recursively the Bauer–Muir transform and obtain clean formulas as we did above for Ramanujan's continued fraction. Let us give the example for $k = 3$, and leave $k = 2$ to the reader (the case $k = 4$, due to G. Rhin and the second author, is slightly more complicated). We only give the final formulas, and the reader can check that they indeed satisfy the recursions of Lemma 7.5.2:

$$a(n, \ell) = 2n^3 - 3n^2 + 3n - 1 + 2(\ell^2 + \ell)(2n - 1) \ ,$$

$$b(n, \ell) = -n^6 \ ,$$

$$r(n, \ell - 1) = -n^3 + 2\ell n^2 - 2\ell^2 n + \ell^3 \ ,$$

$$R(n + 1, \ell - 1) = n^3 + 2\ell n^2 + 2\ell^2 n + \ell^3 \ ,$$

$$d(n, \ell - 1) = \ell^6 \ .$$

After a small computation, we can then obtain Apéry's continued fraction

$$\zeta(3) = \cfrac{6}{a(1) - \cfrac{1^6}{a(2) - \cfrac{2^6}{a(3) - \cfrac{3^6}{a(4) - \cdots}}}} \ ,$$

with

$$a(n) = 34n^3 - 51n^2 + 27n - 5 \ .$$

We have again nonalternating exponential convergence with $S - S(n) \sim C/(1 + \sqrt{2})^{8n}$.

By studying the arithmetic properties of $p(n, n)$ and $q(n, n)$, it is then possible to prove the irrationality of $\zeta(3)$.

7.6. The quotient-difference algorithm

In the computation of L-functions, it is essential to be able to compute *inverse Mellin transforms* and *generalized incomplete gamma functions*. We do not need to give the precise definition for now, but we consider an important special case, the K-Bessel function, and in particular the function $K_0(x)$ defined for $x > 0$ by

$$K_0(x) = \frac{1}{2} \int_0^\infty e^{-(x/2)(t+1/t)} \frac{dt}{t} \ .$$

It is immediate to show that it satisfies the second order linear differential equation $y'' + y'/x - y = 0$, and that its generalized power series expansion at 0 is

$$K_0(x) = \sum_{k \geq 0} \frac{(x/2)^{2k}}{k!^2} (H_k - \gamma - \log(x/2)) \ ,$$

where as usual $\gamma = 0.577\cdots$ is Euler's constant and $H_k = \sum_{1 \leq j \leq k} 1/j$ is the harmonic sum. When x is not too large, one can use this power series to compute $K_0(x)$ very efficiently. However, as soon as x is large, the series becomes difficult to use. Consider for instance $x = 100$, which is not that large! Working at a default precision of 38 decimal digits, we seem to find that $K_0(100) = -311990.519\cdots$, which is totally wrong. We get wrong answers as long as the accuracy is less than

100. This is similar to the phenomenon of evaluating e^{-x} by the power series $\sum_{k\geq0}(-1)^k x^k/k!$.

When x is a little large, it is thus useful to use an *asymptotic expansion*, here:

$$K_0(x) = \left(\frac{\pi}{2x}\right)^{1/2} e^{-x}\left(\sum_{m\geq0}(-1)^m\frac{(2m)!^2}{2^{5m}m!^3}x^{-m}\right).$$

As usual in this kind of expansion, the error made in truncating is of the order of the first neglected term. For a given x this gives an optimal truncation for the sum around $m \approx 2x$, for a relative error of the order of $e^{-2x}/x^{1/2}$ by Stirling's formula. If x is large with respect to the accuracy this is less than 2^{-B} and all is well, but for x of intermediate size this may not be sufficient.

An idea first introduced in this context by T. Dokchitser is to expand the asymptotic series formally as a continued fraction, and try to evaluate the continued fraction. Let us take a typical example: let us compute $K_0(10)$ using the default accuracy of 38 digits. Using the asymptotic expansion we must choose $m = 20$, and we obtain the result to 15 decimal digits, already not that bad (since $K_0(x)$ is of the order of e^{-x} we will have a result with an absolute error of the order of e^{-3x}). On the other hand, using Dokchitser's idea we write

$$\sum_{m\geq0}(-1)^m\frac{(2m)!^2}{2^{5m}m!^3}x^{-m} = 1 + \cfrac{b(0)x}{1+\cfrac{b(1)x}{1+\cdots}},$$

and we compute that $b(0) = -1/8$, $b(1) = 9/16$, $b(2) = 23/48$, etc. Using exactly as many terms (i.e., $m = 20$) already gives 23 decimal digits instead of 15. But, contrary to the asymptotic expansion, there is no reason to stop at $m = 20$, and in fact it *seems* that the continued fraction converges. Choosing $m = 55$ gives perfect accuracy (for $x = 10$ at 38 decimal digits).

There exists an algorithm to compute $b(n)$ knowing the asymptotic expansion, called the *quotient-difference algorithm* (or the QD algorithm for short), due to H. Rutishauser. It is based on the following formulas:

LEMMA 7.6.1. *Consider the formal equality*

$$c(0) + c(1)z + c(2)z^2 + \cdots = c(0) + \cfrac{b(0)z}{1+\cfrac{b(1)z}{1+\cfrac{b(2)z}{1+\cdots}}}.$$

Define two arrays $e(j,k)$ for $j \geq 0$ and $q(j,k)$ for $j \geq 1$ by $e(0,k) = 0$, $q(1,k) = c(k+2)/c(k+1)$ for $k \geq 0$, and by induction

$$e(j,k) = e(j-1,k+1) + q(j,k+1) - q(j,k) \quad \text{for } j \geq 1 \text{ and } k \geq 0,$$
$$q(j+1,k) = q(j,k+1)e(j,k+1)/e(j,k) \quad \text{for } j \geq 1 \text{ and } k \geq 0.$$

Then $b(0) = c(1)$, and for $n \geq 1$, $b(2n-1) = -q(n,0)$ and $b(2n) = -e(n,0)$.

PROOF. The proof is easily done by induction and is left to the reader who can also refer to [**Hen58**]. $\square$

REMARKS 7.6.2.

(1) This algorithm can fail if some $e(j, k)$ becomes 0 or very close to 0. There are methods to overcome this which we will not go into.

(2) If the algorithm is applied with floating point numbers (as opposed to rational numbers if the $c(i)$ are rational), it is rather unstable. There exist more stable versions, but again it is not our purpose to go into details. We explain below how to increase the working accuracy to alleviate the instability problem.

(3) Keeping *arrays* e and q as in the lemma is costly, and we can easily dispense with them and simply keep the latest vectors of e and q as follows:

Quodif.gp

```
1   /* Given a vector c, compute the b[n] for 1 <= n <= N, where
2    * c[1]z + c[2]z^2 + ... = b[1]z / (1 + b[2]z / (1 + ...)) */
3   Quodif(c, N = #c - 1) =
4   { my(b, e, q);
5
6     b = vector(N + 1); b[1] = c[1];
7     e = vector(N);
8     q = vector(N, k, c[k+1] / c[k] * 1.);
9     for (j = 1, N \ 2,
10      my(L = N - 2*j);
11      b[2*j] = -q[1];
12      for (k = 0, L, e[k+1] = e[k+2] + q[k+2] - q[k+1]);
13      b[2*j+1] = -e[1];
14      for (k = 0, L-1, q[k+1] = q[k+2] * e[k+2] / e[k+1]));
15    if (N % 2, b[N+1] = -q[1]);
16    return (b);
17  }
```

REMARK 7.6.3. The above algorithm uses $O(N^2)$ floating point operations. Indeed, line 8 converts the input to floating point numbers and prevents coefficient explosion in case the $c(i)$ are rational. For the exact version, replace this line by

```
q = vector(N, k, c[k+1] / c[k]);
```

Experiments show that if one uses the floating point version, which is orders of magnitude faster, one must take a default accuracy proportional to the number N of desired terms. For instance, for the Bessel functions K_ν, approximately N decimal digits are required. Note also that as for many continued fractions, the behavior of the odd and even convergents may be slightly different. In that case, it is useful to convert the continued fraction obtained by the QD algorithm into its even part by using Corollary 7.3.3.

Now experimentally again for K_0, it seems that $b(n) \approx n/4 - 0.1$ for n even and $b(n) \approx n/4 + 0.35$ for n odd. If we look at Theorem 7.3.7, since $a(n) = 1$ we have $\alpha = 0$, $\beta = 1$, $a_0 = 1$, $b_0 = z/4$ so we are in Case (3), and since in the asymptotic expansion of $K_0(x)$ we have $z = 1/x$, we expect

$$S - S(n) \sim (-1)^n C e^{-4(nx)^{1/2}} .$$

And indeed, for $x = 10$ we find that $S - S(n)$ is of the order of 10^{-25} for $n = 20$ and 10^{-41} for $n = 55$, in perfect accordance with what we have observed above.

More generally, K_ν has the asymptotic expansion

$$\left(\frac{\pi}{2x}\right)^{1/2} e^{-x} {}_2F_0\left(\frac{1}{2} - \nu, \frac{1}{2} + \nu; -\frac{1}{2x}\right) ,$$

and it seems that $b(n) \approx n/4 + \nu/4 - 1/8$ for n even and $b(n) \approx n/4 - \nu/4 + 3/8$ for n odd, so the speed of convergence is essentially the same.

On the other hand, if we try to do the same for inverse Mellin transforms of higher powers of $\Gamma(s/2)$, for instance of $\Gamma(s/2)^3$, the continued fraction is completely irregular, although Dokchitser's method still seems to converge, as we will see in the next chapter.

Exercise: We have seen above that the `Quodif` algorithm (more precisely, its exact version) applied to $\sum_{n \geq 1} B_{2n}/((2n-1)(2n))x^n$ gives an output with complicated fractions. You can check that the same is true for $\sum_{n \geq 1} B_{2n}/(2n)x^n$.

(1) On the contrary, find experimentally the continued fractions corresponding to $\sum_{n \geq 1} B_{2n}x^n$, and to $\sum_{n \geq 1}(2n+1)B_{2n}x^n$.
(2) Deduce continued fractions for $\psi'(x)$ and $\psi''(x)$, where ψ is the logarithmic derivative of the gamma function (very improperly named the digamma function).
(3) Compute their values at $x = 1$, and in analogy with the exercise given at the end of Example 7.4.9, guess an expression for them as a definite integral.

7.7. Evaluation of the quotient-difference output

Once we have performed the QD algorithm, we can compute the value at t of the corresponding series. A simple-minded program is a minor modification of the `CFback` algorithm:

─────────────────────────────── CFeval.gp ───────────────────────────────

```
1  /* Given a continued fraction CF output by
2   * the Quodif program, evaluate its first N terms at t. */
3  CFeval(CF, N, t)=
4  { my(S = 0.);
5
6    if (N > #CF, error("short continued fraction"));
7    forstep (j = N, 1, -1, S = CF[j] * t / (1 + S));
8    return (S);
9  }
```

Note that we need to add $c(0)$ to the result! The time taken by this algorithm is approximately $N(A + M + D)$, where A (resp., M, resp., D) is the time taken for an addition (resp., a multiplication, resp., a division).

This, however, can easily be improved by computing two steps of the continued fraction at once, changing the recursion to

$$S = c(j-1)t/(1 + c(j)t/(1+S)) = c(j-1)t(1+S)/(1 + S + c(j)t) .$$

Written in this way, the time is now approximately $N(2A + 3M + D)/2 = N(A + (3/2)M + D/2)$, and this is already faster than the previous method since in practice $D > M$, in fact $D > 1.5M$. However, we can write this in the more clever way

$$S = c(j-1)(1+S)/((1+S)t' + c(j)) \,,$$

where $t' = 1/t$ is computed once and for all, and this now only requires time $N(2A + 2M + D)/2 = N(A + M + D/2)$. We can still gain a little by computing three or more steps of the continued fraction at once, but we are now in a situation of diminishing returns.

There is, however, a different way to improve the evaluation speed. By Corollary 7.3.3 we have the identity

$$\cfrac{b(1)z}{1 + \cfrac{b(2)z}{1 + \cfrac{b(3)z}{1 + \cfrac{b(4)z}{1 + \cdots}}}} = \cfrac{B(1)z}{1 + A(1)z + \cfrac{B(2)z^2}{1 + A(2)z + \cfrac{B(3)z^2}{1 + A(3)z + \cdots}}} \,,$$

where $A(1) = b(2)$, $A(n) = b(2n-1) + b(2n)$ for $n \geq 2$, $B(1) = b(1)$, and $B(n) = -b(2n-2)b(2n-1)$ for $n \geq 2$. For evaluation purposes, this last continued fraction is better written as

$$\cfrac{B(1)}{1/z + A(1) + \cfrac{B(2)}{1/z + A(2) + \cfrac{B(3)}{1/z + A(3) + \cdots}}} \,,$$

which leads once again to a trivial evaluation program which we do not need to write explicitly, and which now requires only $(N/2)(A + D)$ operations, since this new continued fraction must now be computed only to $N/2$ terms. This is a considerable gain compared to the above. Of course, once again we can improve on this by taking two or more steps of the continued fraction at once. If we take two, this will require $(N/2)(2A + M + D/2)$ operations, and if we take three, it will require $(N/2)(4A/3 + 4M/3 + D/3)$ operations, which in practice is slightly better. We only give the algorithm where we take two steps of the continued fraction at once.

─────────────────── CFEuler.gp ───────────────────

```
1   /* Given a continued fraction CF output by the Quodif program,
2    * convert it into an Euler type continued fraction A(n),B(n),
3    * where
4    * b[1]z / (1 + b[2]z / (1 + b[3]z / (1 + ... + b[lim]z)))
5    * = B[1] / (1/z + A[1] + B[2] / (1/z + A[2]
6    *    + B[3] / (1/z + ... + B[lim\2] / (1/z + A[lim\2])))). */
7   CFtoEuler(CF) =
8   { my(A = vector(#CF \ 2), B = vector(#A));
9
10    for (i = 2, #A,
11      A[i] = CF[2*i] + CF[2*i-1];
12      B[i] = -CF[2*i-1] * CF[2*i-2]);
13    A[1] = CF[2];
```

```
14    B[1] = CF[1]; return ([A, B]);
15  }
16
17  /* Given a continued fraction CFAB output by CFtoEuler,
18   * evaluate its first N terms at 1 / t. */
19  CFEulereval(CFAB, N, t) =
20  { my([A, B] = CFAB, S = 0);
21
22    if (#A < N, error("short continued fraction"));
23    if (N%2, S = B[N] / (A[N] + t); N--);
24    forstep (j = N, 2, -2,
25      S += A[j] + t;
26      S = B[j-1] * S / ((A[j-1] + t) * S + B[j]));
27    return (S);
28  }
```

As an application of the above, we give a program to compute values of the K-Bessel function $K_0(x)$ for reasonably large values of x, using the explicit asymptotic expansion given above. Of course this program is usually slower than the built-in function besselk, but gives a simple example of a general program that we will give in the next chapter for computing inverse Mellin transforms.

────────────────────────────── Besselk0.gp ──────────────────────────────

```
1   Besselk0init() =
2   { my(B = getlocalbitprec(), E, n, T, c);
3
4     localbitprec(32);
5     E = B * log(2); T = E / 25;
6     n = ceil(E^2 / (13 * T));
7     c = vector(n); c[1] = - 1/8;
8     for (m = 2, n, c[m] = -c[m-1] * (2*m-1)^2 / (8 * m));
9     localbitprec(2 * B);
10    return ([CFtoEuler(Quodif(c)), E, T]);
11  }
12
13  /* Data output by Besselk0init(). */
14  Besselk0(x, Data = 0) =
15  {
16    if (!Data, Data = Besselk0init());
17    my([CFAB, E, T] = Data, k);
18    if (x < T, error("x too small for Besselk0"));
19    k = sqrt(Pi / (2 * x)) * exp(-x);
20    if (x < E,
21      my(n = ceil(E^2 / (13 * x)));
22      k *= 1 + CFEulereval(CFAB, n \ 2, x));
23    return (k);
24  }
```

REMARK 7.7.1. Here and for other hypergeometric expansions, it would be more elegant (and a little more efficient) to input directly

$$C = \big(c(1), c(2)/c(1), \dots, c(N)/c(N-1)\big)$$

to Quodif since the elements of C are then simple rational functions evaluated at consecutive small integers instead of huge quotients of factorials. In the present case, the vector c in BesselkOinit would be replaced by

```
C = vector(n, m, - (2*m-1)^2 / (8 * m));
```

We leave the corresponding adaptation of Quodif as an exercise to the reader.

EXAMPLE.

```
? \p500
? Data = BesselkOinit();
time = 8,124 ms.
? Besselk0(50, Data) - besselk(0, 50)
time = 7 ms.
% = 0.E-523 /* Perfect. */
? \p1000
? R = besselk(0, 500)
% = 3.992... E-219 /* Reference value. */
? \p500
? Besselk0(500, Data) - R
time = 1 ms.
% = 5.4... E-519 /* Perfect for absolute accuracy. */
? besselk(0, 500) - R
time = 4 ms.
% = 0.E-719 /* Perfect for relative accuracy. */
? for (n = 1, 100, besselk(0, 50))
time = 71 ms.
? for (n = 1, 100, Besselk0(50, Data))
time = 465 ms. /* Much slower. */
? for (n = 1, 100, besselk(0, 500))
time = 393 ms.
? for (n = 1, 100, Besselk0(500, Data))
time = 78 ms. /* Much faster, see below. */
```

Note that our script computes $K_0(x)$ with *absolute* accuracy less than 2^{-B} which is indeed what is needed in the application of inverse Mellin transform to the computation of L-functions that we will study in the next two chapters, while the Pari/GP implementation computes $K_0(x)$ with *relative* accuracy, so is not surprisingly slower.

CHAPTER 8

Computation of inverse Mellin transforms

8.1. Introduction

This chapter deals with the numerical computation of inverse Mellin transforms. This is not a goal *per se*, but is an essential part of the computation of L-functions that we will study in Chapter 9. We will follow in large part the fundamental paper of T. Dokchitser [**Dok04**].

Recall that the *Mellin transform* of a function f is by definition

$$\mathcal{M}(f)(s) = \int_0^\infty t^s f(t) \frac{dt}{t} ,$$

written in this way to emphasize the fact that dt/t is a Haar measure on the locally compact group $\mathbb{R}_{>0}$, and which simplifies all usual changes of variable such as $t = au^b$ or $t = ae^{bz}$.

We need some reasonable assumptions on f to make this definition useful. Even though far from minimal, we will assume the following:

(1) f is a holomorphic function on some open strip S_β defined by $\Re(t) > 0$ and $|\Im(t)| < \beta$ for some $\beta > 0$.
(2) $f(t)$ tends to 0 faster than any power of $1/|t|$ as $|t| \to \infty$ in the strip S_β (i.e., $|t|^m f(t) \to 0$ for all $m \geq 0$).
(3) $f(t) = O(t^{-\sigma})$ for some σ as $t \to 0$.

These assumptions imply that $g(s) = \mathcal{M}(f)(s)$ is well-defined for $\Re(s) > \sigma$ and is a *holomorphic function* of s in that right half-plane.

The *Mellin inversion formula*, which is a simple variant of the Fourier inversion formula, states that if C is a contour in the half-plane of definition of $g(s)$, one can recover the function f for $x \in S_\beta$ by the formula

$$f(x) = \frac{1}{2\pi i} \int_C x^{-s} g(s) \, ds .$$

Note that since S_β is in the right half-plane $\Re(x) > 0$, we have $x^{-s} = e^{-s \log(x)}$ with the principal determination of the logarithm.

The most famous and important Mellin pair is the gamma function $\Gamma(s)$ defined for $\Re(s) > 0$ by

$$\Gamma(s) = \int_0^\infty t^s e^{-t} \frac{dt}{t} ,$$

so that by Mellin inversion, for $\Re(x) > 0$ we have

$$e^{-x} = \frac{1}{2\pi i} \int_C x^{-s} \Gamma(s) \, ds .$$

Another formula which is occasionally useful is the convolution formula: if $g_i(s) = \mathcal{M}(f_i)(s)$ for $i = 1, 2$, then

$$g_1(s)g_2(s) = \mathcal{M}(f_1 * f_2)(s) \quad \text{where} \quad (f_1 * f_2)(x) = \int_0^\infty f_1(t)f_2(x/t)\,dt/t \ .$$

For computational purposes, it is preferable to write this formula as follows:

$$(f_1 * f_2)(x^2) = \int_1^\infty \left(f_1(xt)f_2(x/t) + f_1(x/t)f_2(xt) \right) \frac{dt}{t} \ .$$

Some immediate formulas for inverse Mellin transforms are as follows: if $f(x) = \mathcal{M}^{-1}(g)(x)$ then

- $\mathcal{M}^{-1}(g(s+a))(x) = x^a f(x)$.
- $\mathcal{M}^{-1}((s+a)g(s))(x) = -f'(x) + af(x)$.
- $\mathcal{M}^{-1}(g(s/a))(x) = af(x^a)$.
- $\mathcal{M}^{-1}(a^s g(s))(x) = f(x/a)$.

The goal of this chapter is the numerical computation of inverse Mellin transforms of the functions $g(s)$ which occur in practice, connected with the evaluation of L-functions.

8.2. Gamma products

8.2.1. The Gamma function. As mentioned above, recall that one of the definitions of the gamma function (and not the one preferred by the authors, see Section 9.6 of [**Coh00**]) is the Mellin transform of the function e^{-x}. This only defines $\Gamma(s)$ for $\Re(s) > 0$. It is well-known and easy that $\Gamma(s)$ can be extended to the whole complex plane into a meromorphic function with simple poles at all negative or zero integers. The standard way to do this is by extending $\Gamma(s)$ strip by strip by using the basic functional equation $\Gamma(s+1) = s\Gamma(s)$, but a more direct way which also immediately gives the residues at the poles is to use the easily proved formula valid for all $s \in \mathbb{C}$

$$\Gamma(s) = \int_1^\infty t^s e^{-t} \frac{dt}{t} + \sum_{n \geq 0} \frac{(-1)^n}{n!(s+n)} \ .$$

The gamma function possesses a huge number of functional properties (see any textbook, including [**Coh00**]), but the main ones that we will use are as follows:

PROPOSITION 8.2.1.

(1) *(Duplication formula.)* We have

$$\Gamma(s/2)\Gamma((s+1)/2) = \pi^{1/2}2^{1-s}\Gamma(s) \ .$$

(2) *(Real Stirling's formula.) As $s \to \infty$, s real, we have the asymptotic formula*

$$\Gamma(s) \sim s^{s-1/2}e^{-s}(2\pi)^{1/2} \ .$$

(3) *(Stirling's formula on vertical strips.) Write $s = \sigma + iT$ with σ and T real. If σ is fixed, as $T \to \pm\infty$ we have*

$$|\Gamma(s)| \sim |T|^{\sigma-1/2}e^{-\pi|T|/2}(2\pi)^{1/2} \ .$$

It follows in particular that $\Gamma(s)$ tends to 0 exponentially fast in bounded vertical strips, which will allow us to shift the contours of integration without worrying about what happens when $|\Im(s)|$ is large.

In view of the functional equations of the Riemann zeta function and of modular forms, it is natural to set the following definitions:

DEFINITION 8.2.2.

(1) We define

$$\Gamma_{\mathbb{R}}(s) = \pi^{-s/2}\Gamma(s/2) \quad \text{and} \quad \Gamma_{\mathbb{C}}(s) = \Gamma_{\mathbb{R}}(s)\Gamma_{\mathbb{R}}(s+1) = 2(2\pi)^{-s}\Gamma(s)$$

(beware of the factor 2 in the definition of $\Gamma_{\mathbb{C}}$).

(2) Let $\mathcal{A} = (\alpha_1, \ldots, \alpha_d)$ be a d-uple of complex numbers. The corresponding *gamma product* is the function

$$\gamma_{\mathcal{A}}(s) = \prod_{1 \le j \le d} \Gamma_{\mathbb{R}}(s + \alpha_j) \,,$$

and its inverse Mellin transform will be denoted $K_{\mathcal{A}}(x)$. The number d of $\Gamma_{\mathbb{R}}$ factors will be called the *degree*.

The reason for introducing these gamma products is that they are the factors "at infinity" which occur in all functional equations of L-functions of order 1, which are the only ones that we will study later (an example of an L-function of order 2 is the *Selberg zeta function*, whose factor at infinity is essentially Barnes's double gamma function $\Gamma_2(s)$).

8.2.2. Elementary cases and reductions. From now on, we will assume that all the α_j's are in $\mathbb{Z}$. This includes almost all gamma factors of usual L-functions, with exceptions such as the L-functions of Maass forms which therefore will not be considered. We can of course always assume that the α_j's are in increasing order. In addition, the functional equation $\Gamma_{\mathbb{R}}(s + 2) = (s/(2\pi))\Gamma_{\mathbb{R}}(s)$ together with the elementary properties of the inverse Mellin transform recalled above implies the induction formula

$$K_{\alpha_1+2,\alpha_2,\ldots,\alpha_d}(x) = (-f'(x) + \alpha_1 f(x))/(2\pi) \,,$$

where we set $f(x) = K_{\alpha_1,\ldots,\alpha_d}(x)$. Thus by induction and doing this for all indices, we may assume that $\alpha_j = 0$ or 1 for all j, although we will not always make this reduction. In addition, using again one of the above formulas, we may assume that not all the α_j are equal to 1. Assuming that computing derivatives of f is "easy", for a given d it is therefore sufficient to study exactly d possibilities, with $\alpha_1 = \cdots = \alpha_m = 0$ and $\alpha_{m+1} = \cdots = \alpha_d = 1$, where $1 \le m \le d$, which we naturally abbreviate as $0^m 1^{d-m}$.

An additional property which can occasionally be useful, and which is an immediate consequence of the duplication formula, is that with evident notation

$$K_{\mathcal{A} \sqcup \mathcal{A}+1}(x) = 2^{d-1-S} K_{2\mathcal{A}}(2^{d/2}x^{1/2}) \,,$$

where $S = \sum_{1 \le j \le d} \alpha_j$. For instance,

$$K_{0^d 1^d}(x) = 2^{d-1} K_{0^d}(2^{d/2}x^{1/2}) \,.$$

For future reference, we note the most important values of $\mathcal{A} = \{\alpha_1, \cdots, \alpha_d\}$ for which the inverse Mellin transform is known explicitly.

(1) For $\mathcal{A} = \{0\}$: $K_{\mathcal{A}}(x) = 2e^{-\pi x^2}$.
(2) For $\mathcal{A} = \{1\}$: $K_{\mathcal{A}}(x) = 2xe^{-\pi x^2}$.
(3) For $\mathcal{A} = \{0,0\}$: $K_{\mathcal{A}}(x) = 4K_0(2\pi x)$.
(4) For $\mathcal{A} = \{0,1\}$: $K_{\mathcal{A}}(x) = 2e^{-2\pi x}$.
(5) For $\mathcal{A} = \{0,0,1,1\}$: $K_{\mathcal{A}}(x) = 8K_0(4\pi x^{1/2})$.
(6) For $\mathcal{A} = \{-1,0,0,1\}$: $K_{\mathcal{A}}(x) = 8K_1(4\pi x^{1/2})/x^{1/2}$.

REMARKS 8.2.3.

(1) We assume no familiarity with the functions K_{ν} occurring in the above formulas, the so-called K-Bessel functions. We shall derive most of what we need from general properties of the functions $K_{\mathcal{A}}$, and shortly recall specific properties.
(2) Note that the occurrence of the same functions for several of the above follows from the duplication formula as explained above.

Recall for future reference that around $x = 0$ we have

$$K_0(x) = \sum_{n \geq 0} \frac{(x/2)^{2n}}{n!^2}\left(H_n - \gamma - \log(x/2)\right)$$

with $H_n = \sum_{1 \leq i \leq n} 1/i$, which converges for all x, and around $x = \infty$ we have

$$K_0(x) = \left(\frac{\pi}{2x}\right)^{1/2} e^{-x}\left(1 - \frac{1^2}{(8x)1!} + \frac{1^2 3^2}{(8x)^2 2!} - \cdots\right),$$

which is an asymptotic expansion converging for no value of x. We will see below the generalization of these formulas to any $\mathcal{A} = (\alpha_1, \ldots, \alpha_d)$.

Note the following convolution formulas:

(1) For $\mathcal{A} = \{0,0,0\}$: $K_{\mathcal{A}}(x) = 8e^{-\pi x^2} * K_0(2\pi x)$.
(2) For $\mathcal{A} = \{0,0,1\}$: $K_{\mathcal{A}}(x) = 4e^{-\pi x^2} * e^{-2\pi x}$.
(3) For $\mathcal{A} = \{0,1,1\}$: $K_{\mathcal{A}}(x) = 4xe^{-\pi x^2} * e^{-2\pi x}$.
(4) For $\mathcal{A} = \{0,0,0,0\}$: $K_{\mathcal{A}}(x) = 16K_0(2\pi x) * K_0(2\pi x)$.
(5) For $\mathcal{A} = \{0,0,0,1\}$: $K_{\mathcal{A}}(x) = 8e^{-2\pi x} * K_0(2\pi x)$.
(6) For $\mathcal{A} = \{0,1,1,1\}$: $K_{\mathcal{A}}(x) = 8e^{-2\pi x} * xK_0(2\pi x)$.
(7) For $\mathcal{A} = \{0,0,0,1,1\}$: $K_{\mathcal{A}}(x) = 16e^{-\pi x^2} * K_0(4\pi x^{1/2})$.
(8) For $\mathcal{A} = \{0,0,1,1,1\}$: $K_{\mathcal{A}}(x) = 16xe^{-\pi x^2} * K_0(4\pi x^{1/2})$.
(9) For $\mathcal{A} = \{0,0,0,0,1,1\}$: $K_{\mathcal{A}}(x) = 32K_0(2\pi x) * K_0(4\pi x^{1/2})$.
(10) For $\mathcal{A} = \{0,0,0,1,1,1\}$: $K_{\mathcal{A}}(x) = 16e^{-2\pi x} * K_0(4\pi x^{1/2})$.
(11) For $\mathcal{A} = \{0,0,1,1,1,1\}$: $K_{\mathcal{A}}(x) = 32xK_0(2\pi x) * K_0(4\pi x^{1/2})$.

These are given for completeness, but in practice they are not useful for the numerical computation of $K_{\mathcal{A}}(x)$.

8.2.3. The asymptotic formula. Recall that S_{β} denotes the strip $\Re(t) > 0$ and $|\Im(t)| < \beta$.

THEOREM 8.2.4. *Fix $\beta > 0$. As $x \to \infty$ in S_{β}, we have the asymptotic expansion*

$$K_{\mathcal{A}}(x) = d^{-1/2} 2^{(d+1)/2} x^A e^{-d\pi x^{2/d}} \sum_{n \geq 0} M_n x^{-2n/d},$$

where $A = (1 - d + \sum_{1 \leq j \leq d} \alpha_j)/d$, $M_0 = 1$, and the M_n satisfy an (explicit) order d linear recurrence relation.

The proof of this theorem can be found in [**Luk69**]. Note that when the α_j are rational, so are the M_n. The explicit recursion for the M_n will be given as a program in the script `Mellininvlargecoefs` below.

COROLLARY 8.2.5. *For $x \in S_\beta$, we have $|x^k K_{\mathcal{A}}(x)| < e^{-U}$ if*

$$d\pi \cdot \Re(x^{2/d}) > U - \log\left(d^{1/2}2^{-(d+1)/2}\right) + (A+k)(d/2)\log(U/(d\pi)) + o(1) .$$

PROOF. Indeed, since $x \in S_\beta$, if $\Re(x)$ is at all large then $|x^{2/d}|$ is comparable with $\Re(x^{2/d})$. We may thus simply use the asymptotic expansion above, together with Lemma 2.3.3 with $a = A+k$, $b = d\pi$, $c = 2/d$, and $B = U-\log\left(d^{1/2}2^{-(d+1)/2}\right)$. $\square$

In view of the computation of L-series, the goal of this chapter will be to compute $x^k K_{\mathcal{A}}(x)$ to *absolute* accuracy e^{-E}, for a given $k \geq 0$ and some large E. This implies two things:

(1) First, if x satisfies the inequality given by the above corollary for $U = E$, we can set $x^k K_{\mathcal{A}}(x) = 0$.
(2) Second, even if x is not that large, we only need to evaluate the asymptotic series with absolute accuracy $d^{1/2}2^{-(d+1)/2}e^{-E+d\pi x^{2/d}}x^{-A-k}$.

8.3. Compendium of possible methods

We come now to the heart of the subject, that of computing $K(x) = K_{\mathcal{A}}(x)$ when no "explicit" formula is known (and even when explicit formulas are known, for instance involving Bessel functions, it may still be useful to find specific methods tailored to our applications).

In view of the formulas that we will give in Chapter 9, we will specifically need to compute $K(nt/\sqrt{N})$ for fixed positive real t (usually of reasonable size), where $N \in \mathbb{Z}_{\geq 1}$ is also fixed, but n varies from 1 to a limit where $K(nt/\sqrt{N})$ becomes negligible compared to the accuracy that we need. It is essential to note that because of the above-mentioned formulas, we need to evaluate these values of $K(nt/N^{1/2})$ with *absolute* accuracy, not relative accuracy, so when it becomes exponentially small, only a few decimals suffice.

A list of possible algorithms to compute these values is as follows:

(1) The DE method to compute the inverse Mellin transform integral directly, with a suitable choice of contour (all the experiments that we have made show that all other integration methods are either much slower or give nonsense results).
(2) The DE method to compute the convolutions occurring in the above cases.
(3) The `IntGaussLaguerreinv` program to compute those convolutions (all other methods fail).
(4) The (generalized) power series expansion of the function K around $x = 0$, which has infinite radius of convergence.
(5) The asymptotic expansion of the function K as $x \to \infty$, either directly or transformed into a continued fraction.
(6) The Taylor expansion of the function K together with the fact that it satisfies a linear differential equation with polynomial coefficients to compute all the values $K(nu)$ for $1 \leq n \leq L$ simultaneously, with $u = t/\sqrt{N}$.

In practice, for reasonably small values of x (such as $|x| \leq 10$) it is best to use the generalized power series, while for larger values we will use the asymptotic expansion transformed into a continued fraction. It seems also reasonable to use the linear differential equation, but we have not studied that method.

8.4. Using the power series around $x = 0$

Recall that

$$K_{\mathcal{A}}(x) = \frac{1}{2\pi i} \int_C x^{-s} \gamma_{\mathcal{A}}(s) \, ds \ ,$$

where C is a suitable contour in the right half-plane $\Re(s) > 0$, for instance a vertical line $\Re(s) = \sigma > 0$. Since the gamma function, and hence also the gamma product $\gamma_{\mathcal{A}}(s)$ tends exponentially to 0 as $|\Im(s)| \to \infty$, it is easy to justify that we can shift the line of integration to the left, catching in passing the residues of the function $x^{-s}\gamma_{\mathcal{A}}(s)$, and by a suitable limiting process prove the formula:

$$K_{\mathcal{A}}(x) = \sum_{s_0 \text{ pole}} \operatorname{Res}_{s=s_0} x^{-s} \gamma_{\mathcal{A}}(s) \ .$$

Since the poles of $\Gamma_{\mathbb{R}}(s)$ are the negative or zero even integers, it follows that the poles s_0 are the numbers of the form $-2n - \alpha_j$ for $1 \leq j \leq d$ and $n \geq 0$. These numbers may be repeated for different values of (α_j, n), and in that case we have multiple poles. The precise formula can be found in Dokchitser's paper and in the script we give below, but for now let us already look at the convergence of the above series.

It is not difficult to see that the convergence is dominated by the degree d of the gamma product, in other words that we can ignore (only for convergence purposes!) the precise values of the α_j. Thus, assume that the α_j are distinct modulo 2, so that the numbers $-2n - \alpha_j$ are all distinct, hence all the poles *simple*.

By the formula given above for $\Gamma(s)$, the residue of $\Gamma_{\mathbb{R}}(s)$ at $s = -2n$ is equal to $2(-1)^n \pi^n / n!$, so that of $x^{-s}\gamma_{\mathcal{A}}(s)$ is bounded in absolute value by $C(d)\pi^{nd} x^{2n} / n!^d$, where $C(d)$ depends only on d, and setting $N = 2n$, up to a small power of N and constants, this is approximately equal to $(2\pi)^{Nd/2} x^N / N!^{d/2}$. In the case of multiple poles, this will be multiplied by $\log(x)^{v-1}$, where v is the largest multiplicity of a pole (see for instance the expansion of $K_0(x)$ given above). Thus, if we want the absolute accuracy to be less than e^{-E}, we need $(d/2)\log(N!) - N((d/2)\log(2\pi) + \log(x)) > E$, hence by Stirling's formula

$$N \log(N) - N \log(2\pi e x^{2/d}) > 2E/d$$

(with $e = 2.718\ldots$), neglecting $(1/2)\log(2\pi)$ and $\log(N)/2$ which will be small compared to E/d.

Now the implicit equation $N \log(N) - AN > B$ for large B and small A has the *exact* solution $N > B/W_0(Be^{-A})$, where W_0 is the Lambert W_0 function defined for $x \geq -1/e$ as the unique solution greater than or equal to -1 of $W_0(x)e^{W_0(x)} = x$ (see Corollary 2.3.4). Thus, we must choose

$$N > \frac{2E/d}{W_0\big((E/d)/(\pi e x^{2/d})\big)} \ .$$

We will see below that a reasonable choice for the threshold between small and large values of x can be expressed in the form $x = E^{d/2}/C^d$, where C is a constant to be determined experimentally, but which does not depend on E, but possibly on d; typically, $C = 9$ or 10 for instance. Thus, $(E/d)/(\pi e x^{2/d}) > C^2/(\pi e)$, so the inequality for N becomes $N > (2E/d)/W_0(C^2/(\pi e))$.

However, this is not the whole story. As can be seen in the expansions of e^{-x} and $K_0(x)$ for instance, the power series has a large amount of cancellation: the largest term is obtained when $n \approx \pi x^{2/d}$, and is approximately equal to $e^{\pi d x^{2/d}}$ which, as we will see, is also the inverse of the approximate size of $K(x)$. We must therefore work with a *relative* accuracy of $e^{-E - \pi d x^{2/d}}$ if we want to obtain an absolute accuracy of e^{-E}. Since we choose $x = E^{d/2}/C^d$ as threshold, this means that we must work with relative accuracy $e^{-E(1 + \pi d/C^2)}$.

The number of arithmetic operations to compute the series is proportional to N and to the square of the accuracy, hence to

$$(E + \pi d x^{2/d})^2 \frac{2E/d}{W_0\big((E/d)/(\pi e x^{2/d})\big)} \ .$$

Even before comparing with other methods, we can write the corresponding program. In what follows, Vga is the vector of the α_j. As usual, we have an initialization phase, and an evaluation. The program is directly taken from T. Dokchitser's paper, with a slight simplification due to the assumption that the α_js are integers. We leave the threshold parameter C as an argument, to allow tuning.

——————————————— | Mellininvsmall.gp | ———————————————

```
1   Mellininvsmallinit(Vga, C = 9) =
2   { my(B = getlocalbitprec(), d = #Vga, x = 'x, LA0, LA1, N, LA);
3     my(P, m, L, G, M);
4
5     localbitprec(32);
6     L = ceil(2*log(2) * B / (d * lambertw(C^2 / (Pi*exp(1)))));
7     localbitprec(B * (1 + d * Pi / C^2));
8     LA0 = vecsort([ a | a <- Vga, a%2 == 0 ]);
9     LA1 = vecsort([ a | a <- Vga, a%2 == 1 ]);
10    LA = if (#LA0 == 0, [LA1],
11            #LA1 == 0, [LA0], [LA0, LA1]);
12    N = #LA;
13    P = vector(N, j, #LA[j]);
14    m = vector(N, j, 2 - vecmin(LA[j]));
15    G = vector(L + 1);
16    M = matrix(N, d + 1);
17    for (j = 1, N,
18      my(v = vector(d, i, (x + Vga[i] + m[j]) / 2));
19      my(T = vecprod(v), p = P[j] + 1);
20      G[1] = prod(i = 1, d, gamma(v[i] + O(x^p)));
21      for (n = 1, L, G[n+1] = G[n] / subst(T, x, x - 2 * n));
22      for (k = 1, P[j],
23        M[j,k] = vector(L, n, 1. * polcoef(G[n+1],-k)) / (k-1)!));
24    return ([P, m, M]);
25  }
26
27  /* sum_{n <= N} v[n] u^n */
```

```
28   Evalvec(v, N, u) =
29   { my(S = 0);
30     forstep (n = min(N,#v), 1, -1, S = u * (v[n] + S));
31     return (S);
32   }
33
34   /* K_Vga( x^(d / 2) ), x < E / C^2 is small */
35   Mellininvsmall(Vga, x, PmM = 0, C = 9) =
36   { my(B = getlocalbitprec(), d = #Vga);
37     my(x2, S, N, Ed, xd, NEWB, v, P, m, M);
38
39     [P, m, M] = if (PmM, PmM, Mellininvsmallinit(Vga, C));
40     localbitprec(32);
41     Ed = log(2) * B / d;
42     xd = max(Pi * abs(x), 1E-11);
43     if (xd > Ed, error("should not be in Mellininvsmall"));
44     N = ceil(2 * Ed / lambertw(C^2 / (exp(1) * xd)));
45     NEWB = ceil(B + d * xd / log(2));
46     localbitprec(NEWB);
47     x = bitprecision(x, NEWB);
48     x = (Pi * x)^(d / 2); x2 = x^2;
49     v = powers(-log(x), vecmax(P) - 1);
50     S = sum(j = 1, #P,
51             x^(-m[j]) * sum(k = 1, P[j],
52                         v[k] * Evalvec(M[j, k], N, x2)));
53     return (bitprecision(S / Pi^(vecsum(Vga) / 2), B));
54   }
```

Note that this program will correctly compute values of $K_A(x)$ with *absolute* bit accuracy less than $2^{-B} = e^{-E}$ for *any* positive x, although it will of course be much less efficient for large x.

8.5. Using the asymptotic expansion

8.5.1. Computing the coefficients M_n. When x is "large", even though the above program will compute $K(x)$ correctly, the computation will be rather slow: first because we will need to use many terms of the generalized power series, and second because we have to increase the working accuracy, at worse double it. It is thus advantageous to use other methods, and the natural method to use is the asymptotic expansion given by Theorem 8.2.4. The following program, adapted from [**Dok04**], computes the coefficients M_n:

──────────────────── Mellininvlargecoefs.gp ────────────────────

```
1   /* Vector of coefficients M_n, n <= N; if asymptotic expansion
2    * is finite, return a possibly shorter polynomial */
3   Mellininvlargecoefs(Vga, N) =
4   { my(d = #Vga, dv = 0, x = 'x, pol, S, A, sh, vp, M);
5
6     if (d == 1, return (Pol(1)));
```

```
7    if (d == 2 && (dv = abs(Vga[1] - Vga[2])) % 2,
8      N = min(N, (dv+1) \ 2));  /* finite expansion */
9    pol = prod(i = 1, d, 1 + d * Vga[i] * x);
10   A = - vecsum(Vga);
11   S = [sum(k = 0, m, A^k * binomial(k + d - m, k)
12                      * polcoef(pol, m - k)) | m <- [0..d]];
13   sh = powers(sinh(x + O(x^(d+2))) / x, d);
14   vp = vector(d, p,
15     my(Sp = [S[m+1] * prod(j = m, p-1, d-j) | m <- [0..d]]);
16     my(w = vector(p\2 + 1, k, polcoef(sh[d-p+1], 2*k-2)));
17     my(D = -(2 * d)^p);
18     vector(N - 1, n,
19       my(c, a = 2*n - p + 1);
20       my(v = vector(p+1, k, a^(k-1) / (k-1)!));
21       sum(m = 0, p,
22           if (c = Sp[m+1],
23             my(q = p - m);
24             c * sum(k = 0, q \ 2, v[q-2*k+1] * w[k+1]))) / D));
25   M = vector(N); M[1] = 1;
26   for (n = 1, N - 1,
27     M[n+1] = sum(p = 2, min(d, n+1), vp[p][n] * M[n+2-p]) / n);
28   return (if (dv, Polrev(M), M));
29 }
```

REMARK 8.5.1. The cases $d = 1$, or $d = 2$ and Vga $= [a, a+1]$ (corresponding to pure exponentials, see above), will give a trivial asymptotic expansion (i.e., $M_0 = 1$, $M_n = 0$ for $n \geq 1$), so we exit immediately. More generally, if Vga $= [a, a + 2k + 1]$ with $k \in \mathbb{Z}$, the asymptotic expansion will only have a finite number of nonzero terms, so we reduce N accordingly if necessary.

We could first try to use it directly: we can hope that the error made by truncating is smaller than the first neglected term, as for most asymptotic expansions. A cursory study (see [**Luk69**]) shows that the M_n occurring in Theorem 8.2.4 are of the order of $n!/(2d\sin(\pi/d))^n$, up to small powers of n. Thus the smallest term is attained for $n \approx 2\pi d \sin(\pi/d)x^{2/d}$, and is approximately equal to $e^{-2\pi d \sin(\pi/d)x^{2/d}}$. Taking into account the factor in front of the asymptotic series, this means that we can directly use the asymptotic expansion as soon as $d\pi x^{2/d} > E/(1 + 2\sin(\pi/d))$ (recall that if $d\pi x^{2/d} > E$ we set the value equal to 0). However, experiments show that even inside its range of applicability this method is slower than the power series program seen above, even when t is large, so we will do things differently.

8.5.2. Using the quotient-difference algorithm. Instead, we will use a *completely heuristic* idea of T. Dokchitser (already present in other contexts, but popularized by his paper in the context of computing inverse Mellin transforms). Note that all the estimates that we have used up to now are rigorous, even if we often make approximations. In the present section, as acknowledged in Dokchitser's paper, nothing is rigorous, but it works remarkably well. We have tried to *prove* some results, but we have been unable to do so except in degree less than or equal to 2 (exponential and K-Bessel functions).

Recall that we have an asymptotic series $f(z) = \sum_{n \geq 0} M_n z^n$, in which we want to plug in $z = 1/(\pi t^{2/d})$. This series never converges (except when it is finite, which happens if and only if Vga is either of the form $[a]$ or of the form $[a, a + 2k + 1]$ with $k \in \mathbb{Z}$), but as we have seen it can nonetheless be used in certain ranges.

The basic idea is this: we are going to *formally* convert this asymptotic series into a *continued fraction*. There are different equivalent ways of writing this continued fraction, but we will choose the following:

$$f(z) = \cfrac{1}{1 + \cfrac{c(1)z^{a_1}}{1 + \cfrac{c(2)z^{a_2}}{1 + \cdots}}} \;,$$

where $a_i \in \mathbb{Z}_{\geq 1}$ and the $c(i)$ are complex numbers (in fact rational numbers in our case since the M_n are rational when the α_j are integers). The proof of existence and uniqueness of this (formal) expansion is trivial and left to the reader.

Two things must be noted about this: first, we almost always have $a_i = 1$ for all i. Second, as we have seen in Section 7.6, the *quotient-difference* algorithm (QD for short) computes the $c(n)$ from the M_n. Since our $c(n)$ are rational numbers, we can proceed in two ways: either *exactly*, using rational numbers, and suffer from coefficient explosion. Or *approximately*, using floating point approximations to the correct values. This version, given in Section 7.6, is orders of magnitude faster, but however suffers from instability. As usual this instability can be compensated by increasing the working accuracy.

Note that if Vga $= [0, 0, 0, 0, 0]$, the vector M starts $[1, -1/5, 1/25, -2/625, \dots]$, and the coincidence $1/25 = (-1/5)^2$ will give a division by 0 because in fact $a_2 = 2$ and not 1: more precisely,

$$\sum_{n \geq 0} M_n z^n = 1/(1 + (1/5)z/(1 + (3/125)z^2/(1 + (13/15)z/(1 + \cdots)))) \;.$$

More generally, this is the case if Vga$=[a, a, a, a, a]$. The simplest way to circumvent this problem is to divide the asymptotic expansion by a linear polynomial of the form $1 - uz$ for some transcendental number u (we have chosen $u = 1/\pi$ in our implementation); the Quodif algorithm will then work without divisions by 0, and after evaluating the corresponding continued fraction it is then sufficient to multiply back numerically by $1 - uz$.

Once we have computed the continued fraction using the Quodif algorithm, to perform the initialization stage for the evaluation of L-functions we will need to *evaluate* the continued fraction for many values of the argument z, and this is the dominant part of our initialization. To improve efficiency, we use the method explained in Chapter 7: we first convert the continued fraction into an Euler-type continued fraction, using CFtoEuler, which we can then evaluate efficiently using CFEulereval, which is up to four times faster than the naïve evaluation using CFeval. Note that these programs are slightly unstable, but as usual we can cheat and compensate this instability by increasing the accuracy.

8.5.3. The main heuristic assumption. As mentioned in the introduction, the evaluation of the continued fraction, either directly on the Quodif output or with the `CFEulereval` program, is *heuristic*, since we do not know how to prove the convergence to the correct value, let alone the *fast* convergence, although it does work very well in practice.

The simplest nontrivial case is for `Vga=[0,0]`, corresponding to the asymptotic expansion of $4K_0(2x)$ with $x = 1/z$. One observes experimentally that the $c(n)$ output by the QD algorithm satisfy $c(n) \approx n/8 - 0.05$ for n even and $c(n) \approx n/8 - 0.0725$ for n odd. Since this is a regular pattern, this should be provable, but we will not attempt to do so. If true, using the results of Chapter 7, this would imply that the continued fraction will converge, and that if $S(n)$ is the nth convergent and $S = 4K_0(2x)$ its limit, we should have

$$S - S(n) \sim (-1)^n e^{-4\sqrt{nx}} .$$

If we now look at the next simplest case $\text{Vga} = [0,0,0]$, corresponding to the gamma factor $\Gamma_{\mathbb{R}}(s)^3$, one observes experimentally that the $c(n)$ vary widely, both in sign and in magnitude. For instance, for $n = 100, \ldots, 105$, we have $c(n) \approx 62.16, -49.84,$ $18.65, -0.41, -51.74, 84.79$, which do not display a regular pattern. Nonetheless, the continued fraction itself *seems* to converge to the correct value, and at exactly the same rate $|S - S(n)| = O(e^{-4\sqrt{nx}})$ as before (the sign is more irregular). Testing other values such as $[0,0,1]$, $[0,0,0,0]$, etc., it again *seems* that the continued fraction always converges, and at the same rate. We repeat that we have no idea how to prove this, if true.

We thus make the fundamental *heuristic assumption* that this is the case and proceed from there. Recall from Theorem 8.2.4 that

$$K_{\mathcal{A}}(t) = P \sum_{n \geq 0} M_n \pi^{-n} t^{-2n/d} \quad \text{with} \quad P = d^{-1/2} 2^{(d+1)/2} t^A e^{-d\pi t^{2/d}}$$

and $A = \left(1 - d + \sum_{1 \leq j \leq d} \alpha_j\right)/d$.

To obtain accuracy less than e^{-E}, as in the case of small t we need to compute the continued fraction with accuracy less than $e^{-E + d\pi t^{2/d} - A\log(t)}$ (since t may be large, it is preferable to keep the term $\log(t)$). Assuming the convergence in $e^{-4\sqrt{nx}}$ with $x = \pi t^{2/d}$, we thus need $4\sqrt{n\pi t^{2/d}} > E - d\pi t^{2/d} + A\log(t)$, in other words

$$n > \frac{(E - d\pi t^{2/d} + A\log(t))^2}{16\pi t^{2/d}} .$$

Now if $d\pi t^{2/d} - A\log(t) > E$, we can set $K_{\mathcal{A}}(t) = 0$. Thus, neglecting the logarithmic term (which for this argument can be done), we may assume that $d\pi t^{2/d} < E$, and it is clear that the function on the right-hand side is a decreasing function of t. Thus the largest n to be used is that for the smallest value of t, again to be determined as a threshold with other methods.

For this, we make the (reasonable) assumption, confirmed by experimentation, that the threshold is of the type $t > E^{d/2}/C^d$, for C a constant to be determined. Under this assumption, again neglecting the logarithmic term, the largest value of n is $(E/(16\pi))(C - d\pi/C)^2 < EC^2/(16\pi)$. We thus set the parameter N in `Mellininvlargecoefs` to this value, and then optimize for C in terms of the speed of the specific implementation. The programs are self-explanatory:

─────────────────────── | Mellininvlarge.gp | ───────────────────────

```
1  Mellininvlargeinit(Vga, C = 9) =
2  { my(B = getlocalbitprec(), M, N, e);
```

```
3
4    localbitprec(32); N = ceil(B * C^2 * log(2) / (16 * Pi));
5    localbitprec(B); M = Mellininvlargecoefs(Vga, N);
6    if (type(M) == "t_POL", return (M)); /* finite expansion */
7    /* Divide expansion by 1 - z/Pi to avoid divisions by 0 */
8    localbitprec(4 * B / 3); e = 1 / Pi;
9    for (i = 2, #M, M[i] += M[i - 1] * e);
10   return (CFtoEuler(Quodif(M)));
11   }
12
13   /* K_Vga( t^(d / 2) ), t > E / C^2 is large */
14   Mellininvlarge(Vga, t, M = 0, C = 9) =
15   { my(B = getlocalbitprec(), d = #Vga, td, T, A, n, P);
16     my(NEWB);
17
18     if (!M, M = Mellininvlargeinit(Vga, C));
19     A = 1 - d + vecsum(Vga);
20     localbitprec(32); td = abs(t);
21     NEWB = max(64, ceil(B+64 - (Pi*d*td - A*log(td)/2) / log(2)));
22     localbitprec(NEWB); t = bitprecision(t, NEWB); T = Pi * t;
23     P = sqrt(2^(d+1) / d * t^A) * exp(-d*Pi*t);
24     /* finite expansion ? */
25     if (type(M) == "t_POL", return (P * subst(M, 'x, 1 / T)));
26     localbitprec(64);
27     n = min(2 * #M[1], ceil((log(2) * B)^2 / (C^2 * td)));
28     localbitprec(B);
29     return (P * CFEulereval(M, n\2, T) * (T - 1 / Pi));
30   }
```

Note that the initialization divides the asymptotic expansion by the artificial $1 - z/\pi$ to avoid divisions by 0, so we need to multiply back at the end.

8.5.4. Putting everything together. Combining the programs using the generalized power series for small t and continued fractions for large t, we can now put things together, leaving the parameter C to be optimized.

—————————————————— Mellininv.gp ——————————————————

```
1    /* Return a closure computing K(x^(d/2)). When K(t) is simple,
2     * give the formula. Otherwise use the power series for t < T,
3     * else the continued fraction. */
4    Mellininvtech(Vga, C = 0)=
5    { my(B = getlocalbitprec(), S, L);
6
7      if (Vga == [0],    return (x -> 2*exp(-Pi*x)),
8          Vga == [1],    return (x -> 2*sqrt(x)*exp(-Pi*x)),
9          Vga == [0, 1], return (x -> 2*exp(-2*Pi*x)));
10     if (!C, C = if (#Vga == 2, 9, 10));
11     localbitprec(32); my(T = log(2) * B / C^2);
12     localbitprec(B);
```

```
13    S = Mellininvsmallinit(Vga, C);
14    L = Mellininvlargeinit(Vga, C);
15    return (x -> if (abs(x) < T, Mellininvsmall(Vga, x, S, C),
16                                 Mellininvlarge(Vga, x, L, C)));
17  }
18
19  /* Return a closure computing K(x) */
20  Mellininv(Vga) = x -> Mellininvtech(Vga)(x^(2 / #Vga));
```

REMARKS 8.5.2.

(1) In the special cases Vga equal to $[0,0]$, $[0,0,1,1]$,or $[-1,0,0,1]$, we have seen above that the inverse Mellin transform can be explicitly expressed in terms of K-Bessel functions. However this is inefficient: Pari/GP's implementation of the K-Bessel functions gives the result with *relative* accuracy 2^{-B}, which is much more precise than what we need, which is absolute accuracy.

(2) It is not difficult to adapt the above programs to compute successive *derivatives* of inverse Mellin transforms: simply differentiate term by term the Taylor series around 0 for small t or the asymptotic series for large t (before applying Quodif as before). The built-in function gammamellininv provides this feature.

After some experimentation, it seems that a value of C between 8 and 10 gives optimal performance, so we have set the default at 9.

8.6. Generalized incomplete Gamma functions

Our main use of inverse Mellin transforms will be in the numerical computation of L-functions that we will study in the Chapter 9. The method that we give in that chapter, due to A. Booker and P. Molin, uses *only* inverse Mellin transforms. However, other methods which can also prove useful need an additional ingredient, generalized incomplete gamma functions. Even though we will not study these other methods, we briefly explain here the necessary computations.

If $\gamma(s)$ is a gamma product and $K(x)$ its inverse Mellin transform, we define the associated (generalized) incomplete gamma function by

$$\gamma(s,x) = \int_x^\infty t^{s-1}K(t)\,dt\,,$$

so that $\gamma(s,0) = \gamma(s)$. Once again, the main question is the computation of $\gamma(s,x)$, both for small x and for large x.

For small x (and reasonable values of s) we write $\gamma(s,x) = \gamma(s) - \int_0^x t^{s-1}K(t)$, and since we know the generalized power series expansion of $K(t)$ around $t = 0$ (generalized because of powers of $\log(t)$), we simply integrate term by term.

For large x, we can once again compute the asymptotic expansion of $\gamma(s,x)$, which will be of the form

$$\gamma(s,x) = A_1 x^{A_2} e^{-d\pi x^{2/d}} \sum_{n\geq 0} N_n x^{-2n/d}$$

for suitable A_i and N_n depending only on s, and as before we use the quotient-difference algorithm to transform this into a continued fraction. The details of all these procedures can be found in [**Dok04**].

CHAPTER 9

Computation of L-functions

9.1. The basic setting and goals

Our goal is the numerical computation of quantities linked to L-functions. We first give a precise definition of the objects that we study, although evidently, more general definitions are possible.

DEFINITION 9.1.1. We will say that a complex-valued function $L(s)$ is a weak L-function if there exist positive integers d, the degree, k, the weight, and N, the conductor, satisfying the following conditions:

(1) It has a Dirichlet series representation $L(s) = \sum_{n\geq 1} a(n)n^{-s}$ with $a(1) = 1$, where $|a(n)| = O_\varepsilon(n^{k-1+\varepsilon})$ for all $\varepsilon > 0$. In particular, this series converges absolutely for $\Re(s) > k$.

(2) It has a meromorphic continuation to the whole complex plane with a finite number of poles, which are all simple with real part strictly greater than $k/2$.

(3) There exist *real numbers* α_j such that if we define the *completed L-function* $\Lambda(s)$ by

$$\Lambda(s) = N^{s/2}\gamma(s)L(s) \quad \text{with} \quad \gamma(s) = \prod_{1\leq j\leq d} \Gamma_{\mathbb{R}}(s+\alpha_j),$$

we have the functional equation

$$\Lambda(k-s) = w\overline{\Lambda}(s),$$

where $\overline{\Lambda}(s) = N^{s/2}\gamma(s)\overline{L}(s)$, $\overline{L}(s) = \sum_{n\geq 1} \overline{a(n)}/n^s$ for $\Re(s) > k$, and w is some complex number, the root number; the function $N^{s/2}\gamma(s)$ is called the *factor at infinity* of the L-function.

We say that the function L is *self-dual* if $a(n)$ is real, so that $\overline{L} = L$ and $\overline{\Lambda} = \Lambda$.

DEFINITION 9.1.2. A weak L-function is an L-function if, in addition, it has an Euler product of the form $L(s) = \prod_p L_p(p^{-s})^{-1}$ where p runs through the set of all prime numbers, $L_p(T)$ is a polynomial of degree equal to the degree d of the L-function for all $p \nmid N$, of degree less than or equal to d otherwise, and such that if we write $L_p(T) = \prod_{1\leq i\leq d}(1 - \alpha_{p,i}T)$, we have $|\alpha_{p,i}| = p^{(k-1)/2}$ when $p \nmid N$, and $|\alpha_{p,i}| = 0$ or $p^{m/2}$ for some integer $m \leq k-1$ when $p \mid N$.

REMARKS 9.1.3.

(1) Note that $\overline{\Lambda}(s) = \overline{\Lambda(\overline{s})}$.

(2) Evidently the root number has modulus 1, and is equal to ± 1 if the L-function is self-dual.

(3) This is not the most general definition of an L-function, but it is the one we will restrict to in this book. A general definition which covers all "reasonable" L-functions has been given by A. Selberg, and is thus called the *Selberg class*, see for instance [**Kac06**] for details.

Our main goal is to compute many values of $L(s)$ as fast as possible. In particular, we may want to compute *special values*, or plot the function and compute its zeros on the critical line $\Re(s) = k/2$, and other possible problems. Until mentioned otherwise, we only assume that our L-functions are weak L-functions, in particular that they do not necessarily have an Euler product.

Note that the methods that we will explain in the main part of this chapter work in reasonable ranges of s, and in particular are *not* well adapted to the computation of L-functions very high in the critical strip. In the important special case of L-functions associated to Dirichlet characters (including the Riemann zeta function), we will study in Section 9.12 the `ZetaRiemannSiegel` and the `Zetafast` algorithms which are extremely efficient for this purpose, but even the Euler–MacLaurin algorithm `ZetaHurwitz` that we gave in Section 4.2.7 is orders of magnitude faster. The main and crucial advantage of the methods that we now explain is that they apply to *all* weak L-functions in the above sense, and not only to those associated to Dirichlet characters, for instance to L-functions associated to modular forms, elliptic curves, or general algebraic varieties.

9.2. The associated Theta function

9.2.1. Definition and basic properties. We denote by $K(t)$ the inverse Mellin transform of the function $\gamma(s)$. This means that $\gamma(s) = \int_0^\infty K(t)t^s \, dt/t$ for $\Re(s) > c$, where $c = \max_j(-\alpha_j)$ (recall that we assume that the α_j are real), and by the Mellin inversion formula, that

$$K(t) = \frac{1}{2\pi i} \int_L t^{-s} \gamma(s) \, ds \;,$$

where L is a suitable contour enclosing on the left all the poles of $\gamma(s)$, for instance a vertical line $\Re(s) = c + \varepsilon$ for any $\varepsilon > 0$. We will set the following definition:

DEFINITION 9.2.1. The theta function associated to $L(s)$ is the function

$$\Theta(t) = \sum_{n \geq 1} a(n) K\big((n/N^{1/2})t\big) \;.$$

Note that we could give a slightly different definition in the case that $L(s)$ has poles, but we will not do so, see the remark below.

Since the α_j are real the function $K(t)$ is real-valued. In addition since $K(t)$ tends to zero exponentially with t as $t \to \infty$ (see Theorem 8.2.4), and since the $a(n)$ are polynomially bounded, it follows that the series for $\Theta(t)$ converges exponentially fast for all $t > 0$, although rather slowly at the beginning if t is very small.

Now note that for $\Re(s) > c$ we have

$$\int_0^\infty t^s K\big((n/N^{1/2})t\big) \, dt/t = N^{s/2} n^{-s} \gamma(s) \;.$$

It follows by uniform convergence that for $\Re(s) > \max(c, k)$ we have

$$\int_0^\infty t^s \Theta(t) \, dt/t = \Lambda(s) \;,$$

and by Mellin inversion, that

$$\Theta(t) = \frac{1}{2\pi i} \int_L t^{-s} \Lambda(s) \, ds \; ,$$

where L is a contour as above, in particular to the right of $\Re(s) = \max(c, k)$.

DEFINITION 9.2.2. We will denote by P^+ the set of poles of $L(s)$, so that $\beta \in P^+$ implies that $\Re(\beta) > k/2$, and by P the set of poles of $\Lambda(s)$. For any $\beta \in P$ we will denote by r_β the residue of $L(s)$ at $s = \beta$, and by R_β the residue of $\Lambda(s)$ at $s = \beta$.

REMARKS 9.2.3.

(1) With evident notation, we have $P = P^+ \sqcup (k - P^+)$. Indeed, by definition and the functional equation we have

$$\Lambda(s) = N^{s/2} \gamma(s) L(s) = w N^{(k-s)/2} \gamma(k-s) \overline{L}(k-s) \; ,$$

so it is clear that if $\beta \in P^+ \cup (k - P^+)$ then $\beta \in P$. Conversely, if $\beta \in P$ and $\beta \notin P^+ \cup (k - P^+)$ then β is a pole both of $\gamma(s)$ and of $\gamma(k-s)$, which is impossible by assumption.

(2) Note also that by the functional equation we have $R_{k-\beta} = -w\overline{R_\beta}$.

PROPOSITION 9.2.4. *For $t \in \mathbb{R}_{>0}$ we have the functional equation*

$$\Theta(1/t) = wt^k \overline{\Theta}(t) + \sum_{\beta \in P} R_\beta t^\beta = wt^k \overline{\Theta}(t) + \sum_{\beta \in P^+} (R_\beta t^\beta - wt^{k-\beta} \overline{R_\beta}) \; .$$

Equivalently, if we set

$$\Theta_1(t) = \Theta(t) + w \sum_{\beta \in P^+} \overline{R_\beta} t^{\beta - k} \quad and$$

$$\Theta_2(t) = \Theta(t) - \sum_{\beta \in P^+} R_\beta t^{-\beta} \; ,$$

we have the functional equations $\Theta_i(1/t) = wt^k \overline{\Theta_i}(t)$ for $i = 1, 2$.

PROOF. Choose for L a vertical line $\Re(s) = \sigma$, where $\sigma > \max(c, k)$, and let $C_T = C_{\sigma,T}$ be the rectangular contour oriented positively $[\sigma - iT, \sigma + iT, k - \sigma + iT, k - \sigma - iT, \sigma - iT]$. By the residue theorem we have

$$I_T = \frac{1}{2\pi i} \int_{C_T} t^s \Lambda(s) \, ds = \sum_{\beta \in P} R_\beta t^\beta \; .$$

On the other hand, since the function $\gamma(s)$ tends to 0 exponentially fast when $\Im(s) \to \pm\infty$, we have

$$\lim_{T \to \infty} I_T = \frac{1}{2\pi i} \int_{\Re(s)=\sigma} t^s \Lambda(s) \, ds - \frac{1}{2\pi i} \int_{\Re(s)=k-\sigma} t^s \Lambda(s) \, ds \; .$$

The first term is by definition equal to $\Theta(1/t)$. In the second term, we change s into $k - s$, so that

$$\frac{1}{2\pi i} \int_{\Re(s)=k-\sigma} t^s \Lambda(s) \, ds = \frac{1}{2\pi i} \int_{\Re(s)=\sigma} t^{k-s} \Lambda(k-s) \, ds$$

$$= wt^k \frac{1}{2\pi i} \int_{\Re(s)=\sigma} t^{-s} \overline{\Lambda}(s) \, ds = wt^k \overline{\Theta}(t) \; ,$$

proving the first part of the proposition, and the second part follows immediately.

□

REMARKS 9.2.5.

(1) It is clear that $\Theta(t)$ tends to 0 exponentially fast as $t \to \infty$ since this is the case for the function $K(t)$.

(2) Since the functional equation for $\Theta_1(t)$ does not involve any extra terms, it is debatable whether to use Θ or Θ_1 as a basic function. The function Θ_2 will be used below.

9.2.2. Computer representation of L-functions. If we want to write computer programs to compute L-functions, the first thing to do is to choose a suitable way to represent them. We have chosen the following: an L-function will always be represented by a seven-component vector

[a, sd, Vga, k, N, w, R]

with the following meaning:

(1) a is the sequence of coefficients of the L-function, which can be given in several ways. We will assume that it is a closure giving the first n coefficients, so that a(M) gives the vector $[a(1), a(2), \ldots, a(M)]$; if for some reason you only have a finite number of coefficients available, so that a is a vector, use n->a[1..min(n,#a)], but this may be dangerous since some programs may need more coefficients than you actually have. Evidently, those programs should issue a warning.

(2) sd is 0 if the L-function is self-dual, i.e., if the $a(n)$ are real, and 1 otherwise. In a general implementation such as the one in Pari/GP, when it is nonzero, sd contains the coefficients for the dual L-function.

(3) Vga is the vector $[\alpha_1, \ldots, \alpha_d]$ of real numbers entering in the gamma product.

(4) k is the integer entering in the functional equation of $\Lambda(s)$.

(5) N is the conductor, so that $N^{s/2}$ enters in the definition of $\Lambda(s)$.

(6) w is the root number occurring in the functional equation of $\Lambda(s)$.

(7) R is the residue of $\Lambda(s)$ at $s = k$, so that the residue r of $L(s)$ is equal to R divided by the factor at infinity in the definition of $\Lambda(s)$ (of course $R = r = 0$ if there is no pole). For simplicity, we will assume that either $L(s)$ is holomorphic in the whole complex plane (it has no pole) or that it has a single *simple* pole at $s = k$. In a more general implementation such as the one in Pari/GP, an arbitrary finite number of poles would be taken into account and multiple poles would be allowed (we would then need to code them differently and require the polar part instead of a residue).

(8) This representation is highly redundant: mathematically, only a should be necessary and determine all the other components; even then, a finite number of coefficients $a(n)$ could be omitted, for instance the $a(p^k)$ for bad primes p dividing the conductor and $k \leq d$, which would be enough to determine the local factor $L_p(T)$ for a true L function with an Euler product (or even fewer k using a local functional equation). This is useful since the theoretical computation of these data, the root number or the conductor for instance, can be very involved. In principle, one could set an unknown component to a placeholder such as oo and write programs

to compute it: replacing all placeholders by unknowns and using the functional equation for Θ (Proposition 9.2.4) at various real or complex t, we can derive as many equations as we please. In full generality, this becomes cumbersome because of the variety of equations involved and also because solving polynomial systems (with approximate coefficients) in a large number of unknowns is hard. Our programs will make provisions to compute N, w and R with various restrictions.

In order to compute $L(s)$ or $\Lambda(s)$, we will need an initialization phase which will both fill in possible missing data in $L(s)$ (the oo placeholders), and compute technical data allowing the fast computation of individual values. The output of this initialization will thus be a two-component vector, where the first component is a *complete* seven-component vector as above, and the second the technical data. We will thus also allow this representation in all programs. Thus for instance, in programs not needing the additional data, we will include the command

```
if (#Ldata == 2, Ldata = Ldata[1]);
/* At this stage, Ldata = [a, sd, Vga, k, N, w, R] */
```

To make our programs more readable, we will setup some magic (see `Lfuntheta` below) so that, given a structure `Ldata` as above, with or without initialization, the syntax `Ldata.a`, `Ldata.Vga`, `Ldata.k`, `Ldata.N`, `Ldata.w`, `Ldata.R` returns respectively the component of the same name in the structure; we also setup `Ldata.d` to return the degree d of the L function, which is also equal to `#Vga`, and `Ldata.r` which is the residue of the L-function at $s = k$. Finally, provided an initialization was performed, `Ldata.prec` return the bitprecision for which it was computed, which will allow us to avoid initializing again the structure until a larger precision is requested.

9.2.3. Computing the Theta function. Recall that $\Theta(t)$ is given by the series $\sum_{n \geq 1} a(n)K(nt/N^{1/2})$.

- From Theorem 8.2.4, we know that $K(x) \sim C(d)x^A e^{-d\pi x^{2/d}}$, where $A = (1 - d + \sum_{1 \leq j \leq d} \alpha_j)/d$ and $C(d) = d^{-1/2}2^{(d+1)/2}$.
- Since $K(x)$ tends to 0 exponentially at infinity, while $a(n)$ has only polynomial growth in n, it is usually possible to neglect the behavior of the factor $a(n)$ in convergence estimates. However, in general we must assume that $a(n) = O(n^{k_1})$ for a suitable k_1. We will discuss the choice of k_1 below (section 9.6).
- We may want to compute values of $\Theta(t)$ *per se*, for instance if we want to compute the root number and/or the residue (see Section 9.2.4 below). If we want an absolute accuracy of e^{-E}, we thus need to find L such that $|\sum_{n>L} a(n)K(nt/N^{1/2})| < e^{-E}$.
- In the application to the computation of $L(s)$, we will instead need to compute $t^\sigma \Theta(t)$ to accuracy e^{-E} for some $\sigma \geq 0$. If σ is small, we can use the same estimate as for $\Theta(t)$ itself, but if σ is large, we will need a specific estimate.
- Finally, note that because of the functional equation relating $\Theta(t)$ and $\Theta(1/t)$ (Proposition 9.2.4), we should only need to compute $\Theta(t)$ for $t \geq 1$ (or possibly for some complex values of t with $|t| \geq 1$). However, as mentioned above, in order to compute unknown quantities such as the root number w and/or the residue r, it will be necessary to use values

of t less than 1, by writing down the functional equation for various pairs $(t, 1/t)$.

For an actual implementation, we must now find precise estimates for the number of terms that we need to use in the sum defining $\Theta(t)$. This is boring and should be skipped on first reading, but of course is essential. As explained above, the general problem is, for given $\sigma \geq 0$, $k_1 \geq 0$, $t > 0$, and N, and an absolute accuracy e^{-E}, to compute L such that $t^\sigma |\sum_{n>L} a(n) K(nt/N^{1/2})| < e^{-E}$. The special case where we want only to compute $\Theta(t)$ of course corresponds to $\sigma = 0$.

We will use reasonable approximations. Using $a(n) = O(n^{k_1})$ and the asymptotics of $K(x)$, the above inequality is approximately equivalent to

$$C(d)t^\sigma \sum_{n>L} n^{k_1}(nt/N^{1/2})^A e^{-d\pi(nt/N^{1/2})^{2/d}} < e^{-E} ,$$

in other words to

$$C(d)t^{\sigma+A}N^{-A/2} \sum_{n>L} n^{k_1+A} e^{-d\pi(nt/N^{1/2})^{2/d}} < e^{-E} .$$

Now this last sum is well approximated by the corresponding integral

$$\int_L^\infty x^{k_1+A} e^{-d\pi(t/N^{1/2})^{2/d}x^{2/d}} \, dx = \frac{d}{2} \int_{L^{2/d}}^\infty x^{d(k_1+A+1)/2-1} e^{-d\pi(t/N^{1/2})^{2/d}x} \, dx .$$

Up to scaling this is an incomplete gamma function, which as $L \to \infty$ is asymptotic to

$$\frac{N^{1/d}t^{-2/d}}{2\pi} L^{k_1+A+1-2/d} e^{-d\pi(t/N^{1/2})^{2/d}L^{2/d}} .$$

Our inequality is thus approximately $L^a e^{-bL^c} < e^{-F}$, where

$$a = k_1 + A + 1 - 2/d, \quad b = d\pi(t/N^{1/2})^{2/d}, \quad c = 2/d ,$$

$$F = E - (A/2 - 1/d)\log(N) + (\sigma + A - 2/d)\log(t) - \log(2\pi/C(d)) .$$

By Lemma 2.3.3, it follows that this inequality is approximately satisfied if

$$L > \max\left(\tau, \beta + \tau\log(\beta)\right)^{1/c}, \quad \text{where} \quad \tau = a/(bc) \quad \text{and} \quad \beta = F/b .$$

We have seen in Chapter 8 that if $d\pi x^{2/d} - A\log(x) > E$, then $K(x) < e^{-E}$, so can be neglected if we want such accuracy. Although this has an exact solution in terms of the Lambert W function, an approximate solution to this inequality which is sufficient for us is $x^{2/d} > E/(d\pi) + A/(2\pi) \cdot \log(E/(d\pi))$.

Since $\Theta(t) = \sum_{n \geq 1} a(n) K(nt/N^{1/2})$, if we assume that $a(n)$ does not grow too fast (otherwise we must compensate), it is sufficient to use the partial sums of the series up to n such that

$$nt > \left(E/(d\pi) + A/(2\pi) \cdot \log(E/(d\pi))\right)^{d/2} N^{1/2} .$$

We can now write down a program computing $\Theta(t)$. As mentioned above, if all the invariants of the L-function are known, we can assume that $|t| \geq 1$ because of the functional equation. In order to be able to determine missing invariants, we do not include the functional equation in the program computing Θ and rather add an optional parameter `tmin` to our initialization function, by default set to 1, which is the minimal value of $|t|$ for which we want to compute $\Theta(t)$.

```
                     ┌─────────────┐
───────────────────── │ Lfuntheta.gp │ ─────────────────────
                     └─────────────┘
1   /* Initialization for Theta computations, |t| >= tmin. */
2   Lfunthetainit(Ldata, tmin = 1, Nmax = Ldata.N)=
3   { my(B = getlocalbitprec(), Vga = Ldata.Vga, d = #Vga, A, E, L);
4     A = (1 - d + vecsum(Vga)) / d;
5     E = log(2) * B / (d * Pi);
6     L = ceil((E + A/(2*Pi) * log(E))^(d/2) * sqrt(Nmax) / tmin);
7     return ([Ldata.a(L + 5), Mellininvtech(Vga)]);
8   }
9
10  /* Theta0(t) = sum_{n > 0} a(n)K(nt). Treat specially the easy
11   * but important cases where #Vga = 1 or Vga = [0,1] in which
12   * case K(t) is an exponential. */
13  Lfuntheta0(Ldata, t, vv = 0)=
14  { my(v, q, q1, q2, S);
15
16    if (!vv, vv = Lfunthetainit(Ldata));
17    my([an, K] = vv, T = min(#an, ceil(#an / real(t))));
18    my(Vga = Ldata.Vga, d = #Vga);
19    if (d == 1,
20      v = vector(T); q1 = q = exp(-Pi * t^2); q2 = q * q;
21      v[1] = q;
22      for (n = 2, T, q1 *= q2; q *= q1; v[n] = q);
23      return (2*sum(n = 1, T, an[n] * (n*t)^Vga[1] * v[n])));
24    if (d == 2 && (Vga = vecsort(Vga)) == [0,1],
25      q = exp(-2 * Pi * t); S = 0;
26      forstep (n = T, 1, -1, S = an[n] + q * S);
27      return (2 * q * S));
28    t = t^(2 / d); v = dirpowers(T, 2 / d);
29    S = sum(n = 1, T, if (an[n], an[n] * K(v[n] * t)));
30    return (S);
31  }
32
33  Lfuntheta(Ldata, t, vv = 0)=
34    Lfuntheta0(Ldata, t / sqrt(Ldata.N), vv);
35
36  /* Accessors to retrieve information from Ldata.
37   * Also support Lfuninit format [Ldataf, VV] */
38  Ldata.a    = if (#Ldata == 2, Ldata[1][1], Ldata[1]);
39  Ldata.Vga  = if (#Ldata == 2, Ldata[1][3], Ldata[3]);
40  Ldata.k    = if (#Ldata == 2, Ldata[1][4], Ldata[4]);
41  Ldata.N    = if (#Ldata == 2, Ldata[1][5], Ldata[5]);
42  Ldata.w    = if (#Ldata == 2, Ldata[1][6], Ldata[6]);
43  Ldata.R    = if (#Ldata == 2, Ldata[1][7], Ldata[7]);
44  Ldata.prec = bitprecision(Ldata[2][1]);
45  Ldata.d    = #Ldata.Vga;
46  /* residue of L at s = k */
```

```
47   Ldata.r = my(k = Ldata.k); \
48     Ldata.R / (Ldata.N^(k/2) * Gammafactor(Ldata.Vga,k));
```

REMARKS 9.2.6.

(1) The program treats specially some easy but important cases where $K(t)$ is an exponential. Other cases such as Vga $= [a, a+1]$ are left as an exercise.

(2) We provide the function $\Theta_0(t) = \sum_{n\geq 1} a(n)K(nt)$ to be used in the Lfunconductor program below. That function guesses the correct conductor when it is unknown, and it does so by evaluating $\Theta_0(t/\sqrt{M})$ for various M. Analogously to the tmin parameter in Lfunthetainit, we provide an optional parameter Nmax, which is an upper bound for the various M. Of course, we set it to N by default outside of Lfunconductor since N is then known and we only need to evaluate $\Theta(t) = \Theta_0(t/\sqrt{N})$.

(3) Given a program to compute the derivatives of K (see Section 8.5.4), one can compute the m-th derivative of Θ using the obvious formula $\Theta^{(m)}(t) = \sum_{n\geq 1} a(n)(n/\sqrt{N})^m K^{(m)}(nt/\sqrt{N})$. It is not difficult to adapt Theorem 8.2.4 to give estimates of $K^{(m)}(x)$ as $x \to \infty$, then to adapt our discussion of cutoffs to the case $m > 0$. The built-in function lfuntheta allows to compute derivatives in this way.

As an immediate application, we can check if a given complete Ldata structure (i.e., containing all the information including the root number and the poles) corresponds to an L-function with functional equation. We do so by checking whether $\Theta_1(1/t_0) = wt_0^k\overline{\Theta_1(\overline{t_0})}$ for some random t_0, using the notations of Proposition 9.2.4 (we have assumed that $P^+ \subset \{k\}$).

$$\boxed{\text{Lfuncheckfeq.gp}}$$

```
1    /* Check if Ldata (assumed complete with root number and poles)
2     * is an L-function with functional equation. If consistent,
3     * the result should be close to 0 to current accuracy. */
4    Lfuncheckfeq(Ldata, vv = 0) =
5    { my(t0 = 355 / 339 + I / 7, a, b);
6
7      if (#Ldata == 2, Ldata = Ldata[1]);
8      my([k, N, w, R] = Ldata[4..7]);
9      if (!vv, vv = Lfunthetainit(Ldata));
10     a = Lfuntheta(Ldata,   1 / t0,   vv) + w * conj(R);
11     b = Lfuntheta(Ldata, conj(t0), vv) + w * conj(R);
12     return (a / (w * t0^k * conj(b)) - 1);
13   }
```

The output should be close to 0 to current bitprecision. In the next subsection we will see how to compute the root number w and the residue R if they are unknown, but note that if the residue *is* known, a trivial modification of the above program approximates the root number. In particular if $L(s)$ has no pole at $s = k$ and $R = 0$, simply replace the last line by

```
return (a / (t0^k * conj(b)));
```

9.2.4. Computing the root number w and the residue R.

PROPOSITION 9.2.7.

(1) *Assume that $\Lambda(s)$ has no pole, i.e., that $R = 0$. Then for any t such that $\overline{\Theta}(t) \neq 0$ we have $w = \Theta(1/t)/(t^k\overline{\Theta}(t))$. In particular, if $\Theta(1) \neq 0$ we have $w = \Theta(1)/\overline{\Theta}(1)$; if $\Theta(1) = 0$ but $\Theta'(1) \neq 0$ we have $w = -\Theta'(1)/\overline{\Theta}'(1)$.*

(2) *Assume that there is a pole and that L is self-dual, i.e., that the $a(n)$ are real. Then for all t such that $t^k \neq w$ we have*

$$R = (\Theta(1/t) - wt^k\Theta(t))/(t^k - w) .$$

Thus, if $w = -1$ we have $R = \Theta(1)$, while if $w = 1$ we have $R = -\big(2\Theta'(1) + k\Theta(1)\big)/k$ by L'Hospital's rule.

(3) *Otherwise, we have the linear equations in w and R for all $t > 0$:*

$$w(t^{2k}\overline{\Theta}(t) - \overline{\Theta}(1/t)) + R(t^{2k} - 1) + t^k(\Theta(t) - \Theta(1/t)) = 0 ,$$

which should be used with t close but not too close to 1, for instance $t = 1.1$.

PROOF. (1) and (2) follow immediately from the functional equation $\Theta(1/t) = wt^k\overline{\Theta}(t) + Rt^k - wR$. Note that for t real, in particular $t = 1$, we have $\overline{\Theta}(t) = \overline{\Theta(t)}$; so $\overline{\Theta}(1) = 0$ if and only if $\Theta(1) = 0$ and similarly for the derivative. For (3), subtracting the functional equation for t_1 from that of t_2 we have

$$w(t_2^k\overline{\Theta}(t_2) - t_1^k\overline{\Theta}(t_1)) + R(t_2^k - t_1^k) - (\Theta(1/t_2) - \Theta(1/t_1)) = 0 ,$$

and (3) follows by choosing $t_1 = 1/t_2$.   □

It is more efficient to compute higher derivatives of Θ at 1 than to use other pairs $t \neq 1/t$, but our simplified program `Lfuntheta` does not allow it (the built-in `lfuntheta` does). An interesting choice, although not as efficient, is $t = \sqrt{2}$ because of the following remark: in the computation of $\Theta(1/\sqrt{2})$, we must compute the quantities $k_n = K(n/\sqrt{2N})$ until they become negligible. But then the computation of $\Theta(\sqrt{2})$ involves only the quantities $K(n\sqrt{2/N}) = k_{2n}$, which thus come for free. If we denote by $x(\varepsilon)$ the smallest u such that $|K(u)| < \varepsilon$ for all $u > x(\varepsilon)$, computing $\Theta(t) = \sum_{n \geq 1} a(n)K(nt/\sqrt{N})$ to accuracy ε requires approximately $x(\varepsilon)N^{1/2}/t$ evaluations of K, so in Cases (1) and (2) we need $1.414 \cdot x(\varepsilon)N^{1/2}$ evaluations (instead of $2.009 \cdot x(\varepsilon)N^{1/2}$ if we evaluate naively at $t = 1.1$ and $1/1.1$).

This yields the following program:

─────────────── Lfunthetaroot2.gp ───────────────

```
/* Lfuntheta at 2^{1/2} and 2^{-1/2} simultaneously.
 * If vv is set keep the current accuracy, otherwise increase
 * it to perform the computation. */
Lfunthetaroot2(Ldata, vv = 0)=
{ my(vK, L, t, Vga, an, K, B = getlocalbitprec(), d);

  localbitprec(if (!vv, B + 32, B));
  if (!vv, vv = Lfunthetainit(Ldata));
  Vga = Ldata.Vga; d = #Vga;
  if (d == 1 || (d == 2 && abs(Vga[1] - Vga[2]) == 1),
    return ([Lfuntheta(Ldata, sqrt(2), vv),
```

```
12            Lfuntheta(Ldata, sqrt(2) / 2, vv)]));
13    [an, K] = vv; L = #an; t = 1 / sqrt(2 * Ldata.N);
14    vK = vector(L, n, K((n * t)^(2/d)));
15    return ([sum(n=1, L \ 2, an[n] * vK[2 * n]),
16            sum(n=1, L, an[n] * vK[n])]);
17  }
18
```

Nonetheless, if we also need to compute values of the *L*-function, we will see below that is is better to use arguments closer to 1, since the precomputations will in large part already have computed the necessary quantities. The following program implements Proposition 9.2.7 using $t = 1$, then $t = \sqrt{2}$ as explained above, then $t = 11/10$:

──────────────────────── Lfunrootres.gp ────────────────────────

```
1   /* Round the root number if close to 1, -1, I or -I. */
2   Roundw(w)=
3   { my(e, W = round(w, &e), B = getlocalbitprec());
4     return (if (e < -B/2, W, w));
5   }
6
7   /* Compute the residue R of Lambda at k and the root number.
8    * Returns [R, w]. If vv is set, keep the current accuracy,
9    * otherwise increase it to perform the computation. */
10
11  /* assume R = 0 */
12  Lfunrootresnopole(Ldata, vv, B) =
13  { my(K1 = Lfuntheta(Ldata, 1, vv), K2);
14
15    if (exponent(K1) > -B / 2,
16      K2 = K1
17    ,
18      [K1, K2] = Lfunthetaroot2(Ldata, vv);
19      K2 *= 2^(-Ldata.k / 2));
20    bitprecision([0, Roundw(K2 / conj(K1))], B);
21  }
22
23  /* [A,B,C] such that A * w + B * R = C from
24   * K1 = theta(t) and K2 = theta(1/t), T = t^(2k) */
25  LfunthetaABC(K1, K2, T) =
26    [ conj(T * K1 - K2), T - 1, T^(1/2) * (K2 - K1) ];
27
28  Lfunrootres(Ldata, vv = 0) =
29  { my(K1, K2, B = getlocalbitprec());
30
31    if (#Ldata == 2, Ldata = Ldata[1]);
32    localbitprec(if (!vv, B + 32, B));
33    if (!vv, vv = Lfunthetainit(Ldata));
```

```
34    my([sd, Vga, k, N, w, R] = Ldata[2..7]);
35    if (!R, return (Lfunrootresnopole(Ldata, vv, B)));
36    if (sd == 0 && w == -1,
37      R = Lfuntheta(Ldata, 1, vv);
38      return (bitprecision([R, w], B)));
39    [K1, K2] = Lfunthetaroot2(Ldata, vv);
40    my([a, b, c] = LfunthetaABC(K1, K2, 2^k));
41    if (w != oo && precision(w) == oo,
42      R = (c - a * w) / b;
43      return (bitprecision([R, w], B)));
44    my(t0 = 11/10);
45    K1 = Lfuntheta(Ldata, t0, vv);
46    K2 = Lfuntheta(Ldata, 1 / t0, vv);
47    my([a2, b2, c2] = LfunthetaABC(K1, K2, t0^(2*k)));
48    [w, R] = [a, b; a2, b2]^(-1) * [c, c2]~;
49    return (bitprecision([R, Roundw(w)], B));
50  }
```

REMARK 9.2.8. As it stands the program fails with a division by zero if $\Theta(1) = \Theta(1/\sqrt{2}) = 0$, and it will lose accuracy if the $\Theta(t)$ or the determinant $ab_2 - ba_2$ occuring in denominators are very small. It could be improved by sampling random values of t in the neighborhood of 1 until all quantities involved in denominators are bigger than $2^{-B/2}$, say, which could require increasing B if $\Theta(1)$ is small.

9.2.5. Computing the conductor N. We now assume that r and w are known, but that the conductor N is unknown. Note that when the conductor is unknown, usually it is the residue r of the L-function which is known, and not $R = N^{k/2}\gamma(k)r$ the residue of the Λ-function. In the latter case, the discussion below simplifies.

Define

$$\Theta_0(t) = -r\gamma(k)t^{-k} + \sum_{n \geq 1} a(n)K(nt) \,,$$

so that with the notation of Proposition 9.2.4 we have $\Theta_2(t) = \Theta_0(t/N^{1/2})$. Let $w^{1/2}$ be one of the complex square roots of w. The functional equation of Θ_2 can thus be written

$$w^{-1/2}t^{-k/2}\Theta_0(1/(tN^{1/2})) = w^{1/2}t^{k/2}\overline{\Theta_0}(t/N^{1/2}) \,.$$

We now borrow an idea from the next section. Define

$$U(z) = w^{-1/2}e^{(k/2)z}\Theta_0(e^z) \quad \text{and} \quad z_0 = -\log(N)/2 \,.$$

Since $|w| = 1$, we thus have for z real $U(z_0 - z) = \overline{U(z_0 + z)}$. It follows that we have the two equations:

$$U(z_0) = \overline{U(z_0)} \quad \text{and} \quad U'(z_0) = -\overline{U'(z_0)}$$

(in other words $\Im(U(z_0)) = 0$ and $\Re(U'(z_0)) = 0$). If $U \neq \overline{U}$ the first formula gives a nontrivial equation for z_0 which usually has only a finite number of solutions, among which probably only a single one is such that $N = e^{-2z_0}$ is close to an integer, while if $U = \overline{U}$ the second formula is $U'(z_0) = 0$, which again usually only has a finite number of solutions.

Note that when L is self-dual, the condition $U = \overline{U}$ means that $w = 1$, so that the condition to be tested is $U'(z_0) = 0$. On the other hand, still when L is self-dual, when $w = -1$ we have $w^{-1/2} = \pm i$ so the condition to be tested is $U(z_0) = 0$.

REMARK 9.2.9. It is quite rare that one needs to compute the conductor of an L-function having a pole. Even when this does happen, it is frequently the case that $L(s)$ is divisible by $\zeta(s)$ and that $L(s)/\zeta(s)$ has no pole and the same conductor. In the even more exceptional case where the residue R of $\Lambda(s)$ is known instead of r, one should choose $\Theta_0(t) = w\overline{R} + \sum_{n \geq 1} a(n)K(nt)$, so that $\Theta_1(t) = \Theta_0(t/N^{1/2})$.

Writing a program to compute the conductor is more delicate, because it is not a priori clear how to solve the equation $\Im(U(z_0)) = 0$ or $\Re(U'(z_0)) = 0$. Nonetheless, we write the following, which must considered *heuristic*; we give the explanations for its use at the end.

─────────────────────── | Lfunconductor.gp | ───────────────────────

```
1   Lfunconductor(Ldata, maxcond=10000, steps=15)=
2   { my(vv, th0, g, w, st);
3
4     if (#Ldata == 2, Ldata = Ldata[1]);
5     if (type(maxcond) != "t_VEC", maxcond = [9/10, maxcond]);
6     my([minN, maxN] = maxcond);
7     my([k, N, w, R] = Ldata[4..7], w2);
8     if (w == oo || R == oo, error("Missing residue/root number"));
9     localbitprec(192);
10    w2 = 1 / sqrt(w);
11    vv = Lfunthetainit(Ldata, 1, ceil(maxN));
12    th0 = (t -> -R + Lfuntheta0(Ldata, t, vv));
13    localbitprec(128);
14    my(Uz = w2 * exp((k/2) * 0.1) * th0(exp(0.1)));
15    if (exponent(imag(Uz)) > -32,
16      g = (N -> imag(w2 * th0(1 / sqrt(N))))  \\ Im(U) != 0, use U
17    ,
18      g = (N -> my(n = 1 / sqrt(N)); \\ use U'
19             real(w2 * (k/2 * th0(n) + th0'(n) * n))));
20    ploth(t = minN, maxN, g(t)); \\ visual control
21    st = (maxN - minN)^(1 / steps);
22    \\ 2: refine until a zero is found; 4: multiplicative search
23    w = solvestep(t = minN, maxN, st, g(t), 2 + 4);
24    return (if (#w == 1, w[1], Vec(w)));
25  }
```

A number of explanations are in order concerning this program; for examples, see Section 9.10.1.

- The program gives a plot corresponding to the algorithm used, and the zeros of this plot are the possible conductors. In addition, the output of this program is either a single number, hopefully the correct conductor, an empty vector if no reasonable value has been found, or a vector containing several plausible conductors (this does happen, see below).

- The different optional parameters serve to help the program find the conductor. The parameter `maxcond` should be set in the ballpark of the conductor. If the conductor is expected to be small, reduce the value, and if large, increase its value. It can also be set to a positive interval, in which case the search and the plot is only done in that interval.
- The program searches the plausible conductors with a geometric progression, and the variable `steps` controls the number of terms used. Thus, if you need a finer mesh to find the conductor, increase `steps`.

Note that this program is given mainly for amusement. In practice the situation is different: the L-function, now assumed to have an Euler product, is known except for the Euler factors and the conductor at a few bad primes. In that case, we must look simultaneously for the missing Euler factors and the bad primes part of the conductor. For instance, in an example that we will see below, we expect that the conductor is a divisor of 64, so it is only necessary to check the (few) divisors of 64 to find the correct conductor.

9.3. Computing $\Lambda(s)$ and $L(s)$

9.3.1. The "approximate" functional equation. The main reason for introducing the associated theta function is the following:

PROPOSITION 9.3.1.

(1) *For $\Re(s) > \max(c, k)$ we have*

$$\Lambda(s) = \int_0^\infty t^s \Theta(t)\, dt/t .$$

(2) *For all $s \neq 0$, k and $A > 0$, we have*

$$\Lambda(s) = \sum_{\beta \in P+} \frac{R_\beta A^{s-\beta}}{s - \beta} - w \sum_{\beta \in P+} \frac{\overline{R_\beta} A^{s+\beta-k}}{s + \beta - k}$$
$$+ \int_A^\infty t^s \Theta(t) \frac{dt}{t} + w \int_{1/A}^\infty t^{k-s} \overline{\Theta(t)} \frac{dt}{t} ,$$

and in particular

$$\Lambda(s) = \sum_{\beta \in P+} \frac{R_\beta}{s - \beta} - w \sum_{\beta \in P+} \frac{\overline{R_\beta}}{s + \beta - k} + \int_1^\infty (t^s \Theta(t) + w t^{k-s} \overline{\Theta(t)}) \frac{dt}{t} .$$

(3) *We have*

$$\int_A^\infty t^s \Theta(t) \frac{dt}{t} = N^{s/2} \sum_{n \geq 1} \frac{a(n)}{n^s} \int_{An/N^{1/2}}^\infty K(t) t^s \frac{dt}{t} .$$

PROOF. (1) has already been used essentially as the definition of Θ and Θ_1, (2) follows for $\Re(s) > k$ by splitting the integral at $t = A$ and using the functional equation of Θ given by proposition 9.2.4, and then for all $s \neq k$, $s \neq 0$ by analytic continuation, and (3) is clear. $\qquad\square$

The formula of (2) for $\Lambda(s)$ is (improperly) called the approximate functional equation, and the last integral $\int_x^\infty K(t)t^s\,dt/t$ is a generalized incomplete gamma function mentioned in Section 8.6.

REMARKS 9.3.2.
  (1) The integral $\int_x^\infty K(t)t^s\,dt/t$ tends exponentially fast to 0 because of the exponential decrease of $K(t)$ at infinity, so the sum giving $\int_1^\infty t^s\Theta(t)\,dt/t$ converges fast and can thus be used to compute numerical values of $\Lambda(s)$, hence of $L(s)$.
  (2) There are several reasons for introducing a parameter A in the approximate functional equation. On the one hand, the programs for computing generalized incomplete gamma functions are not so easy to write, and also lose a number of decimals of accuracy if not done properly, so the fact that the expression should be *independent* of A gives a good check on the correctness of the implementation. On the other hand, as we did with the theta function itself, by varying the parameter A we can compute the root number w and/or the residue R if those are unknown. Since the functions involved are more complicated, it is clearly preferable to do this as we have done, using the theta function, but it is certainly also possible

We do not give any code for the approximate functional equation formulas (or the smoothed version that we mention next), first because it is quite complex and technical to write, and second because in many applications it is better to use the idea of A. Booker and P. Molin explained below in Section 9.4. If you want to write such code, we refer you to the paper of T. Dokchitser [**Dok04**] already mentioned in relation with inverse Mellin transforms, and that of M. Rubinstein [**Rub05**].

9.3.2. The smoothed approximate functional equation. It is, however, possible to modify the approximate functional equation in other ways, for instance if we want to compute $\Lambda(s)$ for $\Im(s)$ large. This is done by *smoothing* the approximate functional equation around the point that we want to compute, by multiplying by elementary exponential factors such as e^{as^2+bs+c} for suitable complex numbers a, b, and c. The following is directly taken from the excellent paper of Rubinstein [**Rub05**]:

THEOREM 9.3.3. *Let* $L(s) = \sum_{n\geq 1} a(n)n^{-s}$ *be a weak L-function, and for simplicity of exposition, assume that $L(s)$ has no poles in $\mathbb{C}$. Let $g(s)$ be an entire function such that for fixed s we have $|\Lambda(z+s)g(z+s)/z| \to 0$ as $\Im(z) \to \infty$ in any bounded strip $|\Re(z)| \leq \alpha$. We have*

$$\Lambda(s)g(s) = \sum_{n\geq 1} \frac{a(n)}{n^s} f_1(s,n) + \omega \sum_{n\geq 1} \frac{\overline{a(n)}}{n^{k-s}} f_2(k-s,n)\,,$$

where

$$f_1(s,x) = x^s \int_{\sigma-i\infty}^{\sigma+i\infty} \frac{\gamma(z)g(z)x^{-z}}{z-s}\,dz \quad and$$

$$f_2(s,x) = x^s \int_{\sigma-i\infty}^{\sigma+i\infty} \frac{\gamma(z)\overline{g(k-\overline{z})}x^{-z}}{z-s}\,dz\,.$$

The parameter σ is any real number greater than the real parts of all the poles of $\gamma(z)$ and than $\Re(s)$.

Several comments are in order concerning this theorem:

(1) The proof is a technical but elementary exercise in complex analysis. In particular, it is easy to modify the formula to take into account possible poles of $L(s)$, see [**Rub05**] once again.

(2) As in the unsmoothed case, the functions $f_i(s, x)$ are exponentially decreasing as $x \to \infty$. Thus this gives fast formulas for computing values of $L(s)$ for reasonable values of s. The very simplest case of this approximate functional equation, even simpler than the Riemann zeta function, is for the computation of the value at $s = 1$ of the L-function of an *elliptic curve* E: if the sign of its functional equation is equal to $+1$ (otherwise $L(E, 1) = 0$), the (unsmoothed) formula reduces to

$$L(E, 1) = 2 \sum_{n \geq 1} \frac{a(n)}{n} e^{-2\pi n/N^{1/2}} ,$$

where N is the conductor of the curve.

(3) It is not difficult to show that as $n \to \infty$ we have a similar behavior for the functions $f_i(s, n)$ as for the inverse Mellin transforms (Theorem 8.2.4), i.e.,

$$f_i(s, n) \sim C \cdot t^a e^{-\pi d (n/N^{1/2})^{2/d}}$$

for some explicit constants a and C (in the preceding example $d = 2$). Thus, we have the so-called $N^{1/2}$-*paradigm*: the series of the theorem indeed converge exponentially fast, but we need at least $\widetilde{O}(N^{1/2})$ terms to obtain any accuracy at all. This is an extremely serious limitation, and probably the most important question in this field of computational number theory: is it possible to do any better? In particular cases (such as $L(E, 1)$ above, or some other special values), there are often other methods using the deeper structure of the problem to find the result, but for instance we believe that nobody has a method faster than $\widetilde{O}(N^{1/2})$ to compute $L(E, \pi)$, say (not that this number has any interest).

    Note, however, that quite surprisingly there are some apparent counterexamples to this paradigm: for instance, in [**Hia14**], Hiary has shown that if the conductor N is far from squarefree, for instance if $N = m^3$, at least in the case of Dirichlet L-functions the computation can be done in time $\widetilde{O}(m) = \widetilde{O}(N^{1/3})$.

(4) The theorem can be used with $g(s) = 1$ to compute values of $L(s)$ for "reasonable" values of s. When s is unreasonable, for instance when $s = 1/2 + iT$ with T large (to check the Riemann hypothesis for instance), one chooses other functions $g(s)$ adapted to the computation to be done, such as $g(s) = e^{is\theta}$ or $g(s) = e^{-a(s-s_0)^2}$; we refer to Rubinstein's paper for detailed examples.

(5) By choosing two very simple functions $g(s)$ such as a^s for two different values of a close to 1, one can compute numerically the value of the root number ω if it is unknown. In a similar manner, if the $a(n)$ are known but not ω nor the conductor N, by choosing a few easy functions $g(s)$ one can find them. But much more surprisingly, if almost nothing is known apart from the gamma factors and N, say, by cleverly choosing a number of functions $g(s)$ and applying techniques from numerical analysis

such as singular value decomposition and least squares methods, one can prove or disprove (numerically of course) the existence of an L-function having the given gamma factors and conductor, and find its first few Fourier coefficients if they exist. This method has been used extensively by D. Farmer in his search for $\mathrm{GL}_3(\mathbb{Z})$ and $\mathrm{GL}_4(\mathbb{Z})$ Maass forms, by Poor and Yuen in computations related to the paramodular conjecture of Brumer–Kramer and abelian surfaces, and by A. Mellit in the search of L-functions of degree 4 with integer coefficients and small conductor. Although a fascinating and active subject, it would carry us too far afield to give more detailed explanations.

9.4. Booker–Molin's idea for computing $\Lambda(s)$: Poisson summation

Until rather recently, the approximate functional equation (smoothed or not) was the main tool for computing many values of $L(s)$. One annoying feature is that one needs to compute the functions $f_i(s, x)$, and this is rather painful. As already mentioned in Section 8.6, in the same paper [**Dok04**] where he explains how to compute inverse Mellin transforms (that we have essentially reproduced in Chapter 8), T. Dokchitser also explains that similar methods (i.e., generalized power series and continued fractions of asymptotic expansions) can be used to compute these functions.

However, A. Booker [**Boo06**] and more recently P. Molin [**Mol10**] have suggested another method for computing L-functions which avoids completely the computation of generalized incomplete gamma functions (but not of inverse Mellin transforms); this method is both simpler and faster than the use of incomplete gamma functions. The basic idea is to use the Poisson summation formula in a slightly modified form, as follows:

LEMMA 9.4.1. *Let ϕ be a piecewise C^∞ function on $\mathbb{R}_{>0}$ with possible simple jump singularities, let $F(s) = \int_0^\infty \phi(t)\, dt/t$ be its Mellin transform defined for $\Re(s) > \sigma_0$, say, and define $f_s(z) = e^{zs}\phi(e^z)$, so that $F(s) = \int_{-\infty}^\infty f_s(z)\, dz$. Then for all $h > 0$ we have for $\Re(s) > \sigma_0$ the identity*

$$\sum_{m \in \mathbb{Z}} f_s(mh) = \sum_{n \in \mathbb{Z}} F(s - 2\pi i n/h) ,$$

if we define $\phi(t) = (\phi(t^+) + \phi(t^-))/2$ for all t including at the jumps.

PROOF. By the standard Poisson summation formula, we have

$$h \sum_{m \in \mathbb{Z}} f_s(mh) = \sum_{n \in \mathbb{Z}} \widehat{f}_s(n/h); .$$

Now

$$\widehat{f}_s(x) = \int_{-\infty}^\infty e^{-2\pi i x z} f_s(z)\, dz = \int_{-\infty}^\infty e^{z(s - 2\pi i x)}\phi(e^z)\, dz = F(s - 2\pi i x) ,$$

proving the lemma. Note in passing that if $\Re(s) > \sigma_0$, we have $\Re(s - 2\pi i n/h) = \Re(s) > \sigma_0$ for all $n \in \mathbb{Z}$. □

From this lemma we deduce the main theorem which underlies the Booker–Molin method:

THEOREM 9.4.2. *For any $h > 0$ we have*

$$\Lambda(s) = \frac{h}{2} \sum_{\beta \in P^+} (R_\beta \coth(h(s-\beta)/2) - w\overline{R_\beta} \coth(h(s+\beta-k)/2))$$

$$+ h \sum_{m \geq 0}{}' (e^{mhs}\Theta(e^{mh}) + we^{mh(k-s)}\overline{\Theta(e^{mh})})) - \sum_{n \in \mathbb{Z} \setminus \{0\}} \Lambda(s + 2\pi i n / h) \,,$$

where $\sum'$ means that the term $m = 0$ must be counted with coefficient $1/2$.

Before giving the proof, note the two crucial aspects of this theorem: first the quantities $\Theta(e^{mh})$ can be *precomputed* once and for all, independently of the argument s. Second, because of the exponential decrease of the gamma function in vertical strips the quantities $\Lambda(s + 2\pi n / h)$ will be *negligible* for $n \neq 0$ if h is chosen suitably small.

PROOF. Set $U(z) = e^{(k/2)z}\Theta(e^z)$, so that the functional equation for Θ reads

$$U(-z) = w\overline{U}(z) + \sum_{\beta \in P^+} (R_\beta e^{(\beta-k/2)z} - w\overline{R_\beta} e^{(k/2-\beta)z}) \,.$$

By the approximate functional equation with $A = 1$, we know that for all s not a pole we have

$$\Lambda(s) = \sum_{\beta \in P^+} \frac{R_\beta}{s-\beta} - w \sum_{\beta \in P^+} \frac{\overline{R_\beta}}{s+\beta-k} + \int_1^\infty (t^s \Theta(t) + wt^{k-s}\overline{\Theta(t)}) \frac{dt}{t} \,.$$

Splitting the integral in two and setting $t = e^z$ (resp. $t = e^{-z}$) gives

$$\int_1^\infty t^s \Theta(t) \, dt/t = \int_0^\infty e^{zs}\Theta(e^z) \, dz = \int_0^\infty e^{z(s-k/2)} U(z) \, dz \,,$$

$$w \int_1^\infty t^{k-s}\overline{\Theta(t)} \, dt/t = w \int_{-\infty}^0 e^{(s-k)z}\overline{\Theta}(e^{-z}) \, dz = w \int_{-\infty}^0 e^{z(s-k/2)}\overline{U}(-z) \, dz \,;$$

using $w\overline{w} = 1$ and the functional equation for U, the second integral becomes

$$\int_{-\infty}^0 e^{z(s-k/2)} \left(U(z) + \sum_{\beta \in P^+} \left(\overline{R_\beta} w e^{(\beta-k/2)z} - R_\beta e^{(k/2-\beta)z} \right) \right) dz \,.$$

Note that since we know that this integral converges absolutely, it is essential to keep the integrand in this way, and not to take out the individual terms.

Denote by $Y(z)$ the step function equal to 0 for $z > 0$ and to 1 for $z < 0$ (thus $Y(z) = H(-z)$, where H is the Heaviside function), and to be in accordance with the lemma, define $Y(0) = 1/2$. We have thus shown the formula valid for all s not a pole

$$\Lambda(s) = \sum_{\beta \in P^+} \frac{R_\beta}{s-\beta} - w \sum_{\beta \in P^+} \frac{\overline{R_\beta}}{s+\beta-k}$$

$$+ \int_{-\infty}^\infty e^{z(s-k/2)} \left(U(z) + Y(z) \sum_{\beta \in P^+} \left(w\overline{R_\beta} e^{(\beta-k/2)z} - R_\beta e^{(k/2-\beta)z} \right) \right) dz \,.$$

If we set

$$\phi(t) = \Theta(t) + Y(\log(t)) \sum_{\beta \in P^+} \left(w\overline{R_\beta} t^{\beta-k} - R_\beta t^{-\beta} \right) \,,$$

with the notation of the lemma we have

$$f_s(z) = e^{z(s-k/2)} \left(U(z) + Y(z) \sum_{\beta \in P^+} \left(w \overline{R_\beta} e^{(\beta-k/2)z} - R_\beta e^{(k/2-\beta)z} \right) \right) ,$$

so by the lemma

$$h \sum_{m \in \mathbb{Z}} f_s(mh) = \sum_{n \in \mathbb{Z}} F(s - 2\pi i n/h) ,$$

where F is the Mellin transform of ϕ. Now by Proposition 9.3.1 the Mellin transform of $\Theta(t)$ is equal to $\Lambda(s)$, and that of

$$Y(\log(t)) \sum_{\beta \in P^+} \left(w \overline{R_\beta} t^{\beta-k} - R_\beta t^{-\beta} \right)$$

is trivially computed to be equal to

$$w \sum_{\beta \in P^+} \frac{\overline{R_\beta}}{s+\beta-k} - \sum_{\beta \in P^+} \frac{R_\beta}{s-\beta} ,$$

so that $F(s)$ is equal to $\Lambda(s)$ with its polar part removed.

REMARK 9.4.3. The use of Poisson summation is justified in our case by the fact that $f_s(z)$ tends to 0 exponentially fast when $z \to \pm\infty$. On the other hand, if $R \neq 0$ the function $f_s(z)$ has a jump discontinuity at $z = 0$, so its Fourier transform will decrease slowly, and this is indeed the case because of the polar terms.

Let us compute the contribution of the "polar terms". First, on the right-hand side, summing symmetrically the terms n and $-n$, we have

$$\sum_{n \in \mathbb{Z}} \frac{1}{s - 2\pi i n/h} = h \sum_{n \in \mathbb{Z}} \frac{1}{hs - 2\pi i n} = \frac{h}{2} \coth(hs/2) .$$

It follows that

$$\sum_{n \in \mathbb{Z}} F(s - 2\pi i n/h) = \sum_{n \in \mathbb{Z}} \Lambda(s + 2\pi i n/h)$$

$$+ \frac{h}{2} \sum_{\beta \in P^+} \left(w \overline{R_\beta} \coth(h(s+\beta-k)/2) - R_\beta \coth(h(s-\beta)/2) \right) .$$

On the other hand, for $m > 0$ we have

$$f_s(mh) = e^{mh(s-k/2)} U(mh) = e^{mhs} \Theta(e^{mh}) ,$$

and for $m < 0$, by the functional equation for $U(z)$ we have

$$f_s(mh) = e^{mh(s-k/2)} \left(U(mh) + \sum_{\beta \in P^+} \left(w \overline{R_\beta} e^{(\beta-k/2)mh} - R_\beta e^{(k/2-\beta)mh} \right) \right)$$

$$= e^{mh(s-k/2)} w \overline{U}(-mh) = w e^{|m|h(k-s)} \overline{\Theta(e^{|m|h})}$$

since $e^{|m|h}$ is real, so that $\overline{\Theta}(e^{|m|h}) = \overline{\Theta(e^{|m|h})}$.

Finally, for $m = 0$, because of the jump discontinuity, we have

$$f_s(0) = \frac{f_s(0^+) + f_s(0^-)}{2} = \frac{\Theta(1) + w \overline{\Theta(1)}}{2}$$

(the same formula is of course obtained by using $Y(0) = 1/2$). It follows that

$$\sum_{m\in\mathbb{Z}} f_s(mh) = \sideset{}{'}\sum_{m\geq 0} \left(e^{mhs}\Theta(e^{mh}) + we^{mh(k-s)}\overline{\Theta(e^{mh})}\right) ,$$

and putting everything together proves the theorem. □

The following corollary is clear:

COROLLARY 9.4.4. *We have*

$$\Lambda^{(j)}(s) = \left(\frac{h}{2}\right)^{j+1} \sum_{\beta\in P^+} \left(R_\beta \coth^{(j)}(h(s-\beta)/2) - w\overline{R_\beta}\coth^{(j)}(h(s+\beta-k)/2)\right)$$

$$+ h^{j+1}\sideset{}{'}\sum_{m\geq 0} m^j\left(e^{mhs}\Theta(e^{mh}) + (-1)^j we^{mh(k-s)}\overline{\Theta(e^{mh})}\right) - \sum_{n\in\mathbb{Z}\setminus\{0\}} \Lambda^{(j)}(s+2\pi in/h) .$$

To use the theorem in computational practice, we rewrite it as follows:

COROLLARY 9.4.5. *We have*

$$\Lambda(s) = P(s) + hS_M(s) + hR_t(M,h) - R_f(h) ,$$

where

$$P(s) = \frac{h}{2}\sum_{\beta\in P^+}\left(R_\beta\coth(h(s-\beta)/2) - w\overline{R_\beta}\coth(h(s+\beta-k)/2)\right) ,$$

$$S_M(s) = \sideset{}{'}\sum_{0\leq m\leq M}\left(e^{mhs}\Theta(e^{mh}) + we^{mh(k-s)}\overline{\Theta(e^{mh})}\right) ,$$

$$R_t(M,h) = \sum_{m>M}\left(e^{mhs}\Theta(e^{mh}) + we^{mh(k-s)}\overline{\Theta(e^{mh})}\right) ,$$

$$R_f(h) = \sum_{n\in\mathbb{Z}\setminus\{0\}}\Lambda(s+2\pi in/h) .$$

In the above, R_t is the truncation error, R_f the Fourier error (both of course also depend on s but we do not need to write explicitly the dependence).

We want to compute $L(s)$ to an accuracy of B bits, i.e., with an error less than $2^{-B} = e^{-B\log(2)}$. In reasonable ranges $L(s)$ is close to 1, so absolute or relative error are essentially the same. Since computing $N^{s/2}\gamma(s)$ can be done easily to high accuracy, this is equivalent to computing $\Lambda(s)$ with relative accuracy less than $e^{-B\log(2)}$. Now when $T = |\Im(s)|$ is large, $|N^{s/2}\gamma(s)|$ is of the order of $e^{-(\pi/4)dT}$. Thus, we need an absolute accuracy less than e^{-E} with $E = B\log(2) + (\pi/4)dT$, and we must perform the computation at accuracy $E/\log(2)$ bits.

9.5. The Fourier error

Consider first the Fourier error $R_f(h)$. Note that $|N^{s+2\pi ik/h}| = |N^s|$, and $|L(s+2\pi ik/h)|/|L(s)|$ is not very small or large. Thus,

$$|\Lambda(s+2\pi in/h)/\Lambda(s)| \approx |\gamma(s+2\pi in/h)/\gamma(s)| .$$

Now by the complex Stirling's formula, we have in a suitable range

$$\left|\frac{\Gamma_\mathbb{R}(s+\alpha_j+2\pi in/h)}{\Gamma_\mathbb{R}(s+\alpha_j)}\right| = \left|\frac{\Gamma\big((s+\alpha_j)/2+\pi in/h\big)}{\Gamma\big((s+\alpha_j)/2\big)}\right|$$

$$\sim \left|\frac{\big((s+\alpha_j)/2+\pi in/h\big)^{(s+\alpha_j-1)/2+\pi in/h}e^{-(s+\alpha_j)/2-\pi in/h}}{\big((s+\alpha_j)/2\big)^{(s+\alpha_j-1)/2}e^{-(s+\alpha_j)/2}}\right|$$

$$= \left|\frac{\big((s+\alpha_j)/2+\pi in/h\big)^{(s+\alpha_j-1)/2+\pi in/h}}{\big((s+\alpha_j)/2\big)^{(s+\alpha_j-1)/2}}\right| .$$

Note that if $s=\sigma+it$ with σ and t real, we have

$$|s^{s-1/2}| = (\sigma^2+t^2)^{(\sigma-1/2)/2}e^{-t\,\mathrm{atan}(t/\sigma)} .$$

To simplify the analysis, we will assume the following. If $|t/\sigma|$ is small (in particular smaller than 1), we neglect the $e^{-t\,\mathrm{atan}(t/\sigma)}$ factor, and approximate by $|\sigma|^{\sigma-1/2}$. On the contrary, if $|t/\sigma|$ is large, we approximate by $|t|^{\sigma-1/2}e^{-(\pi/2)|t|}$, or even simpler by $e^{-(\pi/2)|t|}$.

We apply this approximate analysis to $(s+\alpha_j)/2+\pi in/h$ and to $(s+\alpha_j)/2$. To simplify notation, set $\sigma_1=\sigma+\alpha_j$. Assume first that $|t/\sigma|$ is small. The denominator is approximately equal to $(\sigma_1/2)^{\sigma_1/2-1/2}$. On the other hand, the numerator is approximately equal to

$$(\sigma_1^2/4+\pi^2(n/h)^2)^{(\sigma_1-1)/4}e^{-\pi(n/h)\,\mathrm{atan}(2\pi(n/h)/\sigma_1)} .$$

We choose h so that $\pi/h > |\sigma_1|/2$, in other words $h < 2\pi/|\sigma_1|$, so that this can be approximated by

$$(\pi(|n|/h))^{(\sigma_1-1)/2}e^{-(\pi^2/2)(|n|/h)} = (\pi/h)^{(\sigma_1-1)/2}|n|^{(\sigma_1-1)/2}e^{-(\pi^2/2)(|n|/h)} .$$

We must now take the product of these quantities from $j=1$ to d. Since the α_j are small, this amounts to raising to the dth power and replacing σ_1 by σ, so the product is approximately equal to

$$(\pi/h)^{d(\sigma-1)/2}|n|^{d(\sigma-1)/2}e^{-d(\pi^2/2)(|n|/h)} .$$

The sum for $n=1$ to $+\infty$ of this can be approximated by

$$(\pi/h)^{d(\sigma-1)/2}\int_1^\infty x^{d(\sigma-1)/2+1}e^{-d(\pi^2/2)x/h}\,dx/x$$

$$= (h/\pi)(2/(d\pi))^{d(\sigma-1)/2+1}\int_{(d\pi^2/2)/h}^\infty x^{d(\sigma-1)/2}e^{-x}\,dx$$

$$\sim (2/(d\pi))(\pi/h)^{d(\sigma-1)/2-1}e^{-d\pi^2/(2h)} .$$

Since the product of the denominators is approximately equal to $(\sigma/2)^{d(\sigma-1)/2}$, we obtain finally that $\sum_{n\neq0}|\Lambda(s+2\pi in/h)/\Lambda(s)|$ is approximately equal to

$$\frac{4h}{d\pi^2}\left(\frac{2\pi}{h\sigma}\right)^{d(\sigma-1)/2}e^{-d\pi^2/(2h)} .$$

If we want this to be less than e^{-E}, this can be approximately solved as

$$d\pi^2/(2h) > E + (d(\sigma-1)/2-1)\log(E)$$

(recall that we also need $\pi/h > \sigma/2$).

Assume now that $|t/\sigma|$ is large. The denominator is approximated by

$$|t/2|^{(\sigma-1)/2}e^{-(\pi/4)|t|} ,$$

and the numerator by

$$|t/2 + \pi n/h|^{(\sigma-1)/2}e^{-(\pi/2)|t/2+\pi n/h|} .$$

Assume without loss of generality that $t > 0$. For $n > -th/(2\pi)$ (in particular for $n > 0$) the quotient is thus

$$\left(1 + 2\pi n/(th)\right)^{(\sigma-1)/2}e^{-(\pi^2/2)n/h} .$$

Now note that this is always going to be large if $n < 0$. Since $n = 0$ is excluded, it is sufficient to choose $h < (2\pi)/t$ for $n < 0$ to be excluded. Thus the sum for $n > -th/(2\pi)$ will be

$$\sum_{n \geq 1}(1 + 2\pi n/(th))^{d(\sigma-1)/2}e^{-d(\pi^2/2)n/h} .$$

We approximate this by its first term

$$\left(2\pi/(th)\right)^{d(\sigma-1)/2}e^{-d\left(\pi^2/(2h)\right)} .$$

For $n < -th/(2\pi)$ (in other words, because of our choice of h, for $n \leq -1$), the quotient is

$$\left((\pi|n|/h - t/2)/(t/2)\right)^{(\sigma-1)/2}e^{-(\pi/2)(\pi|n|/h-t)} ,$$

so the sum will be

$$(t/2)^{-d(\sigma-1)/2}e^{d(\pi/2)t}\sum_{n \geq 1}(\pi n/h - t/2)^{d(\sigma-1)/2}e^{-d(\pi^2/2)n/h} .$$

The first term is

$$(2\pi/(ht) - 1)^{-d(\sigma-1)/2}e^{-d(\pi/2)(\pi/h-t)} .$$

Keeping only the exponentials, if we want both the expressions for $n > 0$ and $n < 0$ to be less than e^{-E}, for $n > 0$ we thus need $\pi^2/h > 2E/d$, and for $n < 0$ we need $\pi^2/(2h) - (\pi/2)t > E/d$, in other words $\pi^2/h > 2E/d + \pi t$, which is a stronger condition. Although we have kept only the exponentials, since the exponent of $(2\pi/(ht) - 1)$ is negative, it it not necessary to take it into account.

Combining with the expression obtained when $|t/\sigma|$ is small, we should choose h so that

$$\pi^2/h > 2E/d + \max\left(\pi\max|t|, (2/d)(d(\sigma-1) - 1)\log(E)\right) ,$$

where $\max|t|$ is the maximum of the values of $|\Im(s)|$ that we will want to compute. Note in passing that E also depends on this, since $E = B\log(2) + (\pi/4)d\max|t|$.

9.6. The truncation errors

Recall that

$$\Theta(e^{mh}) = \sum_{n \geq 1}a(n)K((n/N^{1/2})e^{mh}) ,$$

and that as $t \to \infty$ we have $K(t) \sim Ct^{A}e^{-d\pi t^{2/d}}$. We need to find both the limit M of the summation on m, and the number of terms that we take in the sum over n.

Note that although the $a(n)$ are polynomially bounded, they can be rather large (for example Ramanujan's tau function). In the definition, we have $|a(n)| = O_{\varepsilon}(n^{k-1+\varepsilon})$. In view of the case of modular forms, we will make the heuristic assumption that when there is no pole at $s = k$, we have $|a(n)| = O_\varepsilon(n^{(k-1)/2+\varepsilon})$. We will write this as $|a(n)| = O_\varepsilon(n^{k_1+\varepsilon})$, where $k_1 = (k-1)/2$ if there is no pole, and $k_1 = k - 1$ otherwise. Of course, for any *given* class of L-functions, one could use a rigorous bound for $a(n)$, including implicit constants.

Now let m be given. Because of the exponential decrease of K, it is reasonable to require that $n \le L(m)$, where $L(m)$ is such that

$$e^{mh\sigma} L(m)^{k_1} K(L(m)e^{mh}/N^{1/2}) < e^{-E} ,$$

in other words, setting $x(m) = L(m)e^{mh}/N^{1/2}$, that

$$x(m)^{k_1} K(x(m)) = e^{-E-mh\sigma+k_1(mh-\log(N)/2)} .$$

Applying Corollary 8.2.5, this gives approximately

$$L(m) = \left((1/(d\pi)) \left(E + mh\sigma - k_1(mh - \log(N)/2) - \log(d^{1/2}2^{-(d+1)/2}) \right. \right.$$
$$\left. \left. + (A + k_1)d/2 \log \left((E + mh\sigma - k_1(mh - \log(N)/2))/d\pi \right) \right) \right)^{d/2} N^{1/2}e^{-mh} .$$

In view of this formula, we distinguish three cases according to the size of $\sigma - k_1$.

(1) $\sigma - k_1 > 2E/d$. Since $2E/d$ is usually at least equal to 100, this corresponds to an extremely large value of σ. In that case, it is preferable to compute $L(s)$ is a stupid way, approximating it simply by

$$L(s) \approx \sum_{1 \le n \le L} a(n)n^{-s},$$

with for instance $L = e^{E/(\sigma-k_1-1)}$, which will be small.

(2) $2E/d > \sigma - k_1 > 0$. In this case $L(m)$ is a decreasing function of m, so the largest value of $L(m)$ is attained for $m = 0$ and is equal to

$$L := L(0) = \left((1/(d\pi)) \left(E + k_1 \log(N)/2 - \log \left(d^{1/2}2^{-(d+1)/2} \right) \right. \right.$$
$$\left. \left. + (A + k_1)d/2 \log((E + k_1 \log(N)/2)/d\pi) \right) \right)^{d/2} N^{1/2} .$$

It will therefore be sufficient to compute the $a(n)$ for $n \le L$, but we must then compute $L(m)$ for each m, which is fast since there will not be so many values of m (see below).

(3) $\sigma < k_1$. In that case again the largest value of $L(m)$ is attained for $m = 0$ and is equal to L, but here the factor in front of e^{-mh} is also decreasing, so that $L(m) \le Le^{-mh}$, and we do not lose much by choosing Le^{-mh} instead of $L(m)$, which we will do in practice.

Note that we could give a more precise bound using the Lambert function (see Corollary 8.2.5), but the above approximation is sufficient in practice.

To find M, we note that

$$|e^{mhs}\Theta(e^{mh})| \le e^{mh\sigma} \sum_{n \ge 1} |a(n)|K\left((n/N^{1/2})e^{mh}\right) \le e^{mh\sigma} \sum_{n \ge 1} n^{k_1} K\left((n/N^{1/2})e^{mh}\right)$$

approximately. We will need this to be smaller than e^{-E} both for s and $k-s$, so we may assume that $\sigma = \Re(s) \geq k/2$. Since $K(t) \geq 0$, the maximum of $|e^{mhs}\Theta(e^{mh})|$ is thus attained approximately for $m = M$, so we need the inequality

$$e^{Mh\sigma} \sum_{n \geq 1} n^{k_1} K\big((n/N^{1/2})e^{Mh}\big) < e^{-E} \, ,$$

hence by the asymptotic formula for K, approximately

$$e^{Mh\sigma}\big(e^{Mh}/N^{1/2}\big)^A \sum_{n \geq 1} n^{k_1+A} e^{-\pi d(e^{Mh}/N^{1/2})^{2/d}n^{2/d}} < e^{-E}d^{1/2}2^{-(d+1)/2} \, .$$

If $a < bc$ (including the case where $a < 0$), the function $x^a e^{-bx^c}$ is decreasing for $x \geq 1$, so $\sum_{n \geq 1} n^a e^{-bn^c}$ is comparable with the value at $x = 1$, i.e., with e^{-b}. We apply this to $a = k_1 + A$, $b = \pi d(e^{Mh}/N^{1/2})^{2/d}$ and $c = 2/d$, so we deduce that if $k_1 + A < 2\pi(e^{Mh}/N^{1/2})^{2/d}$ the above inequality is approximately

$$X^{\sigma+A}e^{-\pi dX^{2/d}} < N^{-\sigma/2}e^{-E}d^{1/2}2^{-(d+1)/2}$$

with $X = e^{Mh}/N^{1/2}$. By Lemma 2.3.3, the solution X_0 to this inequality satisfies approximately

$$X_0^{2/d} = \frac{1}{\pi d}\left(E' + \frac{A'd}{2}\log\left(\frac{E'}{\pi d}\right)\right) \, ,$$

with $A' = A + \sigma$ and

$$E' = E + (\sigma/2)\log(N) - \log\big(d^{1/2}2^{-(d+1)/2}\big) \, ,$$

so that $Mh > \log(N)/2 + \log(X_0)$. Since h has been determined by the Fourier error, this gives

$$Mh > \log(N)/2 + \max((d/2)\log((k_1+A)/(2\pi)), \log(X_0)) \, ,$$

where X_0 is as above. Of course, in practice, this maximum will be equal to $\log(X_0)$.

We summarize our findings in the following lemma:

LEMMA 9.6.1. *Recall that we set* $E = B\log(2) + (\pi/4)d\max|t|$. *Set* $\sigma = \max(\Re(s), k-\Re(s))$, $A' = A+\sigma$, $E' = E+(\sigma/2)\log(N)-\log(d^{1/2}2^{-(d+1)/2})$, $E'' = E + k_1\log(N)/2 - \log(d^{1/2}2^{-(d+1)/2})$, $X_0 = (E'/(\pi d) + (A'/(2\pi))\log(E'/(\pi d)))^{d/2}$, *and*

$$L_0(m) = \big((E'' - k_1 mh)/(d\pi) + ((A+k_1)/(2\pi))\log((E''-k_1 mh)/(d\pi))\big)^{d/2} \, .$$

We choose

$$h := d\pi^2/(2E + \max(\pi d\max|t|, 2(d(\sigma-1)-1)\log(E))) \, ,$$
$$M := (\log(N)/2 + \max((d/2)\log((k_1+A)/(2\pi)), \log(X_0)))/h \, ,$$
$$L(m) := L_0(m)N^{1/2}e^{-mh} \, ,$$

and we then have with an absolute *error close to* e^{-E} *(hence to a relative error close to* 2^{-B}*)*

$$\Lambda(s) \approx P(s) + h \sum_{0 \leq m \leq M}' \sum_{1 \leq n \leq L(m)} K((n/N^{1/2})e^{mh})(a(n)e^{mhs} + w\overline{a(n)}e^{mh(k-s)}) \, ,$$

where the polar part $P(s)$ *is given by*

$$P(s) = \frac{hR}{2}\coth(h(s-k)/2) - \frac{wh\overline{R}}{2}\coth(hs/2) \, .$$

As already mentioned, note that we can also compute the derivatives $\Lambda^{(j)}(s)$.

REMARK 9.6.2. The reader may notice that the right hand side of the above formula is in fact *periodic* of period $2\pi i/h$, while this is evidently not the case for the left hand side $\Lambda(s)$. This is not a contradiction, since h has been chosen so that we can neglect the sum $\sum_{n\neq 0}\Lambda(s+2\pi in/h)$, in other words so that we can approximate $\Lambda(s)$ by $\sum_{n\in\mathbb{Z}}\Lambda(s+2\pi in/h)$, which is indeed periodic of period $2\pi i/h$.

9.7. Implementation

We have proved the following result:

PROPOSITION 9.7.1. *Let $P(X)$ be the function*

$$P(X) = {\sum_{0\leq m\leq M}}' e^{mhk/2}c(m)X^m + R\left(\frac{1}{e^{-hk/2}X - 1} + \frac{1}{2}\right) ,$$

$$\text{where}\quad c(m) = \sum_{1\leq n\leq L(m)} a(n)K\left(ne^{mh}/N^{1/2}\right)$$

is the truncated Θ series and where we recall that $R = N^{k/2}\gamma(k)r$ is the residue of $\Lambda(s)$ at $s = k$.

(1) *If we set $x = e^{h(s-k/2)}$, we have*

$$\Lambda(s) \approx h\left(P(x) + w\overline{P}(1/x)\right)$$

with a relative error *of the order of 2^{-B}.*

(2) *More generally, for $j \geq 0$ we have*

$$\Lambda^{(j)}(s) \approx h^{j+1}\left(D^j(P)(x) + (-1)^j wD^j(\overline{P})(1/x)\right) ,$$

where $Df(u) = u\,df/du$.

(3) *If $w = -1$ and L is self-dual, at the central point $s = k/2$, we have*

$$N^{k/4}\gamma(k/2)L'(k/2) \approx 2h^2 D(P)(1) ,$$

and more generally if $L^{(j)}(k/2) = 0$ for $j < j_0$ with $(-1)^{j_0} = w$, then

$$N^{k/4}\gamma(k/2)L^{(j_0)}(k/2) \approx 2h^{j+1}D^j(P)(1) .$$

The precomputation of the function P requires approximately

$$\frac{L}{1 - e^{-h}} \approx \frac{L}{h} \approx \frac{2}{d}\frac{LE}{\pi^2}$$

evaluations of the function K. Once precomputed, the evaluation of $\Lambda(s)$ (or its derivatives) requires essentially $2M$ steps, except when s is on the critical line $\Re(s) = k/2$, in which case $\overline{P}(1/x) = \overline{P(x)}$ so the computation requires only M steps.

9.7.1. Improvement of the precomputation speed. The computation time can be improved (see below), but since in any case M is logarithmic both in the conductor and the required accuracy, this is not so important. On the other hand, since L can be extremely large, it is important to improve the speed of the precomputation. This can easily be done using the following trick, analogous to the one we used when evaluating $\Theta(\sqrt{2})$ and $\Theta(1/\sqrt{2})$ simultaneously.

Recall that the choice of h that we have made is an *upper bound*, and that M is then determined by that choice according to the formulas given above. If h_0 is the initial value of h given above, let us choose

$$m_0 = \lceil \log(2)/h_0 \rceil \quad \text{and} \quad h = \log(2)/m_0 \ .$$

We have $h \le h_0$, which is required, and $e^{m_0 h} = 2$. It follows that if $m \ge 0$ is an integer and if $m = qm_0 + r$ is the Euclidean division of m by m_0 we have $e^{mh} = 2^q e^{rh}$. Thus

$$c(m) = \sum_{1 \le n \le L(m)} a(n)K(ne^{mh}/N^{1/2}) = \sum_{1 \le n \le L(m)} a(n)K(2^q ne^{rh}/N^{1/2})$$

and we need only compute values of K for *distinct* values of $2^q n$ over all pairs (q, n). This is done in full generality in the built-in implementation `lfuninit` but for simplicity we now take $L(m) \le Le^{-mh} = Le^{-rh}2^{-q}$. If we set $v(r) = \lfloor \log(Le^{-rh})/\log(2) \rfloor$, we can compute for $0 \le r \le m_0 - 1$ and $0 \le q \le v(r)$ the vectors

$$V(r,q) = \sum_{\substack{1 \le n \le Le^{-rh} \\ v_2(n)=q}} K(ne^{rh}/N^{1/2})[a(n/2^q), \cdots, a(n)]$$

(note that for a given r this again involves Le^{-rh} computations of the function K, simply ordered differently), and so setting $n' = 2^q n$ we have

$$c(m) = \sum_{\substack{1 \le n' \le Le^{-rh} \\ v_2(n') \ge q}} a(n'/2^q)K(n'e^{rh}/N^{1/2}) = \sum_{q \le q' \le v} V(r,q')[q' - q + 1] \ ,$$

where $V[j]$ is the jth component of V starting at 1.

Using this trick, the precomputation only requires

$$L\left(\sum_{0 \le m \le m_0 - 1} e^{-mh} \right) = L(1 - e^{-m_0 h})/(1 - e^{-h}) = L/(2(1 - e^{-h}))$$

evaluations of K, in other words half as many as before.

9.7.2. Improvement of the speed of computation of w and R. We can also use this method to improve the speed of the computation of the root number w and/or of the residue R if one of them is unknown, but only in the context of computing $L(s)$ or $\Lambda(s)$, otherwise the method given in Section 9.2.4 still seems the best. Indeed, first note that for $m \ge 0$ we have $\Theta(e^{mh}) \approx c(m)$, which is thus known, and in particular $\Theta(e^h) \approx c(1)$. Assuming for simplicity that either w or R is known, we thus only need to compute $\Theta(e^{-h})$, which only requires $e^h x(\varepsilon)N^{1/2}$ evaluations of K instead of $1.414x(\varepsilon)N^{1/2}$, using the notation of Section 9.2.4, and note that e^h is typically around 1.05, so already a gain. But we can do even better, since we can write

$$\Theta(e^{-h}) = \sum_{\substack{n \geq 1 \\ n \text{ odd}}} a(n)K(ne^{-h}/N^{1/2}) + \sum_{n \geq 1} a(2n)K(ne^{(m_0-1)h}/N^{1/2}) ,$$

and the non-negligible values of K occurring in the second sum have already been computed, so we gain another factor of 2.

9.7.3. Improvement of the evaluation speed. We will assume in this subsection that the L-function is self-dual, and we will set temporarily $\varepsilon = w = \pm 1$. Recall the following lemma:

LEMMA 9.7.2. *Let T_n and U_n be the Chebyshev polynomials of the first and second kind respectively, i.e., such that $T_n(\cos(t)) = \cos(nt)$ and $U_n(\cos(t)) = \sin((n+1)t)/\sin(t)$. Then:*

(1) *We have*

$$X^m + 1/X^m = 2T_m\big((X+1/X)/2\big) \quad and$$
$$X^m - 1/X^m = (X-1/X)U_{m-1}\big((X+1/X)/2\big) .$$

(2) *We have*

$$T_m(Y/2) = \frac{m}{2} \sum_{k=0}^{\lfloor m/2 \rfloor} (-1)^k \frac{(m-k-1)!}{k!(m-2k)!} Y^{m-2k} \quad and$$

$$U_m(Y/2) = \sum_{k=0}^{\lfloor m/2 \rfloor} (-1)^k \binom{m-k}{k} Y^{m-2k} .$$

COROLLARY 9.7.3. *Recall that*

$$c(m) = \sum_{1 \leq n \leq Le^{-mh}} a(n)K\big(e^{mh}n/N^{1/2}\big) ,$$

let $P_0(X)$ be the polynomial

$$P_0(X) = \sideset{}{'}\sum_{0 \leq m \leq M} e^{mhk/2}c(m)X^m ,$$

so that $P(X) = P_0(X) + R \cdot \big(1/(e^{-hk/2}X - 1) + 1/2\big)$, and set $Y = X + 1/X$.

(1) *(Case $\varepsilon = 1$.) We have*

$$P_0(X) + P_0(1/X) = \sum_{0 \leq n \leq M} \left(\sum_{k=0}^{\lfloor (M-n)/2 \rfloor} (-1)^k \frac{n+2k}{n+k} \binom{n+k}{k} c_{n+2k} \right) Y^n ,$$

where it is understood that $(n+2k)/(n+k) = 2$ if $k = n = 0$.

(2) *(Case $\varepsilon = -1$.) We have*

$$P_0(X) - P_0(1/X) = (X-1/X) \sum_{0 \leq n \leq M-1} \left(\sum_{k=0}^{\lfloor (M-n-1)/2 \rfloor} (-1)^k \binom{n+k}{k} c_{n+2k+1} \right) Y^n .$$

It is clear that thanks to this corollary we only need essentially M steps to evaluate $\Lambda(s)$, which is twice as fast as above. Unfortunately, the coefficients of the Chebyshev polynomials alternate in sign, so there is a large cancellation in the computation of the coefficients of the polynomials in Y and their evaluation, so that it is necessary to increase the working accuracy to compensate for this. Thus it is not clear if we gain in time.

9.8. A possible program for computing $\Lambda(s)$ and $L(s)$

Since all the ideas are explained above, the reader not interested in the specific coding in Pari/GP of the computation of L-functions can skip this section and simply *use* the programs provided on the book's website as a black box.

We have already done a large part of the necessary work: first, in Chapter 8, for the computation of inverse Mellin transforms, certainly the most technical part of the implementation. Second, in Sections 9.2.3 and 9.2.4, for the computation of the associated theta function, and of the root number and/or the residue if unknown. We now tackle specifically the computation of $\Lambda(s)$, hence of $L(s)$.

The basic programs are as follows: as usual there is an initialization program, which can take a considerable amount of time, and a fast evaluation program.

Lfuninit.gp

```
1   /* Initialization: returns a vector [Ldataf,VV], where Ldataf
2    * (f for full) is the initial Ldata component structure with
3    * the root number and/or residues computed if not given
4    * (i.e., set to oo), and VV is a technical vector. Allows to
5    * compute L(s) for |Re(s)| <= maxs and |Im(s)| <= maxt. */
6   Lfuninit(Ldata, maxt = 50, maxs = 10)=
7   { my(B = getlocalbitprec(), h, M, L, tmin, E, sig, m0);
8
9     localbitprec(64);
10    my(NEWB = ceil(B + Pi/4 / log(2) * maxt * Ldata.d));
11    if (#Ldata == 2,
12      my([t, s] = Ldata[2][4..5]);
13      if (t >= maxt && s >= maxs
14                    && Ldata.prec >= NEWB, return (Ldata)));
15      Ldata = Ldata[1]); /* recompute Lfuninit */
16    my([Vga, k, N, w, R] = Ldata[3..7], d = #Vga);
17    maxt = max(abs(maxt), 2);
18    sig = vecmax([1, maxs, k - maxs]);
19    E = log(2) * NEWB;
20
21    localbitprec(NEWB); tmin = 1;
22    if (w == oo || R == oo || bitprecision([R, w]) < NEWB,
23      tmin = 1 / sqrt(2));  /* need to (re)compute w or R */
24    my(vv, [an, K] = vv = Lfunthetainit(Ldata, tmin));
25    L = #an;
26    if (tmin != 1,
27      [R, w] = Lfunrootres(Ldata, vv);
28      L = ceil(L * tmin));
29    h = max(Pi * maxt * d/2, (d*(sig-1) - 1) * log(E));
30    h = d * Pi^2 / (2 * (E + h));
31    m0 = ceil(log(2) / h); h = log(2) / m0;
32    my(Ap = sig + (1 - d + vecsum(Vga)) / d);
```

```
33      my(Ep = (E + (sig / 2) * log(N)) / (d * Pi));
34      M = log(N) / 2 + (d / 2) * log(Ep + (Ap/(2*Pi))*log(Ep));
35      M = ceil(M / h) + 5;
36      my(isqN = 1 / sqrt(N), LL = log(L), v0 = floor(LL / log(2)));
37      my(e = exp(h), ek = powers(e^(k/2), M)); ek[1] = 1/2;
38      my(V = matrix(m0, v0+1), eh = powers(e, m0-1));
39      my(upow = dirpowers(L+1, 2 / d));
40      for (i = 0, m0 - 1,
41        for (j = 0, v0,
42          my(t = eh[i+1] << j);           /* exp(i*h)*2^j */
43          my(U = (floor(L / t) - 1) \ 2);
44          my(C = (t * isqN)^(2 / d));
45          V[i+1, j+1] = sum(u = 0, U,
46            my(v = vector(j+1, a, an[(2*u+1) << (a-1)]));
47            if (v, v *= K(C * upow[2*u+1]))))));
48      my(c = vector(M+1));
49      for (m = 0, M,
50        my([q, r] = divrem(m, m0), v = (LL - r*h) / log(2));
51        c[m+1] = sum(Q = q, v, V[r+1, Q+1][Q-q+1]));
52      my(pol = Polrev(vector(M+1, m, ek[m] * c[m])));
53      Ldata[6] = w;
54      Ldata[7] = R; return ([Ldata, [h, pol, R, maxt, maxs]]);
55    }
```

Since this program is more complex than usual, we give a few explanations:

- If we are in a trivial situation where the necessary technical stuff has been computed to a sufficient accuracy, we return immediately.
- We then look if the root number or the residue of the Λ function are missing: as mentioned before, a parameter is considered missing if it is set to oo.
- We compute sufficiently many coefficients $a(n)$, $n \leq L$, to evaluate $\Theta(t)$ to the requested accuracy for $t \geq 1$ (no missing parameter) or $t \geq 1/\sqrt{2}$ (need to call Lfunrootres). After a possible call to Lfunrootres we can fill in the Ldataf structure which will be the first component of the two-component result. For simplicity, we reuse the generic Lfunrootres program instead of implementing the faster evaluation of w and R from Section 9.7.2.
- We then compute the quantities h, M, m_0 explained above.
- We are then at the heart of the program where almost all the time is spent, the double loop in i and j, lines 40–47. After one loop in i, we can give a reasonable estimate of the time the whole initialization will take, and if it is especially long, we could add a warning to the user estimating the time that the program will need, but we have not done so.
- The end of the program is simple bookkeeping and use of the formulas that we have given, and takes very little time compared to the rest.

The evaluation program is as follows:

———————————————————————— | Lfun.gp | ————————————————

```
1   /* Call lfun init for domain containing s */
2   Lfuninitcheck(Ldata, s)=
3   {
4     if (type(s) == "t_SER", s = polcoef(s, 0));
5     my(re = abs(real(s)), im = abs(imag(s)));
6     return (Lfuninit(Ldata, im, re));
7   }
8
9   /* If flag is 1, return Lambda(s), otherwise L(s);
10   * if s is a t_SER, return Taylor expansion */
11  Lfun(Ldata, s, flag = 0)=
12  { my(S, [Ldataf, vv] = Lfuninitcheck(Ldata, s));
13    my([Vga, k, N, w, R] = Ldataf[3..7], [h, pol] = vv);
14
15    /* First compute Lambda */
16    localbitprec( bitprecision(h) );
17    if (R && s == k, S = R / 'x + O(1), /* pole at k */
18        R && s == 0, S = -w * conj(R) / 'x + O(1), /* pole at 0 */
19      /* else treat as regular value */
20      my(t, t2, z = exp(h * (s - k/2)));
21      t = subst(pol, 'x, z);
22      t2 = if (real(s) == k/2, t, subst(pol, 'x, conj(1/z)));
23      if (R, t += R * (1 / expm1(h * (s - k)) + 1/2);
24            t2+= R * (1 / expm1(-h * conj(s)) + 1/2));
25      S = h * (t + w * conj(t2)));
26    /* Divide by factor at infinity to get L */
27    if (!flag, S /= N^(s/2) * Gammafactor(Vga, s));
28    if (type(s) != "t_SER" && valuation(S, 'x) >= 0,
29        S = polcoef(S, 0));
30    return (S);
31  }
32
33  /* Pi^(-s/2) gamma(s/2) */
34  GammaR(s) =
35  {
36    s /= 2;
37    if (type(s) == "t_INT" && s <= 0, s += 'x / 2 + O('x^2));
38    Pi^(-s) * gamma(s);
39  }
40
41  /* Gamma factor at s. */
42  Gammafactor(Vga, s)= vecprod([GammaR(s + a) | a <- Vga]);
```

REMARKS 9.8.1.

(1) This is a direct translation of Theorem 9.4.2. Note that in fact the *real* work was done in the `Lfuninit` program, in order to compute the polynomial `pol`.

(2) The argument s may be a power series, $s = s_0 + x + O(x^N)$, which allows to recover the expansion of L or Λ around the complex number s_0 (which may be a pole). In particular, this also allows to compute successive derivatives at a regular point.

(3) This convention is also used for outputs: in `GammaR`, evaluating the gamma factor at one of its poles s_0 thus returns a power series where x represents $s - s_0$; and analogously for `Lfun`. This simple trick of introducing Taylor expansions when needed allows to transparently compute regular values of L at an s which is a pole of the gamma factor. Our restriction on the polar part of L and Λ (allowing at most simple poles) makes this simpler, but this can be made to work in full generality, as in the built-in GP function `lfun`.

9.9. Applications

Thanks to the above program, we can now have some real fun!

9.9.1. Direct applications. Our first program computes the Z-function associated to the L-function on the critical line (the Z-function is the natural function associated to $\Lambda(s)$ which is real on $\Re(s) = k/2$ and does not decay exponentially; we leave to the reader to find the easy formula for it).

———————————————— | LfunHardy.gp | ————————————————

```
1   /* Hardy Z function. */
2   LfunHardy(Ldata, t) =
3   {
4     Ldata = Lfuninit(Ldata, abs(t), Ldata.k / 2);
5     my([Vga, k, N, w] = Ldata[1][3..6], d = #Vga);
6     t = bitprecision(t, Ldata.prec);
7     my(k2 = k/2, z = k2 + I*t, E);
8     E = (d * (k2 - 1) + vecsum(Vga)) / 4;
9     return (exp(d * t * atan(t / k2) / 2) / norm(z)^E
10             * real(Lfun(Ldata, z, 1) / sqrt(w)));
11  }
12
13  Lfunplot(Ldata, range = [0,50]) =
14  { my([T, Tmax] = range);
15
16    localbitprec(32);
17    Ldata = Lfuninit(Ldata, vecmax(abs(range)), Ldata.k / 2);
18    ploth(t = T, Tmax, LfunHardy(Ldata, t));
19  }
```

The Z-function vanishes at a real t if and only if $L(k/2 + it) = 0$. Plotting it from $t = 0$ to some moderate height is a nice graphical way to visualize the zeroes on the critical line and see the Riemann hypothesis in action.

We then give the program Lfunorderzero which computes the order of the zero at the center of the critical strip together with the first nonvanishing derivative:

─────────────────────────── Lfunorderzero.gp ───────────────────────────

```
1  /* Assume that L^(j)(k/2) != 0 for some j < N. */
2  Lfunorderzero(Ldata, N = 6) =
3  { my(v, z, B = getlocalbitprec());
4
5    z = Lfun(Ldata, Ldata.k / 2 + 'x + O('x^N));
6    v = valuation(round(2^(B/2) * z), 'x);
7    return ([v, polcoef(z, v)]);
8  }
```

And finally, we give the program Lfunzeros which computes the zeros on the critical line in the interval range, using a reasonably naïve search. Note that if range[1] is equal to 0, we include in the set of zeros the possible zero at the center of the critical strip repeated with multiplicity.

─────────────────────────── Lfunzeros.gp ───────────────────────────

```
1  /* Search and store zeros of Hardy Z-function on critical line,
2   * Im(z) in range = [T,Tmax]. n0 = order of 0 on real axis if
3   * T == 0 (else n0 = 0). For each t, assume there is at most a
4   * single zero in [t, t + Gap(t)] */
5  Zerosearch(Z, Gap, n0, range) =
6  { my([T, Tmax] = range, T0);
7    my(s = sign(Z(T)), w = List(vector(n0)));
8
9    if (n0, T = 2.^(-getlocalbitprec()/ (2*n0)));
10   T0 = T;
11   while (T < Tmax,
12     T = min(T + Gap(T), Tmax);
13     if (s == sign(Z(T)), next);
14     listput(~w, solve(t = T0, T, Z(t)));
15     s = -s; T0 = T);
16   return (Vec(w));
17 }
18
19 /* For an L function of degree d and conductor N, the expected
20  * gap between two consecutive critical zeros at height t and
21  * t2 is 2*Pi / e(t). Return 2*Pi / (divz*e(t)) as a closure */
22 Zerogap(N, d, divz) =
23 { my(C = d + 2 * log(max(1, N / (Pi/2)^d)));
24   t -> my(e = d * log(max(1, t / (2*Pi))) + C);
25       2 * Pi / (divz * e);
26 }
```

```
27
28  /* Naive search for zeros on the critical line for t in
29   * range = [a, b]. Mesh size is controlled by divz, increase
30   * if necessary. */
31  Lfunzeros(Ldata, range = [0,50], divz = 8) =
32  { my(Z, n0, Gap);
33
34    Ldata = Lfuninit(Ldata, vecmax(abs(range)), Ldata.k / 2);
35    my(NEWB = Ldata.prec);
36    Gap = Zerogap(Ldata.N, Ldata.d, divz);
37    n0 = if (range[1] == 0, Lfunorderzero(Ldata)[1]);
38    Z = Zerosearch(t -> LfunHardy(Ldata, bitprecision(t,NEWB)),
39                   Gap, n0, range);
40    return (bitprecision(Z, getlocalbitprec()));
41  }
```

Note that using variants of a method due to A. Turing, it is possible to write an algorithm which *certifies* that all the zeros have been found in a certain interval, see for instance [**Boo06**].

9.9.2. Important classical L-functions. Given a standard mathematical object for which a natural L-function exists (for example Dirichlet L-functions, Dedekind zeta functions, L-functions attached to elliptic curves, modular forms, Hecke characters, etc.), we now need a constructor for the Ldata necessary to work with the Lfunxxx programs. Once the Ldata created, we can use Lfuninit, and then compute L-values using Lfun, or do plots using Lfunplot and LfunHardy, compute zeros using Lfunzeros, etc.

This is quite straightforward in many cases and we now give a few examples. We start with primitive Dirichlet characters given either as a fundamental discriminant D (encoding the Kronecker symbol $\left(\frac{D}{\cdot}\right)$, including Riemann zeta function for $D = 1$) or as a vector of values $(\chi(1), \dots \chi(N))$ where we assume that N is the conductor of χ. It is immediate, but beyond the scope of this section to use more general formats such as Conrey labels or Pari/GP's implementation of Dirichlet characters. As seen in Section 6.2.4, it is more efficient to compute the root number through the functional equation and Lfunrootres than to express it in terms of a Gauss sum.

────────────────────── | Lfunchicreate.gp | ──────────────────────

```
1   /* CHI is a primitive Dirichlet character given either by an
2    * integer D representing the Kronecker symbol (D/.), or by a
3    * vector [CHI[1],...,CHI[N]]. */
4   Lfunchicreate(CHI) =
5   { my(t = type(CHI));
6     my(triv = CHI == 1 || (t == "t_VEC" && #CHI == 1));
7
8     if (triv,            Lfunzetacreate(), /* trivial character */
9         t == "t_INT", Lfunchiquadcreate(CHI), /* quadratic */
10                       Lfunchiveccreate(CHI));
11  }
```

```
12
13   /* Riemann zeta function */
14   Lfunzetacreate() =
15     [n -> vector(n, j, 1), 0, [0], 1, 1, 1, 1];
16
17   /* Quadratic character (D/.) */
18   Lfunchiquadcreate(D) =
19   { my(an, L);
20     an = (n -> vector(n, j, kronecker(D, j)));
21     L = [an, 0, [D < 0], 1, abs(D), oo, 0];
22     L[6] = Lfunrootres(L)[2]; return (L);
23   }
24
25   /* General case chi(i) = v[i] for i = 1..N */
26   Lfunchiveccreate(v) =
27   { my(N = #v, odd, sd, an);
28     odd = exponent(v[N - 1] - 1) > -32; /* is chi odd ? */
29     sd = exponent(imag(v)) < -32; /* is chi self-dual ? */
30     an = (n -> vector(n, j, v[(j - 1) % N + 1]));
31     L = [an, 1 - sd, [odd], 1, N, oo, 0];
32     L[6] = Lfunrootres(L)[2]; return (L);
33   }
```

Next comes the Dedekind zeta function $\zeta_K(s)$, attached to a number field K:

Lfunzetakcreate.gp

```
1    /* 'pol' is an irreducible monic polynomial in Z[X] defining a
2     * number field K. Creates the L function zeta_K(s) */
3    Lfunzetakcreate(pol) =
4    { my(d = poldegree(pol), K = bnfinit(pol), h = K.no, Vga, R);
5      my([r1, r2] = K.sign, N = abs(K.disc), w = K.tu[1]);
6
7      Vga = concat(vector(d - r2), vector(r2, j, 1));
8      R = Gammafactor(Vga, 1) * h * K.reg * 2^r1 * (2 * Pi)^r2 / w;
9      return ([(n -> dirzetak(K, n)), 0, Vga, 1, N, 1, R]);
10   }
11
12   /* Creates zeta_K(s) / zeta(s): reduces the degree and cancels
13    * the pole */
14   Lfunzetakcreate2(pol) =
15   { my(d = poldegree(pol) - 1, K = nfinit(pol));
16     my([r1, r2] = K.sign, N = abs(K.disc), Vga, r, an);
17     Vga = concat(vector(d - r2), vector(r2, j, 1));
18     an = (n -> dirdiv(dirzetak(K, n), vector(n, j, 1)));
19     return ([an, 0, Vga, 1, N, 1, 0]);
20   }
21
```

REMARKS 9.9.1.

(1) We use the analytic class number formula to compute the residue of L at $s = 1$, then multiply by the factor at infinity to get the residue R of Λ, which cancels the term $\sqrt{N}$. We could have used again the Lfunrootres program but it is *much* faster to compute the class number and regulator using bnfinit. Note that this is partly because the latter assumes the Riemann hypothesis, allowing a complexity which is subexponential in $\log N$ instead of $\sqrt{N}$ using Lfunrootres. Without GRH, it would still be faster but both methods would now use $O(\sqrt{N})$ time.

(2) Beware that the residue formula in Lfunzetakcreate may produce inaccurate results: the floating point value is computed on the spot at the current accuracy and not at the accuracy determined later by the Lfuninit program, which depends on the range of complex arguments we want to allow. If the value of R is not sufficiently precise, Lfuninit will have to recompute it. For $K = \mathbb{Q}$ or an imaginary quadratic field, the formula would simplify to a rational number (equal to 2 * K.no / w); but in the general case, the regulator prevents an exact rational representation.

This is a general problem with our Lfuncreate / Lfuninit interface whenever an explicit formula for an entry does not produce an exact result and we use a floating point approximation. This could be solved in the usual way: by allowing a closure instead of hard-coding static arguments; the closure can then be evaluated later, in Lfuninit, once the accuracy is known. For simplicity, we leave the small modifications involved as exercises. The built-in GP function lfuncreate does provide this mechanism.

(3) Note that in the case of Dedekind zeta functions, there is a simpler and more efficient solution, although it remains in general conjectural. According to a special case of the Artin conjecture, $\zeta_K(s)/\zeta(s)$ should be an entire function, which cancels the pole at $s = 1$, hence the need to compute the residue! It also has smaller degree than $\zeta_K(s)$, making the computations faster. This is done in the function Lfunzetakcreate2. It only remains to multiply the result of Lfun by the built-in zeta(s).

(4) More generally, the number field K may contain another number field $k \neq \mathbb{Q}$. Again, the Artin conjecture predicts that $\zeta_K(s)/\zeta_k(s)$ is an entire function and we can again use this to considerably reduce the timings. We leave to the reader to write the corresponding programs. An extreme example where this is very useful is when K is an *abelian* extension of $\mathbb{Q}$: in that case, $\zeta_K(s)$ is a product of Dirichlet L-functions. It is then a simple matter of algebraic number theory to determine exactly the corresponding Dirichlet characters, and then use the Lfunchicreate program above.

And finally, here is code for elliptic curves, and more generally eigenforms with integer coefficients:

———————————————— | Lfunmfcreate.gp | ————————————————

```
1  /* E elliptic curve given by model or Cremona label. */
2  Lfunellcreate(E) =
3  {
4    E = ellinit(E); my(N = ellglobalred(E)[1]);
5    [n -> ellan(E, n), 0, [0, 1], 2, N, ellrootno(E), 0];
```

```
6    }
7
8    /* Assume F eigenform with integer coefficients. */
9    Lfunmfcreate(F) =
10   { my([N, k] = mfparams(F), L, R, w);
11
12     L = [n -> mfcoefs(F, n)[^1], 0, [0, 1], k, N, oo, oo];
13     [R, w] = Lfunrootres(L); L[6] = w; L[7] = R; return (L);
14   }
```

REMARK 9.9.2. In the `Lfunmfcreate` program, the formula for the residue R of Λ at $s = k$ is $r = 2wa(0)$, where w is the root number, but first this root number (equal to ± 1 with our assumptions) must be determined, and second, since the inverse Mellin transform of $\Gamma(s)$ is e^{-x}, so very easily computed, the `Lfunrootres` program is sufficiently fast.

The built-in `Pari/GP` package allows more general functional equations than we do: complex α_j, any (finite) number of poles of any order and arbitrary pairs L/L^* of dual functions (we assumed real α_j, simple poles, and $L^* = \overline{L}$ for simplicity). However, the format used to represent L-functions in this book is close to the one used in `Pari/GP` in the situations they both cover, with the exception of the handling of poles. If you want to use for instance the `lfuninit` program, when there are no poles, instead of writing

> `Lfuninit(LD, maxt, maxs);`

using our script, you would write instead

> `lfuninit(lfuncreate(LD[1..6]), [0, maxs, maxt]);`

The rectangle description `[0,maxs,maxt]` (with `maxs` and `maxt` in the reverse order) *must* be given explicitly since the built-in `lfuninit` has no default values.

9.10. Examples

We now give a number of examples which in particular may better show how to use the programs that we have written.

9.10.1. Computing the conductor.
We have mentioned that the program that we have given for computing the conductor is quite heuristic. We are going to give a number of examples. First, an example where all goes well: for instance

> `LD = Lfunellcreate("5077a1");`
> `Lfunconductor(LD)`

first gives a plot of the functions used to find the conductor, as well as the 2-component vector [34.989..., 5077.000000000000000...], which clearly shows that the only possible conductor is 5077 since the other value is too far from an integer. This is obtained without cheating, i.e., without using the fact that the conductor is in fact already present in the LD structure.

We now give examples where problems may occur. First, a small testing program:

```
test(p)=
{ my(CHI, Ldata);

  localbitprec(64);
  CHI = vector(p, n, kronecker(n, p));
  Ldata = Lfunchicreate(CHI);
  Lfunconductor(Ldata, 2000);
}
```

This creates data associated to the theta function of the Dirichlet character corresponding to the Legendre symbol (n/p), and computes approximations to the conductor. Usually, the result is the correct conductor, p. Let us take other examples:

- For $p = 179$, say, the result is a five-component vector

$[3.011\cdots, 11.00000000000313624, 40.109\cdots, 179.0000000000000000, 798.83\cdots]$.

    The only plausible values, i.e., close to integers, are $N = 11$ and $N = 179$, but it is quite remarkable that $N = 11$ seems also almost perfect.
- For $p = 857$, the result is even worse (or better, depending on one's point of view), it is the two-component vector

$[17.000000000000000000, 857.0000000000000000]$,

    so both 17 and 857 are plausible conductors.
- For $p = 479$, the result is the implausible value 456.163.... To find the conductor, we look at the plot (which is one of its main interests). We see the interesting phenomenon that there is a flex approximately where the curve crosses the real axis. It is thus necessary to zoom in. The flex is around 500, so we write instead Lfunconductor(Ldata,[450,550]), and we see that the flex corresponds in fact to three distinct points of intersection. To be able to find these points, we must increase the number of steps by writing for instance Lfunconductor(Ldata,[450,550],100), which gives $456.1631\cdots$, $479.000000000\cdots$, and $502.980077\cdots$.

9.10.2. Computation of L-functions.

(1) For our first example, we let K be the totally real cubic field of discriminant 10004, defined by the equation $x^3 - 14x - 6 = 0$. We can write

```
LD = Lfunzetakcreate(x^3 - 14*x - 6);
LDV = Lfuninit(LD);
```

    Note that a priori we do not know the residue, but if we ask for LD.R, we see that Lfunzetakcreate has filled in the residue of Λ at $s = 1$, namely $R = 106.986\cdots$. The residue of ζ_K is LD.r, namely $r = 1.069\cdots$. Note that the initialization has required 3.6 seconds.

    We can now ask a number of questions concerning the Dedekind zeta function of the number field. For instance bestappr(Lfun(LDV,-3), 1000) returns 741550066/15: a beautiful theorem of Siegel says that this is indeed a rational number with very small denominator.

    We can ask to plot using ploth(t = 0, 50, LfunHardy(LDV, t)), or the zeros using Lfunzeros(LDV) (or better, Lfunzeros(LDV, [0,20]) since there are many zeros).
(2) We can, however, do better: if we simply write

```
ZK = Lfuninit(Lfunzetakcreate2(x^3 - 14*x - 6));
```

this initialization is done in only 0.7 seconds, if we want the residue of
the Dedekind zeta function we simply ask for `Lfun(ZK, 1)`, and the other
quantities computed above by `Lfun(ZK, -3) * zeta(-3)`, and similarly
for plots and zeros.

(3) For another example, we define and initialize a character χ modulo 101
by the following program:

```
CHI = vector(101);
g = Mod(2, 101);
z = exp(17*Pi*I / 10);
for(j = 0, 99, CHI[lift(g^j)] = z^j);
LC = Lfuninit(Lfunchicreate(CHI));
```

As always, we can now compute quantities related to the L-function of
this character. Note that this is an example of a non self-dual L-function,
as you can see graphically by writing

```
ploth(t = -5, 5, LfunHardy(LC, t))
```

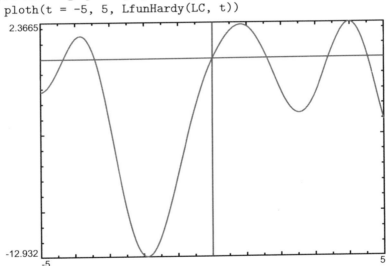

What is particularly spectacular about this example is its smallest zero:
although the plot looks like it passes through the origin, this is not
the case (`LfunHardy(LC, 0) = -0.000274...`), and if we ask at 38D
`Lfunzeros(LC, [0, 2])`, we obtain

```
[5.5425... E-5, 1.5916...]
```

The proximity of the first zero to the real line is absolutely exceptional
for so small a conductor ($N = 101$), and does not occur for any other
character of conductor less than 1000.

(4) For still another example, we define and initialize a character χ modulo
300 by the following more complicated program (note that $(\mathbb{Z}/300\mathbb{Z})^*$ is
isomorphic to the additive group $\mathbb{Z}/20\mathbb{Z} \times \mathbb{Z}/20\mathbb{Z} \times \mathbb{Z}/2\mathbb{Z}$):

```
CHI = vector(300);
G = Mod([101, 151, 277], 300); \\ order 2,2,20
Z = [-1, -1, exp(4*Pi*I / 5)];
forvec(J = [[0,1], [0,1], [0,19]], \
```

```
CHI[lift(factorback(G,J))] = factorback(Z,J));
LC = Lfuninit(Lfunchicreate(CHI));
```

We can plot, or compute zeros, etc. and there is nothing special. Incidentally, we can ask for the root number by asking for LC.w, and we obtain some number w; it can be proved (see [**CZ13**]) that w is a 25th root of unity, and indeed exactly equal to $e^{-22\pi i/25}$.

But this is not the interesting property of this character: what is special is that if we compute Lfuntheta(LC,1), whatever accuracy is used we find a number which is numerically very close to 0, and indeed [**CZ13**] proves that it is exactly equal to 0. Note that this phenomenon is very exceptional: only another related character modulo 600, plus of course the two conjugates of these two characters seem to have this property, among almost a billion characters studied.

(5) As another example, consider now the L-function associated to the Δ function, the normalized cusp form of weight 12 over the full modular group. A simple-minded way to obtain its coefficients (the Ramanujan tau function) is to write

```
DEL = Vec('x * eta('x + O('x^10000))^24);
TAU = n -> DEL[1..min(n, 10000)];
LD = [TAU, 0, [0, 1], 12, 1, 1, 0];
LDV = Lfuninit(LD);
```

Note that the Vga=[0,1] vector, common to all modular forms, corresponds to the gamma factor

$$\Gamma_{\mathbb{C}}(s) = \Gamma_{\mathbb{R}}(s)\Gamma_{\mathbb{R}}(s+1) = 2 \cdot (2\pi)^{-s}\Gamma(s) \,.$$

We can also more simply write LD = Lfunmfcreate(mfDelta()); to obtain LD, which has the advantage both of being more general and second not to have an artificial limitation of 10000 Fourier coefficients. Nonetheless, we keep this simple-minded way since it will be used below for the symmetric square.

Apart from plotting and computing zeros on the critical line $\Re(s) = 6$, we can also compute *special values*: let us write

```
even = vector(5, j, Lfun(LDV, 2 * j, 1));
odd  = vector(6, j, Lfun(LDV, 2 * j - 1, 1));
even /= even[1]; odd /= odd[2];
```

(recall that the last parameter 1 in the Lfun call means that we compute $\Lambda(s)$ instead of $L(s)$). We immediately see that all the components of even and odd are numerically rational numbers (this is less "visible" for the first component of odd), and if we apply the function bestappr(vector,1000), say to both, we find that even is equal to

$$[1, 25/48, 5/12, 25/48, 1]$$

and that odd is equal to

$$[1620/691, 1, 9/14, 9/14, 1, 1620/691].$$

Note that this is a famous theorem of Manin, completing in a fundamental way work of Eichler and Shimura.

(6) It is not difficult, at least in level 1, to define higher symmetric powers of modular forms (in non-squarefree level, it is a pain to find the Euler factors at the bad primes, see below for an example). For Δ, the symmetric square L-function can be defined by the following program:

```
TAUSYM2(n) = sumdiv(n, m, (-1)^bigomega(m)*m^11*DEL[n\m]^2);
LD2 = [n -> vector(n,j,TAUSYM2(j)), 0, [-10,0,1], 23,1,1,0];
LDV2 = Lfuninit(LD2);
```

(in other words we have $k = 23$ (motivic weight 22, double that of Δ), the gamma product vector is [-10,0,1], and we are in level 1 with root number 1). Once again, we have special values, but here either odd integers between 3 and 11, or, by the functional equation $s \mapsto 23 - s$, even integers between 12 and 20. For instance, we can write even = vector(6, j, Lfun(LDV2,10+2*j,1)), and we obtain real numbers which do not seem easy to recognize; however, if we write instead

```
even2 = vector(6, j, even[j] / Pi^(j-1)) / even[1]
```

the result looks rational, and indeed bestappr(even2,10^6) produces

$$[1, 4/7, 2/3, 64/45, 6144/1225, 430080/15893].$$

Note that $15893 = 23 \cdot 691$ involves the numerator of the Bernoulli number B_{12}, in other words it is almost certainly not a random number coming from bestappr. Note in passing that we can immediately recover the Petersson square S of Δ (which we painstakingly computed in Section 3.8.6 using a double integral) by the simple formula S = even[1] / 4096.

(7) As a final example, we write EE = Lfunellcreate("5077a1"), which initializes the L-function of the elliptic curve of rank 3 over $\mathbb{Q}$ with smallest conductor. Apart from doing the usual computations such as plotting, zeros, or special values, we can ask for Lfunorderzero(EE), which will tell us that $s = 1$ has order 3 and that $L'''(1)/3! = 1.7318499 \cdots$, a quantity which can be expressed in terms of the usual quantities occurring in the Birch and Swinnerton-Dyer conjecture.

9.10.3. Computing unknown Euler factors. Another important use of the programs is to find Euler factors at bad primes for true (as opposed to weak) L-functions. In fact, as in Lfuncheckfeq, we only need $\Theta(t)$ and not the more sophisticated functions $\Lambda(s)$ or $L(s)$. We give two examples.

Example 1: Consider the function $(\eta(\tau)\eta(5\tau))^4$, which is a Hecke eigenform of weight 4 on $\Gamma_0(5)$. As we have done above for Δ, we can define its symmetric square, and the corresponding L-function will have data of the type [n->[b(1)...b(n)],0,[-2,0,1],7,25,1,0], where the $b(j)$ are the coefficients of the Dirichlet series. When j is coprime to the level, here 5, the coefficients are given by the same formula as that used for the Δ function, which corresponds to an Euler factor of degree 3. When j is divisible by 5, the degree of the Euler factor will probably decrease, and we will make the assumption (which of course may be wrong) that it has degree 1, hence of the form $1 - yT$ for some y, and we want to determine y. We thus write the following program:

```
FUN = Vec('x*(eta('x+O('x^10000))*eta('x^5+O('x^10000)))^4);
TAU5SYM2(n) = my(v = valuation(n, 5)); n /= 5^v;\
 'y^v * sumdiv(n, m, (-1)^bigomega(m) * m^3 * FUN[n\m]^2);
```

```
LD = [n -> vector(n, j, TAU5SYM2(j)), 0, [-2, 0, 1], 7, 25, 1, 0];
```

Note that we let the variable 'y as an unknown in the definition of TAU5SYM2. If we now ask for Lfuncheckfeq(LD), this will give us a number which should be equal to 0 if the functional equation is satisfied, so we write

```
polroots(numerator(Lfuncheckfeq(LD)))
```

This gives three complex numbers, two large complex ones which we can immediately exclude, and the other equal to 25 to 29 digits of accuracy. It is thus highly believable that the local factor of the symmetric square for $p = 5$ is $1 - 25T$. Once this local factor computed, we can of course do the same kinds of computation that we did with the symmetric square of Δ, such as computing special values.

A final remark concerning this example: if you look at the formulas underlying the Lfuninit program, you will see that even if some Euler factors are wrong, the resulting L-function will still have an exact functional equation $s \mapsto k - s$. However, one can still notice that something is wrong for instance by plotting: for instance, choose 'y=1, which is wrong, and plot the L-function using Lfunplot. You see nothing, because the plot is quite wrong. On the other hand, if you input the correct value 'y=25, you obtain a very reasonable plot.

Example 2. We consider a similar example, where the conductor is now unknown. Consider the function $(\eta(2\tau)\eta(4\tau))^4$, which is a Hecke eigenform of weight 4 on $\Gamma_0(8)$. Once again, we want to determine its symmetric square L-function, and make the same assumption that its bad Euler factor at the prime 2 is of the form $1 - yT$. We thus write as above:

```
FUN2 = Vec('x*(eta('x^2+O('x^10000))*eta('x^4+O('x^10000)))^4);
TAU8SYM2(n) = my(v = valuation(n, 2)); n /= 2^v;\
 'y^v * sumdiv(n, m, (-1)^bigomega(m) * m^3 * FUN2[n\m]^2);
LD = [n -> vector(n, j, TAU8SYM2(j)), 0, [-2, 0, 1], 7, 64, 1, 0];
polroots(numerator(Lfuncheckfeq(LD)))
```

The output of this program is not satisfactory: we find a number of values for y, either too large compared to the Ramanujan bound (and not integral anyway), or even complex. There can be two reasons for this failure: either the Euler factor has degree larger than 1, for instance of the form $1 - yT + zT^2$ with $z \neq 0$, or the conductor $N = 64 = 8^2$ is wrong. Since it is much simpler to test this latter assumption, let us try with other powers of 2. For $N = 32$, the result is again not satisfactory. However, for $N = 16$, we find that one of the roots is equal to 0 to 38 decimals of accuracy. We thus strongly suspect that the Euler factor at $p = 2$ is in fact trivial (just 1), and that the conductor is equal to 16. We can easily make additional checks to see if this is correct: for instance, we can compute $\Theta(1/t) - t^7\Theta(t)$ for a few values of t close to 1, to see if it is very close to 0, or use Lfuninit and then Lfunplot to plot the corresponding L-function on the critical line, and see if the plot is reasonable. All these extra checks of course work.

9.10.4. Additional fun. A first example where the Riemann hypothesis is totally wrong, the L-function associated to the Eisenstein series E_4:

```
LD = Lfunmfcreate(mfEk(4));
Lfunplot(LD);
```

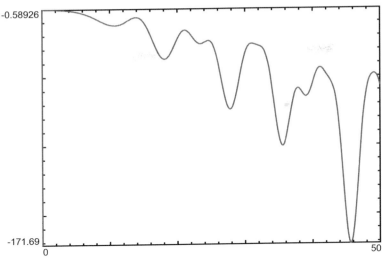

The reason why this plot is so strange is that the L-function associated to E_4 is $\zeta(s)\zeta(s-3)$, and the critical line $\Re(s) = 2$ has nothing to do with the critical lines of $\zeta(s)$ or $\zeta(s-3)$, which are respectively for $\Re(s) = 1/2$ and $\Re(s) = 7/2$.

The second example is more subtle. Consider the L-function attached to the positive definite quadratic form $2x^2 + xy + 3y^2$. It has weight 1, conductor 23, and a pole at $s = 1$. We thus write the following:

```
? an = (n -> 2*Vec(qfrep([4,1;1,6], n, 1)));
? LD = [an, 0, [0, 1], 1, 23, 1, oo];
? LDV = Lfuninit(LD);
? exponent(Lfuncheckfeq(LDV))
% = -127 /* Perfect, functional equation OK */
? LDV.r
% = 1.3101347027385727772685113367482242768 /* Residue */
? Lfunplot(LDV, [0, 50]);
```

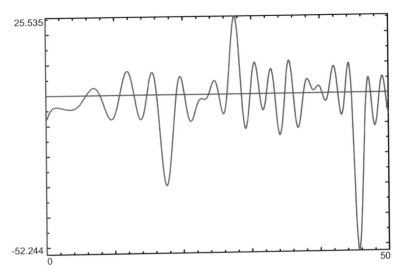

This plot seems quite reasonable, however to the trained eye it is clear that something is going on: near $t = 23$ and $t = 39$ there are little dips which do not cross the real axis: this is an indication that the Riemann hypothesis fails for this L-function, and this comes from the fact that the class number of $\mathbb{Q}(\sqrt{-23})$ is equal to 3, i.e., is not 1. And indeed, after straightforward Newton iterations with initial values $1/2 + 23i$ and $1/2 + 39i$ we find that there are zeros off the critical line at $t = 1.007119 \cdots + 22.56905 \cdots i$ and at $t = 0.820013 \cdots + 39.06517 \cdots i$. In fact, since this L-function does not have an Euler product, it is even quite surprising that *most* of the zeros seem to be on the critical line.

Exercise: Consider the functions

$$f(s) = 2\pi\zeta(s-1) + (s-1)\zeta(s+1) \quad \text{and} \quad F(s) = \pi^{-s/2}\Gamma((s-1)/2)f(s) .$$

(1) Show that $F(1-s) = F(s)$, and deduce that $F(1/2 + it)$ is real for $t \in \mathbb{R}$.
(2) Plot the function $e^{\pi t/4}F(1/2 + it)$ for $t \in [5, 100]$, say, and notice that numerically $f(s)$ seems to satisfy the Riemann hypothesis, although it does not have an Euler product.

What is in our opinion quite extraordinary about this function (and similar ones) is that one can *prove* that the Riemann hypothesis is true for it, i.e., all its nontrivial zeros are on the critical line. This is a theorem of the little-known British mathematician P. R. Taylor just before WWII, see [**Tay45**]. Unfortunately, it is of no help in proving the real Riemann hypothesis.

9.11. Shifting the line of integration in the Booker–Molin method

When computing $L(s)$ for $\Im(s)$ large, because of the exponential decrease of the gamma product $\gamma(s)$ we have seen that we must both work to higher accuracy e^{-E} with $E = B\log(2) + (\pi/4)dT$, and choose a smaller value of h, hence a larger value of M.

A natural idea, already used when using the approximate functional equation, is to find formulas which lessen the exponential decrease of $\gamma(s)$. A simple method which already works very well is as follows. For simplicity of exposition, we explain the method only in the case where the L-function has no pole.

Recall that $K(t) \sim C(d)t^A e^{-\pi dt^{2/d}}$. According to [**Luk69**], this is in fact valid as long as $\Re(t^{2/d})/|t^{2/d}| > \delta > 0$ for some fixed δ. Thus, if $t \in \mathbb{R}_{>0}$, we have

$$K(te^{i\alpha}) \sim C(d)t^A e^{Ai\alpha} e^{-\pi dt^{2/d}e^{2i\alpha/d}}$$

as long as $\cos(2\alpha/d) > \delta > 0$, in other words $\alpha < \pi d/4$.

If this condition is satisfied, we still have $\Theta(te^{i\alpha}) \sim K(te^{i\alpha}/N^{1/2})$, which tends to 0 exponentially fast as $t \to \infty$. It follows that we can still use the Poisson summation formula with this function instead of $\Theta(t)$. More precisely, recall that $\Lambda(s) = \int_{-\infty}^{\infty} e^{zs}\Theta(e^z)\, dz$, Since $\Theta(te^{i\alpha})$ tends to 0 exponentially fast when $z \to \pm\infty$, we can shift the line of integration from $]-\infty, \infty[$ to $]-\infty + i\alpha, \infty + i\alpha[$, so that

$$\Lambda(s) = \int_{-\infty}^{\infty} e^{s(z+i\alpha)}\Theta(e^{i\alpha}e^z)\, dz ,$$

in other words

$$e^{-i\alpha s}\Lambda(s) = \int_{-\infty}^{\infty} e^{zs}\Theta(e^{i\alpha}e^z)\, dz .$$

Thus, by Lemma 9.4.1 we deduce that

$$\sum_{m\in\mathbb{Z}} e^{mhs}\Theta(e^{i\alpha}e^{mh}) = \sum_{k\in\mathbb{Z}} e^{-i\alpha(s-2\pi ik/h)}\Lambda(s - 2\pi ik/h) \ .$$

We now again use the functional equation of Θ, and since there is no pole, this gives

$$\Theta(e^{i\alpha}e^{-mh}) = w(e^{-i\alpha}e^{mh})^k\overline{\Theta}(e^{-i\alpha}e^{mh}) = we^{-ik\alpha}e^{kmh}\overline{\Theta(e^{i\alpha}e^{mh})} \ ,$$

so that

$$\sum_{m\in\mathbb{Z}} e^{mhs}\Theta(e^{i\alpha}e^{mh}) = \Theta(e^{i\alpha})$$

$$+ \sum_{m\geq 1}\left(e^{mhs}\Theta(e^{i\alpha}e^{mh}) + we^{-ik\alpha}e^{mh(k-s)}\overline{\Theta(e^{i\alpha}e^{mh})}\right) \ .$$

We have thus proved the following generalization of Theorem 9.4.2:

THEOREM 9.11.1. *Assume that the L-function has no pole, and let α be such that $0 \leq \alpha < \pi d/4$. Then for any $h > 0$ we have*

$$e^{-i\alpha s}\Lambda(s) = \Theta(e^{i\alpha}) + \sum_{m\geq 1}\left(e^{mhs}\Theta(e^{i\alpha}e^{mh}) + we^{-ik\alpha}e^{mh(k-s)}\overline{\Theta(e^{i\alpha}e^{mh})}\right)$$

$$- \sum_{n\in\mathbb{Z}\backslash\{0\}} e^{-i\alpha s}e^{2\pi n\alpha/h}\Lambda(s + 2\pi in/h) \ .$$

The point of this theorem is that since $\Lambda(s)$ tends to 0 as $e^{-(\pi d/4)|T|}$ as $T = \Im(s) \to \pm\infty$, $e^{-i\alpha s}\Lambda(s)$ will tend to 0 as $e^{-(\pi d/4-\alpha)T}$ as $T \to +\infty$, so we will have considerably diminished the exponential decrease (recall that we must choose $\alpha < \pi d/4$).

We leave to the reader the task of studying the parameters occurring in this method and the corresponding implementation.

9.12. Computing $L(s)$ for large $\Im(s)$

We have already mentioned that the Booker–Molin method explained above is not suited for the computation of L-functions high in the critical strip. For instance, computing $\zeta(1/2 + 10^4i)$ already requires 84 seconds at 38 decimals of accuracy, and to compute $L(\chi_5, 1/2 + 10^4i)$ requires almost 5 minutes, while the programs that we will study below (including the preprogrammed zeta of Pari/GP based on Euler–MacLaurin, i.e., ZetaHurwitz) require milliseconds.

In this section which is completely independent of what we have done up to now, we will describe two important algorithms for computing L-function values for large $T = \Im(s)$, one being the classical Riemann–Siegel formula put into a very precise form by A. de la Reyna [AdR11] which is only applicable to the Riemann zeta function (although a generalization to L-functions of Dirichlet characters due to Siegel exists, see [Sie43]), and the second, much simpler formula discovered recently by K. Fischer [Fis17] which he calls the zetafast algorithm, which is applicable to L-functions of Dirichlet characters.

9.12.1. Computing $\zeta(s)$ by Euler–Maclaurin. We first review the methods that we have seen and which can be applied for computing $\zeta(s)$ for arbitrary complex s, and in particular in the critical strip $0 \leq \Re(s) \leq 1$. The Booker–Molin method that we have just studied in detail works well, but only if $\Im(s)$ is not too large, for instance $|\Im(s)| < 10^3$, and has the unique advantage of being applicable to quite general L-functions. In Chapter 4 we saw that `Sumalt` can compute $\zeta(s)$ outside of its normal range of applicability, but it will be accurate and fast only for real $s > 1$. The `SumMonien` method needs the knowledge of $\zeta(s)$ to compute $\zeta(s)$, so it would be a vicious circle. There remains the most natural method `SumEulerMaclaurin` (which was in fact Euler's initial motivation). Note that `Sumdelta` would also be applicable, but would be less efficient. We have in fact adapted `SumEulerMaclaurin` in the `ZetaHurwitz` program given in Chapter 4, more precisely $\zeta(s)$ is given by the command `ZetaHurwitz(s,1)`. An additional advantage is that it is trivially adaptable to the computation of L-functions of Dirichlet characters when the conductor is reasonably small, see the `Lfunchisimple` program.

The `ZetaHurwitz` algorithm is essentially linear in $T = \Im(s)$, and the timings that we will give below show that it is still very efficient for T as large as 10^7 or 10^8, especially in high accuracy where it becomes the fastest available method.

We will now study algorithms which are essentially linear in $T^{1/2}$, hence much faster when T becomes really large, at least if the accuracy is reasonably small such as 38 decimal digits. For instance, using our rather naïve implementation of the Riemann–Siegel formula given below, we can compute $\zeta(1/2 + iT)$ to 38D for $T = 10^{20}$ in 30 minutes, and more than 80% of that time is spent in simply computing $\sum_{1 \leq n \leq N} n^{-s}$ for N approximately equal to $4 \cdot 10^9$.

9.12.2. Computing $\sum_{1 \leq n \leq N} n^{-s}$. Thus, an essential problem in the Riemann–Siegel formula that we will see below is the computation of the partial sums of $\zeta(s)$. The naïve way would of course simply be to compute the sum as it is written, as `sum(n = 1, N, n^(-s))`. This is clearly wasteful, since once all the p^{-s} are known for $p \leq N$ prime, all the n^{-s} can be computed by factoring n and simple multiplications. In fact, setting up the computation as a sieve avoids factoring n altogether and this is exactly what the GP function `dirpowers` does in order to computes the vector of all n^{-s} for $n \leq N$. So a better command is `vecsum(dirpowers(N, -s))`, which will be faster by a rough factor $\log N$.

Unfortunately `dirpowers` returns a vector with N entries and needs a considerable amount of memory for $N > 10^9$, say. Even storing the p^{-s} for all primes p less than N, which saves a factor $\log N$ by the prime number theorem, becomes unreasonable. It is thus important to look for a method which is almost as fast but uses less memory. The first idea is this: denote by $P(n)$ the largest prime factor of n. If $p = P(n) > \sqrt{N}$ we have $n = pe$ with $e \leq \sqrt{N}$, so that

$$S_1(N) := \sum_{\substack{1 \leq n \leq N \\ P(n) > \sqrt{N}}} n^{-s} = \sum_{\substack{p \text{ prime} \\ \sqrt{N} < p \leq N}} p^{-s} \sum_{1 \leq e \leq \lfloor N/p \rfloor} e^{-s} .$$

To compute this it is sufficient to precompute and store the values of $\sum_{1 \leq e \leq m} e^{-s}$ for $m \leq \sqrt{N}$, which takes little time and space, and then sum over the primes p such that $\sqrt{N} < p \leq N$, which is unavoidable.

We are thus left with the computation of

$$S_2(N) := \sum_{\substack{1 \le n \le N \\ P(n) \le \sqrt{N}}} n^{-s} \ .$$

Since all the p^{-s} for $p \le \sqrt{N}$ have been precomputed, we could simply factor all $n \le N$, keep only those with $P(n) \le \sqrt{N}$ and for those do suitable multiplications. However, we can do better. The second idea is to write uniquely $n = df^2$ with d squarefree. Thus, denoting by $\mathcal{D}_N$ the set of squarefree d such that $P(d) \le \sqrt{N}$, we have

$$S_2(N) = \sum_{\substack{1 \le d \le N \\ d \in \mathcal{D}_N}} d^{-s} F\left(\sqrt{N/d}\right) \ ,$$

where the $F(m) = \sum_{1 \le f \le m} f^{-2s}$ can be precomputed and stored for $m \le \sqrt{N}$.

A third idea is to separate the squarefree d divisible by 2, or 3, or .... For instance, we can write

$$S_2(N) = \sum_{\substack{1 \le d \le N \\ d \in \mathcal{D}_N, \ d \text{ odd}}} d^{-s} F\left(\sqrt{N/d}\right) + 2^{-s} \sum_{\substack{1 \le d \le N/2 \\ d \in \mathcal{D}_N, \ d \text{ odd}}} d^{-s} F\left(\sqrt{N/(2d)}\right)$$

$$= \sum_{\substack{1 \le d \le N \\ d \in \mathcal{D}_N, \ d \text{ odd}}} d^{-s} G(N/d) \ ,$$

where $G(x) = F\left(\sqrt{x}\right) + 2^{-s} F\left(\sqrt{x/2}\right)$ if $x \ge 2$ and $G(x) = F\left(\sqrt{x}\right)$ if $x < 2$. We can do a similar decomposition with the primes 3, 5, etc., but with diminishing returns. We give the script using only 2 and 3:

_____ Sumpow.gp _____

```
/* P = sorted prime divisors of squarefree n;
 * V[k] = k^s for k <= sq = floor(sqrt(N))
 * Return n^s if P^+(n) <= sq, else return 0 */
Smallfact(n, P, sq, V) =
{
  if (!#P, return (1));
  my(p = P[#P]); if (p > sq, return (0)); /* P^+(n) > sqrt(N) */
  my(c = V[p]);
  forstep (i = #P - 1, 1, -1,
    n /= p; if (n <= sq, return (V[n] * c));
    p = P[i]; c *= V[p]);
  return (c);
}

/* Compute sum_{1 <= n <= N} n^s. */
Sumpow(N, s) =
{
  if (N < 10000, return (vecsum(dirpowers(N, s))));

  my(B = getlocalbitprec(), sq = sqrtint(N));
  my(W = vector(sq), F = vector(sq), V, precp, t, S);
```

```
22    s = precision(s, B + 32);
23    localbitprec(B + 32);
24    V = dirpowers(sq, s); W[1] = F[1] = 1;
25    for (n = 2, sq,
26      W[n] = W[n - 1] + V[n];        /* sum_{j <= n} j^s */
27      F[n] = F[n - 1] + sqr(V[n])); /* sum_{j <= n} j^(2s) */
28    precp = t = 1; S = 0;
29    forprime (p = sq + 1, N,
30      t *= (p / precp)^s;
31      S += W[N \ p] * t; precp = p);
32    /* At this point S = S_1(N). Now add S_2(N). */
33    forsquarefree (dF = 1, N,
34      my(d = dF[1]);
35      if (gcd(d, 6) == 1 && t = Smallfact(d, dF[2][,1], sq, V),
36        my(q = N \ d, c);
37        if (q <= 5, c = W[min(q, 4)]
38                  , c = F[sqrtint(q)] + V[2] * F[sqrtint(q\2)]
39                      + V[3] * F[sqrtint(q\3)] + V[6] * F[sqrtint(q\6)]);
40        S += c * t));
41    return (bitprecision(S, B));
42  }
```

REMARKS 9.12.1.

(1) Since the built-in powering program z^s is faster when z is close to 1, we compute p^s using $p^s = pp^s \left(p/pp\right)^s$, where pp is the previous prime and pp^s has already been computed.

(2) The built-in forsquarefree uses a sieve and provides the factorizations of *consecutive* squarefree integers at little additional cost compared to an ordinary loop. By programming at a lower level, in the C implementation, one would include the conditions $(d, 6) = 1$ and $P^+(d) \leq \sqrt{N}$ in the sieve so as not to generate the factorizations of useless indices d. This is the main reason why the built-in dirpowerssum is 2 to 3 times faster than Sumpow.

(3) At bit accuracy B, the program uses $O(B\sqrt{N})$ memory. The built-in GP function dirpowerssum has similar properties.

Another, much simpler, method is to write

$$S = \sum_{0 \leq j \leq e} 2^{js} \sum_{\substack{1 \leq n \leq N/2^j \\ n \text{ odd}}} n^s ,$$

where $e = \lfloor \log_2 N \rfloor$ and the inner sums are of course computed inductively. The following script implements this in a clever way ensuring that n^s is computed with a single multiplication whenever n is not a prime number.

$$\boxed{\text{Sumpow2.gp}}$$

```
Sumpow2(N, s) =
{ my(N3 = N \ 3, D = vectorsmall(N), F = vector(N3));
  my(f2 = 2^s, u = 1, S = 1, ks = 1);

  forstep (k = 3, sqrtint(N), 2,
    if (!D[k], forstep (j = sqr(k), N, 2 * k, D[j] = k)));
  forstep (j = exponent(N), 1, -1,
    /* u = sum_{n odd <= N / 2^j} n^s */
    forstep (k = bitor((N >> j) + 1, 1), N >> (j - 1), 2,
      ks = if (D[k], F[D[k]] * F[k / D[k]]
                   , ks * (k / (k - 2))^s);
      u += ks; if (k <= N3, F[k] = ks));
    S = u + f2 * S);
  return (S);
}
```

REMARKS 9.12.2.
(1) For all odd $3 \leq k \leq N$, the entry $D[k]$ contains 0 if and only if k is a prime number and otherwise the largest proper divisor of k which is $\leq \sqrt{k}$. For all odd $k \leq N/3$, the entry $F[k]$ is k^s.
(2) In low accuracy, say less than 500 decimals, the speed of this program is comparable to that of the more complicated Sumpow given above, but it becomes less competitive in larger accuracies. Another drawback is that it uses $O(NB)$ memory, considerably more than Sumpow.

9.12.3. The Riemann–Siegel formula. We now give without proof the Riemann–Siegel formula, directly taken from [AdR11]. Note that thanks to the functional equation, to compute $\zeta(\sigma + iT)$ we may assume that $\sigma \geq 1/2$ and $T \geq 0$.

THEOREM 9.12.3. Assume $s = \sigma + iT$ with $\sigma \geq 1/2$ and $T > 0$.

(1) Define a double sequence $d_j^{(k)}$ for $0 \leq j \leq 3k/2$ by $d_0^{(0)} = 1$, the induction

$$(6k - 4j)d_j^{(k)} = \frac{1}{2}d_j^{(k-1)} + (1 - 2\sigma)d_{j-1}^{(k-1)} - 2(3k - 2j)(3k - 2j + 1)d_{j-2}^{(k-1)}$$

for $j \neq 3k/2$, where it is understood that $d_j^{(k)} = 0$ if $j < 0$ or $j > 3k/2$, and for $j = 3k/2$:

$$d_{3k/2}^{(k)} = -\sum_{0 \leq i < 3k/2} (-1)^{3k/2-i} d_i^{(k)} \frac{(3k - 2i)!}{(3k/2 - i)!} .$$

(2) Let

$$F(z) = \frac{\exp\left(i\pi(z^2/2 + 3/8)\right) - i\sqrt{2}\cos(\pi z/2)}{2\cos(\pi z)} ,$$

and set

$$C_k(z) = \pi^{-2k} \sum_{0 \leq j \leq 3k/2} (-i\pi/2)^j d_j^{(k)} F^{(3k-2j)}(z) .$$

(3) *Define* $t = T/(2\pi)$, $a = t^{1/2}$, $N = \lfloor a \rfloor$ *and*

$$p = 1 - 2(a - N), \quad U = \exp\left(-i\pi(t\log(t) - t - 1/8)\right).$$

(4) *Finally, set*

$$R_K(s) = \sum_{1 \leq n \leq N} n^{-s} + (-1)^{N-1} U a^{-\sigma} \sum_{0 \leq k \leq K} C_k(p)/a^k.$$

Then for T sufficiently large (see below) we have

$$\zeta(s) = R_K(s) + \pi^{s-1/2}\frac{\Gamma((1-s)/2)}{\Gamma(s/2)} R_K(1 - \overline{s}) + \varepsilon_K,$$

for some small error ε_K.

REMARKS 9.12.4.

(1) Bounds for the error ε_K are given in [**AdR11**]. A reasonable choice ensuring $|\varepsilon_K| < 2^{-B} = e^{-E}$ is

$$K = 2\lceil E/(A - \log(E/A)) \rceil, \quad \text{with} \quad A = \log(1.73T),$$

but in the script we will use a slightly better value.
(2) The condition that T be sufficiently large is satisfied if $T \geq 12B$.
(3) The derivatives $F^{(i)}(p)$, $i \leq 3K$, can be computed all at once using the `derivnum` program. A more complicated solution uses the power series expansion of F in terms of Euler numbers, and it can be shown that it is sufficient to take $3K$ terms of this expansion. In practice the latter is always faster, so our script only includes that method.
(4) For large T, the quantity

$$\frac{\Gamma((1-s)/2)}{\Gamma(s/2)} = \exp\left(\log\Gamma((1-s)/2) - \log\Gamma(s/2)\right)$$

must be computed with care because of enormous underflow, or equivalently because of cancellation when subtracting the $\log\Gamma$ values. We use our standard technique of increasing locally the accuracy to make up for the cancellation.
(5) For large T, by far the most time-consuming part of the algorithm is the simple computation of $S_N(s) := \sum_{1 \leq n \leq N} n^{-s}$ and of $S_N(1 - s)$, so for this we use the built-in `dirpowerssum` command which is the C implementation of the `Sumpow` script given above.

Thanks to the above remarks, it is immediate to write a reasonably simple Riemann–Siegel program:

─────────────────── | ZetaRiemannSiegel.gp | ───────────────────

```
1  FindK(sig, T) =
2  { my(B = getlocalbitprec());
3    localbitprec(32);
4    my(E = (B + 1.5 * sig) * log(2), A = log(T) - 1.528);
5    return (ceil(2 * Solvemulneglog(A, E - (A + 0.5) / 2)));
6  }
7
8  /* F(z) = sum_n C[n+1] z^n; W[n+1] = (-i Pi/2)^n */
9  GetC(K, T) =
```

```
10   { my(B = getlocalbitprec(), M = 3 * K, V, W, W1, W2, VE);
11     my(E, VE, C, c1, c2);
12
13     localbitprec(B + 5 * M);
14     c1 = -I / sqrt(2); c2 = exp(3 * I * Pi / 8) / 2;
15     V = vector(2 * M + 1); V[2 * M + 1] = 1 / factorial(2 * M);
16     forstep (j = 2 * M, 1, -1, V[j] = j * V[j + 1]);
17     W = powers(I * Pi / 2, 2 * M);
18     W1 = vector(M + 1, n, V[2 * n - 1] * W[2 * n - 1]);
19     W2 = vector(M + 1, n, c1 * W1[n] + c2 * V[n] * W[n]);
20     E = eulervec(M);
21     VE = vector(M + 1, n, E[n] * W1[n] << (2 * (n - 1)));
22     C = Vec(Ser(VE) * Ser(W2));
23     forstep (n = 2, #W, 2, W[n] = - W[n]);
24     return ([C, W]);
25   }
26
27   /* D[k+1][j+1] = d_j^{(k)}; K >= 2 */
28   GetD(K, sig) =
29   { my(D = vector(K+1), e = 2.0 - 4*sig);
30     my(A = vector((3*K) \ 2)); /* A[d] = (-1)^(d-1) (2*d)!/d! */
31     A[1] = 2; for (d = 2, #A, A[d] = -A[d-1] * (4*d - 2));
32     D[1] = [1]; D[2] = [1 / 12, e / 4];
33     for (k = 2, K,
34       my(k3 = 3 * k, J = (k3-1) \ 2, Ds = vector(k3 \ 2 + 1));
35       Ds[1] = D[k][1] / (k3 << 2);
36       Ds[2] = (D[k][2] + e * D[k][1]) / ((k3 - 2) << 2);
37       for (j = 2, J - 1,
38         my(c = k3 - 2*j);
39         Ds[j+1] = (D[k][j+1] + e * D[k][j]) / (c << 2)
40                 - (c + 1) * D[k][j-1]);
41       my(c = k3 - 2*J);
42       Ds[J+1] = e * D[k][J] / (c << 2) - (c + 1) * D[k][J-1];
43       if (k % 2 == 0,
44         Ds[J+2] = sum(i = 0, J, A[J-i+1] * Ds[i+1]));
45       D[k+1] = Ds);
46     return (D);
47   }
48
49   /* Re(s) >= 1 / 2, Im(s) >= 12 * B */
50   RiemannSiegel(s) =
51   { my(B = getlocalbitprec(), sig = real(s), T = imag(s));
52     my(K = FindK(sig, T), F, D, S, a, N);
53
54     localbitprec(B + 32);
55     T = bitprecision(T, B + 32);
56     a = sqrt(T / (2 * Pi)); N = floor(a);
57     my([C, W] = GetC(K, T), p = 1 - 2 * (a - N), v, b, P);
```

```
58      v = powers(p^2, #C-1); v = vector(#v, n, v[n] * C[n]);
59      P = powers(1 / p, 3 * K);
60      F = vector(3 * K + 1, j,
61        if (j != 1, v = vector(#v, n, v[n] * (2*n-j)));
62        vecsum(v) * P[j]); /* F[i+1] = F^(i)(p) */
63      b = powers(1 / (Pi^2 * a), K);
64      D = GetD(K, sig);
65      S = sum(k = 0, K,
66              b[k+1] * sum(j = 0, (3*k) \ 2,
67                        W[j+1] * D[k+1][j+1] * F[3*k-2*j+1]));
68      dirpowerssum(N, -s) +
69        (-1)^(N - 1) * S * exp(-s * log(a) + I * (T / 2 + Pi / 8));
70    }
71
72  Getchis(s) =
73  { my(B = getlocalbitprec(), T = imag(s), G);
74
75    localbitprec(32);
76    localbitprec(B + exponent(T * (log(T) - 1)));
77    G = lngamma((1 - s) / 2) - lngamma(s / 2);
78    localbitprec(B + 32);
79    return (exp((s - 1/2) * log(Pi) + G));
80  }
81
82  ZetaRiemannSiegel(s) =
83  { my(B = getlocalbitprec(), X, z1, z2);
84    my(T = imag(s), flconj = T < 0, h = real(s) - 1/2);
85
86    s = bitprecision(s, B + 32);
87    if (flconj, s = conj(s));
88    X = Getchis(s);
89    if (h < 0, s = 1 - conj(s)); /* use functional equation */
90    if (abs(T) < 12 * B, error("T too small in RiemannSiegel"));
91    z1 = RiemannSiegel(s);
92    z2 = if (h, RiemannSiegel(1 - conj(s)), z1);
93    if (h < 0, [z1, z2] = [z2, z1]);
94    z1 += X * conj(z2);
95    return (if (flconj, conj(z1), z1));
96  }
```

This program can be improved in several ways, for instance when s is not on the critical line $\Re(s) = 1/2$ we can modify the Sumpow program (here used in its C-version dirpowerssum for efficiency) so as to compute Sumpow(N, -(1-s)) and Sumpow(N, -s) simultaneously. But it is already remarkably efficient, see below for some comparative timings.

9.12.4. The Zetafast algorithm. Recently, K. Fischer [**Fis17**] invented another method which, as the Riemann–Siegel formula, takes time roughly proportional to the square root of $T = \Im(s)$, but is simpler to explain, and directly applies also to L-functions of Dirichlet characters. We will see that the algorithm is not competitive to compute the Riemann zeta function, but quite useful for Dirichlet L-functions. We first give his main theorem:

THEOREM 9.12.5. *Let χ be a primitive Dirichlet character of conductor q, let $v > 0$ be an integer and $N \geq 1$, set*

$$D(\chi, s) = \sum_{n \geq 1} \chi(n) n^{-s} Q_v(n/N) , \quad where$$

$$Q_v(x) = \frac{1}{(v-1)!} \int_x^\infty t^{v-1} e^{-t} \, dt = e^{-x} \sum_{0 \leq w \leq v-1} \frac{x^w}{w!} ,$$

and for $\varepsilon = \pm 1$ set

$$E_\varepsilon(\chi, s) = \sum_{m \geq 1} \overline{\chi(m)} E_\varepsilon(q, m, s) , \quad where$$

$$E_\varepsilon(q, m, s) = v \binom{v-s}{v} \int_0^Z z^{v-1} (z+m)^{s-v-1} \, dz , \quad where \ Z = i\varepsilon q/(2\pi N)$$

$$= m^{s-1} - \sum_{0 \leq w \leq v-1} \binom{s-1}{w} (m+Z)^{s-1-w} (-Z)^w .$$

All the series are absolutely and uniformly convergent in any strip $0 < \Re(s) < v$, and if $0 \leq \sigma = \Re(s) < 1$ we have

$$L(\chi, s) = D(\chi, s) - \frac{\Gamma(v+1-s)}{(1-s)\Gamma(v)} N^{1-s} \cdot \delta_{q,1}$$

$$+ (2\pi)^{s-1} \Gamma(1-s) q^{-s} \mathfrak{g}(\chi) \sum_{\varepsilon = \pm 1} \chi(-\varepsilon) e^{(i\varepsilon\pi/2)(1-s)} E_\varepsilon(\chi, s) ,$$

where $\mathfrak{g}(\chi) = \sum_{x \bmod q} \chi(x) e^{2\pi i x/q}$ is the usual Gauss sum associated to χ.

The point of giving the integral representations is twofold. First, depending on the efficiency of the integration programs, it may be faster to use them than to use the explicit finite sums given by the theorem. But more importantly, it allows us to give a reasonable error term when truncating the sums in the above formulas.

PROOF. We sketch the proof for the Riemann zeta function itself (the general case is essentially identical), referring to Fischer's paper for details. The main idea, common in analytic number theory, is to introduce a smoothing function $f(n)$ to cut off the sum $\sum_{n \geq 1} n^{-s}$ defining $\zeta(s)$, in other words to consider $\sum_{n \geq 1} f(n) n^{-s}$ instead, where $f(n)$ should be very close to 1 for $1 \leq n \leq N$ for some large N, and tend smoothly to 0 as $n \to \infty$. The choice made by the author is to consider the function $Q_v(x) = e^{-x} \sum_{0 \leq j < v} x^j/j!$ for some large v, and to choose $f(n) = Q_v(n/N)$ for some other large N. It is immediate to check that $f(n)$ is indeed a nice smoothing function, but the additional advantage is that the Mellin transform

$\int_0^\infty t^{z-1} Q_v(t) dt$ is simple:

$$\int_0^\infty t^{z-1} Q_v(t) dt = \sum_{0 \le j < v} \Gamma(z+j)/j! \,.$$

Thus, the Mellin inversion formula tells us that for $x > 0$ we have

$$Q_v(t) = \frac{1}{2\pi i} \sum_{0 \le j < v} \frac{1}{j!} \int_{\Re(z)=x} t^{-z} \Gamma(z+j) \, dz \,,$$

so that after exchanging absolutely convergent sums and integrals

$$D(s) = \sum_{n \ge 1} \frac{Q_v(n/N)}{n^s} = \frac{1}{2\pi i} \sum_{0 \le j < v} \frac{1}{j!} \int_{\Re(z)=x} N^z \Gamma(z+j) \zeta(z+s) \, dz \,.$$

Because of the exponential decrease of $\Gamma(z)$ on vertical strips we can move the line of integration to $\Re(z) = y$ for some y such that $-1 < y < -\sigma$ (which is possible since $0 \le \sigma < 1$). Taking into account the poles at $z = 1 - s$ and (if $j = 0$) at $z = 0$, we obtain

$$\frac{1}{2\pi i} \int_{\Re(z)=x} N^z \Gamma(z+j) \zeta(z+s) \, dz = N^{1-s} \Gamma(j+1-s) + \delta_{j,0} \zeta(s)$$

$$+ \frac{1}{2\pi i} \int_{\Re(z)=y} N^z \Gamma(z+j) \zeta(z+s) \, dz \,.$$

We now use the functional equation for ζ in the form

$$\zeta(z+s) = 2(2\pi)^{z+s-1} \Gamma(1-z-s) \cos\left((\pi/2)(1-z-s)\right) \cdot \zeta(1-z-s) \,.$$

For $\Re(z) = y$ we have $\Re(1-z-s) = 1 - \sigma - y > 1$ since $y < -\sigma$, so we can replace $\zeta(1-z-s)$ by its absolutely convergent expression and

$$\frac{1}{2\pi i} \int_{\Re(z)=y} N^z \Gamma(z+j) \zeta(z+s) \, dz = (2\pi)^{s-1} \sum_{m \ge 1} m^{s-1} \left(I(A_+) + I(A_-)\right) \,,$$

where $A_\pm = \pm 2\pi i N m$ and

$$I(A_\pm) = \frac{e^{\mp i(\pi/2)(1-s)}}{2\pi i} \int_{\Re(z)=y} \Gamma(z+j) \Gamma(1-z-s) A_\pm^z \, dz \,.$$

Now note that $|A_\pm| > 2\pi$ so $A_\pm^z$ tends exponentially to 0 when $\Re(z) \to -\infty$ while $\Gamma(z+j)\Gamma(1-z-s)$ behaves polynomially. It follows that we can shift the line of integration to $y = -\infty$ and capture the poles of $\Gamma(z+j)$. We obtain after a small calculation

$$I(A^\pm) = e^{\pm i(\pi/2)(1-s)} \Gamma(1+j-s)(-A_\pm)^{-j} \left((1 - 1/A_\pm)^{1+j-s} - \delta_{j,0}\right) \,.$$

Using the standard properties of the gamma function proves the theorem. □

To use the above theorem in practice, we must determine suitable values of v and N, and determine where to cutoff the infinite series so as to achieve our usual absolute error bound 2^{-B}. Although very important, this is rather technical, so we will be content with giving some indications. The values given by the author are not optimal (quite far from optimal in certain ranges), so we do not quote them.

We will assume that s is in the critical strip $0 \le \Re(s) \le 1$ (the modifications to be made if s is far from the critical strip are rather complicated). Assume v and N chosen. We can neglect $\chi(n)n^{-s}$, so we must stop the sum giving $D(\chi, s)$ as soon as $Q_v(n/N) < 2^{-B}$. Now when $x > 2v$, say, $Q_v(x)$ is well approximated by the value of the integrand at x, i.e., by $x^{v-1}e^{-x}/(v-1)!$, so we need $x - (v-1)\log(x) > \log(2^B/(v-1)!)$. This inequality can be solved using the Solvedivlog program, and therefore the sum giving $D(\chi, s)$ should be truncated at $n = M = Nx_0$, where x_0 is the smallest solution larger than $2v$.

Concerning the sum giving $E_1(\chi, s)$, first note that by the complex Stirling formula, the factor $\Gamma(1-s)$ in front is of the order of $e^{-(\pi/2)T}$, while the factor $e^{i(\pi/2)(1-s)}$ is of the order of $e^{(\pi/2)T}$, hence the factors essentially cancel (for E_{-1}, the factor would be $e^{-\pi T}$, which is negligible). Concerning $E_1(\chi, s)$ itself, again using the integral representation and the complex Stirling formula the factor in front of the integral is of order $T^v/(v-1)!$. The integrand is dominated by $m^{-v}z^{v-1}$, so the integral is of the order of $|Z|^v/v$, hence $E_1(q, m, s)$ is of the order of $|ZT/m|^v/v!$. For this to be less than 2^{-B} we therefore need $m > |ZT| \cdot (2^B/v!)^{1/v}$, which is therefore where we must truncate the series for E_1.

The above derivations are mathematically precise. On the other hand, the choice of v and N depend in a crucial way on the implementation, and more precisely on the relative speed of computation of Q_v and E_1.

We have found that a choice of v proportional to the bit accuracy B gives reasonable results, and we have chosen $v = B/2$; in fact, varying this "magic constant" $1/2$ between 0.45 or 0.55 does not change much the efficiency of the program. Concerning the choice of N, a choice proportional to $\sqrt{qT}/B$ gives good results, and we have chosen a "magic factor" of 5, again not essential since values between 3 and 8 give a similar efficiency.

The following script incorporates these choices, and the reader is welcome to improve on them. For simplicity it computes $L(\chi_D, s)$, where $\chi_D(n) = \left(\frac{D}{n}\right)$ is the Kronecker symbol, but it can trivially be generalized to any primitive Dirichlet character χ as in the theorem. As mentioned in the analysis, we implicitly assume that σ is not too large for instance $0 \le \sigma \le 1$; for simplicity, we even assume $\sigma \ge 1/2$ instead of applying the functional equation in the script.

──────────────── │ Zetafast.gp │ ────────────────

```
1   /* Compute L(chi_D, s) by the Zetafast algorithm. */
2
3   Zetafastdata(q, T, B) =
4   { my(qT = q * T, v = B \ 2, N, L, M0, M);
5
6       localbitprec(32);
7       N = max(round(5 * sqrt(qT) / B), 1);
8       L = - log(factorial(v) >> B);  /* = log(2^B / v!) */
9       M0 = ceil(qT * exp(L / v) / (2 * Pi * N));
10      M = ceil(N * Solvedivlog(v, L));
11      return ([v, N, M0, M]);
12  }
13
14  Vecsumchi(V, D) =
15  {
16      if (D == 1, return (vecsum(V)));
17      sum(m = 1, #V, kronecker(D, m) * V[m]);
```

```
18    }
19
20    ZetafastE1(D, s, data) =
21    { my([v, N, M0] = data, q = abs(D), Z = I * q / (2 * Pi * N));
22      my(sig = real(s), C, S, VE, E, VZ);
23
24      VZ = powers(-Z, v - 1);
25      VE = vector(M0, m, if (gcd(D, m) == 1, (m + Z)^s, 0));
26      E = vector(M0, m, if (gcd(D, m) == 1, 1 / (m + Z), 0));
27      S = Vecsumchi(dirpowers(M0, s), D);
28      C = 1;
29      for (w = 0, v - 1,
30        S -= C * VZ[w + 1] * Vecsumchi(VE, D);
31        if (w == v - 1, break);
32        C *= (s - w) / (w + 1);
33        VE = vector(M0, m, VE[m] * E[m]));
34      my(G = sqrt(q) * (2*Pi)^s * q^(-s - 1));
35      if (D < 0, G *= -I);
36      /* G = chi(-1) (2 pi)^(s-1) q^(-s) g(chi) */
37      return (G * S * gamma(-s) * exp(-I * Pi/2 * s));
38    }
39
40    ZetafastQ(D, s, data) =
41    { my([v, N, M0, M] = data, VS, VE, P, P0, S);
42
43      VS = dirpowers(M, -s);
44      VE = powers(exp(-1 / N), M);
45      P = vector(v); P[v] = 1;
46      forstep (w = v-1, 1, -1, P[w] = w * N * P[w+1]);
47      P0 = P[1]; P = Polrev(P);
48      S = if (D == 1,
49        sum(n = 1, M, VS[n] * VE[n+1] * subst(P, 'x, n));
50        ,
51        sum(n = 1, M,
52            my(k = kronecker(D, n));
53            if (k, k * VS[n] * VE[n+1] * subst(P, 'x, n))));
54      return (S / P0);
55    }
56
57    /* exp(z) avoiding underflow */
58    Exp(z) =
59    { my(B = getlocalbitprec());
60      localbitprec(32); my(E = (B + 20) * log(2));
61      localbitprec(B); return (if (real(z) > -E, exp(z), 0.));
62    }
63
64    ZetafastG(s, data) =
65    { my([v, N] = data);
```

```
66        Exp(lngamma(s + v) + s * log(N) - log(s * (v - 1)!)));
67    }
68
69    /* Valid when 1/2 <= Re(s) <= 1 */
70    Zetafast(s, D = 1) =
71    { my(B = getlocalbitprec(), la, v, data, S);
72      my(sig = real(s), T = imag(s), flconj = 0);
73
74      if (sig < 1/2, error("please use the functional equation"));
75      if (T == 0 && s == round(s) && s >= 0, return (lfun(D, s)));
76      if (T < 0, flconj = 1; T = -T);
77      [v] = data = Zetafastdata(abs(D), max(T,100), B);
78      if (T < 3*v, return (lfun(D, s)));
79      if (flconj, s = conj(s));
80      localbitprec(B + 32);
81      S = ZetafastQ(D, s, data) - if (D == 1, ZetafastG(1-s, data))
82          + ZetafastE1(D, s-1, data);
83      if (flconj, S = conj(S));
84      return (bitprecision(S, B));
85    }
```

REMARKS 9.12.6.

    (1) **Warning:** Zetafast(s, D) computes $L(\chi_D, s)$. We have exchanged the parameters D and s so as to simply write Zetafast(s) for the Riemann zeta function.

    (2) Since the operations used in this program are very simple, the corresponding C program is almost three times as fast.

    (3) Most of the time is spent in computations which depend only on the size of T. It would therefore possible to modify this program by first writing an initialization part which would be useful for $t \leq T$, and second a faster program using this initialization. The gain being marginal, we have not done so.

    (4) It is not clear how to adapt this program to more general L-functions, although it may not be difficult.

9.12.5. Two fun applications. Thanks to the above programs, we can play with L-functions on the critical strip as we have done above with the Lfun programs, in particular in connection with the Generalized Riemann Hypothesis, but we can now use much larger values of $\Im(s)$. If as usual χ_D denotes the quadratic character for the fundamental discriminant D, the corresponding completed L-function and Hardy Z-function are given by

$$\Lambda(\chi_D, s) = |D|^{s/2} \pi^{-(s+a)/2} \Gamma((s+a)/2) L(\chi_D, s) \quad \text{and}$$

$$Z(\chi_D, T) = \exp((T/2) \operatorname{atan}(2T))(1/4 + T^2)^{(-1)^a/8} \Lambda(\chi_D, 1/2 + iT) ,$$

where $a = 0$ if $D > 0$ and $a = 1$ if $D < 0$. The functional equation $\Lambda(\chi_D, 1 - s) = \Lambda(\chi_D, s)$ implies that $Z(\chi_D, T)$ is real, and we recall that the exponential factor compensates the exponential decrease of the gamma function on vertical strips. The following script which uses our new programs implements this formula by avoiding exponent underflows:

—————————————————— ZetaHardy.gp ——————————————————

```
1  ZetaHardy(T, D = 1) =
2  { my(B = getlocalbitprec(), s = 1/2 + I * T, S, Z, L);
3
4    Z = if (D == 1 && T > 12 * B, ZetaRiemannSiegel(s),
5                                  Zetafast(s, D));
6    if (!Z, return (0.));
7    S = if (D < 0, s + 1, s) / 2; /* (s + a) / 2 */
8    L = -S * log(Pi) + lngamma(S) + (s/2) * log(abs(D)) + log(Z);
9    return (exp((T / 2) * atan(2 * T) + real(L)) * cos(imag(L))
10           * (1/4 + T^2)^(sign(D) / 8));
11 }
```

A command such as

```
? \pb 32 /* Set to low accuracy for plotting */
? ploth(t = 10^10, 10^10 + 3, ZetaHardy(t))
```

gives in 24 seconds a plot of the Hardy Z-function associated to the Riemann zeta function at height 10^{10}, showing in particular the position of the zeros.

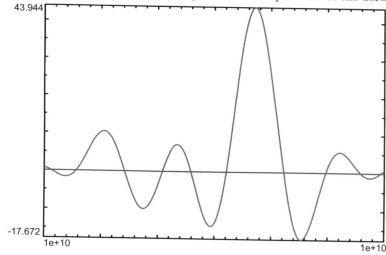

This directly leads us to our second application, the location of these zeros. The following script uses the naïve zero searching program already used in the more general Lfunzeros program:

—————————————————— Zetazeros.gp ——————————————————

```
1  Zetazeros(range, D = 1, divz = 8) =
2  { my(Gap = Zerogap(abs(D), 1, divz));
3    return (Zerosearch(t -> ZetaHardy(t, D), Gap, 0, range));
4  }
```

For instance, this script shows that the first zero occurring in the above plot occurs at height $10^{10} + 0.060634346\cdots$. For another example, an immediate program shows that for negative fundamental discriminants, the smallest zero is minimal with respect to the previous ones for $D = -43$, -67, -148, and -163, and it is of course not a coincidence that three of these four discriminants have class number $h = 1$: indeed, $h = 1$ implies that $L(D, 1) = \pi h/\sqrt{|D|}$ is comparatively small, so it is not unreasonable to expect that the lowest zero is also quite small.

9.12.6. Sample timings for $\zeta(1/2+iT)$. We first give a table of the times for the three programs ZetaHurwitz, ZetaRiemannSiegel, and Zetafast, at 38, 115, and 346 decimal digits, with $s = 1/2 + 10^k i$ for $4 \le k \le 18$ (we do not include the Lfun program which, as already mentioned, is orders of magnitude slower for these values of s). Here, as in other tables, $\otimes$ means that the program is not applicable or gives nonsense results, and ∞ that waiting for 5 minutes was not enough to get a result.

k	Hur	RS	fast	Hur	RS	fast	Hur	RS	fast
4	**0.01**	**0.01**	**0.01**	**0.01**	0.26	0.03	**0.04**	$\otimes$	0.21
5	0.03	**0.00**	0.03	0.07	**0.06**	0.08	**0.27**	2.14	0.54
6	0.23	**0.00**	0.07	0.57	**0.03**	0.25	2.07	**0.76**	1.58
7	1.89	**0.00**	0.23	4.35	**0.02**	0.79	16.9	**0.37**	5.02
8	16.4	**0.00**	0.73	35.2	**0.02**	2.54	143.	**0.24**	16.4
9	151.	**0.02**	2.29	∞	**0.04**	8.22	∞	**0.25**	54.9
10	∞	**0.04**	7.30	∞	**0.08**	26.7	∞	**0.38**	180.
11	∞	**0.12**	23.2	∞	**0.22**	85.8	∞	**0.83**	∞
12	∞	**0.30**	73.6	∞	**0.62**	277.	∞	**2.13**	∞
13	∞	**0.83**	235.	∞	**1.74**	∞	∞	**5.93**	∞
14	∞	**2.36**	∞	∞	**5.00**	∞	∞	**17.0**	∞
15	∞	**6.96**	∞	∞	**14.4**	∞	∞	**49.3**	∞
16	∞	**20.8**	∞	∞	**40.8**	∞	∞	**144.**	∞
17	∞	**62.5**	∞	∞	**116.**	∞	∞	∞	∞
18	∞	**189.**	∞	∞	∞	∞	∞	∞	∞

Dependence on $\Im(s) = 10^k$ at 38D, 115D, and 346D.

On this table we first see clearly that the Zetafast algorithm is never competitive. Second, that for larger values of $T = \Im(s)$, ZetaRiemannSiegel is orders of magnitude faster than ZetaHurwitz, as expected since the latter is essentially linear in T while the former is linear in $T^{1/2}$. Finally, we note that in small accuracy ZetaRiemannSiegel is already best as soon as T is large enough to be accepted by the program; but gets slower than ZetaHurwitz as the accuracy increases while T remains (small and) fixed.

The second table that we give shows the dependence of the three programs with respect to the bit accuracy B which we choose equal to a power of 2, and where $\zeta(s)$ is computed for $s = 1/2 + 10^4 i$, $s = 1/2 + 10^6 i$, and $s = 1/2 + 10^8 i$ (the times for ZetaHurwitz are given after Bernoulli initialization):

B	Hur	RS	fast	Hur	RS	fast	Hur	RS	fast
2^6	0.01	**0.00**	0.01	0.17	**0.00**	0.04	11.7	**0.01**	0.41
2^7	**0.01**	0.01	0.01	0.23	**0.00**	0.07	16.2	**0.01**	0.71
2^8	**0.01**	0.06	0.02	0.38	**0.01**	0.15	24.7	**0.01**	1.47
2^9	**0.02**	0.77	0.04	0.73	**0.06**	0.37	47.3	**0.04**	3.71
2^{10}	**0.03**	$\otimes$	0.15	1.71	**0.50**	1.19	117.	**0.17**	12.5
2^{11}	**0.09**	$\otimes$	0.85	**4.96**	5.23	5.32	∞	**1.23**	52.9
2^{12}	**0.29**	$\otimes$	6.14	**16.4**	48.0	27.2	∞	**11.0**	299.
2^{13}	**1.03**	$\otimes$	314.	**56.1**	∞	133.	∞	**111.**	∞
2^{14}	**3.73**	$\otimes$	∞	**193.**	∞	∞	∞	∞	∞
2^{15}	**15.7**	$\otimes$	∞	∞	∞	∞	∞	∞	∞
2^{16}	**75.7**	$\otimes$	∞	∞	∞	∞	∞	∞	∞

Dependence on $B = 2^k$ at $T = 10^4$, 10^6, and 10^8.

The main conclusion from this table is that, with our tuning choices, the Riemann–Siegel algorithm scales badly with increasing accuracy, contrary to the zetafast algorithm, although the latter is again not competitive: either Riemann–Siegel (large T) or Euler–Maclaurin (large accuracies) turn out to be faster.

Note that the zetafast algorithm also applies to L-functions of Dirichlet characters. There exists an analogous Riemann–Siegel formula for such L-functions (see [**Sie43**]), but we have not found statements as precise as in [**AdR11**]. Recall that Euler–Maclaurin can also be applied to such L-functions, see the Lfunchisimple program. We thus give a small table giving the time to compute $L(\chi_{4 \cdot 10^k + 1}, 1/2 + Ti)$ with $T = 10^2$, 10^4, and 10^6 to 38 decimal digits, using Lfunchisimple and Zetafast, as well as the general Lfun program (applied to Lfunchicreate) given above (the built-in lfun program is only 20% faster):

k	Lfunchi	fast	Lfun	Lfunchi	fast	Lfun	Lfunchi	fast	Lfun
0	0.01	**0.00**	**0.00**	0.13	**0.02**	267.	12.2	**0.17**	∞
1	0.06	**0.01**	**0.01**	1.26	**0.06**	∞	122.	**0.56**	∞
2	0.64	**0.02**	0.03	12.6	**0.18**	∞	∞	**1.78**	∞
3	6.60	**0.08**	0.09	126.	**0.57**	∞	∞	**5.70**	∞
4	58.9	**0.21**	0.26	∞	**1.67**	∞	∞	**16.7**	∞
5	∞	**0.66**	0.81	∞	**5.26**	∞	∞	**53.2**	∞
6	∞	**2.36**	2.86	∞	**18.2**	∞	∞	**183.**	∞
7	∞	**7.50**	9.04	∞	**58.0**	∞	∞	∞	∞
8	∞	**24.1**	29.2	∞	**186.**	∞	∞	∞	∞
9	∞	**74.1**	89.6	∞	∞	∞	∞	∞	∞
10	∞	**221.**	269.	∞	∞	∞	∞	∞	∞

Dependence on $q = 4 \cdot 10^k + 1$ at $T = 10^2$, 10^4, and 10^6.

If $q = 4 \cdot 10^k + 1$ denotes the conductor of the character, we clearly see the approximate linear dependence in q for Lfunchisimple, and in $q^{1/2}$ for Zetafast and Lfun. On the other hand, we see that Lfun (as well as the built-in lfun) is not at all designed to handle large imaginary parts (even as small as 10^4), and in that case the `Zetafast` algorithm is by far the best choice, at least if we restrict to Dirichlet characters (as opposed to more general L-functions). A straightforward implementation of Siegel's generalization of the Riemann–Siegel formula to Dirichlet characters shows that it is much faster than `Zetafast`, but only for reasonably small conductors.

9.13. Explicit formulas

9.13.1. The basic theorem. Up to now, we have mainly considered weak L-functions, in other words we have not assumed that they have an Euler product (except in some examples above). In this section, we now assume that $L(s)$ is a true L-function, and in particular has an Euler product of the type described at the beginning of this chapter. In other words, for $\Re(s)$ sufficiently large, we have

$$L(s) = \prod_{p \in P} L_p(p^{-s})^{-1} \,,$$

where L_p is a polynomial of degree less than or equal to the degree d of the L-function, with additional properties given in Definition 9.1.2.

In a different direction, it is not difficult to prove that a weak L-function is a function of *order* 1, meaning that $|(s-k)^v L(s)| = O(e^{|s|^{1+\varepsilon}})$ for all $\varepsilon > 0$ as $|s| \to \infty$, where v is the order of the possible pole of L at $s = k$. An important theorem of complex analysis is that functions of finite order have *Hadamard products*. Here, since the order is 1 this product is particularly simple: we have

$$(s - k)^v L(s) = e^{as+b} \prod_{\rho,\ L(\rho)=0} \left(1 - \frac{s}{\rho}\right) e^{s/\rho} = e^{a's+b'} \prod_{\rho,\ L(\rho)=0} \left(1 - \frac{s}{\rho}\right) \,,$$

where a, b, a', and b' are constants, and in the second product it is understood that we take the limit as T tends to infinity of the product for $|\Im(\rho)| \le T$.

Thus, if L is a true L-function, we have two different representations of $L(s)$ as an infinite product. Taking logarithms and equating, this gives a relation between on the one hand a sum over prime numbers (in fact, as we will see, better expressed as a sum over prime powers), and on the other hand a sum over the zeros of $L(s)$. This is the first manifestation of the so-called "explicit formulas", and was the amazing insight of B. Riemann which ultimately led to the proof of the prime number theorem.

An additional essential ingredient is to introduce a suitable *test function* $F(x)$, so as to obtain a formula depending on F and a suitable transform (essentially the Fourier transform) of F. One can write the explicit formulas in many ways, but we will closely follow the presentation of J.-F. Mestre [**Mes86**] (generalizing G. Poitou's version of A. Weil's original results).

THEOREM 9.13.1. *Let $L(s)$ be an L-function of degree d, conductor N, with functional equation $s \mapsto k - s$, and keep all the notation that we have used up to now. On the other hand, let $F(x)$ be a function defined on $\mathbb{R}$ such that, for some $\varepsilon > 0$:*

(1) *$F(x)e^{(k-1/2+\varepsilon)x}$ is absolutely integrable on $\mathbb{R}$;*
(2) *$F(x)e^{(k-1/2+\varepsilon)x}$ has bounded variation and its value at any point is the average of its left and right limits;*
(3) *The function $(F(x) - F(0))/x$ has bounded variation.*

Define

$$\Phi(s) = \int_{-\infty}^{\infty} F(x)e^{(s-k/2)x}\, dx\ .$$

Then we have the equality

$$\sum_{\rho} \Phi(\rho) - \delta(\Phi(k) + \Phi(0)) = (\log(N) - d\log(\pi))F(0)$$

$$+ \int_0^{\infty} \left(dF(0)\frac{e^{-x}}{x} - \frac{F(x/2) + F(-x/2)}{2}\frac{e^{-kx/4}}{1 - e^{-x}}\sum_{1 \le j \le d} e^{-\alpha_j x/2} \right) dx$$

$$- \sum_{p,m,j} \frac{\log(p)}{p^{km/2}}\left(\alpha_{p,j}^m F(m\log(p)) + \overline{\alpha_{p,j}}^m F(-m\log(p)) \right)\ ,$$

where ρ runs through the zeros of Λ with $0 \le \Re(\rho) \le k$ repeated with their multiplicities, $\delta = 0$ if $\Lambda(s)$ has no pole, and $\delta = 1$ otherwise.

PROOF. The proof is an exercise in complex integration, but it still seems worthwhile to give it. First note that $\Phi(s)$ can be expressed in many different ways: for instance, $\Phi(k/2 - it)$ is the Fourier transform of F, and $\Phi(s)$ is the Mellin transform of the function $x^{-k/2}F(\log(x))$. We will keep the above normalization.

Consider the integral

$$I = \frac{1}{2\pi i}\int_{C_{a,T}} \Phi(s)\frac{\Lambda'(s)}{\Lambda(s)}\, ds\ ,$$

where $C_{a,T}$ is the rectangular contour boundary of $[-a, k+a] \times [-T, T]$ for some $a > 0$ and large T. On the one hand, by the residue theorem since $\Phi(s)$ has no poles and Λ has only possible simple poles at $s = k$ and $s = 0$, the contribution of the poles of $\Lambda(s)$ is $-(\Phi(k) + \Phi(0))\delta$, where $\delta = 0$ if $\Lambda(s)$ has no pole, otherwise $\delta = 1$. On the other hand, if ρ is a zero of $\Lambda(s)$ of multiplicity v, its contribution to I is $v\Phi(\rho)$. We have therefore

$$I = \sum_{\rho} \Phi(\rho) - (\Phi(k) + \Phi(0))\ ,$$

where by convention the sum on ρ is with multiplicity. This is the weak L-function part of the proof.

Now by our assumptions on Φ and the fact that $\Lambda(s)$ tends exponentially fast to 0 as $|\Im(s)| \to \infty$, we have

$$I = \int_{\Re(s)=k+a} \Phi(s)\frac{\Lambda'(s)}{\Lambda(s)}\, ds - \int_{\Re(s)=-a} \Phi(s)\frac{\Lambda'(s)}{\Lambda(s)}\, ds$$

$$= \int_{\Re(s)=k+a} \left(\Phi(s)\frac{\Lambda'(s)}{\Lambda(s)} + \Phi(k-s)\frac{\overline{\Lambda}'(s)}{\overline{\Lambda}(s)} \right) ds$$

using the functional equation of $\Lambda(s)$.

We now use the Euler product in the following form: we have for $\Re(s) > k$:

$$L(s) = \prod_{p \in P}\prod_{1 \le j \le d} (1 - \alpha_{p,j}p^{-s})^{-1}\ .$$

Thus,

$$\frac{L'(s)}{L(s)} = -\sum_{p \in P} \log(p) \sum_{1 \le j \le d} \sum_{m \ge 1} \frac{\alpha_{p,j}^m}{p^{ms}} \, ,$$

and

$$\frac{\Lambda'(s)}{\Lambda(s)} = \frac{\log(N)}{2} - \frac{d \log(\pi)}{2} + \frac{1}{2} \sum_{1 \le j \le d} \psi\left(\frac{s + \alpha_j}{2}\right) + \frac{L'(s)}{L(s)} \, .$$

It is not difficult to justify the fact that we can integrate term by term, so we look at each term separately.

(1) The constants: By Mellin inversion, $\frac{1}{2\pi i} \int_{\Re(s)=k+a} \Phi(s) \, ds$ is the inverse Mellin transform of $\Phi(s)$ at $x = 1$, and since $\Phi(s)$ is the Mellin transform of $x^{-k/2} F(\log(x))$, this is equal to $F(0)$.

(2) The terms p^{-ms}: similarly, $\frac{1}{2\pi i} \int_{\Re(s)=k+a} p^{-ms} \Phi(s) \, ds$ is the inverse Mellin transform of $\Phi(s)$ at $x = p^m$, hence is equal to $p^{-km/2} F(m \log(p))$.

(3) The terms $\psi((s + \alpha_j)/2)$: this is slightly more delicate. Recall the well-known formula

$$\psi(s) = \int_0^\infty \left(\frac{e^{-x}}{x} - \frac{e^{-sx}}{1 - e^{-x}}\right) dx \, .$$

Using (1) and (2) and an easy justification of the exchange of integration, we deduce that

$$\frac{1}{2\pi i} \int_{\Re(s)=k+a} \psi((s + \alpha_j)/2) \Phi(s) \, ds = \int_0^\infty \left(F(0)\frac{e^{-x}}{x} - F(x/2)\frac{e^{-(\alpha_j+k/2)x/2}}{1 - e^{-x}}\right) .$$

For the other integral involving $\Phi(k - s)$, we see by an immediate change of variable that $\Phi(k - s)$ is the Mellin transform of $x^{-k/2} F(-\log(x))$, so in the formulas that we have obtained above we simply replace $F(X)$ by $F(-X)$. Putting everything together, we obtain

$$I = (\log(N) - d\log(\pi))F(0)$$

$$+ \int_0^\infty \left(dF(0)\frac{e^{-x}}{x} - \frac{F(x/2) + F(-x/2)}{2} \frac{e^{-kx/4}}{1 - e^{-x}} \sum_{1 \le j \le d} e^{-\alpha_j x/2}\right) dx$$

$$- \sum_{p,m,j} \frac{\log(p)}{p^{km/2}} (\alpha_{p,j}^m F(m \log(p)) + \overline{\alpha_{p,j}}^m F(-m \log(p))) \, ,$$

proving the theorem. $\square$

9.13.2. Application I: Sums over nontrivial zeros. There are many important applications of the explicit formula, and we give three, of which the first uses it only implicitly.

PROPOSITION 9.13.2. *Let $L(s)$ be a weak L-function, and keep all the usual notation. Let*

$$(s(k - s))^v \Lambda(s) = ae^{bs} \prod_\rho \left(1 - \frac{s}{\rho}\right)$$

be the Hadamard product of $\Lambda(s)$, where $v = 0$ or 1 is the order of its pole at $s = k$, and where it is understood that the product is the limit as $T \to \infty$ of the product for $|\Im(\rho)| \le T$.

(1) *We have*

$$a = \begin{cases} w\overline{\Lambda(k)} & \textit{if } \Lambda(s) \textit{ has no pole} \\ -wk\overline{R} & \textit{if } \Lambda(s) \textit{ has a simple pole of residue } R \ . \end{cases}$$

(2) b *is purely imaginary, and if $L(s)$ is self-dual, $b = 0$.*

(3) *If $L(s)$ is a true L-function self-dual or not, we always have $b = 0$.*

PROOF. (1). We could prove this proposition by using the explicit formula with $F(x) = (e^{-ax} - 1)e^{(k/2)x}$ when $x < 0$, and $F(x) = 0$ when $x \geq 0$, and this is left as an excellent exercise for the reader. However, in this simple case we prefer working directly with the Hadamard product.

Since $L(s)$ is a weak L-function, it has order 1, so it has a Hadamard product as above. We could also write an absolutely convergent product $\prod_\rho (1 - s/\rho)e^{s/\rho}$, changing the constants a and b, but it is more natural to write it in this way

If we set $s = 0$ in the Hadamard product, the right-hand side is equal to a. If $\Lambda(s)$ has no pole, the left-hand side is equal to $\Lambda(0) = w\overline{\Lambda(k)}$ by the functional equation; otherwise the left-hand side is equal to $k\operatorname{Res}_{s=0}\Lambda(s) = -kw\overline{R}$. We have just proved (1).

For (2), note that the functional equation implies that if $\Lambda(\rho) = 0$ we have $\Lambda(k - \overline{\rho}) = w\overline{\Lambda(\rho)} = 0$, so zeros come in pairs $(\rho, k - \overline{\rho})$ (note that the generalized Riemann hypothesis, for L-functions, would imply that $k - \overline{\rho} = \rho$).

When s is changed into $k - s$, the left-hand side is changed into

$$(s(k - s))^v w\overline{\Lambda(\overline{s})} = w\overline{a}e^{\overline{b}s}\prod_\rho(1 - s/\overline{\rho}) = w\overline{a}e^{\overline{b}s}\prod_\rho(1 - s/(k - \rho)) \ ,$$

and the right-hand side becomes

$$ae^{b(k-s)}\prod_\rho(1 - (k - s)/\rho) \ .$$

Computing the quotient of this latter expression by the former, we obtain

$$1 = \frac{a}{w\overline{a}}e^{b(k-s)-\overline{b}s}\prod_\rho\frac{(\rho - k + s)/\rho}{(k - \rho - s)/(k - \rho)} = \frac{a}{w\overline{a}}e^{bk-s(b+\overline{b})}\prod_\rho\left(1 - \frac{k}{\rho}\right) \ .$$

It follows that the coefficient of s in the exponential must vanish, so that $b + \overline{b} = 0$, in other words b is purely imaginary.

If, in addition, $L(s)$ is self-dual, we have $\overline{\Lambda(\overline{s})} = \Lambda(s)$, which implies that its zeros come in pairs $(\rho, \overline{\rho})$ (or a single value if ρ is real). Thus, changing s into $\overline{s}$ and conjugating, we deduce that $ae^{bs} = \overline{a}e^{\overline{b}s}$, in other words that a and b are real. Since b is also purely imaginary, it must vanish.

(3). This is more subtle, and apparently not so well-known. We heartily thank "Lucia" from the MathOverflow forum for showing us this proof, which we only sketch.

Denote by $N^+(T)$ (resp. $N^-(T)$) the number of zeros of $\Lambda(s)$ with $0 < \Im(s) \leq T$ (resp $-T \leq \Im(s) < 0$). By the principle of the argument from complex analysis, since $L(s)$ is a true L-function it is easy to show that as $T \to +\infty$ we have

$$N^\pm(T) = \frac{T}{2\pi}\log\left(N\left(\frac{T}{2\pi e}\right)^d\right) + O(\log(N(T + 1))^d) \ ,$$

where as usual d is the degree of the L-function and N its conductor. This implies first that $N^+(T) - N^-(T) = O(\log(N(T+1))^d)$, and second that the number of zeros of imaginary part at height around T in an interval of length 1, say, is also $O(\log(N(|T|+1))^d)$.

We can now prove that $b = 0$: let R be some large real number. From the Hadamard product we have

$$(\Lambda'/\Lambda)(R) = b + \lim_{T \to \infty} \sum_{|\Im(\rho)| \leq T} 1/(R - \rho),$$

and since we know that $\Re(b) = 0$ we only look at imaginary parts. We have trivially $\Im((L'/L)(R)) = O(2^{-R})$ which tends exponentially to 0 when $R \to \infty$. Since we have assumed that the α_j occurring in the gamma factors are real, the imaginary part of the other components of $(\Lambda'/\Lambda)(R)$ vanish. We leave to the reader to check that even if some α_j is complex, $\Im((\Lambda'/\Lambda)(R))$ will still tend to 0, although not exponentially.

It is thus sufficient to consider the sum over the zeros ρ. Write $\rho = \beta + i\gamma$ with β and γ real. Since the zeros are all in the critical strip we have $0 \leq \beta \leq k$, so

$$\Im(1/(R - \rho)) = \gamma/((R - \beta)^2 + \gamma^2) = \gamma/(R^2 + \gamma^2) + O(\gamma R/((R^2 + \gamma^2)^2)) \ .$$

We consider the sum of the main term and the error term separately. Consider first the error term. Separating the zeros ρ with $|\gamma| \leq R$ and $|\gamma| > R$, and using the result recalled above about the number of zeros in an interval of length 1 we immediately obtain $\sum_{|\gamma| \leq T} |\gamma| R/(R^2 + \gamma^2)^2 = O(\log(NR)/R)$.

The remaining sum is that of the main terms, i.e., $\sum_{|\gamma| \leq T} \gamma/(R^2 + \gamma^2)$, and the result on $N^\pm(T)$ recalled above implies by an easy partial summation left to the reader that $\sum_{|\gamma| \leq T} \gamma/(R^2 + \gamma^2) = O(\log(NR)/R)$ once again, proving (3) since everything tends to 0 as $R \to \infty$. □

DEFINITION 9.13.3. We denote by $[n]f_{s=a}$ the nth coefficient of the power series expansion of a function f at $s = a$, in other words $[n]f_{s=a} = f^{(n)}(a)/n!$.

COROLLARY 9.13.4.

(1) *With the same convention as in the proposition, we have*

$$\sum_\rho \frac{1}{s - \rho} = -b + v\left(\frac{1}{s} + \frac{1}{s - k}\right) + \frac{\Lambda'(s)}{\Lambda(s)} \ .$$

(2) *In particular, for all $m \geq 1$ we have*

$$\sum_\rho \frac{1}{\rho^m} = b\delta_{m,1} + \frac{v}{k^m} + (-1)^{m-1}[m-1]\overline{\left(\frac{\Lambda'}{\Lambda}\right)}_{s=k} ,$$

where δ is the Kronecker symbol, and where we recall from the proposition that $b = 0$ when L is a true L-function.

PROOF. (1). Simply take the logarithmic derivative of both sides of the proposition.

For (2), we expand both sides around $s = 0$: since $\Lambda(s)$ has a pole of order v the terms in $1/s$ disappear, while the other terms give the equality

$$-\sum_\rho 1/\rho^m = -b\delta_{m,1} - v/k^m + [m-1](\Lambda'/\Lambda)_{s=0} \ .$$

Expanding $(\Lambda'/\Lambda)(s) = \sum_{m \geq -1} a_m s^m$ around 0, it follows that $(\Lambda'/\Lambda)(k-s) = \sum_{m \geq -1} (-1)^m a_m (s-k)^m$, hence $\overline{(\Lambda'/\Lambda)}(s) = \sum_{m \geq -1} (-1)^{m+1} a_m (s-k)^m$ by the functional equation, so

$$[m-1](\Lambda'/\Lambda)_{s=0} = (-1)^m [m-1]\overline{(\Lambda'/\Lambda)}_{s=k} \ ,$$

proving the corollary.   □

In the same way it is immediate to prove the following:

COROLLARY 9.13.5. *Assume that L is self-dual and satisfies the Generalized Riemann Hypothesis, in other words that the zeros of $\Lambda(s)$ are all of the form $k/2 + i\gamma$ with $\gamma \in \mathbb{R}$.*

(1) *For $m \geq 2$ even, we have*

$$\sum_{\gamma} \frac{1}{\gamma^m} = (-1)^{m/2}\left(\frac{2v}{(k/2)^m} - \left([m-1]\frac{\Lambda'}{\Lambda}\right)_{s=k/2}\right) \ .$$

(2) *We have*

$$\sum_{\gamma} \frac{1}{k^2/4 + \gamma^2} = \frac{2v}{k} + 2\left([0]\frac{\Lambda'}{\Lambda}\right)_{s=k} \ .$$

Note that in the above, we sum over all γ, positive, negative, or 0, since there may be a zero at $s = k/2$.

Using the above corollaries, the functional equation, and standard properties of the gamma function and its logarithmic derivatives, it is easy to specialize to any classical L-function. The proofs of the following results for the Riemann zeta function are left as exercises for the reader.

PROPOSITION 9.13.6. *Denote by $\rho = \beta + i\gamma$ the nontrivial zeros of $\zeta(s)$, and define the Stieltjes constants γ_m by $\zeta(s) = 1/(s-1) + \sum_{m \geq 0} (-1)^m (\gamma_m/m!)(s-1)^m$, so that for instance $\gamma_0 = \gamma$.*

(1) *We have*

$$\sum_{\rho} \frac{1}{\rho} = \frac{1}{2}(2 + \gamma - \log(4\pi)) \ , \quad \sum_{\rho} \frac{1}{\rho^2} = 1 - \frac{\pi^2}{8} + 2\gamma_1 + \gamma^2 \ ,$$

and more generally $\displaystyle\sum_{\rho} \frac{1}{\rho^k} = 1 - (1 - 1/2^k)\zeta(k) + \delta_k \ ,$

where the δ_k are defined by induction by

$$\delta_{k+1} = (k+1)\frac{\gamma_k}{k!} + \sum_{j=0}^{k-1} \frac{\gamma_j \delta_{k-j}}{j!} \ .$$

(2) *We have*

$$\prod_{\rho}\left(1 - \frac{1}{\rho^2}\right) = \frac{\pi}{3} \ , \quad \prod_{\rho}\left(1 - \frac{4}{\rho^2}\right) = \zeta(3) \ , \quad \text{and more generally}$$

$$\prod_{\rho}\left(1 - \frac{m^2}{\rho^2}\right) = \begin{cases} (-1)^{(m-1)/2} 2m(m-1)\pi B_{m+1}\zeta(m) & \text{when m is odd} \ , \\ (-1)^{m/2-1} m(m^2-1) B_m \zeta(m+1) & \text{when m is even} \ , \end{cases}$$

where $(m-1)\zeta(m)$ is to be interpreted as 1 for $m = 1$.

(3) *Assume the Riemann hypothesis. We have*

$$\sum_{\gamma>0} \frac{1}{\gamma^2} = \frac{1}{2}\left([1]\frac{\zeta'}{\zeta}\right)_{s=1/2} + \frac{\pi^2}{8} + G - 4 , \quad \text{and more generally}$$

$$\sum_{\gamma>0} \frac{1}{\gamma^m} = (-1)^{m/2-1}\left(\frac{1}{2}\left([m-1]\frac{\zeta'}{\zeta}\right)_{s=1/2} + (2^m-1)\frac{\zeta(m)}{4}\right.$$
$$\left. + 2^{m-2}L(\chi_{-4},m) - 2^m\right) \quad \text{for } m \geq 2 \text{ even .}$$

(4) *Assume the Riemann hypothesis. We have*

$$\sum_{\gamma>0} \frac{1}{(1/4+\gamma^2)^k} = \sum_{j=1}^{k} \binom{2k-j-1}{k-1} S_j ,$$

where $S_j = \sum_\rho 1/\rho^j$ is given by (1).

(5) *We have*

$$\frac{\zeta'(1/2)}{\zeta(1/2)} = \left([0]\frac{\zeta'}{\zeta}\right)_{s=1/2} = \frac{1}{2}\left(\log(8\pi) + \gamma + \frac{\pi}{2}\right) ,$$

$$\left([2]\frac{\zeta'}{\zeta}\right)_{s=1/2} = \frac{7}{2}\zeta(3) + \frac{\pi^3}{8} , \quad \text{and more generally}$$

$$\left([m]\frac{\zeta'}{\zeta}\right)_{s=1/2} = \left(2^m - \frac{1}{2}\right)\zeta(m+1) + \frac{(-1)^{m/2}}{4}\frac{E_m \pi^{m+1}}{m!}$$

for $m \geq 2$ even, where $E_0 = 1$, $E_2 = -1$, $E_4 = 5$, $E_6 = -61,\dots$ are the Euler numbers. Note that these correspond to odd values of m in Corollary 9.13.5 (1), for which the left-hand side trivially vanishes.

Much more difficult (and as far as the authors are aware, unsolved), are similar problems when we only sum over the zeros of positive imaginary part (except of course for even functions of γ such as (3) above). Assume that L is self-dual, for instance that L is the Riemann zeta function. The zeros then come in pairs $(\rho, \overline{\rho})$. Corollary 9.13.4 gives us the value of

$$S_m = \sum_\rho 1/\rho^m = \sum_{\Im(\rho)>0} (1/\rho^m + 1/\overline{\rho}^m) = 2\sum_{\Im(\rho)>0} \Re(1/\rho^m) .$$

But how do we compute (to hundreds of decimals, say) the corresponding *imaginary* part, or equivalently $\sum_{\Im(\rho)>0} 1/\rho^m$?

Similarly, assume the Riemann hypothesis for our L-function, so write $\rho = k/2 + i\gamma$. Corollary 9.13.5 gives us the value of $\sum_\gamma 1/\gamma^m$ for m even, hence also of $\sum_{\gamma>0} 1/\gamma^m$ since zeros come in pairs $(k/2+i\gamma, k/2-i\gamma)$. But since $\sum_\gamma 1/\gamma^m = 0$ for m odd, how do we compute $\sum_{\gamma>0} 1/\gamma^m$ when m is odd ?

For both these problems, using some clever manipulations and the knowledge of 10^{10} zeros of the Riemann zeta function, one can give approximately 28 decimals for these problems, see for instance [BPT20], but as far as the authors are aware, it is completely out of the question to compute 100 decimals, say. We would of course be happy to be contradicted on this subject.

9.13.3. Application II: $b = 0$ for true L-functions. Using the explicit formula we can give an alternative proof of the fact that $b = 0$ for true L-functions as in Proposition 9.13.2, at least for L-functions of Dirichlet characters, although it should be easy to generalize to L-functions satisfying the above assumptions.

Let us choose $F(x) = e^{kx/2}$ for $x < 0$, $F(x) = 0$ for $x > 0$, and since F is a Fourier transform, $F(0) = (F(0^+) + F(0^-))/2 = 1/2$. It is clear that F satisfies the necessary assumptions, and that $\Phi(s) = 1/s$, which is what we want. Thus the explicit formula gives us another expression for $\sum_\rho 1/\rho$, where as usual this is understood as the limit as $T \to \infty$ of the sum for $|\Im(\rho)| \le T$.

To fix ideas, assume that $L(s) = L(\chi, s)$, where χ is an odd primitive character modulo N (the proof is essentially the same for even characters). We thus have $k = d = 1$, $\alpha_1 = 1$, $\alpha_{p,j} = \chi(p)$. The integral term is

$$\frac{1}{2} \int_0^\infty \left(\frac{e^{-x}}{x} - \frac{e^{-x}}{1 - e^{-x}} \right) dx$$

which is equal to $-\gamma/2$, so the explicit formula gives us

$$\sum_\rho \frac{1}{\rho} = \frac{1}{2}(\log(N/\pi) - \gamma) - S$$

with

$$S = \sum_{p \in P} \log(p) \sum_{m \ge 1} \frac{\overline{\chi(p)}^m}{p^m} = \sum_{p \in P} \log(p) \frac{\overline{\chi(p)}}{p - \overline{\chi(p)}} .$$

Now by differentiation, it is immediate to see that $S = -(L'/L)(\overline{\chi}, 1)$, so that

$$\sum_\rho \frac{1}{\rho} = \frac{1}{2}(\log(N/\pi) - \gamma) + \frac{L'(\overline{\chi}, 1)}{L(\overline{\chi}, 1)} .$$

Thus, by Corollary 9.13.4, since we assume that χ is an odd character we have $\Lambda(\overline{\chi}, s) = N^{s/2} \pi^{-(s+1)/2} \Gamma((s+1)/2) L(\overline{\chi}, s)$, so

$$(\Lambda'/\Lambda)(\overline{\chi}, 1) = (\log(N/\pi) - \gamma)/2 + (L'/L)(\overline{\chi}, 1) ,$$

and we deduce from the corollary that $b = 0$, as desired.

9.13.4. Application III: Discriminant bounds. Perhaps the most famous application of the explicit formula is to discriminant bounds of number fields, due to Stark, Odlyzko, Poitou, Serre, Diaz y Diaz, etc. Such a theorem is as follows:

THEOREM 9.13.7. *Let K be a number field of degree n, signature (r_1, r_2) with $r_1 + 2r_2 = n$, and discriminant $D(K)$. Assume the Generalized Riemann Hypothesis, i.e., that the real part of the nontrivial zeros ρ of $\zeta_K(s)$ satisfy $\Re(\rho) = 1/2$. Then as $n \to \infty$ we have*

$$\log(|D_K|) > (\log(8\pi) + \gamma - o(1))n + (\pi/2)r_1 .$$

Note that there exist versions of this theorem with worse constants if we do not assume the GRH, and that the $o(1)$ term can be made explicit, both asymptotically, and for given number fields. We refer to the abundant literature on the subject for details.

PROOF. We give two proofs. One is a short proof, due to N. Elkies, which does not use the full force of the explicit formula but only the Hadamard product. The second is the standard proof, due in essence to H. Stark but optimized by A. Odlyzko.

Elkies's proof goes as follows. We begin by the following identity:

PROPOSITION 9.13.8. *Let K be a number field, $D(K)$ its discriminant, (r_1, r_2) its signature (number of real embeddings and half the number of complex embeddings). For all* real *s, we have*

$$\log(|D(K)|) = r_1(\log(\pi) - \psi(s/2)) + 2r_2(\log(2\pi) - \psi(s))$$
$$- 2\frac{\zeta_K'(s)}{\zeta_K(s)} - \frac{2}{s} - \frac{2}{s-1} + 2\sum_\rho \Re\left(\frac{1}{s-\rho}\right) ,$$

where ζ_K is the Dedekind zeta function of K and ρ ranges among all the nontrivial zeros of ζ_K.

PROOF. We know that

$$\Lambda_K(s) = |D(K)|^{s/2}\Gamma_\mathbb{R}(s)^{r_1+r_2}\Gamma_\mathbb{R}(s+1)^{r_2}\zeta_K(s)$$
$$= |D(K)|^{s/2}\Gamma_\mathbb{R}(s)^{r_1}\Gamma_\mathbb{C}(s)^{r_2}\zeta_K(s)$$
$$= |D(K)|^{s/2}2^{r_2}\pi^{-r_1s/2}(2\pi)^{-r_2s}\Gamma(s/2)^{r_1}\Gamma(s)^{r_2}\zeta_K(s) .$$

It follows that

$$2\frac{\Lambda_K'(s)}{\Lambda_K(s)} = \log(|D(K)|) - (r_1\log(\pi) + 2r_2\log(2\pi)) + r_1\psi(s/2) + 2r_2\psi(s) + 2\frac{\zeta_K'(s)}{\zeta_K(s)} ,$$

and since $\zeta_K(s)$ is self-dual and has a simple pole, we have $b = 0$ and $v = 1$, so that

$$\sum_\rho \frac{1}{s-\rho} = \frac{1}{s} + \frac{1}{s-1} + \frac{\Lambda_K'(s)}{\Lambda_K(s)}$$

as given above, proving the proposition since when s is real, all the terms are real except possibly the sum on ρ, so we can replace it by its real part. □

Using this proposition, we first give a wrong proof of the theorem, and then modify this wrong proof to obtain a correct one.

Since for $s > 1$ we have $\zeta_K(s) = \prod_p \prod_{\mathfrak{p}|p}(1 - p^{-f(\mathfrak{p}/p)s})^{-1}$, it follows that

$$-\frac{\zeta_K'(s)}{\zeta_K(s)} = \sum_p \sum_{\mathfrak{p}|p} \sum_{m\geq 1} f(\mathfrak{p}/p)\log(p)p^{-mf(\mathfrak{p}/p)s} .$$

Thus, if $s > 1$ is real, $-\zeta_K'(s)/\zeta_K(s)$ as well as all its derivatives of even order are positive, and those of odd order are negative, since this is the case for each individual $p^{-mf(\mathfrak{p}/p)s}$.

Now assume that this is not only true for $s > 1$ but also for $s = 1/2$, which for now is of course not justified. We thus have

$$\log(|D(K)|) \geq r_1(\log(\pi) - \psi(1/4)) + 2r_2(\log(2\pi) - \psi(1/2))$$
$$- 4 + 4 + 2\sum_\rho \Re(1/(1/2 - \rho)) .$$

Since we assume GRH we have $1/2 - \rho = -it$ for some real t, whose real part vanishes. We now use the known values $\psi(1/2) = -(2\log(2) + \gamma)$ and $\psi(1/4) = -(3\log(2) + \gamma + \pi/2)$, giving

$$\log(|D(K)|) \geq r_1(\log(8\pi) + \gamma + \pi/2) + (n - r_1)(\log(8\pi) + \gamma)$$
$$= n(\log(8\pi) + \gamma) + r_1\pi/2 ,$$

"proving" the theorem. At least, we now see where the strange constants occurring in the theorem come from.

We now give Elkies's rigorous proof of the theorem. We begin in the same way, noting that if $s > 1$ is real, $-\zeta_K'(s)/\zeta_K(s)$ as well as all its derivatives of even order are positive, and those of odd order are negative. We still want to push this to $s = 1/2$. For this, we want to use Taylor's formula in the form

$$f(1/2 + \varepsilon) = \sum_{m \geq 0} ((-1)^m/2^m) f^{(m)}(1 + \varepsilon)/m! ,$$

as long as this series converges absolutely.

We thus differentiate m times and multiply by $(-1)^m$ the formula of the proposition, using the known sign of the derivatives of $\zeta_K'(s)/\zeta_K(s)$. Using the Kronecker symbol $\delta_{m,0}$, we obtain

$$\delta_{m,0} \log(|D(K)|) > (-1)^m \left(r_1 \frac{d^m}{ds^m}(\log(\pi) - \psi(s/2)) + 2r_2 \frac{d^m}{ds^m}(\log(2\pi) - \psi(s)) \right.$$
$$\left. + 2 \cdot m! \left(\Re\left(\frac{1}{(s-\rho)^{m+1}}\right) - \frac{1}{s^{m+1}} - \frac{1}{(s-1)^{m+1}} \right) \right) .$$

We would now like to sum this divided by $2^m m!$ for all $m \geq 0$, to obtain the expansion around $s - 1/2$, but we have to be careful. Set $s_0 = 1 + \varepsilon$ with ε small, and let $s = s_0 - 1/2 = 1/2 + \varepsilon$. We will choose a large integer M satisfying the following two conditions:

(1) The sum of the first M terms of the Taylor expansion of $\psi(s_0 - 1/2)$ and $\psi(s_0/2 - 1/4)$ is less than ε from their value. This is possible since the radii of convergence of $\psi(x)$ and of $\psi(x/2)$ around $x = 1$ are both exactly equal to $1 > 1/2$.

(2) The sum of the first M terms of the Taylor expansion of $\Re(1/(s-1/2-\rho))$ around s_0 at $s = s_0 - 1/2$ is positive for all complex numbers ρ of real part $1/2$. This is indeed possible: write $\rho = 1/2 + it$ with t real, so that

$$\Re(1/(s_0 - 1/2 - \rho)) = \Re(1/(\varepsilon - it)) = \varepsilon/(\varepsilon^2 + t^2) .$$

The Taylor expansion around $s = s_0$ without the real part is

$$1/((s - s_0) + s_0 - 1 - it) = (1/(s_0 - 1 - it))/(1 + (s - s_0)/(s_0 - 1 - it))$$
$$= \sum_{m \geq 0} (s_0 - s)^m/(s_0 - 1 - it)^{m+1} ,$$

so the sum of the terms for $m > M$ is equal to $R_M = ((s_0 - s)/(s_0 - 1 - it))^M/(s - 1 - it)$, so

$$\Re(R_M) \leq |R_M| = |s_0 - s|^M/(\varepsilon^2 + t^2)^{M/2}/((s-1)^2 + t^2)^{1/2} .$$

and for $s = s_0 - 1/2$ this gives

$$\Re(R_M) = 2^{-M}/(\varepsilon^2 + t^2)^{M/2}/((\varepsilon - 1/2)^2 + t^2)^{1/2} ,$$

and this is clearly less that $\varepsilon/(\varepsilon^2 + t^2)$ uniformly in t for M sufficiently large, so the sum of the first M terms is indeed positive, as claimed.

We thus sum from $m = 0$ to M the formula that we have given above divided by $2^m m!$, which corresponds to the Taylor formula at $s = s_0 - 1/2$, and we obtain

$$\log(|D(K)|) > r_1(\log(\pi) - \psi(1/4) - \varepsilon) + 2r_2(\log(2\pi) - \psi(1/2) - \varepsilon) - A(M, \varepsilon)$$
$$= n(\log(8\pi) + \gamma - \varepsilon) + r_1\pi/2 - A(M, \varepsilon)$$

for some (possibly large) constant $A(M, \varepsilon)$ depending on M and ε. Thus, if we want $\log(|D(K)|) > n(\log(8\pi) + \gamma - \eta) + r_1\pi/2$, we choose $\varepsilon = \eta/2$, and this inequality will be valid for $n > A(M, \varepsilon)/(\eta/2)$, proving the theorem.

We now give the standard proof. Consider the function

$$F(x) = \begin{cases} (1 - |x|)\cos(\pi x) + \sin(\pi|x|)/\pi & \text{for } x \in [-1, 1] , \\ F(x) = 0 & \text{otherwise .} \end{cases}$$

An uninteresting computation shows that

$$\int_{-\infty}^{\infty} F(x)e^{(s-1/2)x}\,dx = 4\pi^2 \frac{1 + \cosh(s - 1/2)}{((s - 1/2)^2 + \pi^2)^2} .$$

Fix some positive λ, and apply the explicit formula to the function $F(x/\lambda)$. We have

$$\Phi(s) = 4\pi^2\lambda \frac{1 + \cosh(\lambda(s - 1/2))}{(\lambda^2(s - 1/2)^2 + \pi^2)^2} .$$

Since we assume the Generalized Riemann Hypothesis, if ρ is a nontrivial zero of $\zeta_K(s)$ we have $\rho = 1/2 + i\gamma$ for some real number γ, so

$$\Phi(\rho) = 4\pi^2\lambda(1 + \cos(\lambda\gamma))/(\pi^2 - \lambda^2\gamma^2)^2 \geq 0$$

for all ρ. Thus, the left hand side of the explicit formula given in Theorem 9.13.1 is greater than or equal to

$$-(\Phi(1) + \Phi(0)) = -8\pi^2\lambda(1 + \cosh(\lambda/2))/(\pi^2 + \lambda^2/4)^2 .$$

Let us now consider the infinite sum occurring on the right-hand side. Note that the Euler factors of the Dedekind zeta function are all of the form $(1 - p^{-fs})^{-1}$ for some $f \geq 1$, so $\alpha_{p,j}$ ranges through all fth roots of unity. Now $\sum_{0 \leq i < f} \zeta_f^m = 0$ if $f \nmid m$, and is equal to f otherwise, but in any case is a nonnegative real number. Since, as is easily checked, the function F is also nonnegative, it follows that the sum on p, m, j is also nonnegative, which is in the correct direction for the inequality we are aiming for, in other words we have

$$\log(|D(K)|) \geq n\log(\pi) - C(\lambda) - I(\lambda) ,$$

where $C(\lambda) = 8\pi^2\lambda(1 + \cosh(\lambda/2))/(\pi^2 + \lambda^2/4)^2$ depends only on λ, and $I(\lambda)$ is the integral

$$I(\lambda) = \int_0^{\infty} \left(n\frac{e^{-x}}{x} - F(x/\lambda)\frac{e^{-x/4}}{1 - e^{-x}}(r_1 + r_2 + r_2 e^{-x/2})\right) dx$$

since the gamma factor of $\zeta_K(s)$ is equal to $\Gamma_{\mathbb{R}}(s)^{r_1+r_2}\Gamma_{\mathbb{R}}(s + 1)^{r_2}$.

Now it is not difficult to show that, as $\lambda \to \infty$, the integral $I(\lambda)$ tends to the same integral with $F(x/\lambda)$ replaced by the function 1, i.e., to

$$J = \int_0^\infty \left(n\frac{e^{-x}}{x} - \frac{e^{-x/4}}{1-e^{-x}}(r_1 + r_2 + r_2 e^{-x/2}) \right) dx$$

$$= (r_1 + r_2) \int_0^\infty \left(\frac{e^{-x}}{x} - \frac{e^{-x/4}}{1-e^{-x}} \right) dx + r_2 \int_0^\infty \left(\frac{e^{-x}}{x} - \frac{e^{-3x/4}}{1-e^{-x}} \right) dx$$

$$= (r_1 + r_2)\psi(1/4) + r_2\psi(3/4) ,$$

since

$$\psi(a) = \int_0^\infty \left(\frac{e^{-x}}{x} - \frac{e^{-xa}}{1-e^{-x}} \right) dx .$$

It follows that $I(\lambda) = (r_1 + r_2)\psi(1/4) + r_2\psi(3/4) + n\varepsilon(\lambda)$, where $\varepsilon(\lambda)$ tends to 0 when $\lambda \to \infty$, so that

$$\log(|D(K)|) \geq n\log(\pi) - (r_1 + r_2)\psi(1/4) - r_2\psi(3/4) - n\varepsilon(\lambda) - C(\lambda) .$$

Since $\psi(3/4) = \psi(1/4) + \pi$, we have as in the theorem

$$n\log(\pi) - (r_1 + r_2)\psi(1/4) - r_2\psi(3/4) = n(\log(\pi) - \psi(1/4)) - r_2\pi$$

$$= n(\log(8\pi) + \gamma + \pi/2) - (n - r_1)\pi/2$$

$$= n(\log(8\pi) + \gamma) + r_1\pi/2 ,$$

proving the theorem as with Elkies's proof.　　　　□

APPENDIX A

List of relevant GP programs

In this appendix, we give a list and short description of the built-in `Pari/GP` programs corresponding or related to the scripts given in the main part of the book, referring to the online help manual for detailed explanations of the commands. An (init) following a command simply means that there exists a corresponding initialization program.

- `asympnum`: asymptotic expansion, assumed rational.
- `asympnumraw`: asymptotic expansion, no rationality assumption.
- `bernfrac, bernpol, bernreal, bernvec`: Bernoulli numbers and polynomials.
- `besselj, besselk, besseli, besseln`: Bessel functions.
- `contfrac`: continued fraction expansion of a real number.
- `contfraceval`: evaluate a continued fraction in Euler form.
- `contfracinit`: transform a power series into a continued fraction in Euler form, for use with `contfraceval`.
- `contfracpnqn`: partial quotients of a continued fraction.
- `derivnum`: numerical derivatives.
- `dirpowers`: vector of n^s from $n = 1$ to N.
- `dirpowerssum`: sum from $n = 1$ to N of n^s.
- `eulerfrac, eulerpol, eulerreal, eulervec`: Euler numbers and polynomials.
- `gamma`: complex and p-adic gamma function.
- `gammamellininv(init)`: inverse Mellin transform.
- `gammamellininvasymp`: asymptotic expansion at infinity of inverse Mellin transform.
- `getlocalbitprec`: local bit precision.
- `hypergeom`: general hypergeometric function.
- `incgam, incgamc`: incomplete gamma functions.
- `intcirc`: curvilinear integral around a circle.
- `intnum(init)`: numerical integration using double-exponential integration.
- `intnumgauss(init)`: numerical integration using Gauss–Legendre.
- `intnumromb`: numerical integration using Romberg.
- `lambertw`: Lambert W-function.
- `lfun(init)`: L-function numerical computation.
- `lfuncheckfeq`: using functional equation, check if L-function data is plausible.
- `lfunconductor`: find conductor of L-function.
- `lfuncreate`: general constructor for L-functions.
- `lfunhardy`: Hardy Z-function corresponding to a given L-function.

- lfunlambda: completed L-function with factors at infinity.
- lfunartin, lfunetaquo, lfunmf, lfunqf: L-functions associated to an Artin L-function, an eta quotient, a modular form, a quadratic form.
- lfunorderzero: order of zero at the center of the critical strip.
- lfunrootres: root number and residues at poles.
- lfuntheta(init): theta function associated to an L-function.
- lfunzeros: zeros of an L-function on the critical strip.
- limitnum: limit of a sequence.
- lngamma: log gamma function.
- localbitprec: set local bit precision.
- polchebyshev, polhermite, pollaguerre, pollegendre: corresponding orthogonal polynomials.
- polroots, polrootsreal: complex or real roots of a polynomial.
- polylogmult: multiple polylogarithms.
- prodeulerrat: infinite Euler product of rational function at p^{-s}.
- prodnumrat: infinite product of rational function.
- psi: logarithmic derivative of the gamma function.
- quodif (must be installed with install(quodif,GL)): quotient-difference algorithm.
- sumalt: numerical summation of alternating series.
- sumeulerrat: infinite Euler sum of rational function at p^{-s}.
- suminf: naïve infinite summation of rapidly convergent series.
- sumnum(init): numerical summation using discrete Euler–Maclaurin.
- sumnumap(init): numerical summation using Abel–Plana.
- sumnumlagrange(init): numerical summation using Lagrange extrapolation.
- sumnummonien(init): numerical summation using Monien.
- sumnumrat: infinite sum of rational function.
- sumpos: numerical summation using Van Wijngaarden.
- zeta: Riemann zeta function.
- zetahurwitz: Hurwitz zeta function.
- zetamult, zetamultall, zetamultconvert, zetamultdual: Multiple Zeta Value functions.

Bibliography

[AdR11]   Juan Arias de Reyna, *High precision computation of Riemann's zeta function by the Riemann-Siegel formula, I*, Math. Comp. **80** (2011), no. 274, 995–1009, DOI 10.1090/S0025-5718-2010-02426-3. MR2772105

[Akh17]   P. Akhilesh, *Double tails of multiple zeta values*, J. Number Theory **170** (2017), 228–249, DOI 10.1016/j.jnt.2016.06.020. MR3541706

[BCM15]   Frits Beukers, Henri Cohen, and Anton Mellit, *Finite hypergeometric functions*, Pure Appl. Math. Q. **11** (2015), no. 4, 559–589, DOI 10.4310/PAMQ.2015.v11.n4.a2. MR3613122

[BEW98]   Bruce C. Berndt, Ronald J. Evans, and Kenneth S. Williams, *Gauss and Jacobi sums*, Canadian Mathematical Society Series of Monographs and Advanced Texts, John Wiley & Sons, Inc., New York, 1998. A Wiley-Interscience Publication. MR1625181

[Boo06]   Andrew R. Booker, *Artin's conjecture, Turing's method, and the Riemann hypothesis*, Experiment. Math. **15** (2006), no. 4, 385–407. MR2293591

[Bou51]   Nicolas Bourbaki, *Éléments de mathématique. IX. Première partie: Les structures fondamentales de l'analyse. Livre IV: Fonctions d'une variable réelle (théorie élémentaire). Chapitre I: Dérivées. Chapitre II: Primitives et intégrales. Chapitre III: Fonctions élémentaires* (French), Actualités Sci. Ind., no. 1074, Hermann et Cie., Paris, 1949. MR0031013

[BPT20]   Richard Brent, David Platt, and Timothy Trudgian, *Accurate estimation of sums over zeros of the riemann zeta-function*, 2020, preprint, arXiv:2009.13791.

[BZ11]   Richard P. Brent and Paul Zimmermann, *Modern computer arithmetic*, Cambridge Monographs on Applied and Computational Mathematics, vol. 18, Cambridge University Press, Cambridge, 2011. MR2760886

[CKR20]   Edgar Costa, Kiran Kedlaya, and David Roe, *Hypergeometric L-functions in average polynomial time*, 2020, preprint, arXiv:2005.13640.

[Coh]   Henri Cohen, *Exponential sums can (usually) be computed explicitly*, preprint.

[Coh93]   Henri Cohen, *A course in computational algebraic number theory*, Graduate Texts in Mathematics, vol. 138, Springer-Verlag, Berlin, 1993, DOI 10.1007/978-3-662-02945-9. MR1228206

[Coh00]   Henri Cohen, *Advanced topics in computational number theory*, Graduate Texts in Mathematics, vol. 193, Springer-Verlag, New York, 2000, DOI 10.1007/978-1-4419-8489-0. MR1728313

[Coh07a]   Henri Cohen, *Number theory. Vol. I. Tools and Diophantine equations*, Graduate Texts in Mathematics, vol. 239, Springer, New York, 2007. MR2312337

[Coh07b]   Henri Cohen, *Number theory. Vol. II. Analytic and modern tools*, Graduate Texts in Mathematics, vol. 240, Springer, New York, 2007. MR2312338

[CRVZ00]   Henri Cohen, Fernando Rodriguez Villegas, and Don Zagier, *Convergence acceleration of alternating series*, Experiment. Math. **9** (2000), no. 1, 3–12. MR1758796

[CS17]   Henri Cohen and Fredrik Strömberg, *Modular forms: A classical approach*, Graduate Studies in Mathematics, vol. 179, American Mathematical Society, Providence, RI, 2017, DOI 10.1090/gsm/179. MR3675870

[CZ13]   Henri Cohen and Don Zagier, *Vanishing and non-vanishing theta values* (English, with English and French summaries), Ann. Math. Qué. **37** (2013), no. 1, 45–61, DOI 10.1007/s40316-013-0003-x. MR3117737

[Dok04]   Tim Dokchitser, *Computing special values of motivic L-functions*, Experiment. Math. **13** (2004), no. 2, 137–149. MR2068888

[ERS19]   Salma Ettahri, Olivier Ramaré, and Léon Surel, *Fast multi-precision computation of some Euler products*, 2019, preprint, `arXiv:1908.06808`.

[Fis17]   Kurt Fischer, *The Zetafast algorithm for computing zeta functions*, 2017, preprint, `arXiv:1703.01414v7`.

[FS87]   William F. Ford and Avram Sidi, *An algorithm for a generalization of the Richardson extrapolation process*, SIAM J. Numer. Anal. **24** (1987), no. 5, 1212–1232, DOI 10.1137/0724080. MR909075

[Gou93]   Xavier Gourdon, *Algorithmique du théorème fondamental de l'algèbre*, Rapport de recherche 1852, INRIA, 1993.

[Hec22]   Erich Hecke, *Über analytische Funktionen und die Verteilung von Zahlen mod. eins* (German), Abh. Math. Sem. Univ. Hamburg **1** (1922), no. 1, 54–76, DOI 10.1007/BF02940580. MR3069388

[Hen58]   Peter Henrici, *The quotient-difference algorithm*, Nat. Bur. Standards Appl. Math. Ser. **49** (1958), 23–46. MR94901

[Hia14]   Ghaith A. Hiary, *Computing Dirichlet character sums to a power-full modulus*, J. Number Theory **140** (2014), 122–146, DOI 10.1016/j.jnt.2013.12.005. MR3181649

[JM18]   Fredrik Johansson and Marc Mezzarobba, *Fast and rigorous arbitrary-precision computation of Gauss-Legendre quadrature nodes and weights*, SIAM J. Sci. Comput. **40** (2018), no. 6, C726–C747, DOI 10.1137/18M1170133. MR3880259

[Joh15]   Fredrik Johansson, *Rigorous high-precision computation of the Hurwitz zeta function and its derivatives*, Numer. Algorithms **69** (2015), no. 2, 253–270, DOI 10.1007/s11075-014-9893-1. MR3350381

[Joh17]   Fredrik Johansson, *Arb: efficient arbitrary-precision midpoint-radius interval arithmetic*, IEEE Trans. Comput. **66** (2017), no. 8, 1281–1292, DOI 10.1109/TC.2017.2690633. MR3681746

[Kac06]   Jerzy Kaczorowski, *Axiomatic theory of L-functions: the Selberg class*, Analytic number theory, Lecture Notes in Math., vol. 1891, Springer, Berlin, 2006, pp. 133–209, DOI 10.1007/978-3-540-36364-4_4. MR2277660

[KZ98]   Masanobu Kaneko and Don Zagier, *Supersingular j-invariants, hypergeometric series, and Atkin's orthogonal polynomials*, Computational perspectives on number theory (Chicago, IL, 1995), AMS/IP Stud. Adv. Math., vol. 7, Amer. Math. Soc., Providence, RI, 1998, pp. 97–126, DOI 10.1090/amsip/007/05. MR1486833

[Lev73]   David Levin, *Development of non-linear transformations of improving convergence of sequences*, Internat. J. Comput. Math. **3** (1973), 371–388, DOI 10.1080/00207167308803075. MR359261

[Lou02]   Stéphane R. Louboutin, *Efficient computation of class numbers of real abelian number fields*, Algorithmic number theory (Sydney, 2002), Lecture Notes in Comput. Sci., vol. 2369, Springer, Berlin, 2002, pp. 134–147, DOI 10.1007/3-540-45455-1_11. MR2041079

[LS95]   Stephen K. Lucas and Howard A. Stone, *Evaluating infinite integrals involving Bessel functions of arbitrary order*, J. Comput. Appl. Math. **64** (1995), no. 3, 217–231, DOI 10.1016/0377-0427(95)00142-5. MR1365426

[Luk69]   Yudell L. Luke, *The special functions and their approximations, Vol. I*, Mathematics in Science and Engineering, Vol. 53, Academic Press, New York-London, 1969. MR0241700

[Mes86]   Jean-François Mestre, *Formules explicites et minorations de conducteurs de variétés algébriques* (French), Compositio Math. **58** (1986), no. 2, 209–232. MR844410

[Mol10]   Pascal Molin, *Intégration numérique et calculs de fonctions L*, Ph.D. thesis, Université Bordeaux 1, 2010.

[Mon10]   H. Monien, *Gaussian quadrature for sums: a rapidly convergent summation scheme*, Math. Comp. **79** (2010), no. 270, 857–869, DOI 10.1090/S0025-5718-09-02289-3. MR2600547

[OM91]   Takuya Ooura and Masatake Mori, *The double exponential formula for oscillatory functions over the half infinite interval.*, J. Comput. Appl. Math. **38** (1991), no. 1-3, 353–360 (English).

[OM99]   Takuya Ooura and Masatake Mori, *A robust double exponential formula for Fourier-type integrals*, J. Comput. Appl. Math. **112** (1999), no. 1-2, 229–241, DOI 10.1016/S0377-0427(99)00223-X. Numerical evaluation of integrals. MR1728462

[Pin18]   Iosif Pinelis, *An alternative to the Euler-Maclaurin summation formula: approximating sums by integrals only*, Numer. Math. **140** (2018), no. 3, 755–790, DOI 10.1007/s00211-018-0978-y. MR3854359

[PTVF07]   William H. Press, Saul A. Teukolsky, William T. Vetterling, and Brian P. Flannery, *Numerical recipes: The art of scientific computing*, 3rd ed., Cambridge University Press, Cambridge, 2007. MR2371990

[Ram19]   Olivier Ramaré and Aled Walker, *Products of primes in arithmetic progressions: a footnote in parity breaking* (English, with English and French summaries), J. Théor. Nombres Bordeaux **30** (2018), no. 1, 219–225. MR3809717

[Rub05]   Michael Rubinstein, *Computational methods and experiments in analytic number theory*, Recent perspectives in random matrix theory and number theory, London Math. Soc. Lecture Note Ser., vol. 322, Cambridge Univ. Press, Cambridge, 2005, pp. 425–506, DOI 10.1017/CBO9780511550492.015. MR2166470

[Sid79]   Avram Sidi, *Some properties of a generalization of the Richardson extrapolation process*, J. Inst. Math. Appl. **24** (1979), no. 3, 327–346. MR550478

[Sid82a]   Avram Sidi, *An algorithm for a special case of generalization of the Richardson extrapolation process*, Numer. Math. **38** (1981/82), no. 3, 299–307, DOI 10.1007/BF01396434. MR654099

[Sid82b]   Avram Sidi, *The numerical evaluation of very oscillatory infinite integrals by extrapolation*, Math. Comp. **38** (1982), no. 158, 517–529, DOI 10.2307/2007286. MR645667

[Sid88]   Avram Sidi, *A user-friendly extrapolation method for oscillatory infinite integrals*, Math. Comp. **51** (1988), no. 183, 249–266, DOI 10.2307/2008589. MR942153

[Sie43]   Carl Ludwig Siegel, *Contributions to the theory of the Dirichlet L-series and the Epstein zeta-functions*, Ann. of Math. (2) **44** (1943), 143–172, DOI 10.2307/1968761. MR7760

[Sug97]   Masaaki Sugihara, *Optimality of the double exponential formula—functional analysis approach*, Numer. Math. **75** (1997), no. 3, 379–395, DOI 10.1007/s002110050244. MR1427714

[SV14]   Burhan Sadiq and Divakar Viswanath, *Finite difference weights, spectral differentiation, and superconvergence*, Math. Comp. **83** (2014), no. 289, 2403–2427, DOI 10.1090/S0025-5718-2014-02798-1. MR3223337

[Sze75]   Gábor Szegő, *Orthogonal polynomials*, 4th ed., American Mathematical Society, Providence, R.I., 1975. American Mathematical Society, Colloquium Publications, Vol. XXIII. MR0372517

[Tay45]   P. R. Taylor, *On the Riemann zeta function*, Quart. J. Math. Oxford Ser. **16** (1945), 1–21, DOI 10.1093/qmath/os-16.1.1. MR12626

[TB86]   Ian J. Thompson and A. Ross Barnett, *Coulomb and Bessel functions of complex arguments and order*, J. Comput. Phys. **64** (1986), no. 2, 490–509, DOI 10.1016/0021-9991(86)90046-X. MR845195

[vzGG99]   Joachim von zur Gathen and Jürgen Gerhard, *Modern computer algebra*, Cambridge University Press, New York, 1999. MR1689167

[Wat95]   George Neville Watson, *A treatise on the theory of Bessel functions*, Cambridge Mathematical Library, Cambridge University Press, Cambridge, 1995. Reprint of the second (1944) edition. MR1349110

[WW96]   Edmund Taylor Whittaker and George Neville Watson, *A course of modern analysis*, Cambridge Mathematical Library, Cambridge University Press, Cambridge, 1996. An introduction to the general theory of infinite processes and of analytic functions; with an account of the principal transcendental functions; Reprint of the fourth (1927) edition, DOI 10.1017/CBO9780511608759. MR1424469

Index of Programs

General Index